Scientific Bases of Human Movement

THIRD EDITION

Scientific Bases of Human Movement

THIRD EDITION

Barbara A. Gowitzke, Ph.D.

F.A.A.H.P.E.R.D.
Associate Professor of Physical Education
School of Physical Education and Athletics
Faculty of Social Sciences
McMaster University,
Hamilton, Ontario, Canada

Morris Milner, Ph.D., P.Eng., C.C.E.

F.I.E.E.
Director, Rehabilitation Engineering & Research Departments
Hugh MacMillan Medical Centre
Professor and Chairman,
Department of Rehabilitation Medicine
Faculty of Medicine
University of Toronto,
Toronto, Ontario, Canada

Editor: John P. Butler
Associate Editor: Victoria M. Vaughn
Copy Editor: Lindsay E. Edmunds
Design: Saturn Graphics
Illustration Planning: Wayne Hubbel
Production: Theda Harris

Printed in the United States of America

First Edition, 1972
Reprinted 1973, 1976
Second Edition, 1980
Reprinted 1981, 1982, 1984

Library of Congress Cataloging-in-Publication Data

Gowitzke, Barbara A.
Scientific bases of human movement.

Rev. ed. of: Understanding the scientific bases of human movement. 2nd ed. c1980.
Bibliography: p.
Includes index.
1. Human mechanics. 2. Kinesiology. I. Milner, Morris. II. Gowitzke, Barbara A. Understanding the scientific bases of human movement. III. Title.
[DNLM: 1. Movement. WE 103 G723u]
QP303.G63 1988 612′.76 87-6287
ISBN 0-683-03593-2

88 89 90 91 92
10 9 8 7 6 5 4 3 2

Cover Illustration: The display on the cover was obtained with Lise Arsenault-Goertz fitted with illuminating souces at the wrist, shoulder, hip, knee and ankle, then executing a back walkover while being tracked optoelectronically in one plane by use of a Selspot system coupled to a PDP 11/10 computer. A maximum of 711 frames at a sampling frequency of 161 Hz were collected, the data were processed using a 5 Hz low-pass digital filter and the display was generated by taking every seventh sample set, showing the pertinent stick diagram and generating the trajectories of each light source. The walkover commenced on the right side of the display with the subject upright. (J. A. Wallace provided technical assistance and Dr. H. deBruin attended to the computational aspects of the display. We are grateful to both of them.)

Dedication

To my husband, David B. Waddell, who encouraged me to write the second edition, and who provided the moral support, patience, and assistance through this, the third edition.

B. A. Gowitzke

Foreword

Writing a foreword for the previous edition of this excellent book some years ago was a great pleasure that remains fresh to this day. Like a fine wine, this work ages well and how I savor the pleasure of a second tasting. The passing of time can mature a book or a vintage into a classic or can reduce either of them to sour uselessness. An essential factor in ensuring quality is meticulous nurturing that enhances the original. In books, the greatest opportunity for nurture (or for ruin) can come with a new edition.

In the previous edition of this classic, the writing team of Barbara Gowitzke and Morris Milner blended their efforts to give fresh new life to the earlier work. Dr. Gowitzke undertook the primary responsibility for this edition with the support of three contributing authors. She has blended old and new into a fresh and palatable result.

The third edition of the *Scientific Bases of Human Movement* remains a distinguished textbook of knowledge for all students and teachers who want to know more than cookbook facts. It provides rationale and up-to-date scientific explanation in a clear and interesting manner. I am delighted to be given this fresh opportunity to hail its appearance.

John V. Basmajian, M.D., F.A.C.A., F.R.C.P.(C)
Professor Emeritus of Medicine and Anatomy
McMaster University
Hamilton, Ontario, Canada

Preface

In presenting the third edition, the goal has been to make improvements on material found in the second edition and to update information. Revising and rearranging the contents of the second edition led to new chapter titles, addition of new material, substantial reworking of illustrations as well as adding new ones, and the welcome input of three contributing authors.

There is continued dedication to the fundamental idea that mechanics of the human body cannot be dealt with adequately unless the neurophysiology of the human body is also discussed. This dedication stems, in part, from my own heritage and that of the previous authors, and from some herculean efforts on the part of some of my colleagues to define what should be included in a kinesiology/biomechanics course. In February 1980, a task force of leaders of the Kinesiology Academy of the National Association of Sport and Physical Education (NASPE) proposed guidelines and standards for undergraduate kinesiology. This report was adopted by the Academy and NASPE (Kinesiology Academy, 1980). Course content recommended by the task force includes not only mechanical considerations, i.e., kinematics and kinetics, but also very specific anatomical considerations related to joint structure and function, muscular function, and neuromuscular concepts. Competencies in anatomy and mathematics are considered as prerequisites to an undergraduate kinesiology course.

About the same time, an Ad Hoc Committee of the International Society of Electrophysiological Kinesiology (ISEK), was attempting to standardize technical terminology, especially in the reporting of electromyographic research (ISEK, 1980). Their report, first made available in August 1980, is titled, "Units, Terms and Standards in the Reporting of EMG Research," and is available from the Newsletter Editor of ISEK.

In addition, physical therapy recognized the need to standardize terms, especially biomechanical ones, and the American Physical Therapy Association published a glossary of terms, concepts, and units (Rodgers and Cavanagh, 1980). It is quite clear just how eclectic the field of kinesiology/biomechanics really is when studying the groupings employed in their glossary; they include kinematics, kinetics, a special section on forces, computational methods, muscle mechanics, mechanics of materials, instrumentation, body segment variables, and even units of measurement and conversion factors. I am aware also of the work of a committee within the Canadian Society of Biomechanics charged with the task of providing a document for standardization of terms used in kinesiology/biomechanics. This material may also be available by the time the present textbook is published.

With all of these guidelines for standardization available, it is important that this textbook follow the recommendations of the reports cited. It is not difficult to follow the recommendation of the NASPE Committee report since the textbook always has included neurophysiological considerations. Also, attempts are made to incorporate the recommendations of the ISEK Committee and the Physical Therapy Association. In the process, the Imperial system and United States customary units are dropped and only units of measurement derived from the *Système International d'Unités* (the International System of Units or SI system) are used.

Chapter 1 is a short introduction defining what the textbook intends, as well as defining a few basic terms. Chapter 2 is the previous Chapter 1, and concentrates on the skeletal framework of the human body. Chapters 3 and 4, Kinematics and Kinetics, have taken on new dimensions. Under new titles, they have been reorganized with new content, and in each, many more examples are provided. In addition, the reader is invited to attempt solving sample problems. The solutions and answers are provided in Appendix 1. Chapter 5 adds additional examples of concepts discussed in Chapters 3 and 4, and offers an even more extensive set of problems for the reader to solve. Having dealt with the mechanics of motion, Chapter 6 deals with the forces that power movement, the skeletal muscle system, and combines the material that was offered formerly in Chapters 4 and 7. Chapters 7, 8, and 9 (formerly chapters 8, 9, and 10) attempt to explain the neural mechanisms that control the motor elements, including basic neurophysiology, a review of the central nervous system, and the overall organization of the elements of the neuromuscular system. Chapter 10 on proprioceptors remains essentially unchanged except for updating and new information. Chapter 11 is the concluding chapter and deals with instrumentation. Most of the material is elaborated and synthesized on the basis of measurement schemes. The textbook concludes with selected references and appendices that include body segment tables not easily found elsewhere, as well as solutions and answers to problems.

Barbara A. Gowitzke

Acknowledgments

I am greatly indebted to Morris Milner for his considerable contribution to the second edition and for allowing his name to be used on this edition for continuity. The present edition also takes advantage of his contribution as well as contributions from three invited authors who provided special expertise in specific areas.

Hubert deBruin

Dr. Hubert deBruin, Assistant Professor of Medicine at McMaster University, rewrote the last chapter on Measurement and Instrumentation. His very substantial experience as a biomedical engineer and researcher in the field of locomotion, especially of the handicapped, offered the opportunity to synthesize a very large and widely variant number of topics into digestible form. Concepts, ideas and applications relating to these topics that are of particular interest to the therapy-oriented reader, are supplied by two other colleagues.

Kevin Olds

Kevin Olds, member of the Faculty of Medicine in the Department of Rehabilitation Medicine at the University of Toronto, supplied numerous suggestions for several segments of the textbook and contributed problems with solutions, found principally in Chapters 3, 4 and 5.

Paul Stratford

Paul Stratford, Teaching Master at Mohawk College's Department of Physiotherapy as well as part-time Assistant Professor at McMaster University, offered suggestions for most sections of the text; his contributions are found principally in Chapters 2, 3, 4, 5 and 10 and include sample problems with solutions.

My very special thanks go to David B. Waddell, my husband, who took charge of the illustrations and single-handedly reworked those from the previous edition as well as producing many new graphics. David also functioned as technical editor. Other graphics were produced by Diane Commins of Ottawa and artists in Instructional Media Services of the University of Toronto. The input of Dr. John V. Basmajian is warmly recognized. Even in retirement, he found the time and with enthusiasm wrote the foreword for this new edition. Finally, all my students provided the motivation and impetus to write another edition.

B. A. G.

Contents

CHAPTER 3
KINEMATICS

CHAPTER 4
KINETICS

CHAPTER 5
APPLICATION OF MECHANICAL PRINCIPLES TO THE HUMAN BODY

CHAPTER 6
THE SKELETAL MUSCLE SYSTEM

CHAPTER 7
NEURAL BASIS OF BEHAVIOR

CHAPTER 8
BASIC NEUROANATOMY: THE CENTRAL NERVOUS SYSTEM

CHAPTER 9
BASIC ORGANIZATION OF THE NEUROMUSCLAR SYSTEM

CHAPTER 10
PROPRIOCEPTORS AND ALLIED REFLEXES

CHAPTER 11
ANALYSIS OF MOVEMENT: MEASUREMENT AND INSTRUMENTATION

CHAPTER 1

Introduction

Studying human movement can be as casual as sitting on the front porch and making mental notes of the techniques used by runners (and would-be runners) as they respond to the universal push for physical fitness. Or it can be as intense as examining the mechanisms of injury by arthroscopy. It can be concerned with gross motor patterns or it can be confined to the details of fine motor elements. Whatever may be the purpose of a human movement study, scientific aspects are usually involved. Quite often, studies will involve anatomy, neurophysiology, and/or mechanics. It is the latter term which usually moves the study into the purview of a kinesiology or biomechanics study.

Mechanics is the science which deals with the motion (or nonmotion) of material objects. Conventional division of mechanics into two areas of study, "at rest" and "in motion," has given rise to the terms statics and dynamics. **Statics** is concerned with those conditions under which objects remain at rest or in equilibrium. **Dynamics** refers to that phase of mechanics in which objects are moving. "Statics deals with the determination of the forces acting on objects which are at rest or moving in a straight line with constant speed; i.e., statics is the study of objects under the action of balanced forces. Dynamics, on the other hand, is concerned with the study of objects under the action of unbalanced forces" (Watkins, 1983:2).

The area of dynamics is divided into two branches, kinematics and kinetics. **Kinematics** refers to a geometric description of motion in terms of displacement, velocity, and acceleration against time, without regard for the source of the motion, i.e., the forces which cause the motion to occur. **Kinetics**, on the other hand, is concerned with the forces which are responsible for the motion, or change or even stop the motion.

Applying mechanics to the human body is difficult since mechanics deals with material objects, which are quite often rigid bodies. But "all solid materials change shape to some extent when forces are applied to them. Nevertheless, if the movements associated with the changes in shape are very small compared with the overall movements of the body as a whole, then the ideal concept of rigidity is quite acceptable" (Meriam, 1975:197). In the context of the human body, it is customary to regard skeletal parts as the rigid bodies upon which forces are exerted. Thus, it is obvious why in-depth knowledge of the skeletal system is needed.

The human body can be subjected to any of the forces encountered by material objects. They include the forces that are external to the human body as well as those that are present within the human body. The muscles provide the main internal forces. Therefore, knowledge of the mechanics of muscle and understanding of how muscle acts on the skeletal system is absolutely essential to the study of movement.

Finally, the mechanisms which "drive" the muscles to exert their forces, as well as the systems of monitoring performance and modifying signals to muscle, make knowledge of the nervous system and all of its components integral to the scientific bases of human movement.

CHAPTER 2

The Skeletal System

INTRODUCTION

The skeletal system is the bony framework that supports the body organs, protects many of them, and forms the hard core of all body segments. Its many articulations provide mobility, and it is the functions of these mobile articulations that are of concern in facilitating descriptions of human movement.

The driving forces for skeletal components or links derive from muscular actions, which in turn are controlled by the nervous system. Overall movements are thus the manifestation of integrated activities within a complex which is often termed the neuromusculoskeletal system. For the sake of convenience and to facilitate our understanding of the component parts of the neuromusculoskeletal system, we tend to dissect the system and look at its parts in relative isolation, and simplify the descriptions of overall performance or behavior of these parts. For example, the skeletal system may be considered as a system of interconnected links (Fig. 2.1).

LINKS IN THE BODY

The link concept, originally designed by engineers, was used by Dempster (1955), the first kinesiologist to apply the concept to those problems involving kinetic and kinematic treatments of movements of the human body. Since engineering links involve overlapping articulating members held together by pins which act as axes of rotation, a **link** is considered to be a straight line of constant length running from axis to axis. A system of links is essentially a geometric entity for analysis of motion by geometric or kinematic methods. ". . . In engineering mechanisms the links move in relation to a framework, and this framework itself forms a link in the system. Thus, to transmit power, the links of machinery must form a closed system in which the motion of one link has determinate relations to every other link in the system" (Reuleaux, quoted by Dempster (1955)).

Reuleaux (1875) also introduced the term **"kinematic chain"** to refer to a mechanical system of links. In engineering, the chain forms a closed system where, as quoted earlier, "the motion of one link has determinate relations to every other link in the system," and "the closed system assures that forces are transmitted in positive predetermined ways." Thus, in engineering a kinematic chain is a closed system of links joined in such a manner that if any one is moved on a fixed link, all

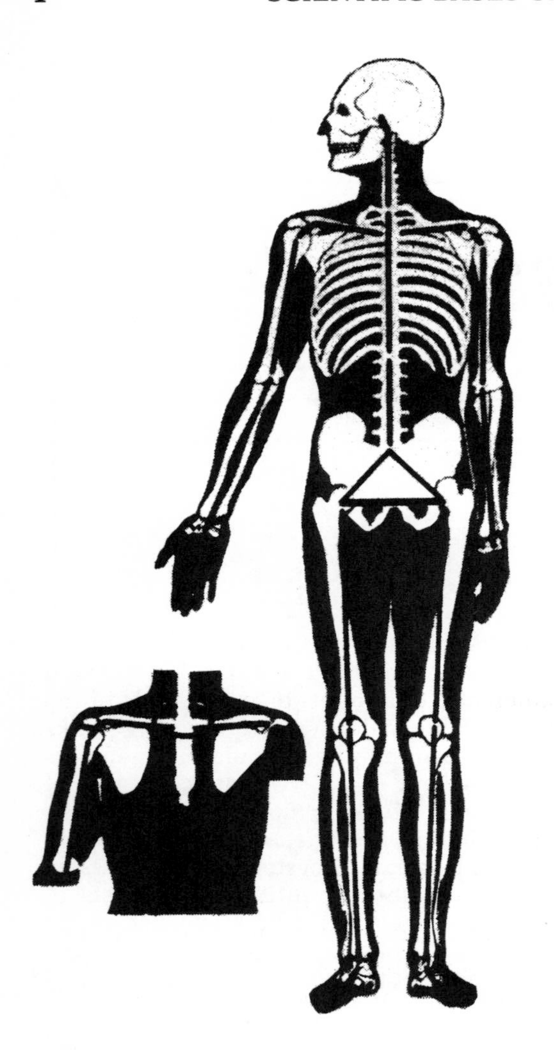

Figure 2.1. The human skeleton as a system of links. Note the simplified link segments drawn between major joints that constitute anatomical links. (From Dempster, W.T., 1955. *Space Requirements for the Seated Operator.* Wright Air Development Center Technical Report 55159.)

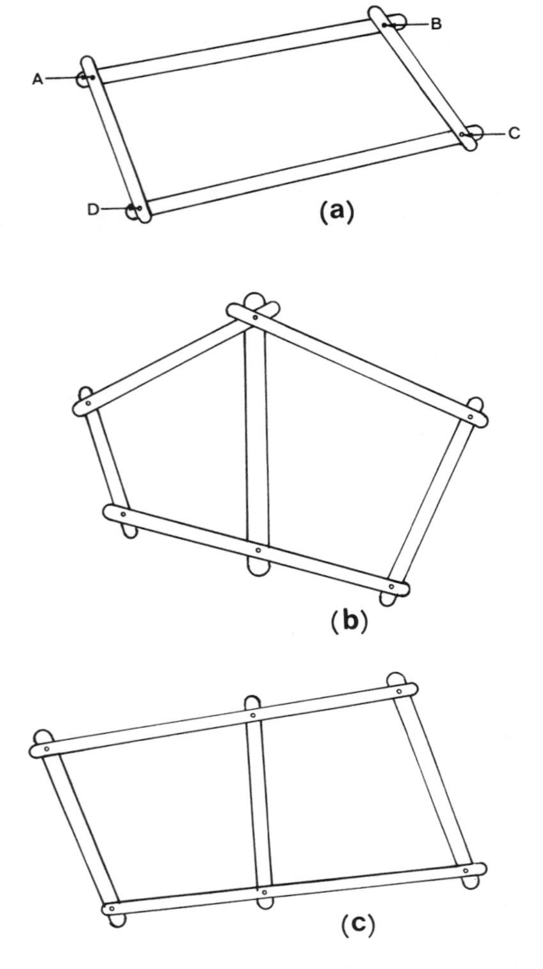

Figure 2.2. Types of closed kinematic chains. All joints are pin-centered and free to move. If for example in (a), link AB (pinned at A and B) is held fixed, links BD, CD, and DA will, when influenced by externally applied forces on one or more of them, be able to move in a predictable pattern which can be determined by constructing circles with A and B as centers and respective radii AD and BC . The initial distance between C and D must be maintained throughout the ranges of movement of C and D to describe the resulting possible movement pattern of the link system. The interested student might care to attempt such constructions and make simple models to obtain appropriate visualizations.

of the other links will move in a predictable pattern (Fig. 2.2). With a few exceptions the system of skeletal links in the human body is generally not composed of closed chains, but of open ones, as the peripheral ends of the extremities are free (Fig. 2.1). Forces may be transmitted in positive ways, predetermined by the central nervous system, but the central nervous sytem is notorious for never accomplishing the same act in exactly the same way from one time to the next, even though the external results may appear similar. Thus, when speaking of a living kinematic chain, we are usually referring to a series of links arranged in an open system, whose dimensions are determined by the linear distance from joint axis to joint axis, ignoring muscle mass, bone structure, and type of articulation between body segments.

Although most living kinematic chains are open, Brunnstrom (1983) defines two closed kinematic chains in the body. The first is the pelvic girdle, which is made up of three bony segments united at the two sacroiliac joints and at the symphysis pubis. This can hardly be classified as a kinematic chain because normally no movement occurs at the joints mentioned. Dempster classes the pelvis as a single triangular link (Fig. 2.1). The second closed kinematic chain in the body according to Brunnstrom is the thorax, where the upper 10 ribs are jointed to the vertebral column and sternum. The rib cage, however, does constitute a system of closed kinematic chains because the upper 10 ribs of the left side cannot move without similar movement by the upper 10 ribs of the right side when they lift the sternum on inhalation. Steindler[a] (1973) considers a closed living chain (which he terms kinetic rather than kinematic) to exist in "all situations in which the peripheral joint of the chain meets with overwhelming resistance" (Fig. 2.3).

The rehabilitation specialist can make use of this concept when prescribing a self-assisted mobilizing exercise. Consider the following example, where the goal of treatment is to increase knee flexion in an individual with limited range of motion. The starting position is with the subject seated on the floor (or an exercise table) so that the involved knee is flexed as far as possible and the foot is flat on the floor.

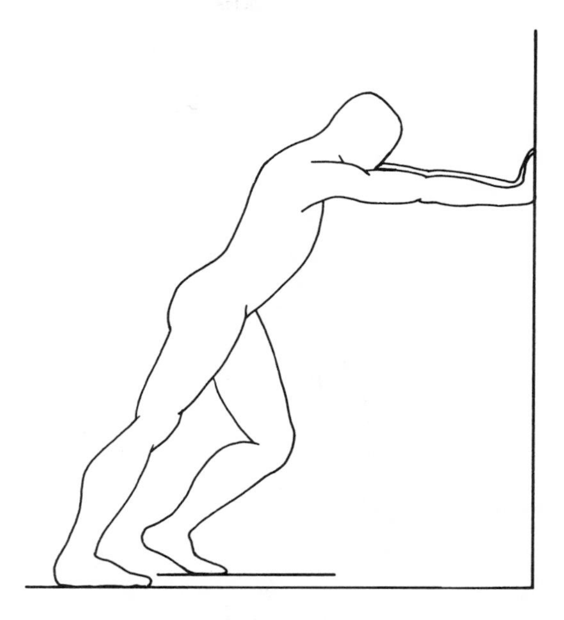

Figure 2.3. Example suggested by Steindler of a human closed kinematic chain. The wall and floor provide "overwhelming" resistance at the peripheral joints.

The subject is instructed to dorsiflex at the ankle while maintaining heel contact with the floor. Since the floor acts as an overwhelming resistance, a closed system exists. By dorsiflexing the ankle, forces are transmitted to the knee joint, with the result being increased knee flexion.

Because the bones of the body rarely overlap at joints as at the ankle and, except for the atlantoaxial joint, have no pin-centered axes, and because at many joints movement can occur in different directions and planes, the engineering concept of links must be redefined to fit the need of the kinesiologist. Dempster proposed the use of the term "link" in kinesiology as the distance between joint axes; e.g., the leg link becomes the linear distance between the joint axes passing through the distal end of the femur and the proximal end of the talus (or through the two malleoli), thus spanning both the knee and ankle joints (Fig. 2.4).

Because they are attached to each other

[a] While Brunnstrom and Steindler used the term "chain" modified by kinematic or kinetic, respectively, when referring to a series of body segments, they did not use the term "links" in their discussions.

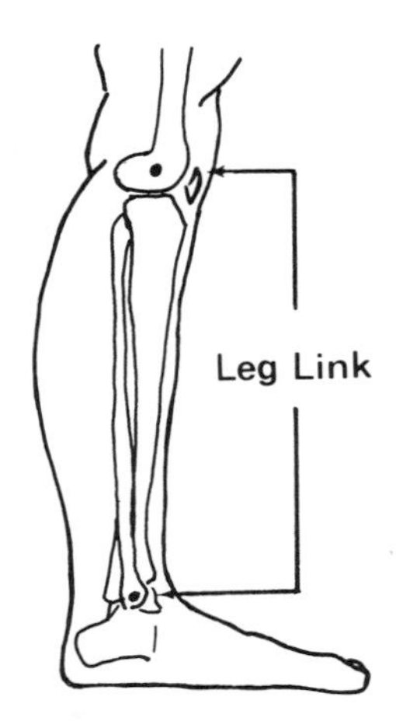

Figure 2.4. Leg link.

by a rigid link, these two joint centers are said to be constrained to keep a fixed distance from each other. This is called a **kinematic constraint.**

In the appendicular skeleton a body segment consists of a hard core made up of one or more bones enclosed in an irregular mass of soft tissue (muscle, connective tissue, and skin). Ligaments and muscle tendons cross the joint between contiguous body segments, anchoring to the adjacent bone or bones and holding the segments together. The instantaneous axis around which one segment moves on another generally passes through one of the bones through an area near the joint.

JOINT AXES AND DEGREES OF FREEDOM

Anatomists and kinesiologists speak of joints as being uniaxial, biaxial, or multiaxial and as having certain degrees of freedom (Steindler (1973); Brunnstrom (1983); Terry and Trotter (1955)). A joint with only one axis **(uniaxial)** has one degree of freedom; that is to say, the articulating bones can move only in one plane. Examples in the human body include hinge and pivot joints. **Hinge** joints occur at the elbow, knee, interphalangeal, and ankle joints. The **pivot** joints are the atlantooccipital in the vertebral column and the radioulnar joints in the forearm. Joints that can move about two axes **(biaxial)** have two degrees of freedom and so produce movement in two different planes. The wrist, the metacarpophalangeal, and the metatarsophalangeal joints are biaxial. Joints that can permit movement in all three planes have three degrees of freedom but are called **multiaxial** rather than triaxial. This is because movement can occur in oblique planes as well as in the three major planes, which are defined as three mutually perpendicular planes. Examples of multiaxial joints include the **ball and socket** joints at the hips and shoulders and the numerous **plane** joints of the axial skeleton. In this instance the term "plane" is an adjective referring to the almost flat articular surfaces which can glide over one another, with movement being limited only by ligaments or by the joint capsule. Examples include those joints between the articular processes of the vertebrae and between the ribs and the vertebrae. These joints have such a limited amount of movement at any one articulation that total movement of the torso occurs only because of the combined action of many or all of the joints and their degrees of freedom.

For the kinesiologist there is a distinct advantage in using the term "degree of freedom." While no one joint can have more than three degrees of freedom, the degrees at adjacent joints can be summed to express the total amount of freedom of motion of a distal segment relative to a proximal one. For instance, the distal phalanges of a pianist enjoy 17 degrees of freedom relative to his trunk: one degree at each of the distal and proximal phalangeal joints; two degrees at the metacarpophalangeal joints; two degrees at the wrist joint; one degree in the forearm at the radioulnar joints; one degree at the elbow; three degrees at the shoulder; three degrees at the acromioclavicular joint; three degrees at the sternoclavicular joint. Observation of many pianists might, however, lead us to add three more degrees of freedom arising from the motion in the vertebral column. This would express the freedom of the phalanges relative to the pelvis, which is resting on the piano bench, rather than relative to his torso, making a total of 20 degrees of freedom available at the fingertips.

The number of degrees of freedom represents the number of angles one must monitor when studying human movement. It should be noted, however, that there are several kinematic constraints within the system. Summing up the degrees of freedom of the pianist's fingertips involves listing the joints occurring between the distal phalanges and the pelvis. These joints unite the various body segments, which move upon each other in the manner of links in a chain. Thus, we have a further example of an open, living kinematic chain.

To illustrate the impact of altering the natural number of degrees of freedom required to perform a given task, the reader may wish to attempt the following exercise. First, walk several hundred meters and climb several flights of stairs. Then, stiffen the right knee in full extension and repeat the task. While one is probably able to complete the second task, there is an obvious reduction in the fluidity of the movement, which is accompanied by an increase in the energy expended to perform the task.

LIMITATIONS OF MOVEMENT

The type and range of movement about any given joint depend upon the structure of the joint and the number of its axes, the restraints imposed by ligaments and muscles crossing the joint, and the bulk of adjacent tissue. Because of its structure, a joint with three degrees of freedom may have a very limited range of motion, as was indicated earlier for the intervertebral joints, while a joint with only one degree of freedom may have a large range of motion. For example, the forearm can move through an average range of 150° from the position in line with the arm to the fully bent position (full flexion). The range may be increased by from 5° to 15° in the individual who has a smaller than average olecranon process or a deeper than average olecranon fossa that permits the forearm to move beyond the position in line with the arm (i.e., *hyperextend*). Conversely, in an individual with overdeveloped biceps brachii and brachialis muscles or with excessive adipose tissue, flexion may be limited by the very bulk of the soft tissue of the arm. Similar factors may also affect the range of mobility of other joints.

Factors which limit passive range of motion fall into two categories.

1. Factors associated with normal limitation of movement: (a) bone on bone, e.g., hyperextension at the elbow; (b) capsular tightness, e.g., external rotation at the shoulder; (c) apposition of soft tissues, e.g., knee flexion; (d) two-joint muscles, e.g., angle dorsiflexion performed with the knee extended, thereby stretching the gastrocnemius; (e) extracapsular ligaments, e.g., knee extension.
2. Factors associated with pathological or abnormal restrictions of movement: (a) bone on bone, e.g., extension of the knee in which there is degenerative joint disease; (b) capsular stiffness due to prolonged immobilization, e.g., frozen shoulder; (c) capsular stiffness due to intra-articular swelling, e.g., swollen knee; (d) noncapsular adhesion formation, e.g., adhesions of the anterior talofibular ligament at the ankle restricting plantar flexion and inversion *or* adhesions between the quadriceps femoris and the femur restricting knee flexion; (e) loose bodies in the joint, e.g., bony or cartilaginous chips jammed between articulating surfaces.

In the above examples, it should be noted that restraints refer to restrictions on joint ranges of motion. Restraints do not affect degrees of freedom, unless they reduce the range of movement to zero.

REFERENCE POSITIONS AND PLANES

For convenience in specifying the positions of anatomical members in space, a reference system must be established. This system may be established in various locations depending upon the details of the issue being investigated. For example, while whole body movements might be referred to a reference system whose origin is located at the center of gravity of the body, finger movements might more conveniently be defined by a reference system whose origin is chosen to be at a selected metacarpal joint. Traditionally, kinesiologists have made use of an orientation system which defines three *cardinal* mutually perpendicular planes of orientation having a common intersection at the *center of gravity*[b] of the body while it occupies the anatomical position exemplified in Figure 2.5. From a practical standpoint, three essential, mutually perpendicular planes of concern are the *sagittal, transverse,* and *frontal* planes. Views in these respective planes are depicted in Figure 2.6(a), (b), and (c). Each of the cardinal planes may be referred to as primary when it passes through the center of gravity of the body because it divides the body into two equal sections. Thus, the cardinal planes are the **sagittal**, which divides the body into right and left sections, the **frontal plane**, which divides the body anterioposteriad into front and back sections, and the **transverse plane**, which divides the body into upper and lower sections. Figure 2.7 is a diagrammatic representation of the primary cardinal reference planes. Location of any anatomical members can be specified by noting their distances from the three mutually perpendicular axes which pass through the origin (in this case the center of gravity of the body).

[b] This is the point in or about the body through which the resultant body force will act due to the gravitational pull of the earth (or other environment) upon the masses of the various body parts. The location of the center of gravity changes as limbs vary their relative positions. From a practical standpoint it is not the best choice of a reference origin since a great deal of anatomical information must be supplied to specify its loction exactly. The location of the center of gravity can be determined using the methods depicted in Chapter 5 under "Locating the Center of Gravity.

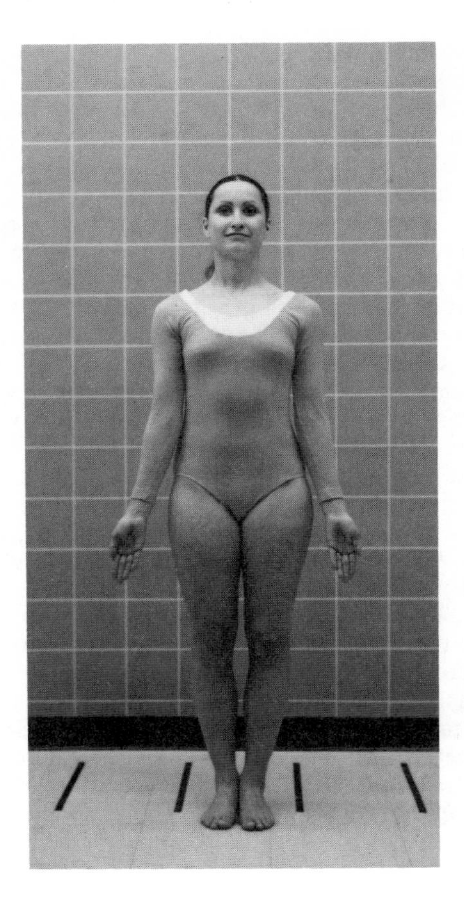

Figure 2.5. Anatomical position.

Figures 2.8 through 2.16 show the joint axes for major movements about the shoulder, elbow, forearm and wrist, fingers and thumb, the hip joint, the knee joint, the ankle and foot, and finally the axial skeleton. Knowledge of the movement descriptors discussed in the following paragraphs will be helpful in interpreting the aforementioned figures.

MOVEMENT DESCRIPTORS

To illustrate the explanations allied with the pertinent movement descriptors, photographs are utilized in what follows.

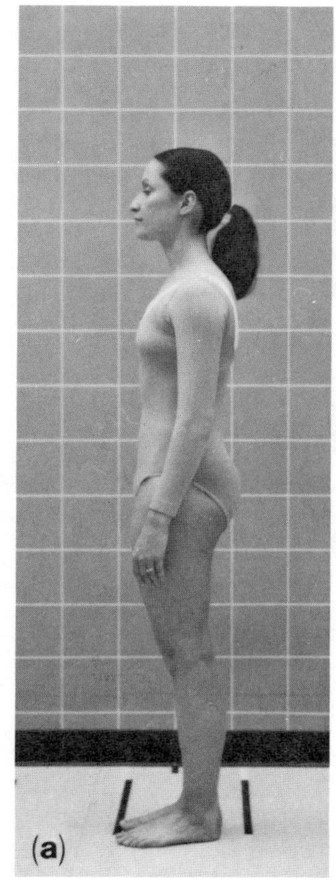

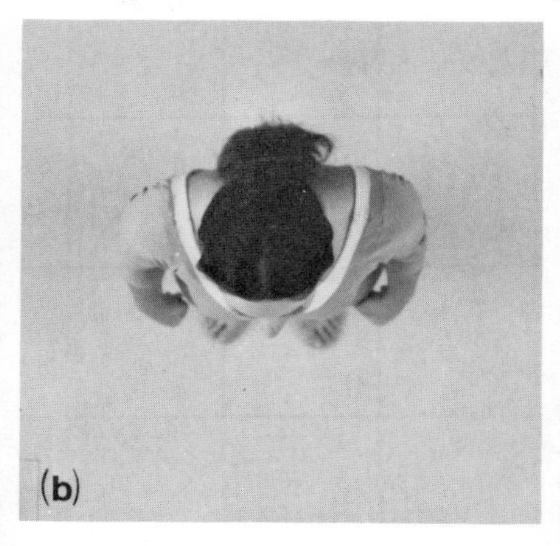

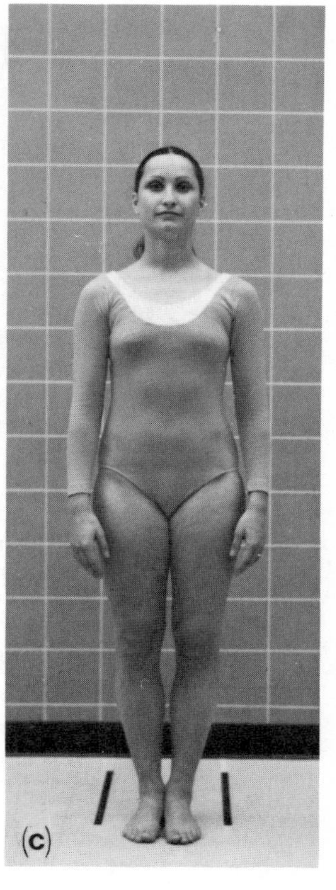

Figure 2.6. Viewing planes of the body: (a) sagittal plane; (b) transverse plane; (c) frontal plane.

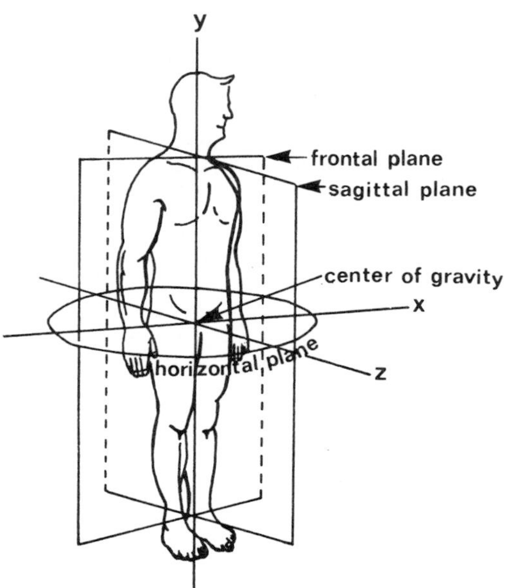

Figure 2.7. Diagrammatic representation of the primary cardinal reference planes.

It should be noted that movements in the frontal plane are best depicted by frontal views, while those in the sagittal plane rely chiefly on side views. It should be recognized that the positions depicted are reached from the anatomical position. This implies that movement occurs before the particular pose shown is achieved.

Flexion-Extension

Flexion is popularly considered to be a movement which decreases the angle between the moving part and the adjacent segment (as in elbow or finger flexion), and **extension** is considered to be a movement which increases this angle (Fig. 2.17).

For the purpose of defining joint movement, it is convenient to assume that the body is in the anatomical position (Fig. 2.5). Then flexion and extension are movements in which the moving segments

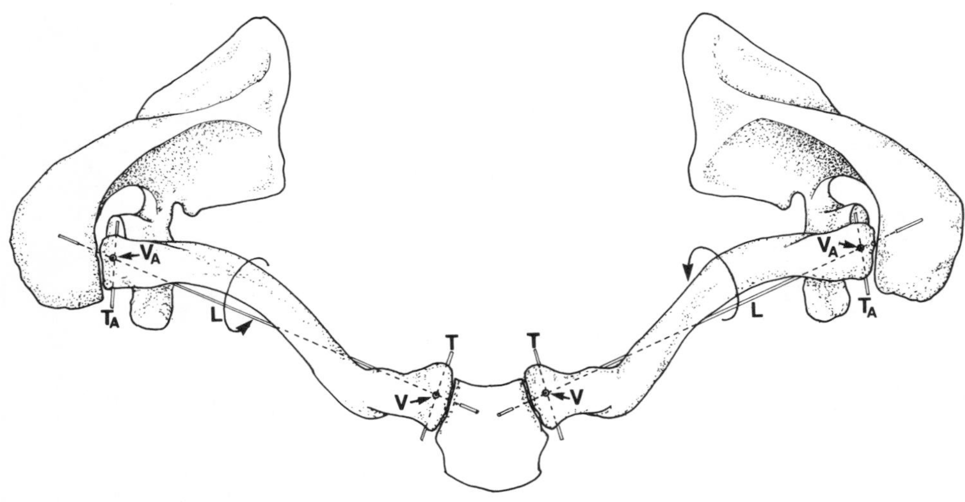

Figure 2.8. Major axes of the shoulder girdle as seen from above. V, vertical axis for protraction and retraction of the shoulder girdle; T, transverse axis for elevation and depression of the shoulder girdle; L, longitudinal axis for the limited rotational movements of the clavicle; V_A, vertical axis, and T_A, transverse axis, at the acromial end of clavicle for scapular motion. (Joint axes in Figure 2.8 to 2.16 drawn after Grant, J.C.B. and Smith, C.G., 1953. In *Morris' Human Anatomy*, edited by J.P. Schaeffer. New York: The Blakiston Company.)

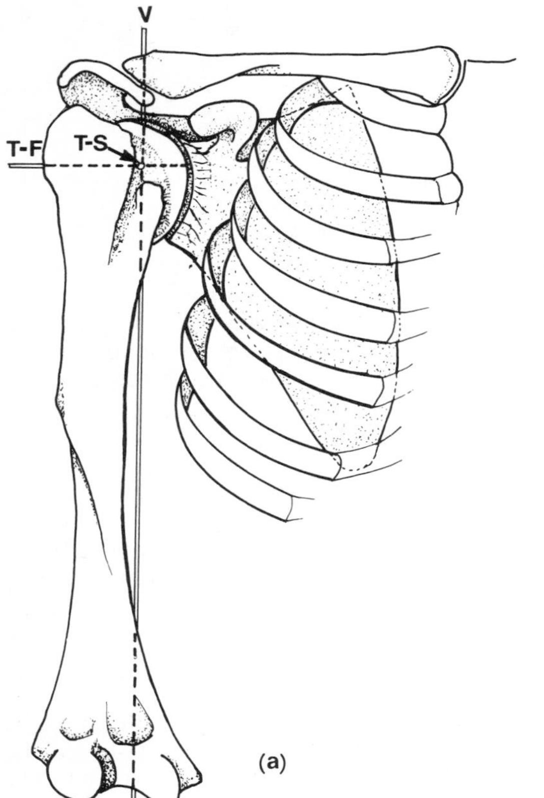

Figure 2.9. Axes for movements at the shoulder joint. (a) from the anatomical position. T-F, transverse axis in the frontal plane for movements of flexion and extension; T-S, transverse axis in the sagittal plane for movements of abduction and adduction; V, vertical axis running the length of the humerus for movements of inward and outward rotation of the arm. (b) Same axes but arm is abducted 90°. Note altered positions of T-F and T-S axes.

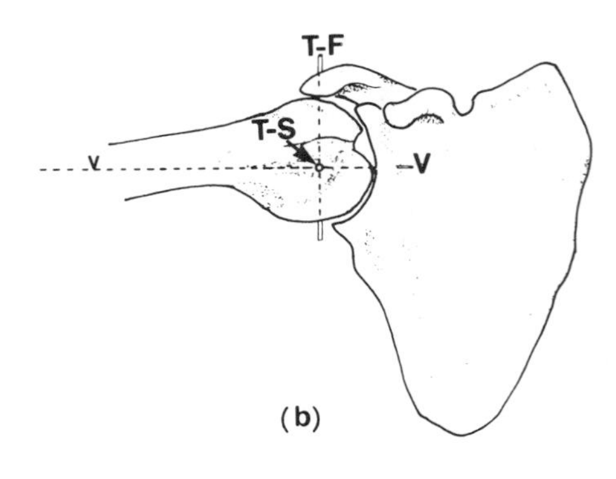

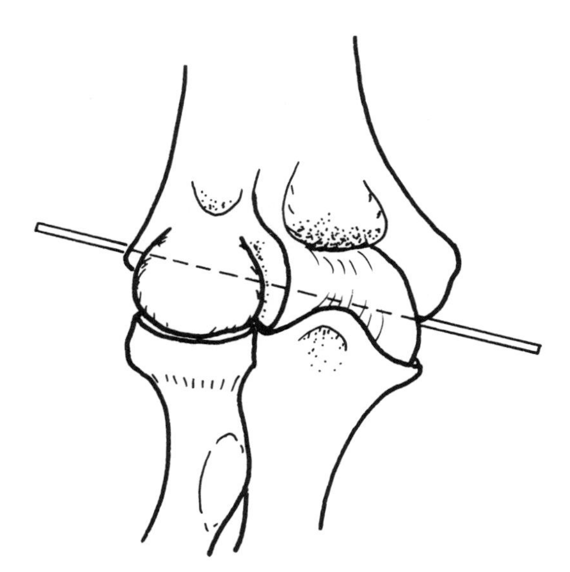

Figure 2.10. Transverse axis through the elbow joint. Movements are flexion and extension.

travel in a sagittal plane around a horizontal axis defined by anatomical frontal and transverse planes through the axis. While this definition is adequate to a degree, it is not applicable to all joints. A more satisfactory one which may be applied to all except the shoulder joints is based on the anatomical concept that flexion is the approximation of ventral or *volar* surfaces. This concept is based on the embryological development of the human fetus. Soon after the limb buds first appear in the embryo (Fig. 2.18(a)), they project laterally with the thumbs and great toes uppermost (Fig. 2.18(b)). As the limbs develop, they bend ventrad at the elbows and knees so that the apices of these joints are pointed outward and the palms of the hands and soles of the feet (the volar surfaces) face the torso (Fig. 2.18(c)). Finally, both pairs of limbs rotate 90° but in opposite directions, the rotations taking place about the long axes of the limbs (Fig. 2.18(d)). The upper extremities rotate laterad so that the elbow points backward, the thumbs are outward, and the ventral and volar (palmar) surfaces face forward. The lower extremity rotates mediad so that the knees

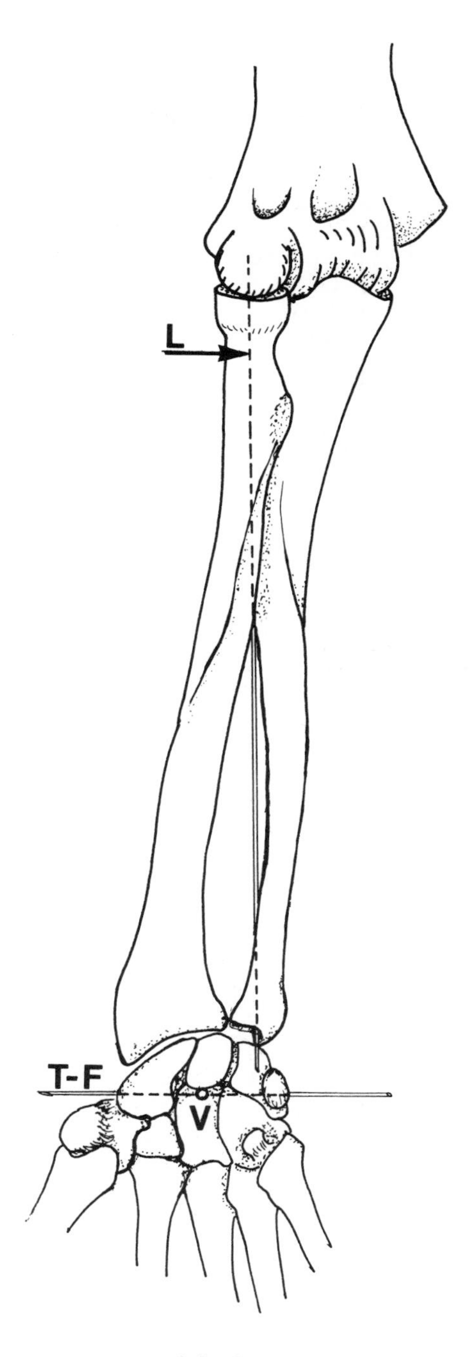

Figure 2.11. Axes of the forearm and wrist. L, long axis of the forearm for pronation and supination: T-F, compromise transverse axis in the frontal plane for wrist flexion and extension; V, volardorsal axis for radial and ulnar deviation of the hand at the wrist.

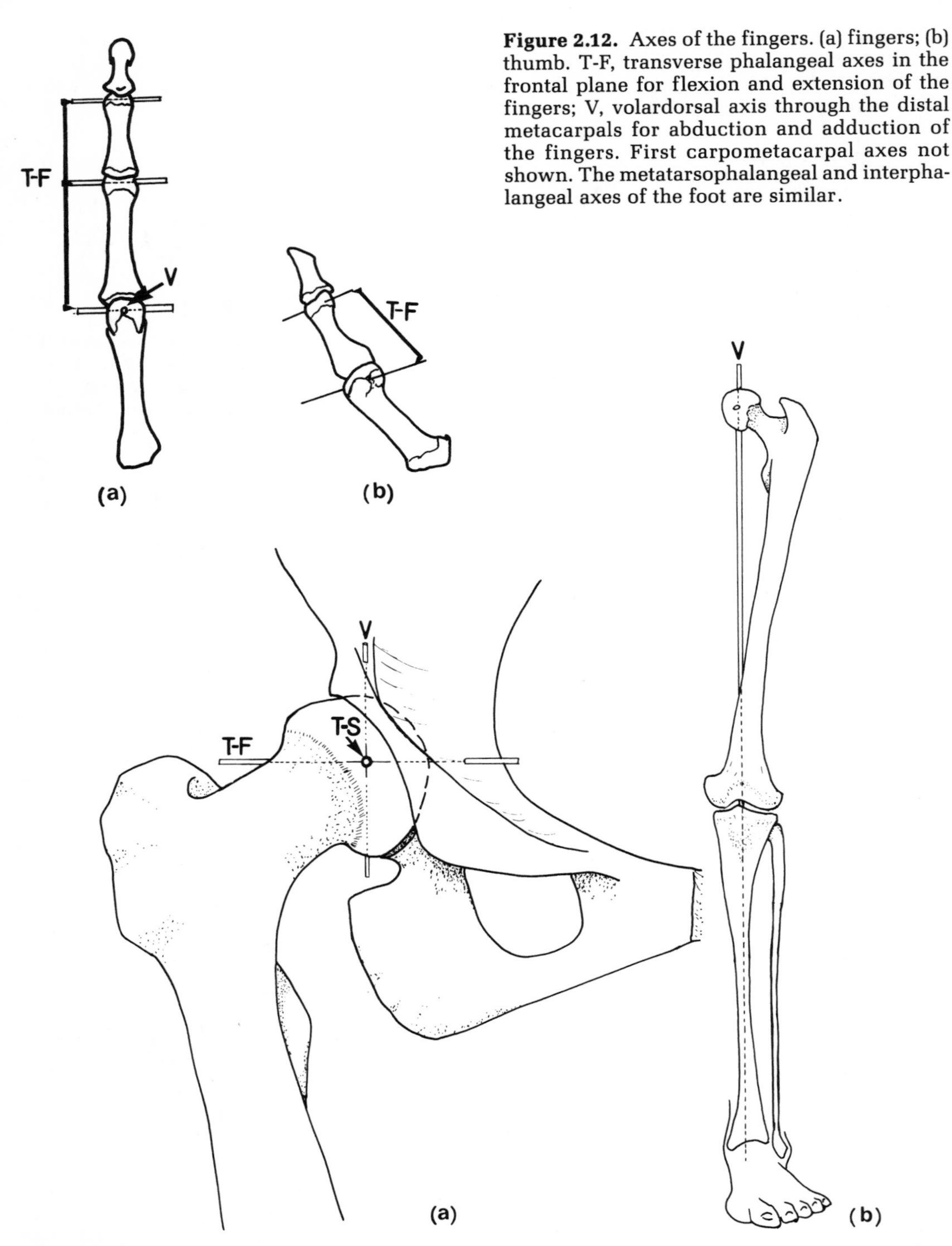

Figure 2.12. Axes of the fingers. (a) fingers; (b) thumb. T-F, transverse phalangeal axes in the frontal plane for flexion and extension of the fingers; V, volardorsal axis through the distal metacarpals for abduction and adduction of the fingers. First carpometacarpal axes not shown. The metatarsophalangeal and interphalangeal axes of the foot are similar.

Figure 2.13. Axes for movements at the hip joint. T-F, transverse axis in the frontal plane for flexion and extension of the thigh; T-S, transverse axis in the sagittal plane for abduction and adduction of the thigh; V, vertical axis for inward and outward (lateral and medial) rotation of the thigh. Note location of axis through the lower limb in (b).

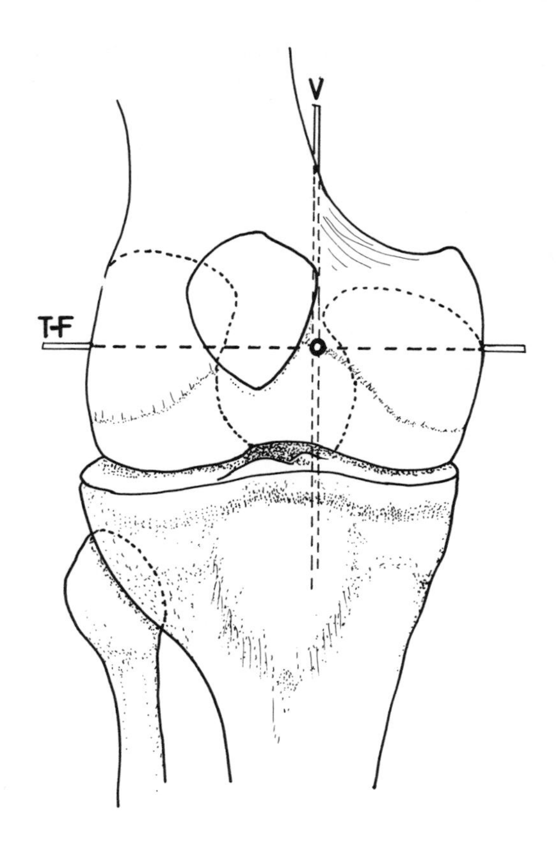

Figure 2.14. Axes through the knee joint. T-F, transverse axis in the frontal plane for flexion and extension; V, vertical axis around which the tibia can rotate when the knee is flexed.

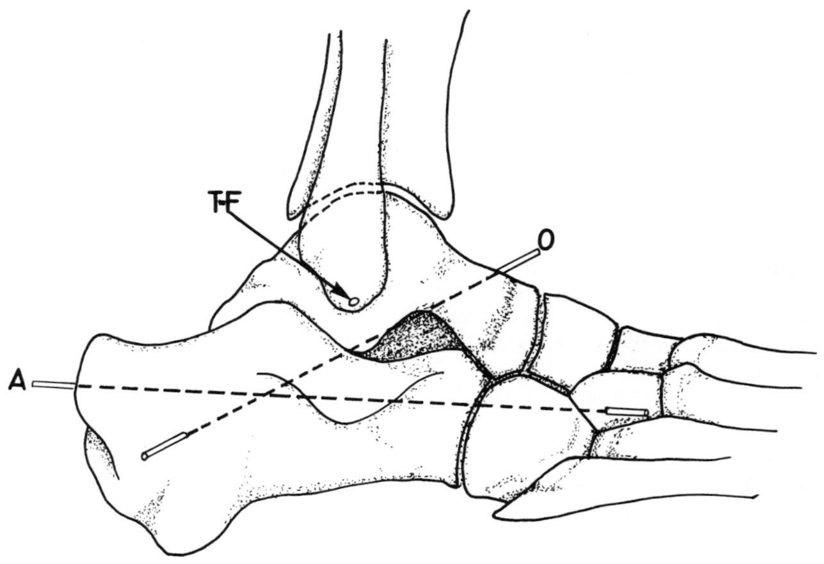

Figure 2.15. Axes of the ankle and foot. T-F, transverse axis in the frontal plane passing through both malleoli and the talar head. The only movements are dorsiflexion and plantar flexion. A and O are compromise axes for inversion and eversion at the intertarsal joints, A through Chopart's joint and O through the talocalcaneal and talonavicular joints.

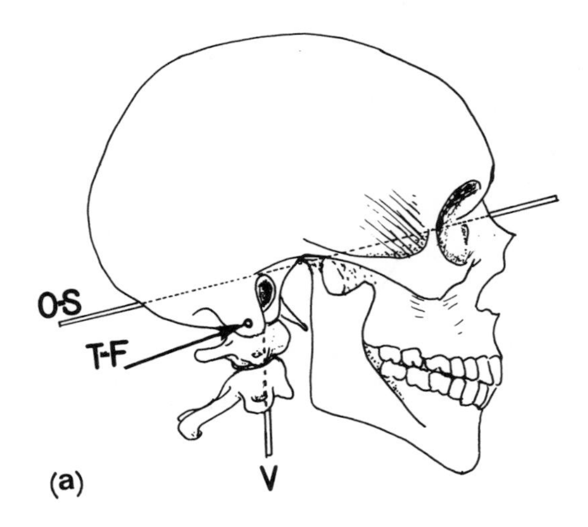

Figure 2.16(a). Joint axes of the axial skeleton axes for movement of the skull on the vertebrae. O-S, oblique axis in the sagittal plane for lateral flexion of the head; T-F, transverse axis in the frontal plane for dorsi- and ventriflexion of the head; V, vertical axis for head rotation, right and left.

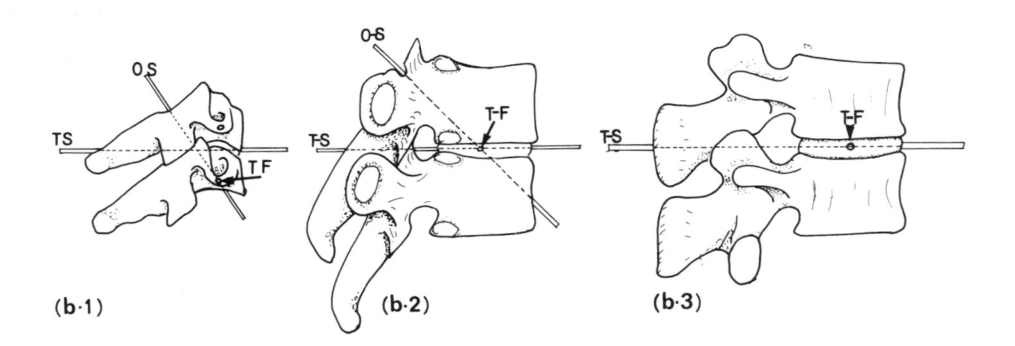

Figure 2.16(b). Axes for movement of one vertebra on the adjacent one below: 1, cervical; 2, thoracic; 3, lumbar vertebrae. O-S, oblique axis in the sagittal plane for movements of rotation combined with abduction; T-F, transverse axis in the frontal plane for movements of flexion and extension; T-S, transverse axis in the sagittal plane for movements of lateral flexion (abduction) right and left.

point forward, the great toes are inward, and the ventral surfaces face backward, as do the soles of the feet (volar surface) when one is standing on the toes. The proximal surface of the limbs retains its embryological orientation in the regions of the axillae and groin. Because of this situation a large portion of the upper part of the thigh still presents some ventral surface on the anterior aspect; therefore, a movement of the lower extremity forward and upward at the hip joint (Fig. 2.17(b)) is an approximation of ventral surfaces and conforms to the latter definition of flexion. Because the rotation is complete at the knee, flexion of this joint also meets

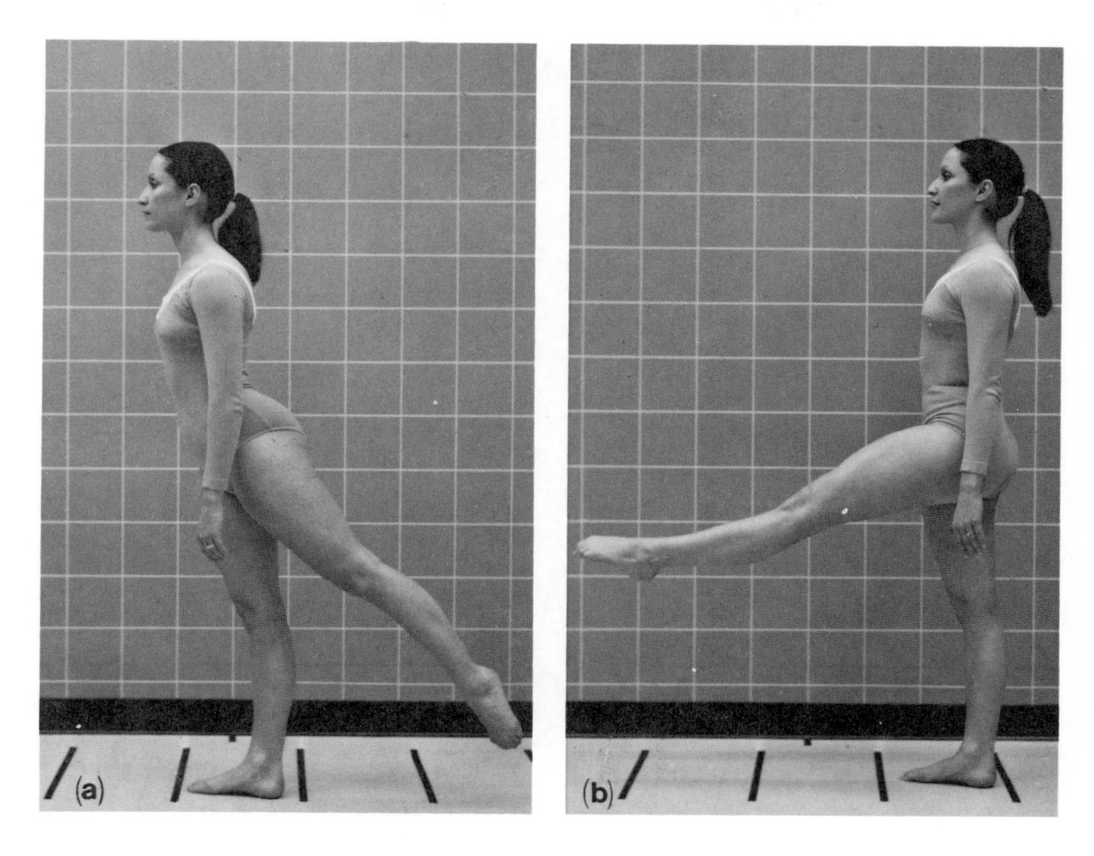

Figure 2.17. Viewing left side in sagittal plane. (a), hip extension; (b), hip flexion.

the anatomical definition. Shoulder flexion and extension are not easily reconciled to either definition, so these movements are correlated with the direction of the movements at the hip: flexion of the arm at the shoulder is defined as movement forward and upward in the sagittal plane and extension as movement downward and backward in the same plane (Fig. 2.19). Figure 2.19(c) depicts hyperflexion of the shoulder. The prefix "hyper" always describes an exceptional continuation of a movement. In hyperextension this usually means a continuation of extension beyond the anatomical position.

Flexion of the elbow, wrist, fingers, toes, and vertebral column all conform to both concepts, i.e., the anatomical concept of approximating the ventral or volar surfaces and the popular concept of decreasing the angle between the body segments. Extension of these same joints is, of course, movement in the opposite direction. However, this agreement between anatomical and popular definitions breaks down when the concepts on which the definitions are based are applied to the movements occurring at the ankle joint.

Study of Figure 2.20 illustrates the divergence. Decrease of the angle between the foot and the leg, anatomical extension, is popularly called ankle flexion; "pointing the toes," anatomical flexion, is popularly known as ankle extension. If Figure 2.21 is consulted one sees how the anatomical terminology is derived. Because of this paradoxical situation the term **dorsiflexion** has been adopted for anatomical

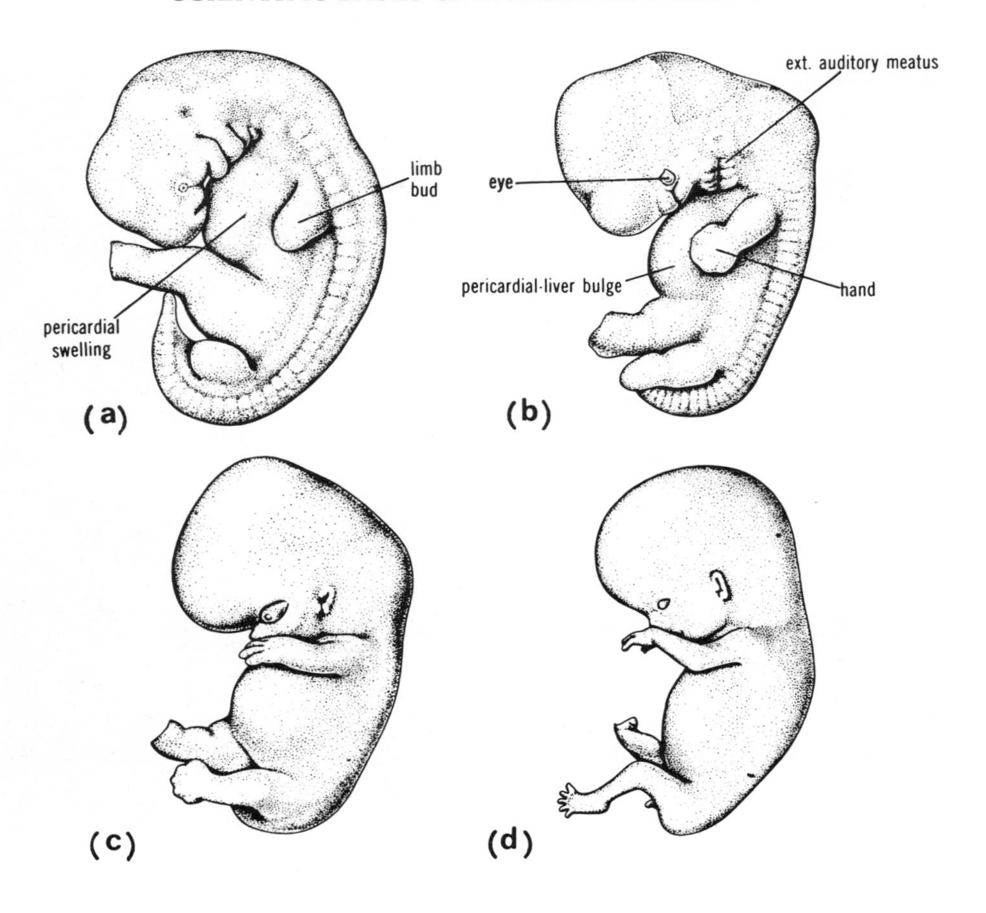

Figure 2.18. The human embryo: (a) at 5 weeks; (b) at 6 weeks; (c) at 7 weeks; (d) at 8 weeks. (From Langman, J., 1969. *Medical Embryology: Human Development, Normal and Abnormal*. Baltimore: The Williams & Wilkins Company.)

extension/popular flexion, and **plantarflexion** is the term applied for anatomical flexion/popular extension.

Abduction-Adduction

This pair of movements takes place in the frontal plane and occurs at biaxial (metacarpophalangeal and metatarsophalangeal) joints and at multiaxial (shoulder, hip, and first carpometacarpal) joints. **Abduction** of the fingers and toes is movement away from the middle digit, while **adduction** is movement toward that digit.

Abduction at a ball and socket joint (shoulder or hip) is movement of the limb upward and away from the midline (Fig. 2.22). At the glenohumeral joint the arm can be raised only 90° before the greater tuberosity of the humerus contacts the acromion process. Further abduction is accomplished by upward rotation of the glenoid fossa of the scapula (as depicted in Fig. 2.23). As a result, the total range of abduction of the upper extremity can be as much as 180°. Further details on movements of the scapula are presented later. In adduction of either the upper or lower extremity the limb may be drawn across the midline of the body (Fig. 2.24).

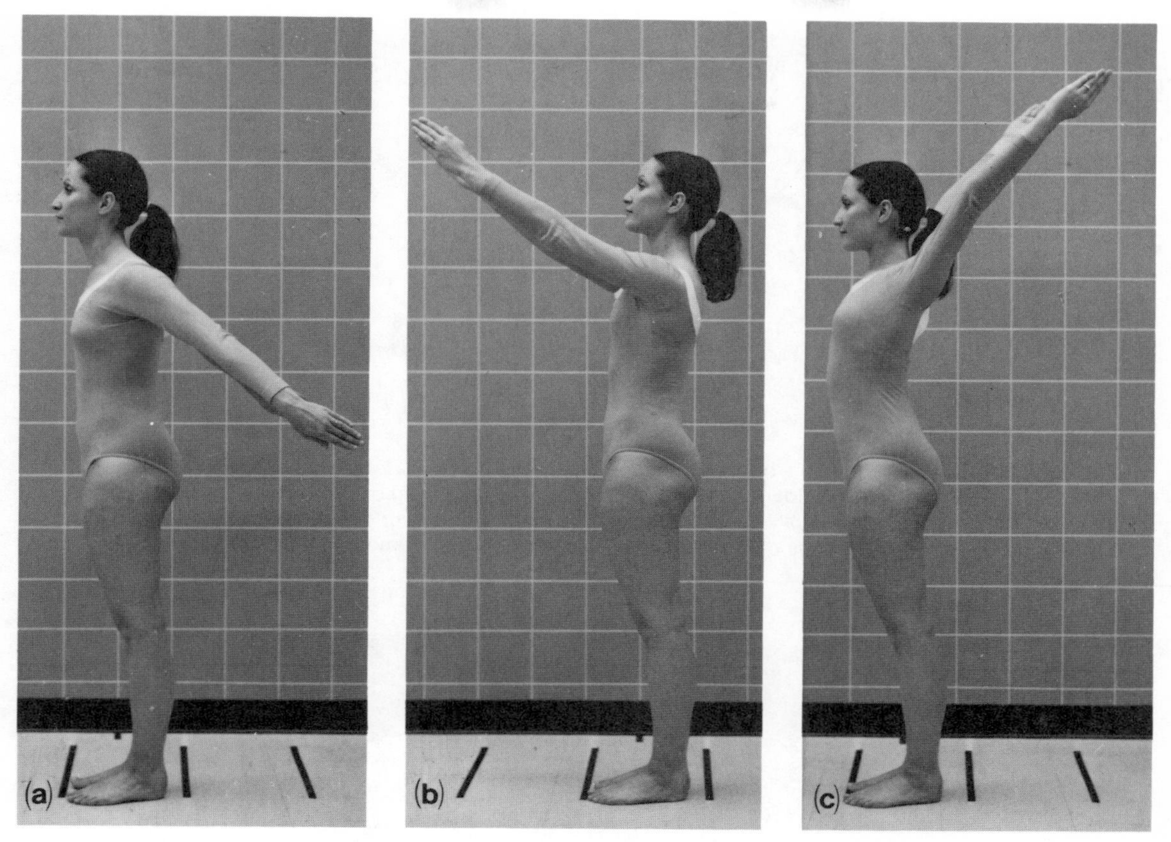

Figure 2.19. Viewing the left side in sagittal plane. (a) shoulder extension; (b) shoulder flexion; (c) shoulder hyperflexion.

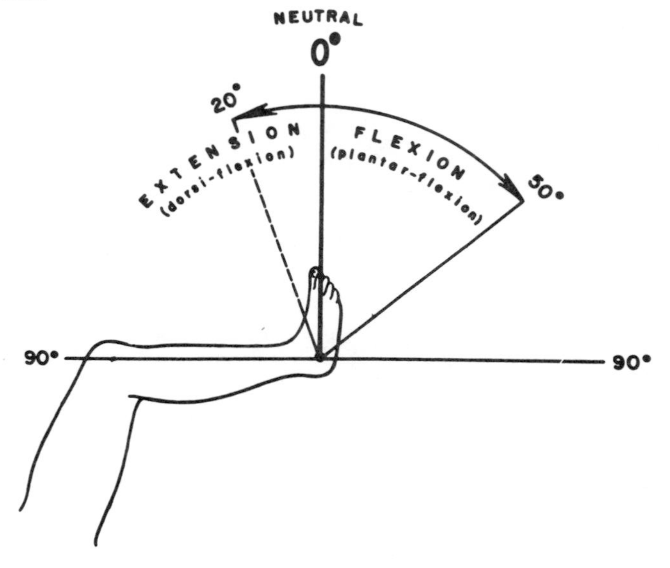

Figure 2.20. Movements at the ankle joint. Dorsiflexion and plantar flexion. (From American Academy of Orthopaedic Surgeons, 1965. *Joint Motion—Method of Measuring and Recording*).

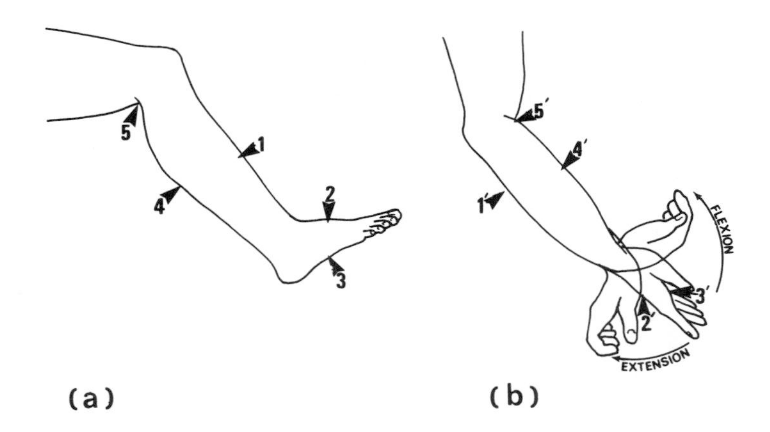

Figure 2.21(a). Lateral aspect of right lower extremity; (b) lateral aspect of upper extremity. 1, anterior surface of leg, homologous to back of forearm, 1′. 2, instep or dorsum of the foot, homologous to the back of the hand, 2′. 3, sole of foot, homologous to palm of the hand, 3′. 4, back of leg, homologous to ventral surface of forearm, 4′. 5, popliteal fossa homologous to the cubital fossa of the forearm, 5′.

Figure 2.22. Abduction. Both shoulders and the left hip are abducted.

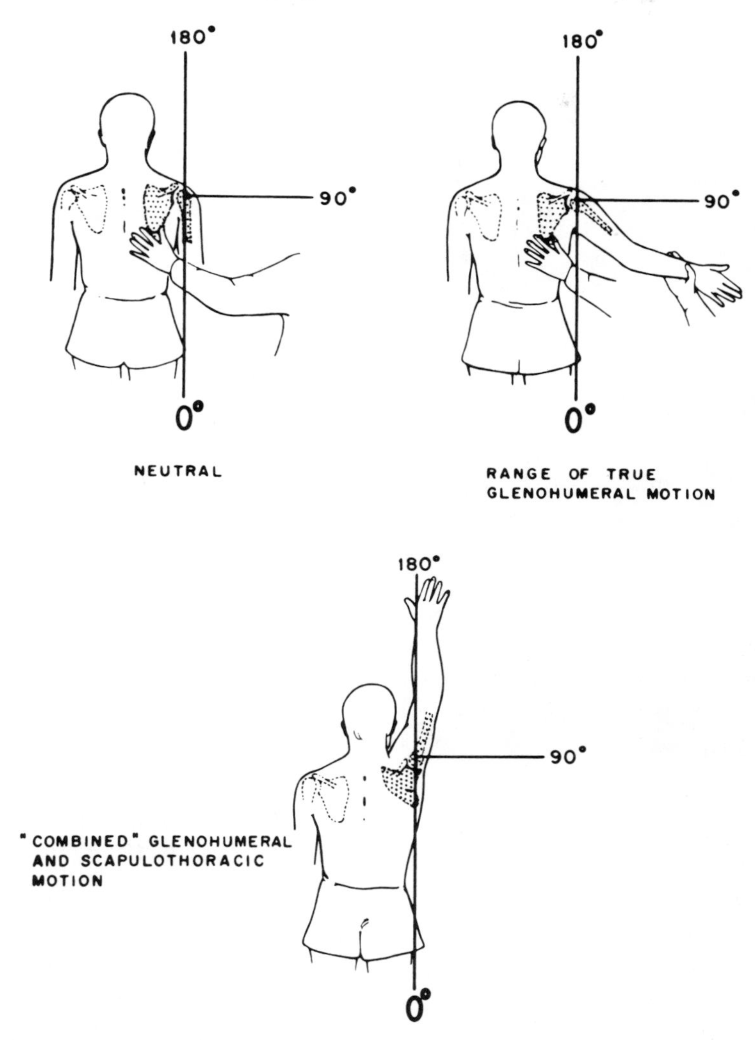

Figure 2.23. Glenohumeral motion. Note the upward rotation of the scapula as abduction is completed. (From American Academy of Orthopaedic Surgeons, 1965. *Joint Motion—Method of Measuring and Recording*).

Horizontal Flexion-Extension

In making reference to Figure 2.25 it will be seen that **horizontal flexion** is performed by the upper extremity from a position of abduction; the motion of the extremity is in a horizontal plane and the limb is carried forward across the front of the body. **Horizontal extension** is similar movement in the opposite direction. There has been a tendency among some professionals to use the term horizontal adduction for horizontal flexion because the arm is moved across the midline of the body. It should be noted, however, that those horizontal movements occur around the same axis through the head of the humerus as do flexion and extension of the shoulder performed from the anatomical position. The abduction of the arm has merely changed the orientation of this axis through the head of the humerus from the horizontal to the vertical (Fig. 2.9).

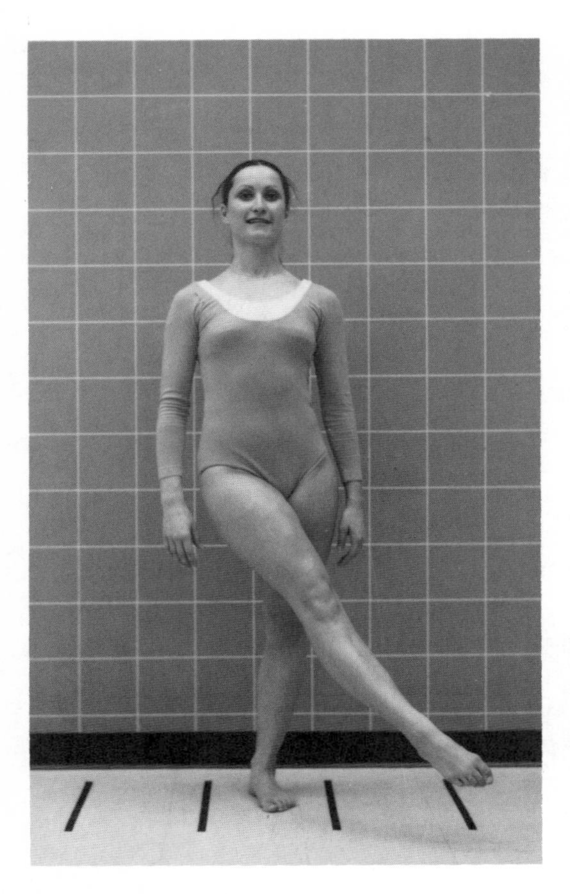

Figure 2.24. Adduction at the right hip.

Medial (Inward) and Lateral (Outward) Rotation

Rotation movements occur at multiaxial joints around a longitudinal axis running the length of the bone of the rotating segment but not necessarily within the shaft of the bone (Fig. 2.9, 2.11, and 2.13). With the exception of the long axis of the clavicle (which rotates slightly with most movements of the upper extremity, see below under "Movements of the Shoulder Girdle"), the rotation axes are vertical or near vertical when the body is in the anatomical position. As vertical axes are defined by the intersection of the frontal and sagittal planes, it follows that the movement of rotation around these axes must occur in the transverse plane. Rotation of an upper extremity in either direction is seen most clearly when the elbow is flexed, as this has the effect of visually magnifying the amount of rotation because of the length of the forearm and hand (Fig. 2.26). Elbow flexion also eliminates the addition of pronation (or supination) to the position of the hand as an indication of the amount of shoulder rotation. **Medial** or **lateral rotation** of the humerus (Fig. 2.27) or femur (Fig. 2.28) may take place with the bone in any starting position, but rotation at the hip is more limited than at the shoulder.

Circumduction

This may occur at any biaxial or multiaxial joint and is a combination of flexion-abduction-extension-adduction or the reverse, and it may involve rotation of the limb concerned. The extremity travels in a cone-shaped path with the apex at the fulcrum of the joint at which the movement originates (Fig. 2.29).

The hip and shoulder joints, described as multiaxial joints, differ in the amount of flexibility permitted by their structures. In the hip joint, which requires stability, the socket of bone and cartilage is exceptionally deep. In the shoulder, which requires more flexibility, the bony and soft tissue parts of the socket are shallow. The resulting looseness of the upper extremity makes it possible to move through large ranges of flexion/extension and abduction/adduction.

Codman's Paradox

An interesting paradox results when large ranges of movement are attempted. Concomitant with any of the movements of flexion, extension, abduction, and adduction through large ranges is the rotation of the upper extremity around its long axis. Figure 2.30 illustrates the fact that approximately 180° of lateral rotation has automatically occurred as a direct result

Figure 2.25. Horizontal flexion/extension at the right shoulder. (a) horizontal flexion or shoulder adduction; (b) horizontal extension or shoulder abduction.

of moving the right upper extremity, first, through 180° of shoulder flexion until the upper arm rests beside the ear (Fig. 2.30(a), (b), and (c)); and second, through 180° of shoulder adduction until the upper extremity rests beside the body (Fig. 2.30(d), (e), and (f)). Notice the position of the hand at the start and finish of the series (Fig. 2.30(a), and (f)). The reader is invited to demonstrate personally the fact that 180° of medial rotation occurs at the shoulder joint when combinations of abduction and extension are used: from anatomical position with the hand in a neutral position, i.e., palm facing the thigh), abduct the upper extremity 180° until the upper arm rests beside the ear; then, move the upper extremity through 180° of extension until the upper extremity is vertical beside the body; the back of the hand will now be facing the thigh. Ranges of movement less than 180° result in smaller amounts of rotation occurring, commensurate with the amount of flexion, extension, abduction, or adduction. Clearly, none of the shoulder joint sagittal and frontal plane movements can be functionally classified as purely flexion/extension or abduction/adduction, respectively, because the skeletal framework requires that rotation at the shoulder joint must also occur at the same time.

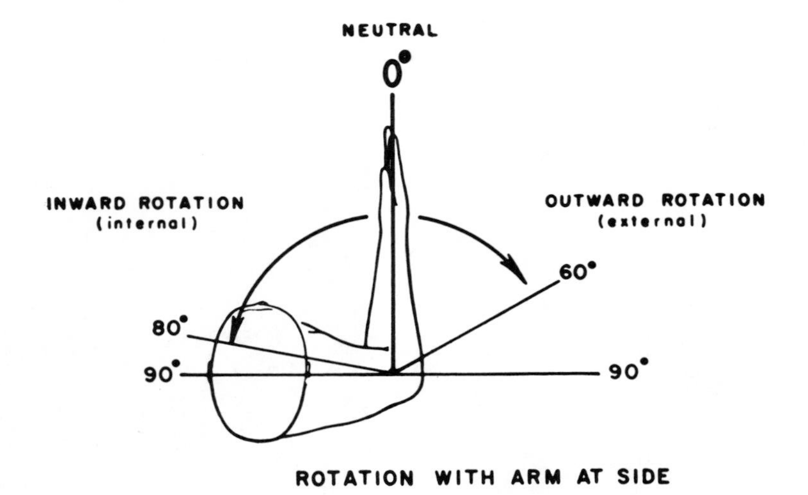

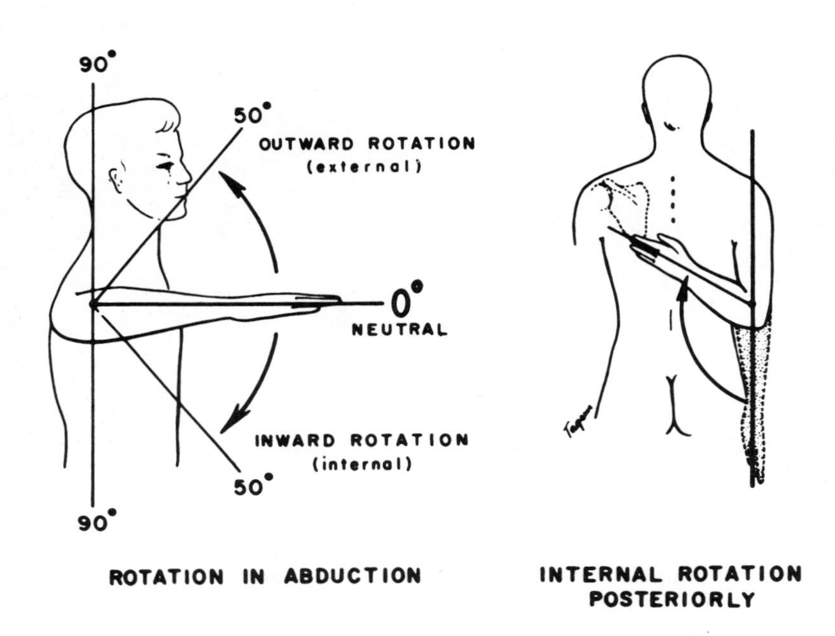

Figure 2.26. Inward and outward (medial and lateral) rotation at the shoulder. (From American Academy of Orthopaedic Surgeons, 1965. *Joint Motion—Method of Measuring and Recording*).

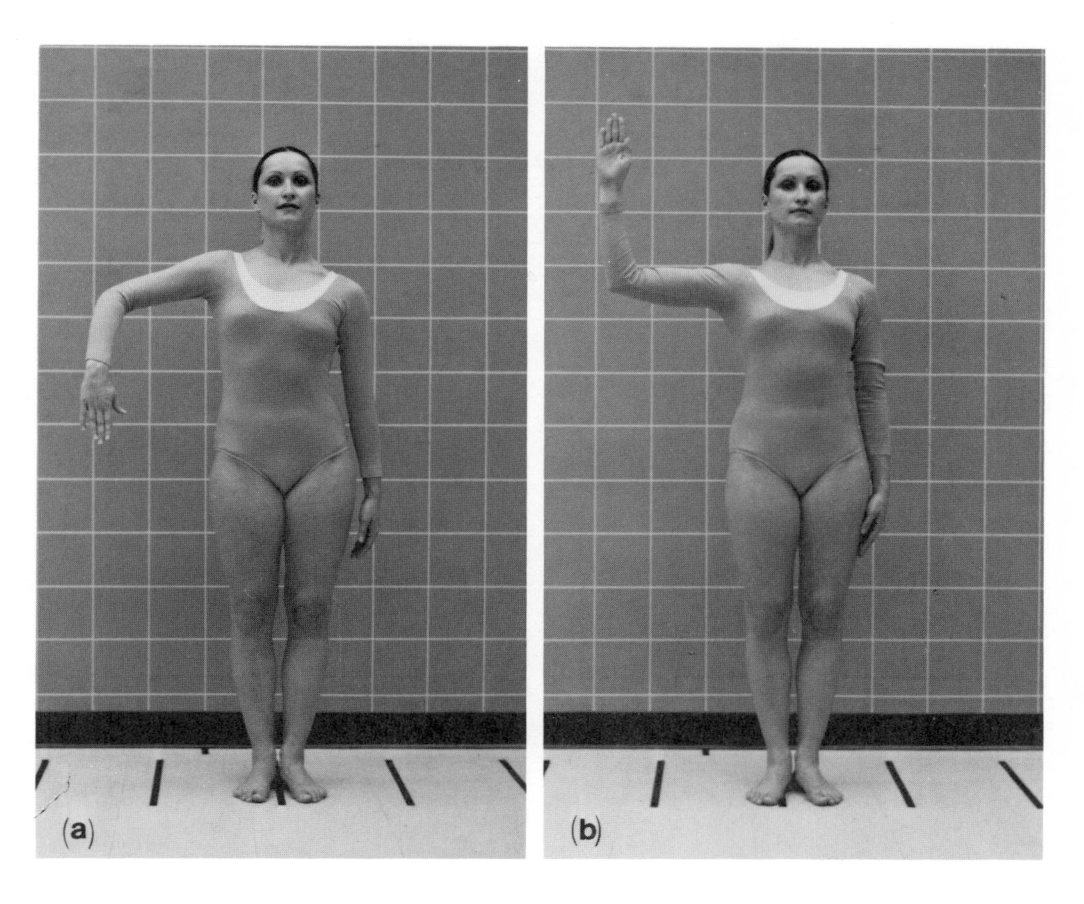

Figure 2.27. Rotation of the humerus at the right shoulder. (a) medial rotation with upper extremity abducted about 90° at the shoulder and flexed 90° at the elbow. (b) lateral rotation with upper extremity abducted 90° at the shoulder and flexed 90° at the elbow.

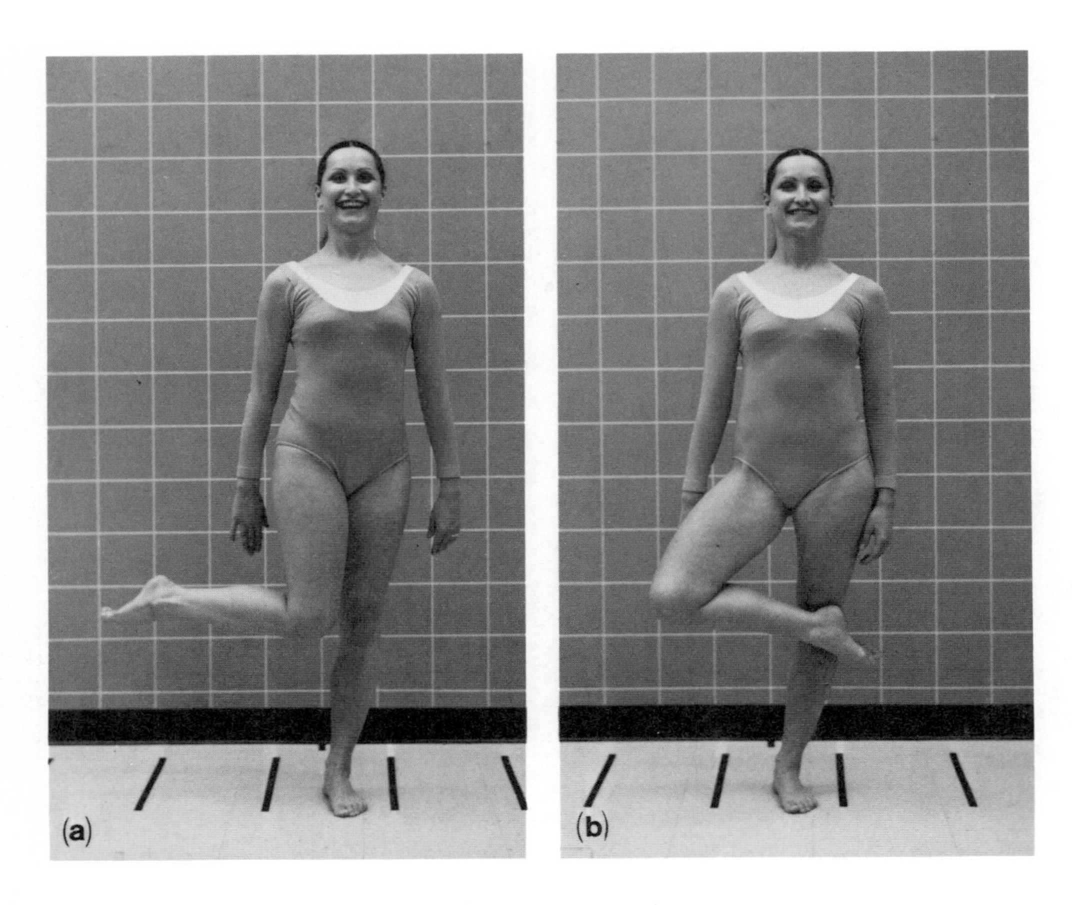

Figure 2.28. Rotation of the right thigh at the hip. (Knee is flexed to exaggerate the movement.) (a) medial (inward) rotation; (b) lateral (outward) rotation.

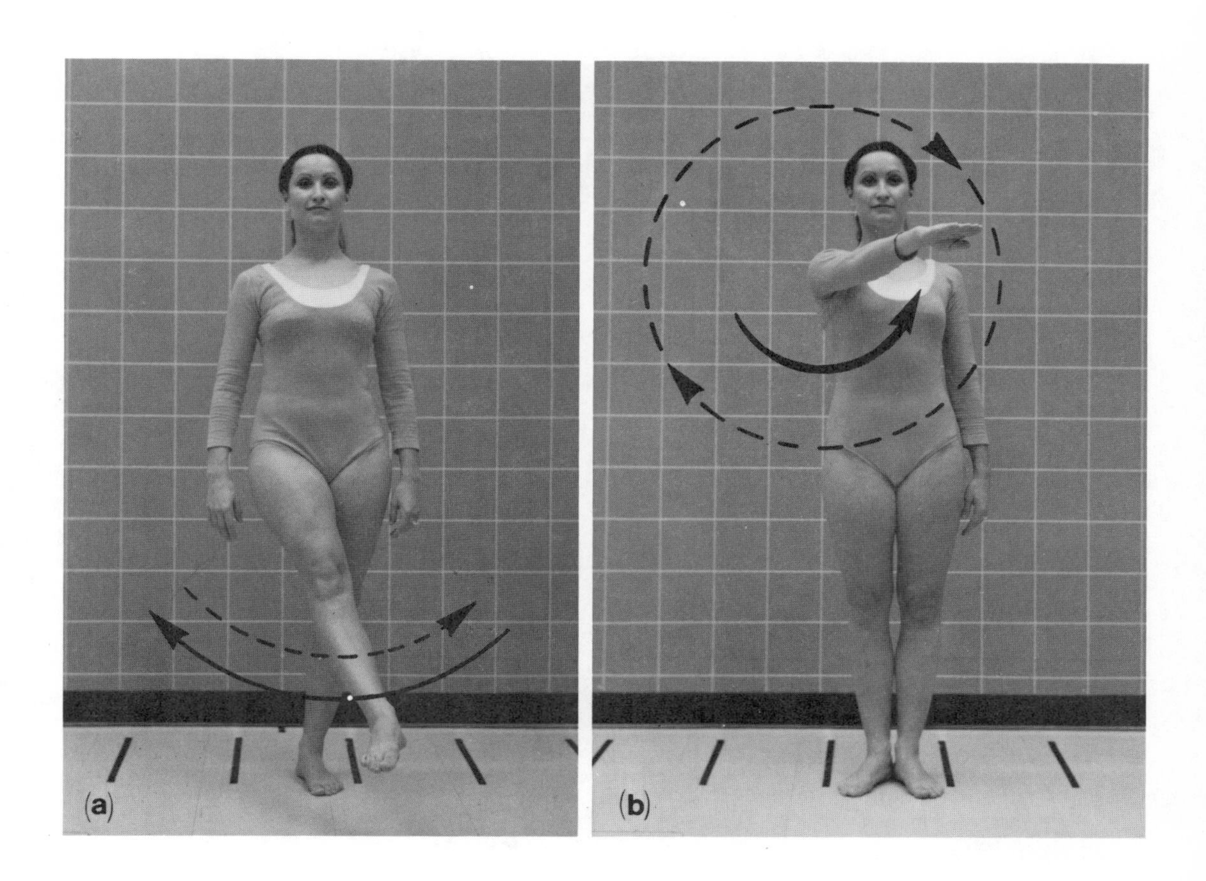

Figure 2.29. Circumduction. (a) of right lower extremity; (b) of right upper extremity.

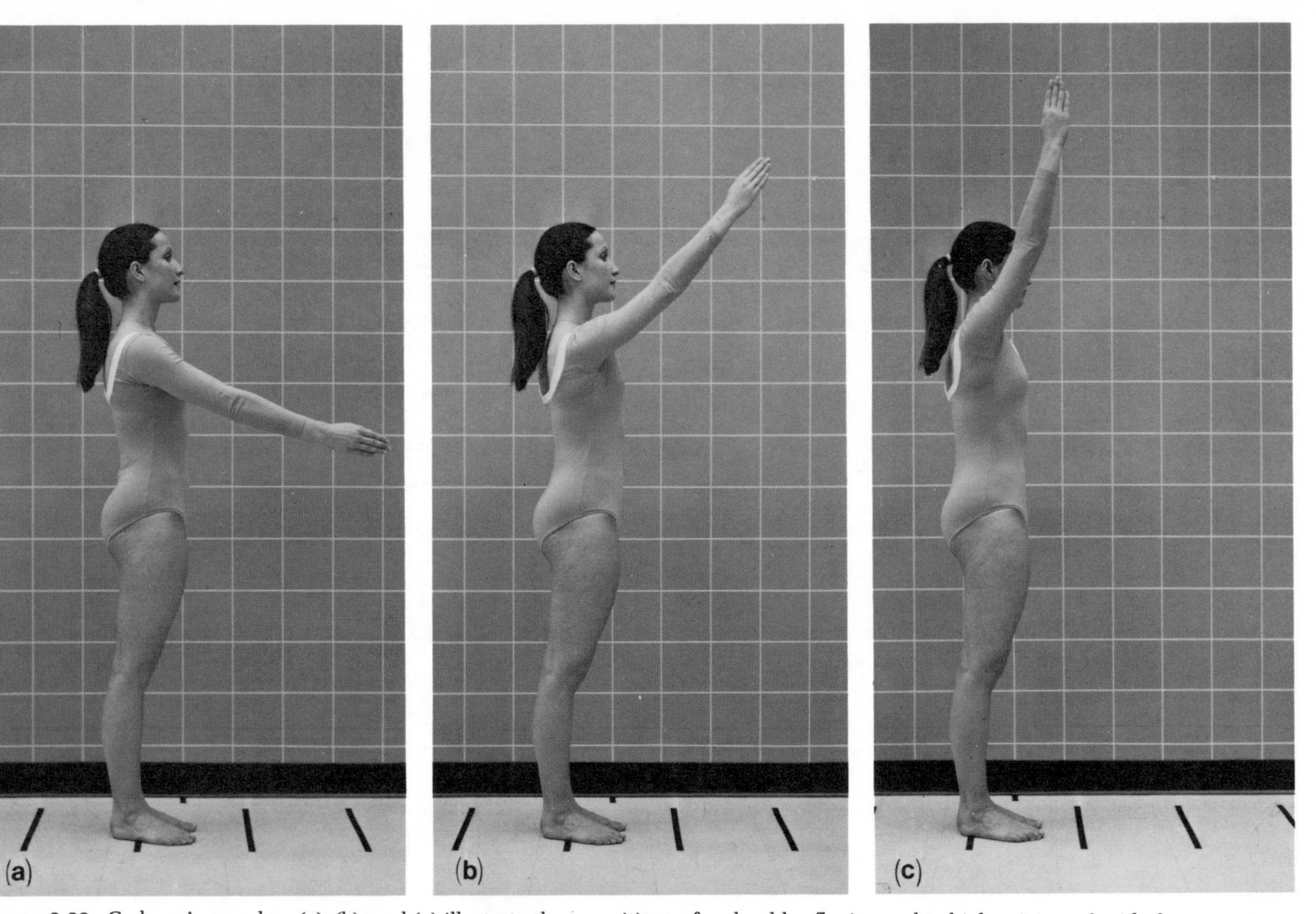

Figure 2.30. Codman's paradox. (a), (b), and (c) illustrate three positions of a shoulder flexion task which originated with the arm resting beside the body. (d), (e), and (f) illustrate three positions of a shoulder adduction task which originated with the arm beside the ear, as in (c). See text for explanation.

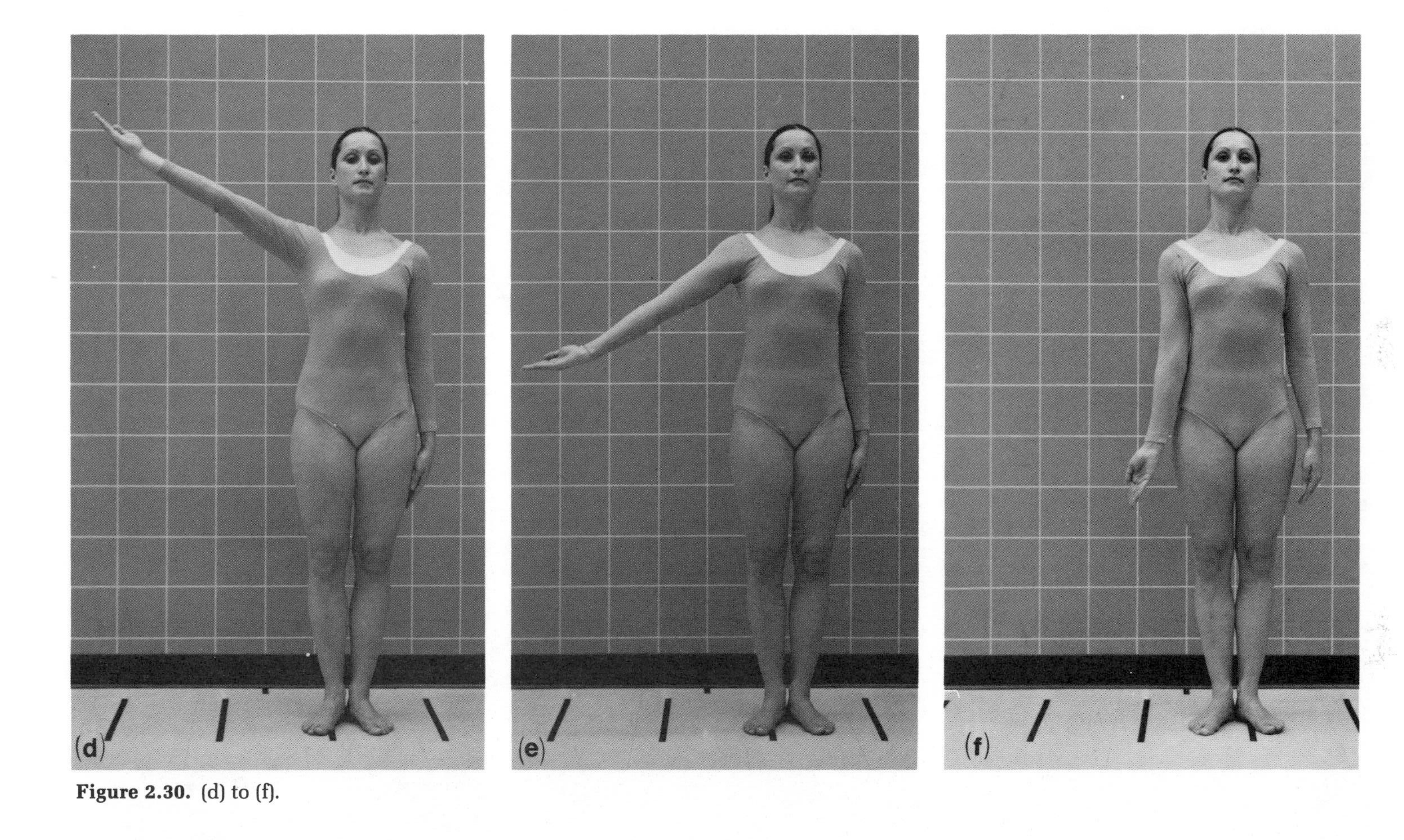

Figure 2.30. (d) to (f).

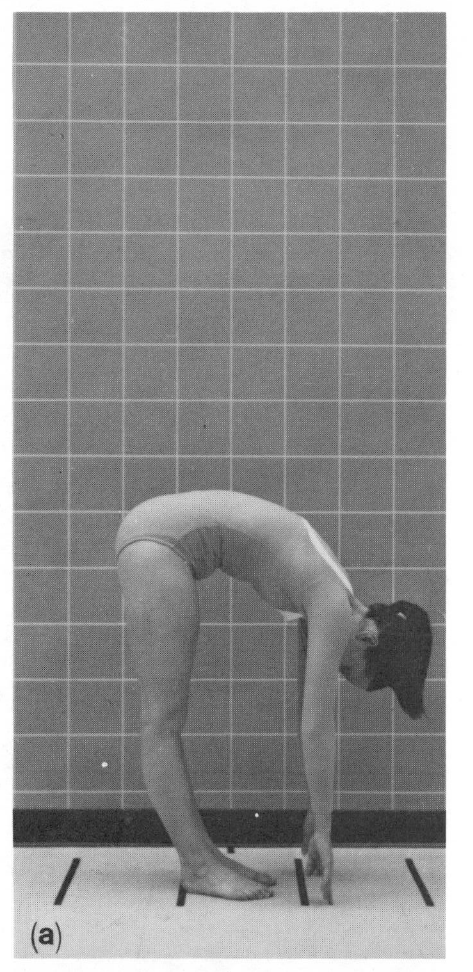

Figure 2.31. Movements of the vertebral column. (a) flexion; (b) flexion combined with hip hyperflexion; (c) extension; (d) hyperextension; (e) lateral flexion (abduction) to the right; (f) rotation to the left.

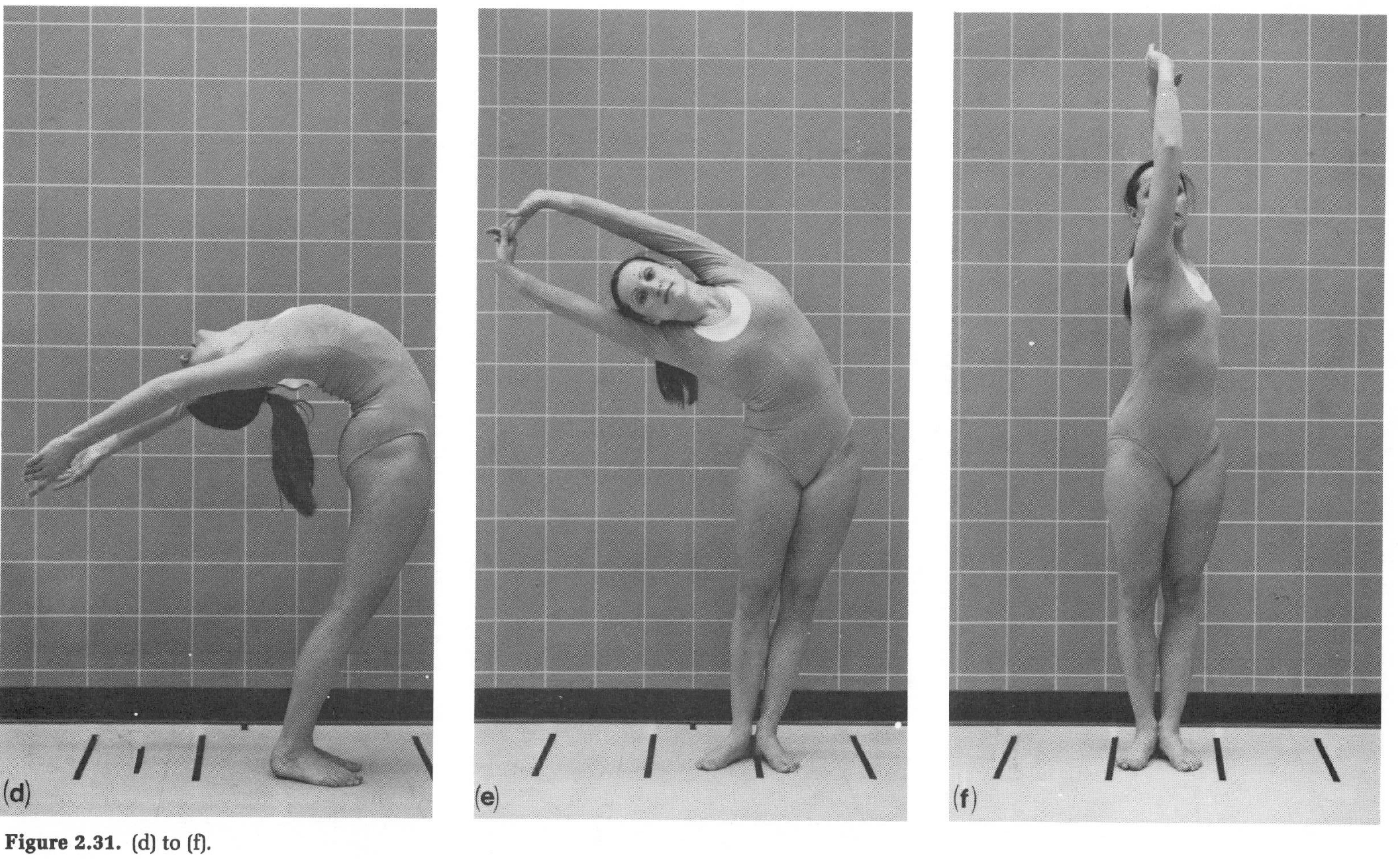

Figure 2.31. (d) to (f).

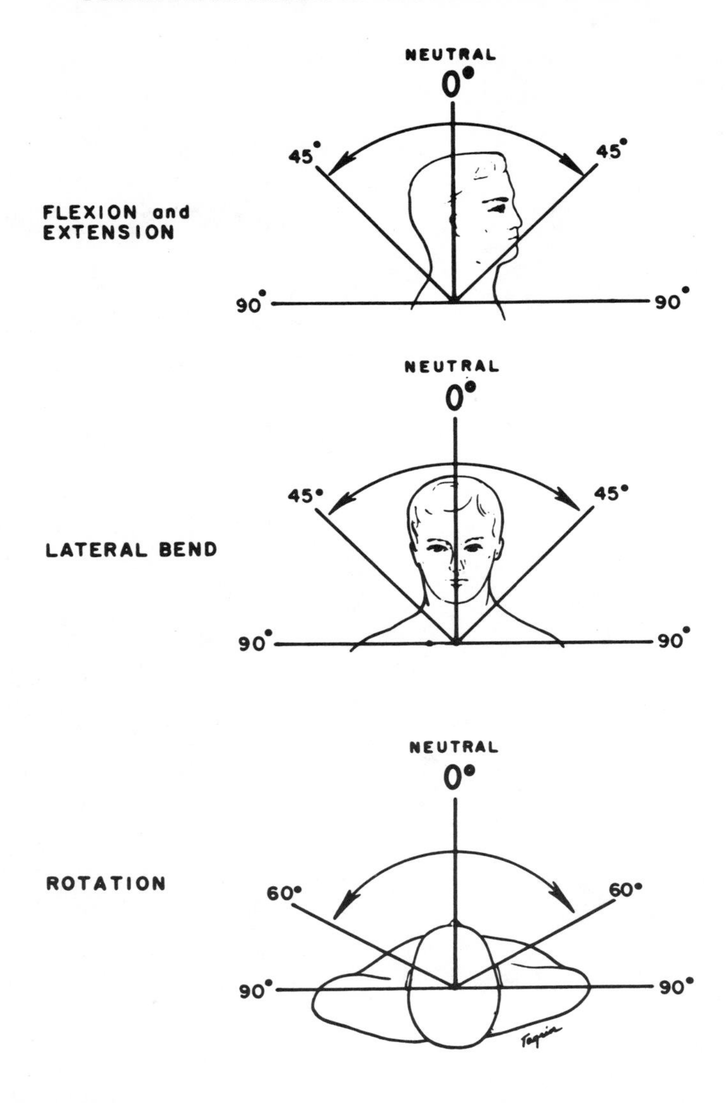

Figure 2.32. Actions at the cervical spine. (From American Academy of Orthopaedic Surgeons, 1965. *Joint Motion—Method of Measuring and Recording*).

MOVEMENTS OF THE AXIAL SKELETON (VERTEBRAL COLUMN)

The appearances of major movements of the vertebral column are reflected in Figure 2.31.

The definition of flexion as an approximation of ventral surfaces and of extension as movement in the opposite direction is appropriate here (Fig. 2.31(a) to (d)). Optional terms are **ventriflexion** and **dorsiflexion**, respectively. Movement of the trunk to either side is called lateral bending by the orthopaedic surgeons and **lateral flexion** (or sometimes abduction) by anatomists (Fig. 2.31(e)). **Rotation to right** or **left** can occur around the lengthwise axis of the vertebrae (Fig. 2.31(f)).

The way that these same movements affect the head when they occur in the cervical vertebrae is illustrated in Figures 2.32 and 2.33. Figure 2.34 demonstrates atlantooccipital movements. Most head movements require a combination of movement about both the cervical intervertebral and atlantooccipital axes.

MOVEMENTS AT JOINTS WITHIN THE FOREARM

Pronation and Supination

These movements occur at the pivot joints within the forearm, not at the wrist or elbow. The axis runs from the center of the head of the radius to the center of that of the ulna (Fig. 2.11). In the anatomical position the palm faces forward because the forearm is supinated. When the elbow is flexed 90° with the palm perpendicular to the floor and the thumb uppermost, the forearm is in the neutral or midposition (Fig. 2.35). Rotating the forearm medially so that the palm turns downward is the movement of **pronation**; rotating the forearm in the opposite direction, i.e., laterally, turns the palm upward and is the movement of **supination**. In both of these movements the ulna remains comparatively stationary while the radius is the bone that moves. When the forearm is supinated the radius lies almost parallel to the ulna; when the forearm is pronated, the radius lies across the ulna (Fig. 2.36). Pronation and supination are illustrated in Figure 2.37; both extremities are abducted at the shoulder joints and flexed at the elbow joints.

Radial and Ulnar Deviation of the Hand at the Wrist

These actions at the wrist joint occur in the frontal plane as the hand is moved toward the radial aspect of the forearm **(radial deviation)** or toward the ulnar side **(ulnar deviation)** (Fig. 2.38). Radial deviation replaces the older term of radial abduction, as ulnar deviation replaces that of ulnar abduction.[c]

MOVEMENTS AT JOINTS WITHIN THE FOOT

Inversion and Eversion

Inversion and **eversion** (Fig. 2.39) are a result of many small gliding movements between the intertarsal and tarsometatarsal joints. Hicks (1953) located as many as 12 different axes in the foot around which these movements occur, each contributing to an end result of either inversion or eversion. Compromise axes for these movements are indicated in Figure 2.15.

The first of these movements, inversion, has been defined as raising the medial border of the foot and/or rotating the sole of the foot inward. Eversion is movement in the opposite direction and involves raising the lateral border of the foot and/or rotating the sole of the foot outward. Walking over rough, uneven, or sloping terrain would be difficult, if not impossible, without the availability of these movements.

---

[c] The terms radial and ulnar flexion are sometimes used to describe these movements, but as radial and ulnar movements do not meet the anatomical criteria for flexion (see under "Flexion-Extension," above), the term "deviation" is preferred.

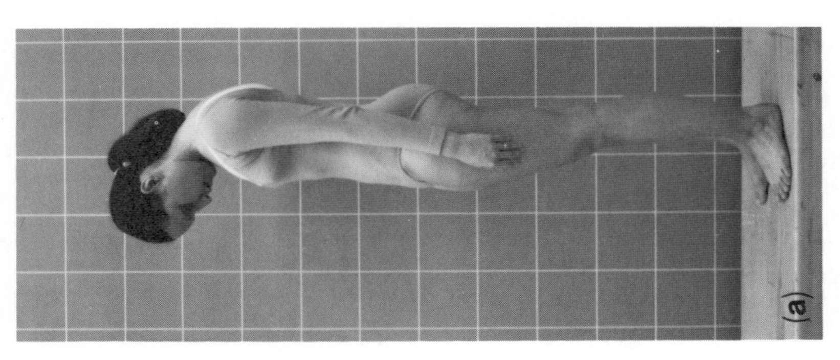

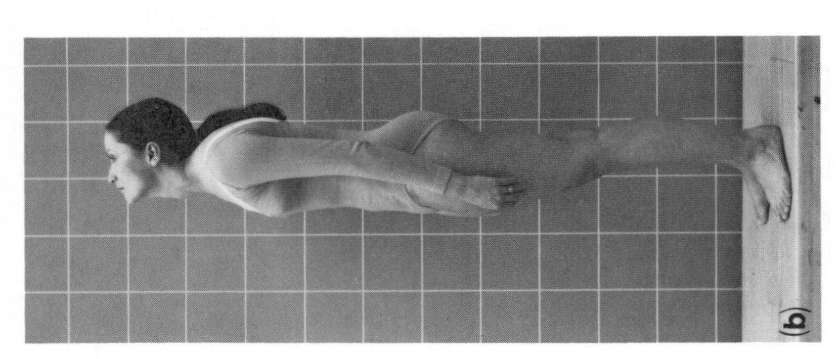

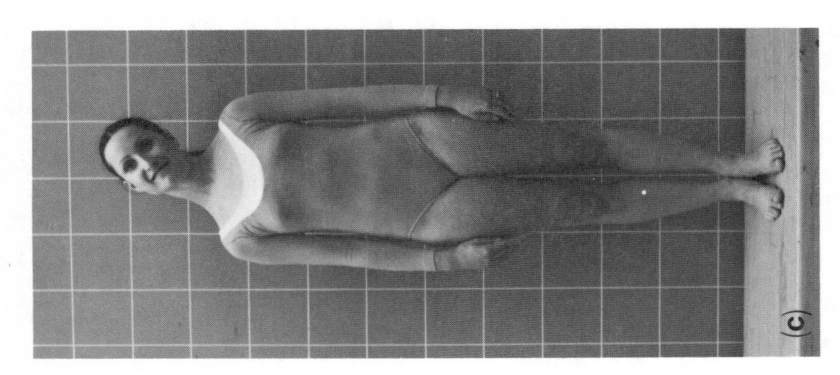

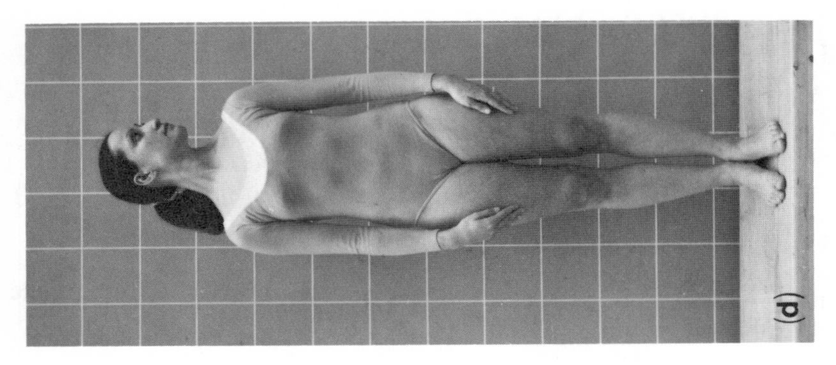

Figure 2.33. Cervical spine movements. (a) ventriflexion; (b) dorsiflexion; (c) lateral flexion; (d) rotation to the left.

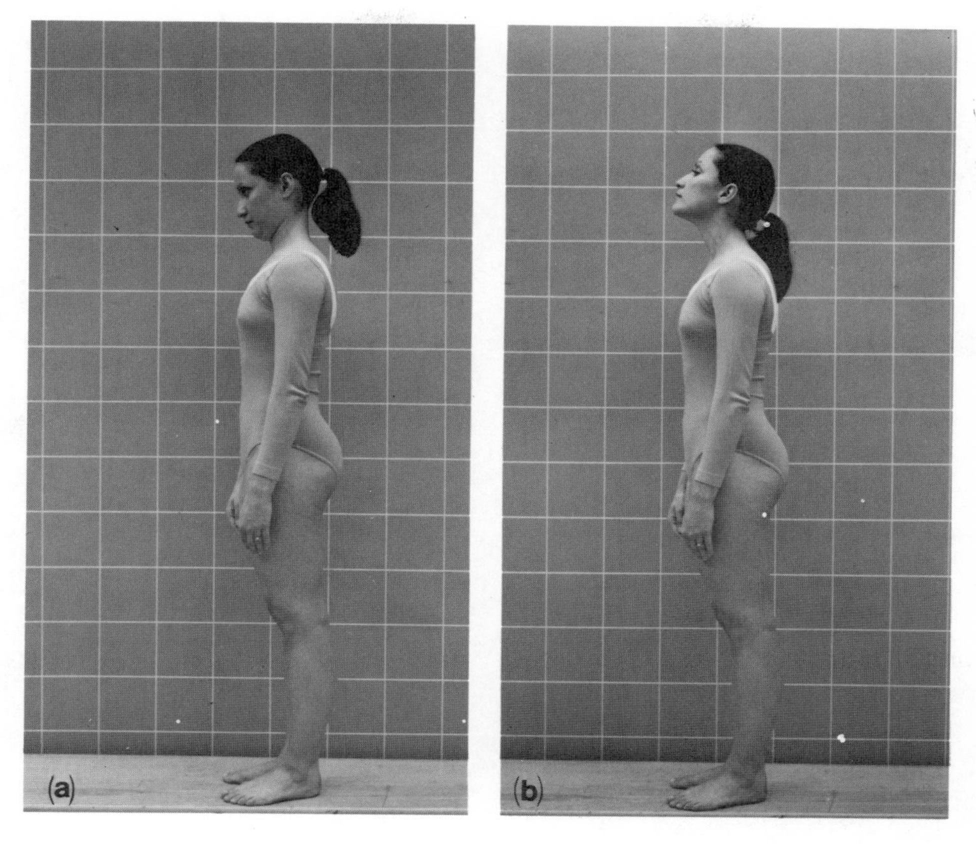

Figure 2.34. Atlantooccipital movements. (a) ventriflexion; (b) dorsiflexion.

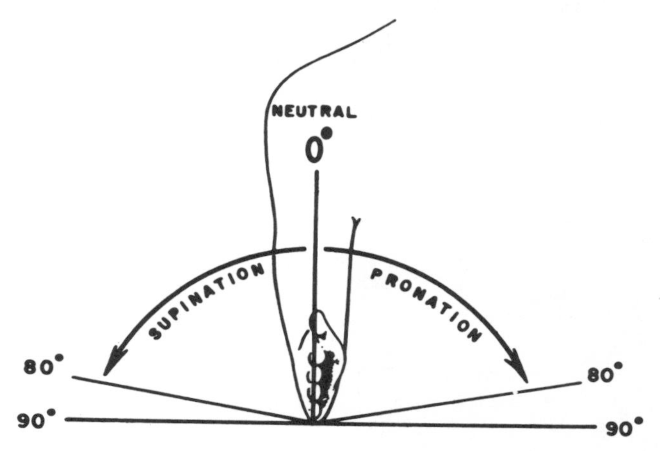

Figure 2.35. Pronation and supination at the radioulnar joints. (From American Academy of Orthopaedic Surgeons, 1965. *Joint Motion—Method of Measuring and Recording*).

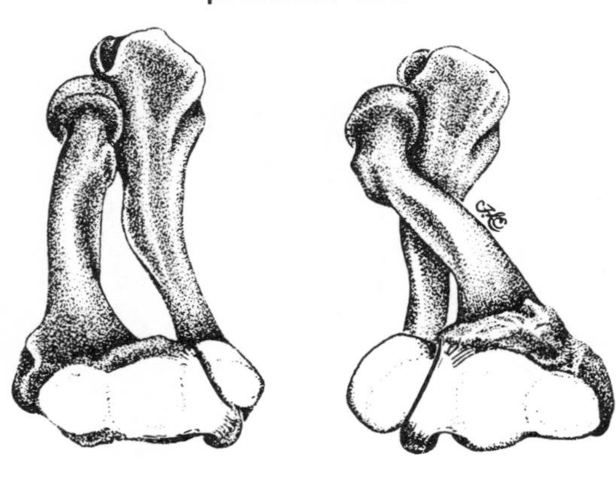

Figure 2.36. The bones of the right forearm as viewed from the volar surface: left, in supination; right, in pronation. (From MacConaill, M.A., and Basmajian, J.V., 1977. *Muscles and Movements: A Basis for Human Kinesiology*. Baltimore: The Williams & Wilkins Company.)

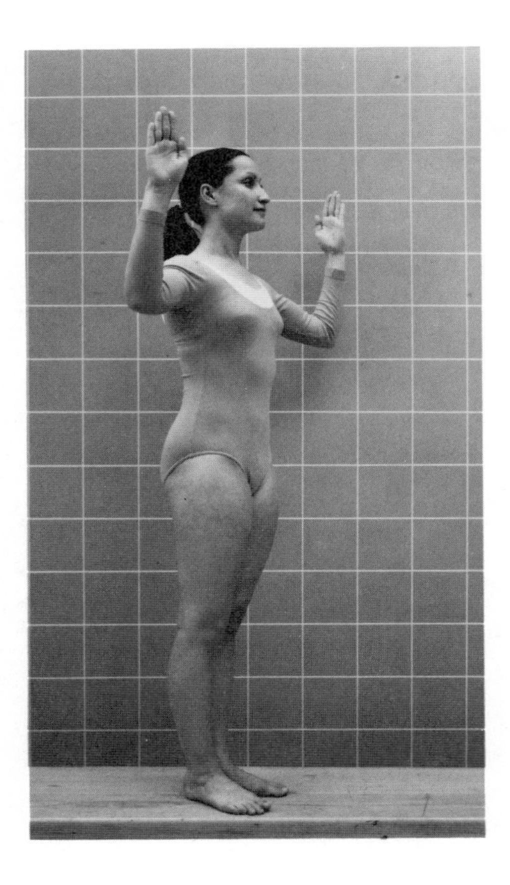

Figure 2.37. Pronation and supination. Right hand is pronated; left hand supinated.

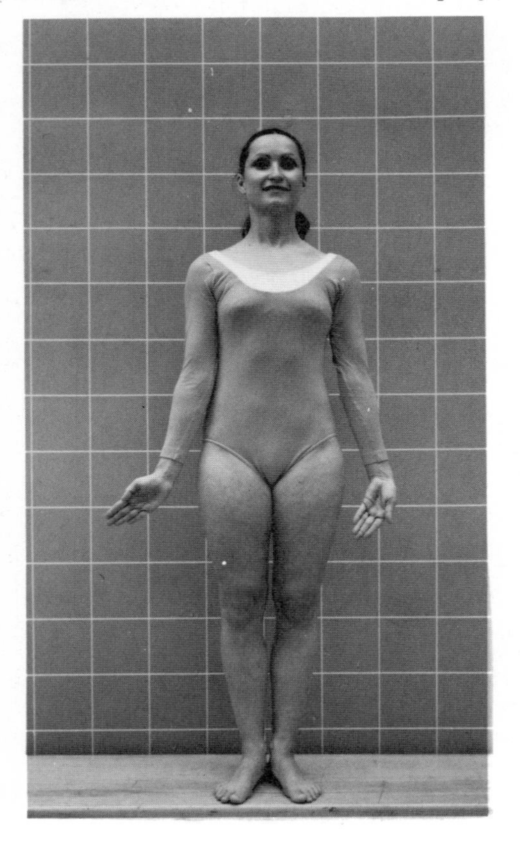

Figure 2.38. Deviation of hand on forearm. Subject exhibits right radial and left ulnar deviations of hand on forearm.

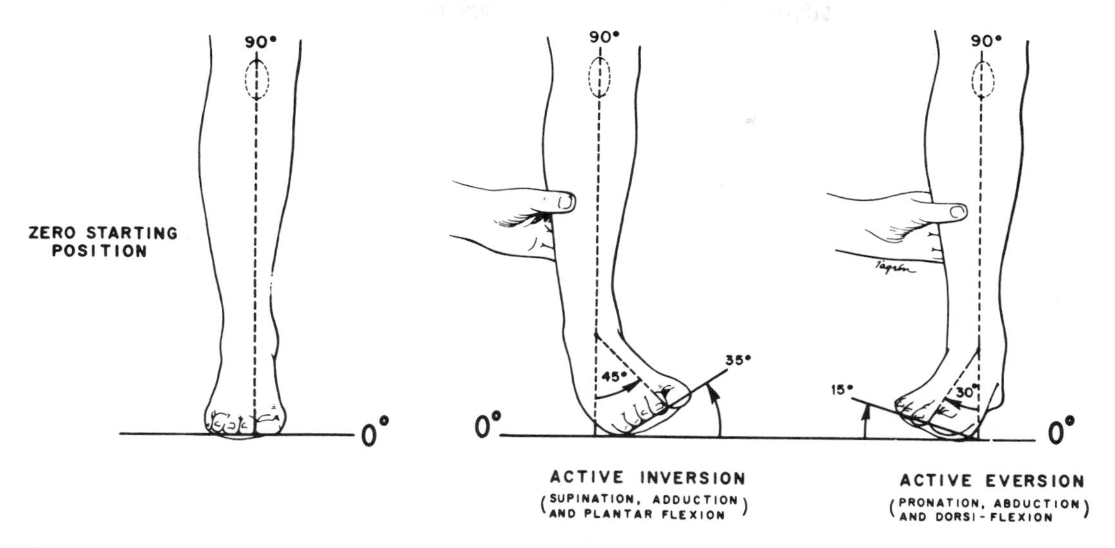

Figure 2.39. Movements of inversion and eversion of the foot. (From American Academy of Orthopaedic Surgeons, 1965. *Joint Motion—Method of Measuring and Recording*).

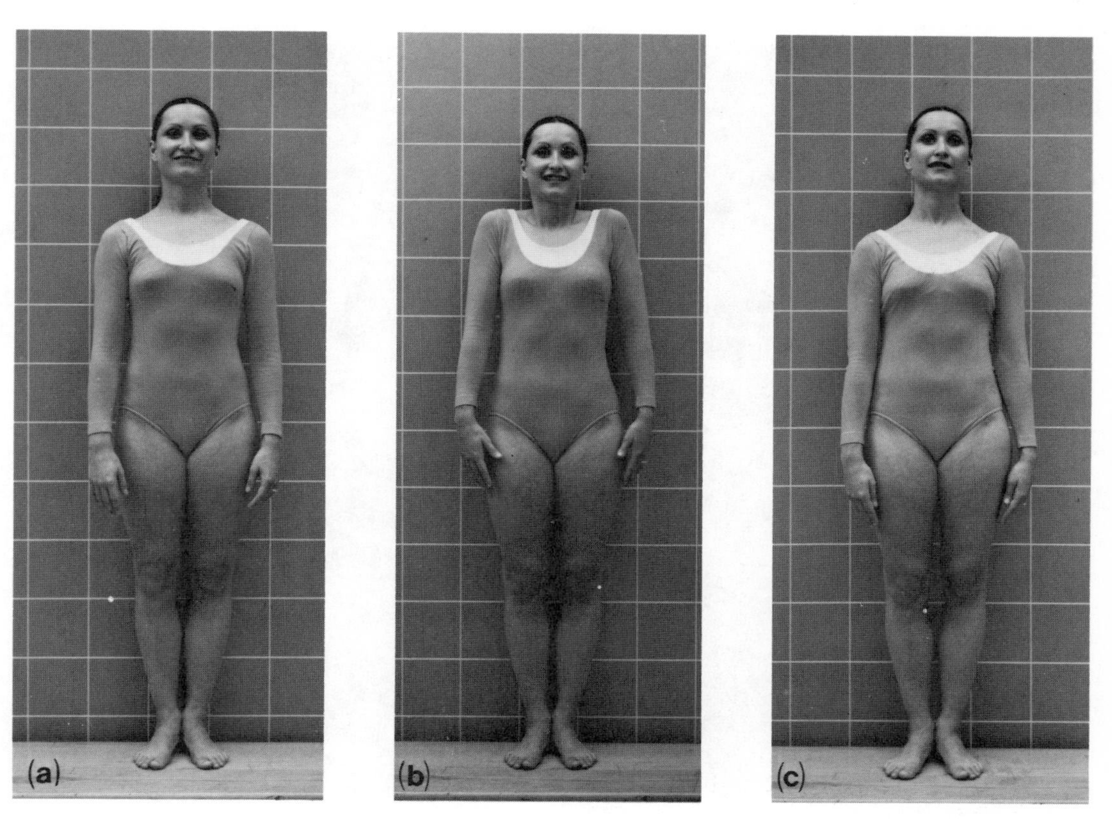

Figure 2.40. Movements of the shoulder girdle at the sternoclavicular joint (frontal views). (a) neutral; (b) elevation; (c) depression.

MOVEMENTS OF THE SHOULDER GIRDLE

The shoulder and pelvic girdles are comparable in that they both provide support and attachment for the limbs. While the innominate bones of the pelvis are anchored securely to the axial skeleton at the sacrum, the shoulder girdle is tied to the axial skeleton only at the two small sternoclavicular joints. The sacroiliac joint is so firmly bound together by fibrous ligaments that only a negligible amount of movement is possible, while the sternoclavicular joints allow a limited range of movement around three different axes passing through the proximal end of the clavicle (Fig. 2.8). These movements, discussed below, are elevation-depression, protraction-retraction, inward and outward rotation, and circumduction.

Elevation-Depression

These movements, illustrated in Figure 2.40, occur around a transverse axis slightly oblique to the sagittal plane (T in Fig. 2.8). The total angular excursion of the clavicle from maximal **depression** to maximal **elevation** has been estimated at 60° (Steindler (1973)).

Protraction-Retraction (Flexion-Extension)

These movements shown in Figure 2.41 take place around a vertical axis through

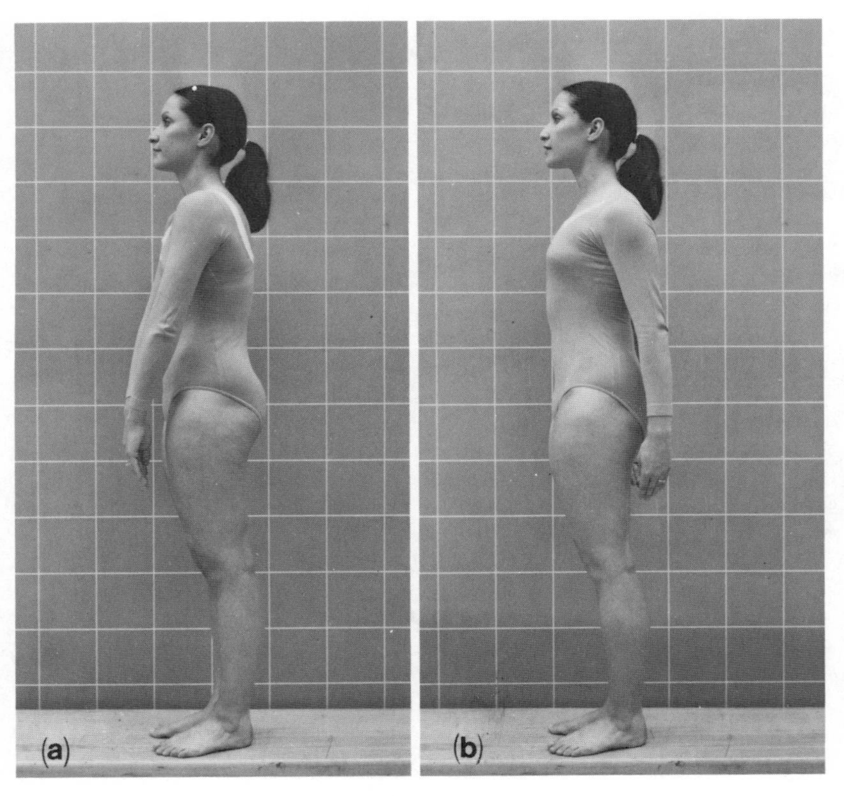

Figure 2.41. Movements of the shoulder girdle at the sternoclavicular joint (sagittal views). (a) protraction (flexion); (b) retraction (extension).

the proximal (sternal) end of the clavicle (V in Fig. 2.8). Anatomists prefer the terms **protraction** for a forward movement of the shoulders and **retraction** for a backward movement, although the flexion-extension terminology selected by the orthopaedic surgeons meets the criteria presented earlier for these movements (see under "Flexion-Extension").

Rotation

Rotation occurs around the long axis L in Figure 2.8. **Inward (or upward) rotation** of the clavicle around this axis is largely responsible for the movement of the glenoid fossa forward during shoulder flexion, while **outward (or downward) rotation** moves the glenoid fossa downward and backward during extension of the humerus at the shoulder.

Circumduction

When the upper extremity is circumducted the range of the movement is increased by **circumduction of the clavicle.**

Acromioclavicular Movements

While the range of movement at this joint is limited in comparison to many others, the joint is multiaxial and it does have three degrees of freedom. It is this freedom that enables the scapula to move on the clavicle so that the glenoid fossa can be further rotated upward and sideward, upward and forward, and downward, thus making possible the extreme ranges of motion of the upper extremity.

Movements of the Scapula

The movements described above of the clavicle on the sternum are caused not as much by muscles acting directly on the clavicle as they are by muscles which act indirectly by holding and moving the scapula on the thorax. For this reason many authorities prefer to omit discussion of sternoclavicular actions and simply to describe the observed behavior of the scapula in anatomical terms as follows.

Elevation moves the scapula higher on the rib cage, while **depression** moves it downward.

Abduction of the scapula is movement away from the vertebral column, while **adduction** of the scapula draws it closer to the spine.

Rotation upward: the inferior angle of the scapula moves laterad and upward while the glenoid fossa also turns upward.

Rotation downward: the inferior angle moves mediad as the glenoid fossa shifts slightly laterad and faces downward.

"MICRO" ASPECTS OF HUMAN MOVEMENT

The study of *osteokinematics* and *arthrokinematics* is of particular interest to kinesiologists and health professionals who are interested in understanding the "microkinematics" of human movement.

Most of the work in this field has been produced by MacConaill (1946, 1950, 1958). Further, a comprehensive review of this work is provided in the text by MacConaill and Basmajian (1977).

Osteokinematics

MacConaill and Basmajian (1977) refer to **osteokinematics** as the study of bone movements. They suggest that any bone movement may be described in terms of spin and swing. **Spin** is defined as a rotation about the long axis of a bone. An example of spin occurs in the humerus during internal and external rotation at the shoulder (starting position is with the shoulder abducted to 90°. **Swing** refers to any movement other than pure spin. Swing is further divided into: (1) **pure** or **cardinal swing**, which occurs without spin (Fig. 2.42(a)), and (2) **impure swing** or **arcuate movement**, which combines the simultaneous occurrence of swing and spin (Fig. 2.42(b)).

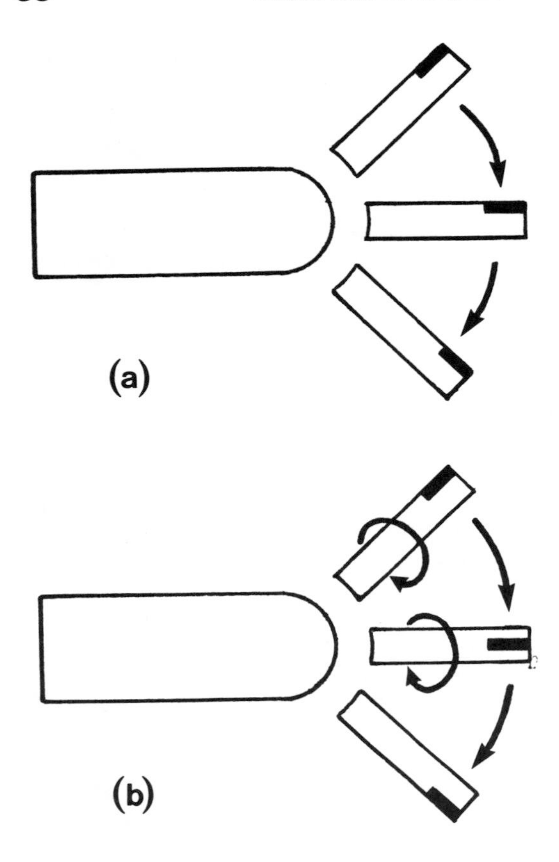

Figure 2.42. Examples of swing. (a) pure swing; (b) impure swing.

Arthrokinematics

While osteokinematics is concerned with the study of bone movements, **arthrokinematics** focuses on describing joint movements. These movements are taking place in joints that are freely moving, i.e., **synovial joints**. MacConaill and Basmajian (1977) advanced the following classification scheme for synovial joints.

1. **Unmodified ovoid.** This type of joint has three degrees of freedom. Examples include the shoulder and hip joints.
2. **Modified ovoid.** This type of joint has two degrees of freedom. Examples include the metacarpophalangeal joints and the radioscaphoid articulation.
3. **Unmodified sellar.** This type of joint has two degrees of freedom. An example is the trapezium-first metacarpal articulation.
4. **Modified sellar.** This type of joint has one degree of freedom. Examples include the knee joint and interphalangeal joints of the digits.

The elements of a synovial joint are referred to as being either male or female. In an ovoid joint, the male element is identified as the component with the convex surface, while the female element has the concave surface. In a sellar joint, the male element is the component with the larger articulating surface.

Of additional interest is the concept of **joint congruency**, the degree to which the elements conform to each other. The position of maximum congruency is referred to as the **close-packed position**, as it is in this position that the curvature of the male element most closely resembles that of the female element (MacConaill (1950)). It is also in this position that the joint capsule and collateral ligaments are most taut. For these reasons, there is little or no joint play in the close-packed position. In all other joint positions, the curvature of the male element is less than that of the female element. This leads to a reduced congruency and is referred to as the **loose-packed position** (MacConaill (1950)).

Joint movements are identified as spin, roll, slide, and rock. The **spin** movement which occurs at the joint is analogous to the spin described in the osteokinematics section. To illustrate the terms roll, slide, and rock, two examples are presented. The first case examines abduction at the glenohumeral joint, i.e., unmodified ovoid. For the purpose of this example, the female element (glenoid fossa) is the fixed component and the male element (head of humerus) is the moving component. In order for abduction to occur, a combination of roll and slide must take place. **Roll** occurs when new points on the male element come into contact with new points on the female element (Fig. 2.43). From Figure 2.43(a), it can be seen that if roll was the only movement to occur at the shoulder, the head of the humerus would roll off the top of the glenoid fossa—or more correctly, it would jam against the inferior aspect of the acromion process. **Slide** occurs when a constant point on one bone comes into contact with new points on the second bone (Fig. 2.43(b)). Thus, in order to successfully achieve abduction, the male humeral head must roll cranially and slide caudally.

The second case examines the movement of flexion at the fifth metacarpophalangeal joint, i.e., modified ovoid. Here, the male metacarpal element will be considered to be fixed and the female phalangeal element to be in motion. MacConaill and Basmajian (1977) state that when a female element moves onto a fixed male element, initially slight **rock** occurs, which is followed by slide (sometimes accompanied by a slight rock) and finally, rock occurs at the end of the range, (Fig. 2.44). During the movement of a female element on a fixed male element, the rock and slide are always in the direction of the movement.

MOVEMENTS OCCURRING AT MORE THAN ONE JOINT

Diagonal Patterns

Seldom, if ever, does the human being isolate movement to one joint unless it be flexion/extension at one of the distal phalangeal joints in the upper or lower extremity. All of us are aware of the vari ety of multiple joint movement combinations which can occur at any one instant in time. Some of the multiple combinations have captured the interests of therapists because of the predictable patterns

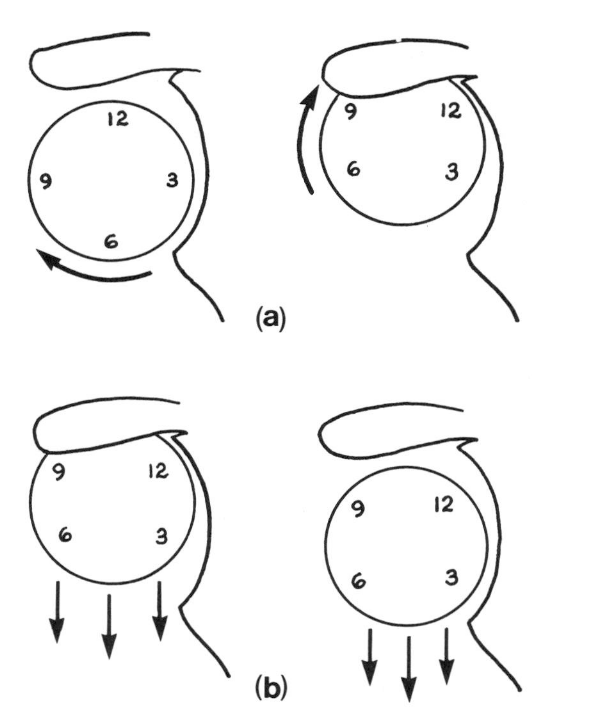

Figure 2.43. Schematic of the glenohumeral articulation. In order to accomplish abduction, the humeral head must (a) roll cranially and (b) slide caudally.

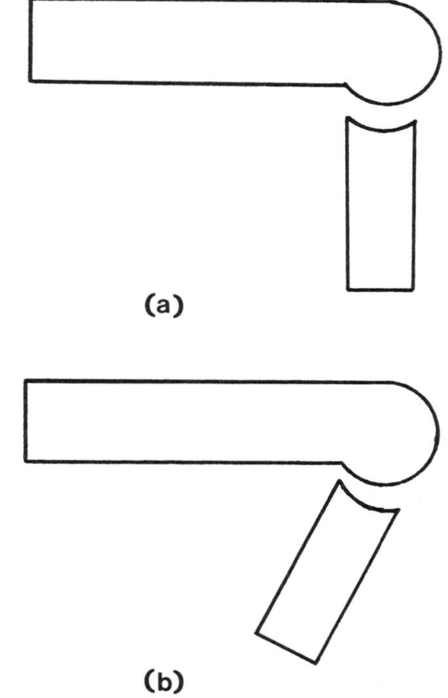

Figure 2.44. Schematic of the fifth metacarpophalangeal joint. (a) male and female apposition part way through the range; (b) example of rock of the female element at the end of the range.

which seem to be assured when a particular joint movement is involved. These have been identified by Knott and Voss as the **diagonals of movement** (Knott and Voss (1968)). They are included in this chapter because their existence is directly associated with the anatomical arrangements and functional abilities of the musculoskeletal system.

The alignment of muscle associated with shoulder and hip actions is rotatory and diagonal in direction suggesting that seldom, if ever, are joint movements performed in the pure cardinal planes of the body.

Two diagonal patterns of motion have been identified. The diagonal patterns for the extremities are described according to the three degrees of freedom, hence the three components of motion possible at the proximal joints—the shoulder and the hip. Each pattern is defined by one of the two movements associated with each of the three components; thus, a pattern includes a component of flexion or extension for sagittal plane movement, abduction or adduction for frontal plane movement, and medial or lateral rotation for transverse plane movement.

The intermediate joints—the elbow and knee—each having essentially one degree of freedom, may be in a position of flexion or extension. The distal joints are in a position consistent with the proximal joints regardless of the action at the intermediate joints.

The **first diagonal pattern** is the same for both the upper and lower extremities. The proximal joint—shoulder or hip—is flexed, adducted, and laterally rotated. The intermediate joints are either flexed or extended. The distal joints show patterns similar to the proximal joints. The reciprocal actions at the shoulder and hip are extension, abduction, and medial rotation.

The **second diagonal pattern** differs between upper and lower extremities. The upper extremity second diagonal pattern is composed of shoulder flexion, abduction, and lateral rotation with consistent patterning at the distal joints. The reciprocal pattern consists of shoulder extension, adduction, and medial rotation. The elbow is flexed or extended.

The second diagonal pattern for the lower extremity consists of hip flexion, abduction, and medial rotation. The reciprocal actions are extension, adduction, and lateral rotation. Again, the distal joint actions are similar in pattern to those of the proximal joints and the knee is either flexed or extended.

The first and second diagonal patterns are always described in relation to the sagittal plane components of the proximal joints, e.g., first diagonal flexion pattern, or first diagonal extension pattern.

Examples of diagonal patterns operating in a sport skill may be seen in the execution of a badminton smash (see Fig. 11.27). The movements of the right upper extremity at the shoulder joint, especially from frame 20 to completion, are viewed as extension, adduction, and medial rotation, hence all of the ingredients for the second diagonal extension pattern in the upper extremity. The left upper extremity also meets the requirements of the second diagonal extension pattern, but with one noticeable difference—the elbow is flexed. Interestingly, and in contrast, the movements at the left hip from frame 1 through frame 10 are primarily extension and medial rotation, with some slight abduction visible, making it a prime example of the first diagonal extension pattern. It would appear that the right lower extremity, especially from frame 12 to completion, fits into the second diagonal flexion pattern primarily because of the hip flexion and abduction. The reader might wish to explore several sport skills in similar fashion, in an attempt to confirm (or refute) the presence of diagonal patterns.

Skilled Patterns of Movement

After acquiring the necessary vocabulary, it is then possible to describe movement patterns even without illustrations. Figure 2.45 depicts a skilled gymnast in two different poses of a one-leg balance. With the exception of the right upper extremity which is used to brace the body

Figure 2.45. Skilled patterns of human movement. (a) gymnastic forward needle: one leg balance with left upper extremity supporting left lower extremity. (b) gymnastic grand battement: one-leg kick to momentary balance with left lower extremity unsupported.

mechanically, all major joint positions are visible. It is assumed that the movement commenced from the anatomical position in each of the two cases depicted.

As an exercise, the reader is invited to describe all of the major joint actions taking place in Figure 2.45(a) and (b). A tabular format is recommended, listing the appropriate joints under each of the body parts involved, i.e., right and left lower extremity, vertebral column, cervical spine, left scapula, and left upper extremity. The appropriate descriptors should be listed for each of the two cases. Where there is no joint movement, it is appropriate to indicate "anatomical position".

When comparing Figure 2.45(a) and (b) particular attention should be given to the position of the left upper extremity, the amplitude of the left lower extremity movement and the position of the support of the right lower extremity. After completing the exercise, Appendix 1 should be consulted for the authors' description of the movements depicted.

In the exercise above, it is also appropriate to make note of the planes of the movements and/or position of body parts with respect to the cardinal planes. Additionally, it is of value to note whether or not diagonal patterns are observable, and to note if there are any distinctive features in this gymnast's style, particularly in the dynamic pose depicted in Figure 2.45(b).

After completing a number of exercises similar to the above, the reader should be able to describe human movement patterns completely and accurately. The skill of describing movement in kinesiological terms is essential to the human movement specialist.

THE CHALLENGE TO FOLLOW

The skeletal system, its series of links, and their interactions with each other have been discussed. Common descriptors have been provided as a kind of motor alphabet to ease the task of communication about human movement. Photographs have been used to illustrate the movement descriptions and readers have been urged to imagine movement occurring before and after the instants in time when the photographs were taken. It is essential that the proper nomenclature is mastered in order to describe as well as to comprehend human movement in kinesiological terms. The level of success in comprehension will be tested in the ensuing chapters where photographs may not be provided and the reader's attention is focussed on motion and complex movement skills.

CHAPTER 3

Kinematics

INTRODUCTION

The skeletal system is described in Chapter 2, and a wide variety of static body positions are indicated. To move into or out of these positions, muscular forces must act upon the skeletal system. At this juncture, however, we are not yet concerned with such forces. Rather, we are concerned with describing the time sequences of events which would be noted by an external observer of these body movements. Recordings made by such an observer may or may not be aided by technological devices such as photographic cameras, video recorders, or even more sophisticated devices which might be interfaced directly with a digital computer. To facilitate our descriptions it is of value to classify the various basic types of relevant motion and to define important terms including displacement, velocity, and acceleration, i.e., the kinematics.

We are considering only a planar analysis of motion, meaning that all parts of the body are analyzed as though they are moving in parallel planes. Since the motion can be projected "onto a single plane parallel to the motion called the **plane of motion**," (Meriam, 1975:198), it follows that any description of movement of body parts in that plane may be accomplished by use of a standard two-dimensional analysis with the customary X and Y designations. Planar or two-dimensional analysis is also more typical than three-dimensional analysis because all photos or figures on a page are automatically constrained to a planar environment. The reader who is interested in three-dimensional analysis of three-dimensional motion should refer to the literature which deals specifically with this topic.

Examples of problems with application to human motion are included and the reader is invited to attempt solutions by "inserting" real numbers. Solutions and answers for these problems are found in Appendix 1. At the end of the chapter, facets of applied mathematics are introduced which are generally helpful with regard to human movement studies.

TYPES OF MOTION

Motion *per se* always involves a changing of place or position of a body. This change may be described by any one or more of four different categories and is represented in Figure 3.1. "**Translation** is defined as any motion in which every line fixed in the body remains parallel to its original position at all times" (Meriam, 1975:198). Such examples as the wheelchair and its occupant, or the ski jumper including his skiis (Fig. 3.1(a) and (b)) fit the definition as long as attention is focused on the fact that every line in the body remains fixed and parallel to its original position.

Rectilinear or **linear translation** (Fig. 3.1(a)) is translation in which all parts of the body move in straight lines. So, with the exception of the wheels, and ruling out any movement on the part of the patient, the wheelchair and occupant meet the requirements of this definition. **Curvilinear translation** (Fig. 3.1(b)) is translation in which all parts of the body move in curves, so that if each curve were superimposed on the other, they would coincide exactly. The assumption here, of course, is

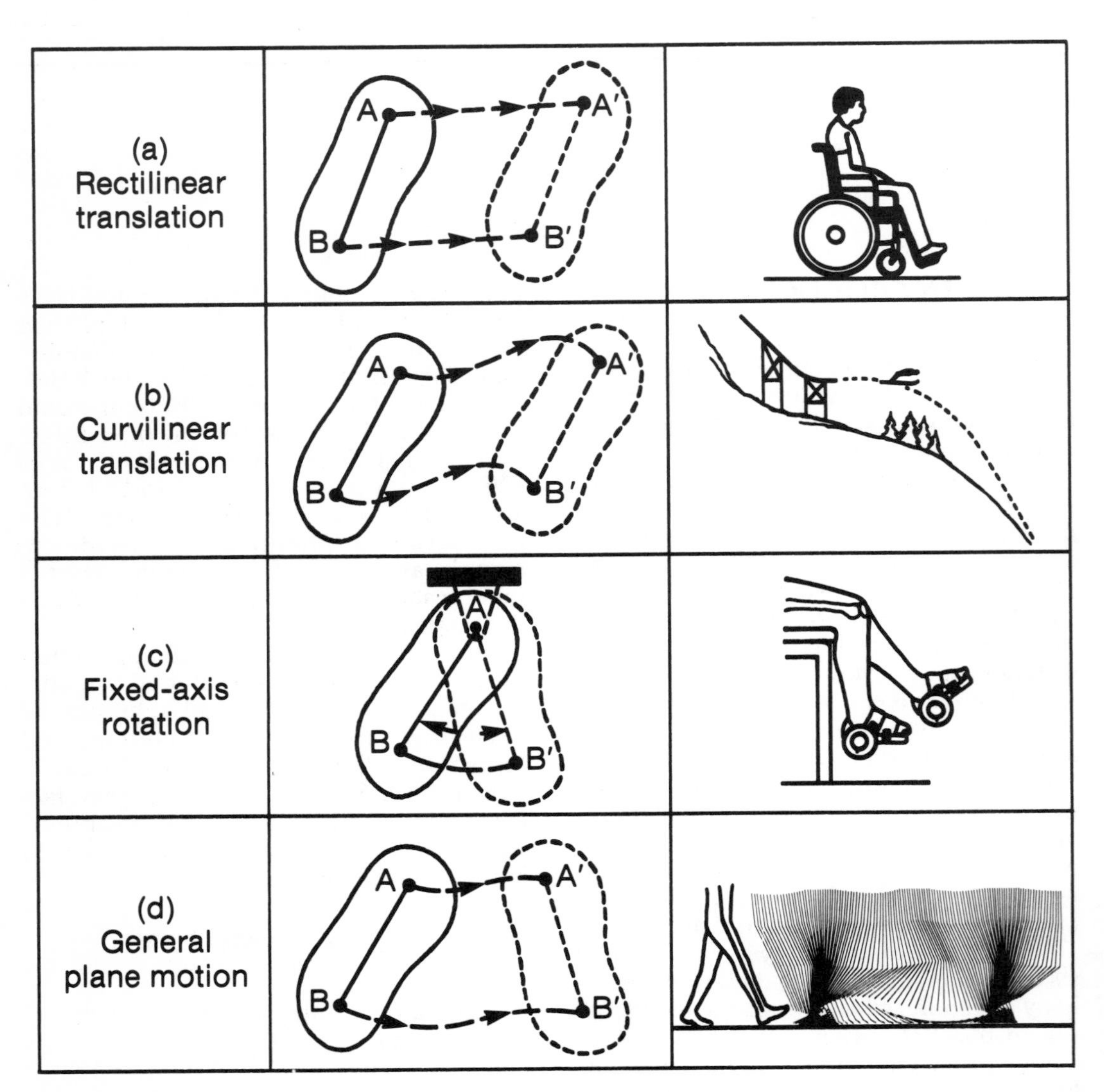

Figure 3.1. Types of rigid-body plane motion with examples. (Modified from Meriam, 1975, p. 198.)

that the ski jumper will maintain the posture in which he is shown in the illustration if his jump is to meet all of the requirements of curvilinear translation. "It should be noted that in each case of translation the motion of the body is completely specified by the motion of any point in the body, since all points have the same motion" (Meriam, 1975:198).

Fixed-axis rotation (Fig. 3.1(c)) is angular or rotational movement about an axis fixed in space. This means that all parts of the body, such as points on the leg, foot, and boot in the example, move in curved paths about the axis at the knee; and all lines such as the tibia or fibula rotate through the same angle in the same amount of time.

General plane motion (Fig. 3.1(d)) combines rotation and translation, and may be illustrated by the movement of a lower extremity in a walk. The leg is free to rotate about an axis at the knee while, at the same time, the knee axis is experiencing curvilinear translation.

Some Examples

From the foregoing it may be seen that translatory motion of a body part must involve movement at more than one joint as the parallel lines must be straight or curved rather than circular. Thus, reciprocating actions of three or more segments at two or more joints can displace the distal end of the chain in translatory motion.

A passenger in an automobile traveling along a straight highway is undergoing rectilinear translation. When the highway curves, the translation of the car and passenger becomes curvilinear. A shussing skier, or one sliding down a track of fixed slope before taking off from a ski jump, would also undergo rectilinear translation.

When a skier swerves to avoid a tree or enters a slalom gate, there is curvilinear translation of at least some parts of the body (some movement will be rotatory as weight shifts and arms and poles swing). On the other hand, when a ski jumper makes his leap up and out from the jump track and assumes his forward lean, he is undergoing essentially curvilinear motion as he travels outward and down to his landing.

All human locomotion, i.e., walk, run, hop, jump, or leap, involves translatory motion which has attributes of rotatory, rectilinear, and curvilinear motions. During the swing phase of the walking gait depicted in Figure 3.2, a lower extremity swings forward, moving about an axis in the head of the femur (proximal end). At heel contact, the foot becomes weight bearing as the body rides over it during the support or stance phase. During this portion of the step there are movements at a number of axes at the distal end of the lower extremity (in the toes, at the metatarsophalangeal joints, and at the ankle). The movements occur as the thigh and leg rotate over these axes while carrying the rest of the body forward. As the lower limbs are undergoing rotatory motions at alternating ends, the trunk is progressing curvilinearly. This latter motion is clearly exhibited in the sagittal plane.

To fully appreciate the different types of motions and their areas of applicability in human movement, it is necessary to become familiar with the fundamental principles involved, as well as with the anatomical and physiological factors presented in this text.

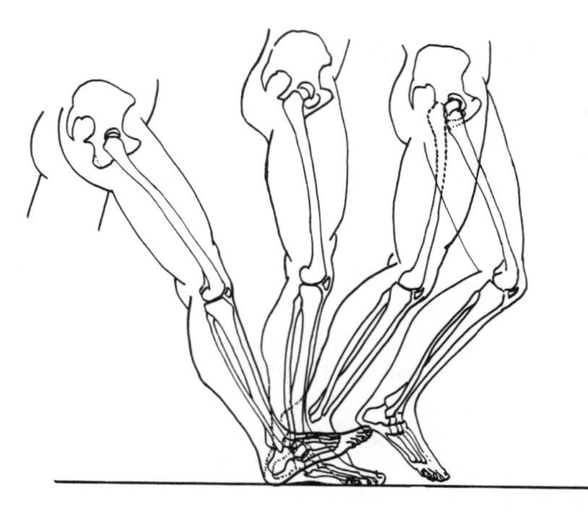

Figure 3.2. Single stride of right lower extremity: walking.

MECHANICS OF MOTION

Vectors and Scalars

Vectors and scalars are important in the description of the characteristics of motion. **Vector quantities** are defined as having both magnitude and direction. A **scalar quantity** is one which has magnitude only. Thus, a vector is described by two scalars: one indicating magnitude, the other direction.

Scalar quantities may be added arithmetically and include such quantities as distance, mass, volume, area, time, temperature, energy, and speed. Speed *per se* has no direction and is the magnitude of a vector quantity termed velocity.

Vector quantities, since they possess both magnitude and direction, may be represented graphically by segments of straight lines. The length of the line segment can be scaled to represent magnitude; the orientation of the line together with an arrowhead specify direction. Such quantities include *velocity* (where speed is the magnitude and the direction is known), *displacement* from one point to another (e.g., from point A to point B, from Boston to New York), and *force*, which might be a push, or a pull, or a weight.

The following example will serve to distinguish some important differences in the properties of scalars and vectors.

If an individual (a) rides a bicycle 5 kilometers, (b) walks 2 kilometers, and (c) rides a skateboard 3.3 kilometers, 10.3 kilometers has been traveled. These distances have magnitude only and so are scalar quantities and may be added as noted above. The destination of the traveler is not known since no *directions* of travel are given. If the directions for each of the above magnitudes are known, then vector quantities are involved, i.e., **displacement,** and the destination of the traveler can be determined.

Given that the distances traveled are (a) 5 kilometers northwest, (b) 2 kilometers east, and (c) 3.3 kilometers northeast, vectors may be added graphically as in Figure 3.3 by laying out vectors to scale and in proper directions. In this figure, vector $\vec{AB}$ represents the 5 kilometers northwest, $\vec{BC}$ the 2 kilometers east, and $\vec{CD}$ the 3.3 kilometers northeast. Note that the arrows above $\vec{AB}$, $\vec{BC}$, and $\vec{CD}$ indicate that these are vector quantities. For $\vec{AB}$, the length *AB* indicates the magnitude of the vector which is directed from *A* to *B*. By measuring the vector $\vec{AD}$ we find that the traveler ends the trip 5.9 kilometers north-northeast of the starting point. By drawing vector $\vec{AD}$ the polygon of displacements is closed to produce the resultant of the summation of vectors $\vec{AB}$, $\vec{BC}$, and $\vec{CD}$.

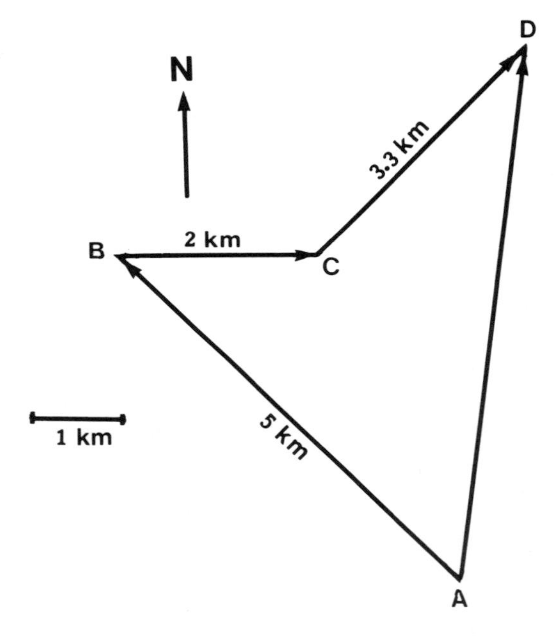

Figure 3.3. Vector diagram of distances traveled: $\vec{AD} = \vec{AB} + \vec{BC} + \vec{CD}$.

A graphical solution for the summation of distance vectors is provided. It is also possible to solve this type of problem using trigonometry. In this method, each vector is resolved into its horizontal and vertical components with regard to a set of mutually perpendicular reference axes. In choosing the horizontal axis to be positive from west to east and the vertical axis to

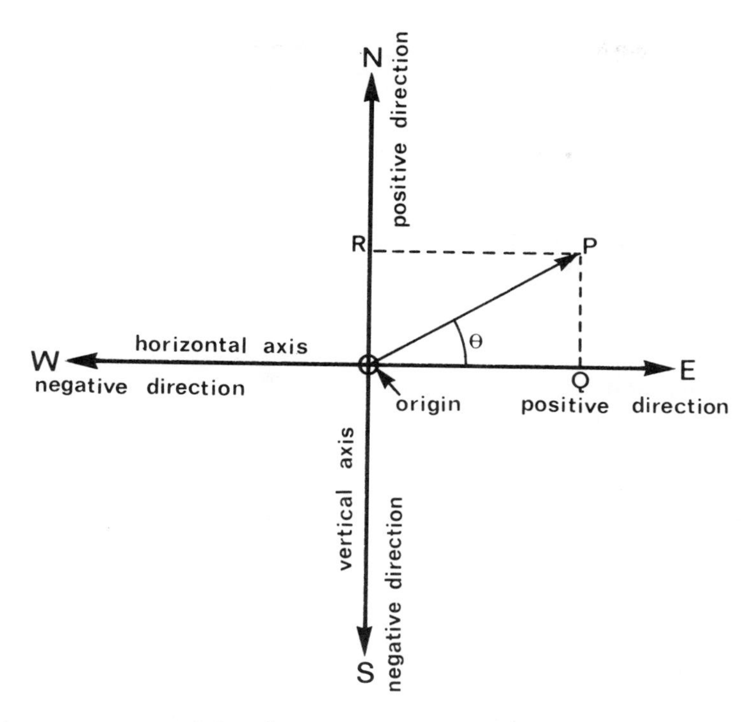

Figure 3.4. Vector components: OQ and OR are, respectively, the horizontal and vertical components of OP.

$$OQ = OP \cos \theta$$
$$OR = OP \sin \theta$$

Note:

$$OP^2 = OQ^2 + OR^2$$
$$OP = \sqrt{OQ^2 + OR^2}$$
$$\tan \theta = PQ/OQ = OR/OQ$$

be positive from south to north, the requisite components may be resolved by recognizing that horizontal components are the product of the vector magnitude and the cosine of the angle between the vector and the horizontal axis, and vertical components are the product of the vector magnitude and the sine of the angle between the vector and the horizontal axis (Fig. 3.4).

Each vector and its respective components are listed in Table 3.1.

Referring to Figure 3.4:

$$\text{Resultant} = \sqrt{(\text{Horiz.})^2 + (\text{Vert.})^2}$$
(from Table 3.1)

$$\text{Resultant} = \sqrt{(0.79)^2 + (5.87)^2}$$
$$= \sqrt{0.6241 + 34.4569}$$
$$= \sqrt{35.081}$$
$$= 5.92 \text{ kilometers}$$

Table 3.1
Vectors and Components for Trigonometric Analysis

Vector	Magnitude (km)	Direction θ^a	Cos θ	Horizontal Component	Sin θ	Vertical Component
$\vec{AB}$	5	135°	−0.70711	−3.54	+0.70711	+3.54
$\vec{BC}$	2	0°	+1.0	+2.0	0	0.0
$\vec{CD}$	3.3	45°	+0.70711	+2.33	+0.70711	+2.33
Sum of Horizontal Components				+0.79		
Sum of Vertical Components						+5.87

[a]Angle with respect to horizontal axis

The angular disposition, θ_R, of the resultant can be determined since

$$\tan \theta_R = \frac{5.87}{0.79} = 7.4304$$

$$\theta_R = \arctan 7.4304 = 82° \ 20'$$

The traveler ends the tour 5.92 kilometers and 82° 20′ north of east of the starting point.

Vector $\vec{AD}$ can be completely specified by writing

$$\vec{AD} = 5.92 \text{ kilometers } \underline{|82° \ 20'}$$

where 5.92 kilometers is the magnitude of $\vec{V}_{VA}$ and $\underline{|82° \ 20'}$ signifies that the vector is disposed at 82° 20′ with respect to the chosen horizontal reference axis.

To solve this kind of problem the arithmetic may be greatly facilitated by using a hand-held calculator having trigonometric capabilities. In the absence of the latter capabilities, reference must be made to pertinent trigonometric tables.

Linear Kinematics

Velocity

Suppose that in the foregoing example the 5 kilometers were traversed on a bicycle in 10 minutes, the 2-kilometer walk took 40 minutes, and the skateboard ride (all downhill) took 10 minutes, so that the time elapsed was 1 hour. Cognizance must now be taken of time which is a scalar quantity. The velocity in traversing $\vec{BC}$ (Fig. 3.3) can be calculated from the definition of velocity:

$$\vec{V}_{BC} = \vec{BC}/t \tag{3.1}$$

where $\vec{BC}$ is the displacement of magnitude, 2 kilometers, that takes place in an easterly direction in a time t = 40 minutes. This displacement occurs at a speed of (2 × 60)/40 = 3 km/h also in an easterly direction. Thus, **velocity** $\vec{V}_{BC}$ is completely specified in magnitude and direction. Its magnitude is the scalar, speed, and its direction follows from the displacement vector (BC). Its units (e.g., km/h) reflect a time rate of change of displacement. As defined in Equation 3.1, velocity is the quotient of a vector quantity and a scalar quantity and thus retains the vector property.

The average velocities of each of the three aforementioned modes of transport are $\vec{V}_{AB}$ = 30 km/h $\underline{|135°}$, $\vec{V}_{BC}$ = 3 km/h $\underline{|0°}$, and $\vec{V}_{CD}$ = 20 km/h $\underline{|45°}$. The average velocity for the complete trip $\vec{V}_{AD}$ = 5.92 km/h $\underline{|82° \ 20'}$.

Velocity as a vector is always in a straight line. The example given places each velocity in its own line but, as each is in a different direction, they can only be added by using the quantitive displacement as calculated by vector summation. When all velocities (or any vectors) are in the same direction or straight line they may be added arithmetically.

Let's look at another example, wheelchair basketball. The reader is invited to solve the problem and check the answer with that given in Appendix 1. Imagine that a wheelchair is travelling forward at 2.0 m/s and that a basketball is thrown sideways from the wheelchair, at a 90° angle to the path of the chair, at 5.0 m/s.

What velocity does the ball achieve and in what direction does it travel relative to its velocity vector?

In human movement studies the velocity units usually used are meters or centimeters per second (m/s or cm/s). Only when discussing large distances are kilometers per hour (km/h) used.

Average and Instantaneous Velocities

In the discussion of the traveler who used three modes of transport, the *average velocities* for each segment of the trip were calculated. Only the average value could be obtained, since the details of each segment from instant to instant were not given. Thus, we might speculate on how the trip might have been made. For example, the 5 kilometers traversed on a bicycle in 10 minutes could have been done by riding steadily at 30 km/h all the way, or at 45 km/h for 6 minutes and at 7.5 km/h for the remaining 4 minutes. Clearly there are an infinite number of possibilities unless we know the record of velocities from instant to instant.

The instantaneous velocity-time graph for a jogger might appear as in Figure 3.5.

The jogger spends the first minute getting up (accelerating) to the steady velocity level of 10 km/h. The speed is maintained for 6 minutes, then decreases over a period of a minute (decelerates) to a standstill when there is a rest of 2 minutes and then a repeat of the previous performance. The average velocity for the 18 minutes recorded is obtained by taking the area under the instantaneous velocity-time graph and dividing by the baseline length of 18 minutes, i.e.,

$$
\begin{aligned}
v_{av} = {} & [(\tfrac{1}{2} \times 1 \times 10) + (6 \times 10) \\
& + (\tfrac{1}{2} \times 1 \times 10) \\
& + (2 \times 0) + (\tfrac{1}{2} \times 1 \times 10) \\
& + (6 \times 10) \\
& + (\tfrac{1}{2} \times 1 \times 10)]/18 \\
= {} & 140/18 = 7.77 \text{ km/h.}
\end{aligned}
$$

So, the average velocity is 77.7% of the maximum steady velocity. This value would be reduced further by a longer rest period, slower rises to, and rapid declines from the steady level.

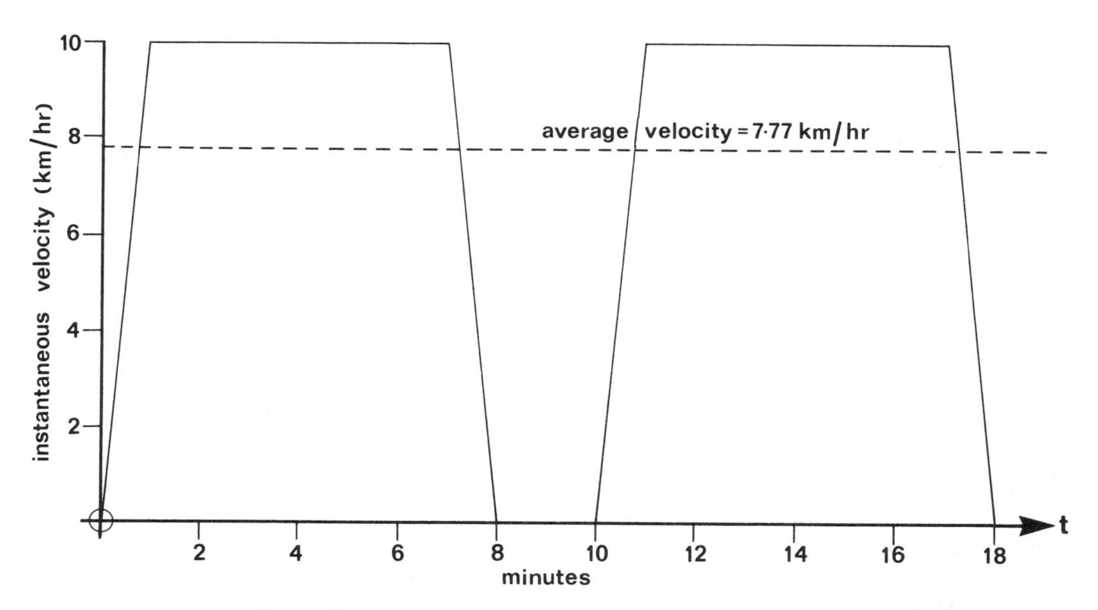

Figure 3.5. Instantaneous velocity-time graph.

Acceleration

Accleration is defined as the time rate of change of velocity. It should be borne in mind that velocity depends upon both speed and direction, and therefore acceleration may be affected by changes in either or both of these parameters.

If the direction of the velocity vector is fixed, then:

$$a = \frac{\text{change in velocity}}{\text{time taken for change}} \quad (3.2)$$

Let the velocity of a body moving along a straight line be v_1 m/s at a time t_1 s and consider that the velocity undergoes a change to velocity v_2 m/s in a time interval Δt s such that $\Delta t = t_2 - t_1$. Then:

$$a = \frac{(v_2 - v_1)}{(t_2 - t_1)} \frac{\text{m/s}}{\text{s}} \quad (3.3)$$

The dimensions of a in this case are (m/s)/s which may be written m/s/s, m/s^2 or m·s^{-2}.

Let

$$\Delta v = v_2 - v_1 \quad (3.4)$$

and

$$\Delta t = t_2 - t_1$$

where the prefix Δ (Greek letter delta) is read as "the change in" then

$$a = \frac{\Delta v}{\Delta t} \quad (3.5)$$

If v_2 exceeds v_1 for positive values of t, a will be positive; if v_1 exceeds v_2 for positive values of t, v and hence a will be negative.

Referring to the example illustrated in Figure 3.5, it will be seen that the jogger has uniform accelerations, of 0.167 km/m^2, i.e., 0.0464 m/s^2, regardless of whether velocity is increasing or decreasing.

Relative Motion

It is important to examine cases where human motion takes place upon a base that is also moving, as in rectilinear translation. A good example may be found in wheelchair basketball (Fig. 3.6). Player A attempts to throw the ball to a teammate so that it will not be intercepted by the upstretched arms of opponent C. Player A, whose wheelchair is moving toward opponent C at a velocity of 2.0 m/s, takes aim to throw the ball over the opponent who is stationary. Player A's perception is that the planned trajectory of the ball will clear easily the fingertips of opponent C. Unfortunately, the selected trajectory was calculated in relation only to player A and not in relation to the floor. The velocity of the wheelchair must be taken into account. The wheelchair velocity must be added as a vector to the velocity of the ball relative to player A, i.e.,

$$\vec{V}_{B/G} = \vec{V}_{B/A} + \vec{V}_{A/G}$$

where: B refers to the ball; G refers to the ground; and A refers to player A. By not considering the absolute velocity of the ball (in relation to the floor or ground), player A in fact throws the ball directly into the hands of opponent C.

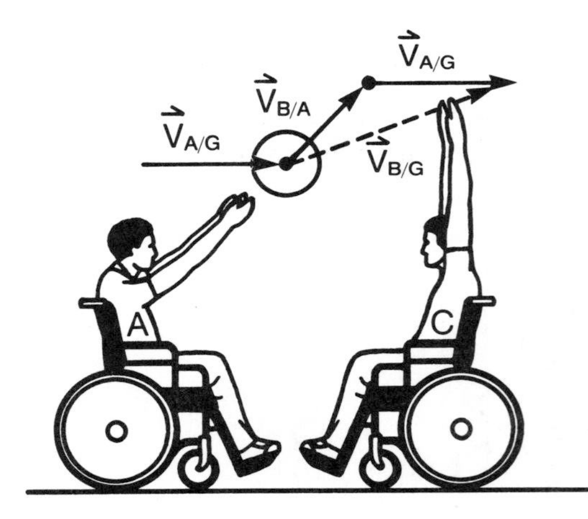

Figure 3.6. Wheelchair basketball. The illustration shows the players at the instant of ball release. $\vec{V}_{A/G}$ = velocity of player A with respect to the ground; $\vec{V}_{B/A}$ = velocity of the ball with respect to player A; $\vec{V}_{B/G}$ = velocity of the ball with respect to the ground.

The foregoing example emphasizes the need to specify all aspects of a movement, whether they be translatory, fixed-axis rotation, or general plane motion. In this case, a type of fixed-axis rotation movement (i.e., throwing the basketball) takes place *relative* to a linear translation movement (i.e., wheelchair motion).

Another good example of relative motion may be found in a person's attempt to ascend an "up" and a "down" escalator. If the walking velocity of a person who is ascending an "up" escalator (i.e., with respect to the escalator) is 0.5 m/s and the velocity of the escalator with respect to the ground is 1.0 m/s, what is the velocity of the person with respect to the ground? Vector mathematics shows the problem as follows:

$$\text{Given: } \vec{V}_{P/E} = 0.5 \text{ m/s}$$
$$\vec{V}_{E/G} = 1.0 \text{ m/s}$$

Find: $\vec{V}_{P/G}$ where P = velocity of the person; E = velocity of the escalator; and G = velocity of the ground.

Solution: (N.B.: Notice the cancellation of subscripts.)

$$\vec{V}_{P/G} = \vec{V}_{P/E} + \vec{V}_{E/G}$$
$$\vec{V}_{P/G} = 0.5 + 1.0$$
$$\therefore \vec{V}_{P/G} = 1.5 \text{ m/s}$$

But what is the solution if the person attempts to ascend a "down" escalator? The same approach is taken as follows:

Solution: (Notice that the "up" orientation has been arbitrarily taken to be positive.)

$$\vec{V}_{P/G} = \vec{V}_{P/E} + \vec{V}_{E/G}$$
$$\vec{V}_{P/E} = 0.5 \text{ and } \vec{V}_{E/G} = -1.0$$

thus

$$\vec{V}_{P/G} = 0.5 + (-1.0)$$
$$\therefore \vec{V}_{P/G} = -0.5 \text{ m/s}$$

In this case, although the person walks up, the direction of travel is down, as signified by the negative answer. It is important to notice that velocity may be negative.

The orientation of positive and negative for direction is arbitrary; notice that once the orientation for the system is established, it must be strictly followed for all parts of the problem. It should be emphasized also that when vectors function in the same direction (as in the first part of the problem), they may be added mathematically.

An Important Case

In cases of **constant acceleration** in a straight line (the situation encountered by bodies acted upon by gravity) the velocity after any elapsed time t is given by

$$v = v_0 + at. \qquad (3.6)$$

where v_0 is the initial velocity or the velocity at $t = 0$, and a is the acceleration.

It should be evident that for acceleration in a straight line, the velocity will be directed along that same line as will the displacement. Displacement s may be calculated by realizing that

$$s = s_0 + vt \qquad (3.7)$$

where s_0 is the displacement measured from the origin at time $t = 0$, and v is the velocity that obtains for time t.

The following example, illustrated by a graphical solution, shows the determination of displacement as a function of time from a knowledge of acceleration. Refer to Figure 3.7, in which there are graphs (top to bottom) of acceleration, velocity, and displacement. Given that acceleration a is constant for a body initially at rest (i.e., $v = 0$) at the origin (i.e., $s = 0$) at time $t = 0$, the graph of v against time may be constructed by recalling Equation 3.6. Since $v_0 = 0$, $v = at$. Substituting values of a (2) for the range of values of t of concern leads to a number of discrete points through which the straight line can be drawn as shown. As expected, for a constant acceleration the velocity of a body increases linearly.

From Equation 3.7, $s = s_0 + vt$. Initially, $s = 0$ (i.e., $s = 0$ for $t = 0$) and therefore $s = vt$.

For $t = 1$, $v = 2$ and the average speed in the time interval $t = 0$ to $t = 1$ is $v_{av} = 1$. Thus, in this interval $s = 1 \times 1 = 1$. The point (A) can therefore be plotted. In the interval between $t = 1$ and $t = 2$ we expect s to have increased by twice as much as in the preceding 1-unit interval. This increase must, according to Equation 3.7,

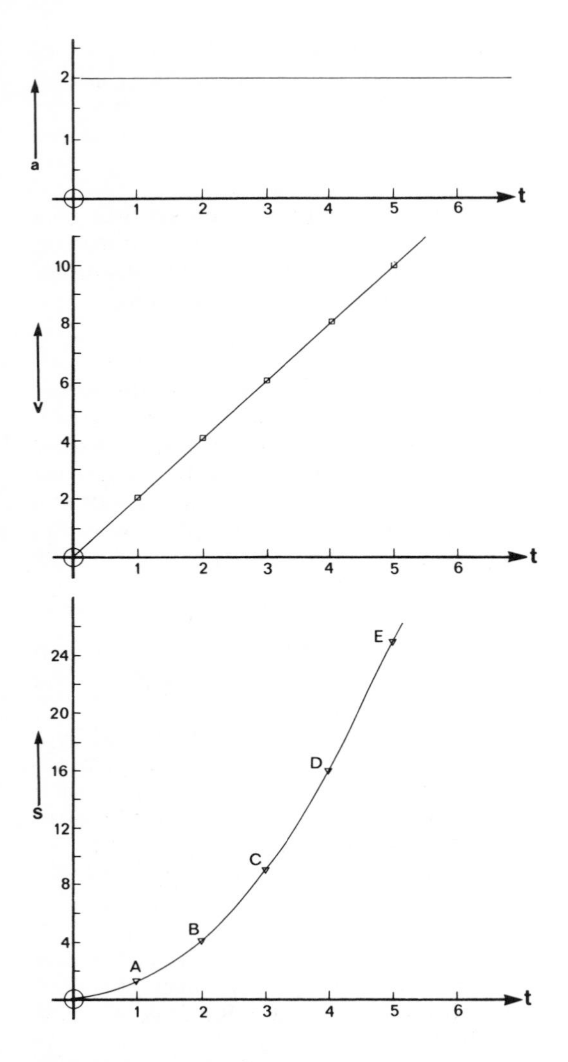

Figure 3.7. Acceleration, a , velocity, v , and displacement, s , as functions of time t for a body initially at rest at the origin of measurement.

be added to the displacement or distance already traversed. The average velocity in the interval of concern is 3 units and the distance traversed in the interval is 3 units. The accumulated distance is therefore 4 units corresponding to point (*B*).

Points (*C*), (*D*), and (*E*) follow in a similar fashion. If a smooth line is drawn through these points (and others which may be constructed as desired), the resulting curve turns out to be a parabola as shown. The equation of the parabola is

$$s = t^2$$

If a had been chosen initially to be 1 unit instead of 2 all of the s values would be reduced by a factor of 2. If a had been doubled, all of the values of s would be doubled. Thus, we can write s in terms of a and t as follows

$$s = \tfrac{1}{2}at^2 \qquad (3.8)$$

For a body with initial velocity v_0 at $t = 0$, subjected to an acceleration a thereafter,

$$s = v_0t + \tfrac{1}{2}\,at^2 \qquad (3.9)$$

The simplest example of constant acceleration for a body is that due to the gravitational pull of the earth, which accelerates a freely falling body at $g = 9.81\ m/s^2$. The direction of g is always vertically downward (being directed toward the center of the earth).

Sample Problem in Linear Kinematics

To test the reader's understanding of the foregoing, a sample problem concerning a skier is provided (Fig. 3.8). Racing down a steep incline at an angle θ, a skier experiences uniform (constant) acceleration. However, the acceleration is modified by the angle of the slope. Note that Equation 3.9 applies where s is the displacement down the hill during period t seconds; v_0 is the initial velocity at the beginning of the period t seconds; and a is the acceleration along the surface of the hill. Therefore,

$$a = g \sin \theta$$

It follows that if $a = g \sin \theta$, and if $\theta = 11.8°$, then $a = 2\ m/s^2$. Equation 3.9 can then be reduced to the following:

$$s = V_0t + t^2$$

It is now possible to solve various problems concerned with the skier, and to confirm answers with the benefit of Figure 3.6, which has an acceleration of 2 units. As well, the numerical solutions are shown in Appendix 1.

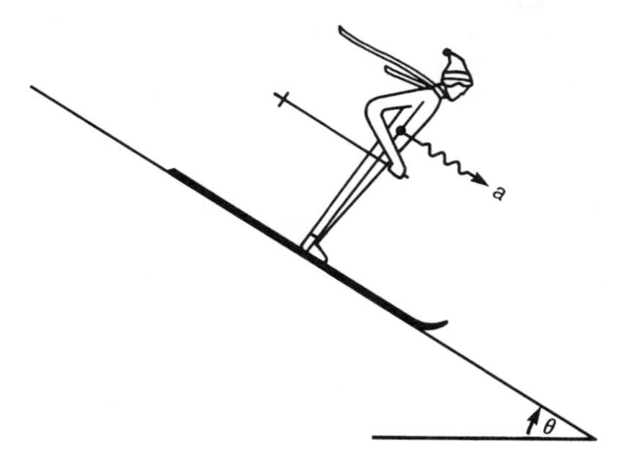

Figure 3.8. Schematic of downhill skier.

Problem 1. Given that the initial velocity is 2 m/s and time is 2 s, find displacement. A check on your answer may be made by consulting Figure 3.7. On the velocity graph, at the point on the curve where initial velocity is 2 units, the time is 1 second. At that same instant on the displacement graph and for 2 seconds afterward, the displacement curve has traversed an area from approximately 1 to 9 units, or a net displacement of 8.

Problem 2. Given that the initial velocity is 4 m/s and time is 3 s, find displacement. Once again, check your answer with the velocity and displacement graphs in Figure 3.7 as well as with Appendix 1.

Problem 3. Given that the initial velocity is 6 m/s and time is 1 s, find displacement.

Note that all of the above problems are set to find displacement, given initial velocity, time and acceleration, or

Find s, given V_0, t and a.

Additional practice may be gained by setting new unknowns. For example:

Find V_0, given s, t and a.
Find t, given V_0, s and a.
Find a, given V_0, t and s; i.e., θ becomes a variable.

In addition, all of the above problems may be verified graphically using Figure 3.7.

Variable Acceleration

When acceleration increases or decreases, as the speed of an escalator might vary, then the velocity graph is not a straight line. Let's examine the problem of the escalator on the way up; i.e., positive acceleration. If the positive acceleration increases, then the person is moving at an increasing velocity. If the velocity continues to increase, but less quickly now, and finally levels off at some maximum, we say that the change in velocity, while it is still an increase, represents a decrease in the acceleration; i.e., a decreasing positive acceleration. Finally, if some maximum velocity is reached and maintained, there is no acceleration, i.e., acceleration is zero.

Let's assume that a malfunction causes the escalator to slow down and then stop. The slowing down, i.e., decreasing velocity, is a negative acceleration. That is to say the negative acceleration is increasing and reaches some maximum. Just as quickly as the escalator comes to a halt, it starts moving again, but this time, backwards, i.e., moving downward. The orientation of the downward-moving escalator causes a negative velocity which increases to some maximum and then decreases. Accordingly, the negative acceleration, which peaked somewhere near the change in escalator direction, now decreases toward zero; the instant at which the velocity does not change, we have zero acceleration again. And, finally, as the negative velocity continues to decrease, there is positive acceleration once more.

In summary then, the escalator trip demonstrates, in order, increasing positive acceleration, maximum positive acceleration, decreasing positive acceleration, zero acceleration, increasing negative acceleration, maximum negative acceleration, decreasing negative acceleration, zero acceleration, and then positive acceleration again.

Rotational Kinematics

Rotation of a body about a fixed point or axis causes any portion of that body to travel in a circular path as it undergoes angular displacement. As this occurs when any body segment moves on another (rotatory motion), it behooves the kinesiologist to have some understanding of the laws governing such action.

Concepts relating to circular motion are of fundamental importance. Angular displacements, velocities, and accelerations as they relate to bodies moving in circular paths are examined. The same laws that apply to a ball on a string or a space capsule in orbit are also applicable to the movement of a distal end of a body kinematic chain moving in an arc about an axis through a proximal joint.

Angular Displacement and Velocity

Imagine a weightless object *OP* pivoted at and rotating about *O* (Fig. 3.9). *O* is the origin of the horizontal (*X*) and vertical (*Y*) axes shown. At time t_1 s, *OP* makes an angle θ_1 radians with the *X* axis and at t_2 s this angle is θ_2 radians.

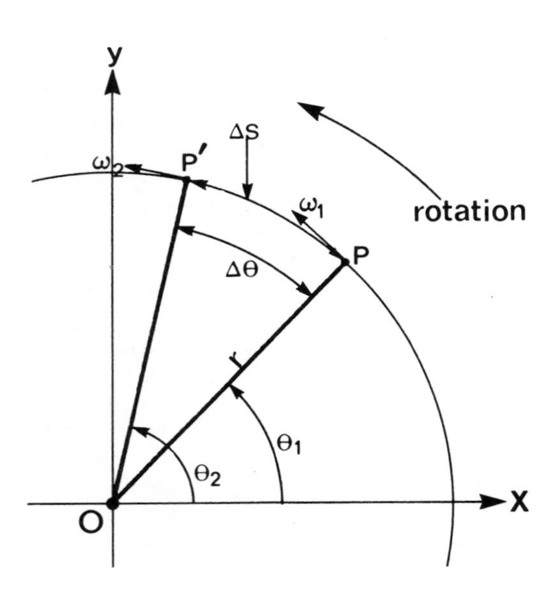

Figure 3.9. Circular motion:

$OP = r$

$PP' = \Delta s$

Now $\Delta\theta = \theta_2 - \theta_1$ is the change in angular displacement and $\Delta t = t_2 - t_1$ is the time over which that change takes place.

$$\text{Angular velocity } \omega = \frac{\Delta\theta}{\Delta t} \text{ (radians/s)} \quad (3.10)$$

and the velocity is directed tangentially at the instantaneous position occupied by *P*.

If the value of ω is constant, the angular displacement in a time *t* is $\theta = \omega t$. This angular displacement corresponds to a linear distance *s* traversed by *P*.

It is well known that

$$s = r\theta \quad (3.11)$$

where r is the radius of the movement. (For a complete traverse of the circle, $\theta = 2\pi$ radians, and $s = 2\pi r$.)

It follows from Equation 3.11 that since *r* is a constant

$$\Delta s = r\Delta\theta \quad (3.11a)$$

Dividing both sides by Δt

$$\frac{\Delta s}{\Delta t} = r\frac{\Delta\theta}{\Delta t}$$

recalling Equation 3.10 and knowing that $v = \Delta s/\Delta t$ leads to

$$v = \omega r \quad (3.12)$$

(It will be helpful to remember that 2π radians = 360°.)

Angular Acceleration

As indicated in Figure 3.9, the angular velocity at any point on the circular path is in a straight line tangential to the path (perpendicular to the radius). While the magnitude of ω may be fixed, its line of action changes from moment to moment and position to position. **Angular acceleration**, α, is defined by

$$\alpha = \frac{\omega_2 - \omega_1}{t_2 - t_1} = \frac{\Delta\omega}{\Delta t} \quad (3.13)$$

(This result is analogous to that for linear acceleration. (See Equations 3.3 and 3.5.))

Sample Problem in Rotational Kinematics

Consider the problem of a tennis player, or squash player, who is hitting a forehand drive and is viewed from above similar to the schematic in Figure 3.10. In the schematic, notation is used to provide information about the movement, as follows:

$\vec{V}_S = \vec{V}_{S/G} =$ velocity of the shoulder with respect to the ground
$\vec{V}_H = \vec{V}_{H/G} =$ velocity of the hand with respect to the ground
$\vec{V}_B = \vec{V}_{B/G} =$ velocity of the ball with respect to the ground
$\omega =$ rotational velocity of the upper extremity about the shoulder joint
$r_1 =$ length of the upper extremity
$r_2 =$ length of the racquet

Problem 1. If the length of the upper extremity is 0.8 m, the angular velocity is 10 rad/s, and the velocity of the shoulder with respect to the ground is 2 m/s, find the velocity of the hand with respect to the ground. Hint: the velocity of the hand with respect to the shoulder is $r_1\omega$. You may confirm your answer in Appendix 1.

Problem 2. If the length of the racquet is 0.7 m, and all other values are the same as for Problem 1, find the velocity of the ball with respect to the ground.

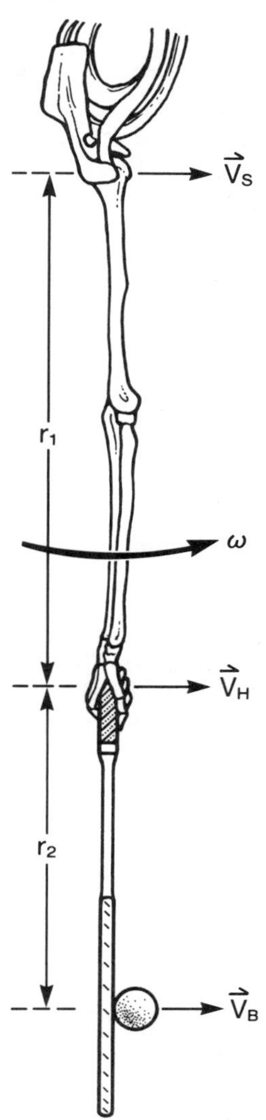

Figure 3.10. Schematic of a tennis forehand drive.

Radial Acceleration

Figure 3.11 shows a graphic method for deriving instantaneous acceleration in magnitude and direction when a body *OP* is rotated at constant angular velocity ω in a clockwise direction.

OP is the initial position occupied by the body and *OP′* is its final position. The line *PY* is constructed such that *PY* is perpendicular to *OP* and therefore in the direction of the instantaneous ω. Similarly line *XP′Z* is constructed tangential to *OP′* at *P′* to represent the direction of ω at *P′*. *X* is the intercept with *PY*; *XY* is made the magnitude of ω to a suitably chosen scale, and likewise for *XZ*.

Thus, *XY* represents $v_1 = \omega r$ and *XZ* represents $v_2 = \omega r$ so that

$$v_2 - v_1 = \vec{XZ} - \vec{XY} = \vec{XZ} + \vec{YX} = \vec{YZ}$$

It will be seen that *YZ* is parallel to *OX*. Thus, in the angle $\Delta\theta$ which occurs in time $(\Delta\theta/\omega)$, v is directed radially toward the origin. This renders the direction for an acceleration component a_R, its magnitude is

$$a_R = \omega^2 r \tag{3.14}$$

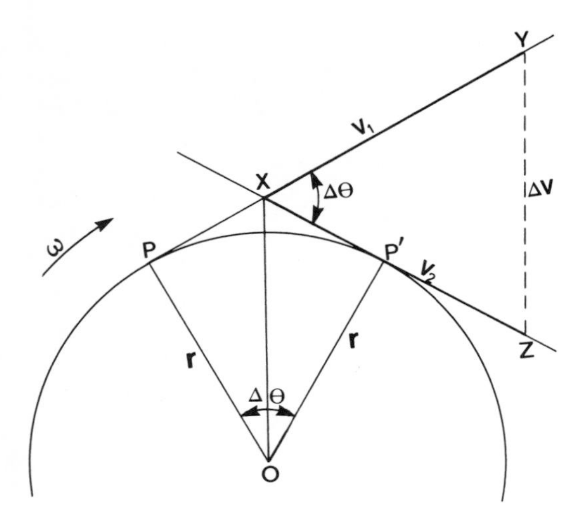

Figure 3.11. Graphical determination of radial acceleration.

(This follows by considering that $\Delta v = \omega r \Delta\theta$ (refer to Fig. 3.11 and Equations 3.11a and 3.12). In Figure 3.11

$$\Delta v = YZ$$

$$\Delta v = \omega r \Delta\theta$$

Since

$$a_R = \frac{\Delta v}{\Delta t},$$

which follows from the definition for linear acceleration, and

$$\Delta t = \frac{\Delta\theta}{\omega},$$

then by substituting in the expression for a_R, Δv and Δt just described, it follows that

$$a_R = \frac{\omega r \Delta\theta}{\Delta\theta} \cdot \omega = \omega^2 r.$$

Since

$$v = \omega r$$

$$a_R = \frac{v^2}{r} \qquad (3.15)$$

a_R is a **radial** or **centripetal acceleration**.

Tangential Acceleration

When motion in a circular path is not constant, the velocity vectors will have different lengths, as in Figure 3.12(a). If the vectors of v_1 and v_2 are drawn as before (Fig. 3.12(b)) with the angle $\Delta\theta$ between them, the vector Δv represents the resultant change in velocity. In Figure 3.12(c) this change of velocity has been resolved into two components, the change in radial velocity Δv_R which results from the *change in direction*, and the change in tangential velocity Δv_T which results from the *change in the magnitude* of the velocity. The tangential acceleration a_T is:

$$a_T = \frac{\Delta V_T}{\Delta t} \qquad (3.16)$$

where **tangential acceleration** is equal to the change in tangential velocity divided by the change in time, and it will be in the direction of the changing velocity (Fig. 3.12(d)).

As a_R and a_T form two sides of a right triangle, when $\Delta\theta$ is very small (practically zero) the actual resultant acceleration is:

$$a = \sqrt{a_T{}^2 + a_R{}^2} \qquad (3.17)$$

EXAMPLE–Rotational Velocity and Acceleration Components

A figure skater initially at rest uniformly increases her rotational speed about a vertical axis until after 15 s she is spinning at 300 rpm. Find the average angular acceleration of the skater and the final linear velocity and linear acceleration of her shoulder tip if it is 20 cm from the axis of spin. Then discuss the relative values of radial and tangential acceleration components as a function of speed.

A speed of 300 rpm corresponds to

$$2\pi \times 300/60 \text{ rad/s} = 10\pi \text{ rad/s}$$

Recalling Equation 3.13:

$$\alpha_{av} = (\omega_2 - \omega_1)/\Delta t = (10\pi - 0)/15$$

$$= 2\pi/3 \text{ rad/s}^2 = 2.095 \text{ rad/s}^2$$

From Equation 3.12 the instantaneous linear velocity at the shoulder has a magnitude

$$v = \omega r = 10\pi \times 20 = 200\pi \text{ cm/s}$$
$$= 628.2 \text{ cm/s}$$

It is directed at right angles to the line drawn from shoulder to shoulder, and in the direction of spinning.

The centripetal acceleration a_R follows from Equation 3.15:

$$a_R = v^2/r = \omega^2 r$$
$$= 100\pi^2 \cdot 20 = 2000\pi^2$$
$$= 19739.2 \text{ cm/s}^2$$

The tangential acceleration a_T is 0 since a constant angular velocity is maintained when the final spinning speed is reached.

In progressing from rest and accelerating to 300 rpm there will, of course, be both radial and tangential acceleration components. At 150 rpm, the radial component

$$a_R = 19739.2/4 = 4934.80 \text{ cm/s}^2$$

(since a_R depends on v^2).

To find a_T, reference must be made to Equation 3.16 and Figure 3.12. At 150 rpm

$$\omega_1 = 5\pi \text{ rad/s}$$

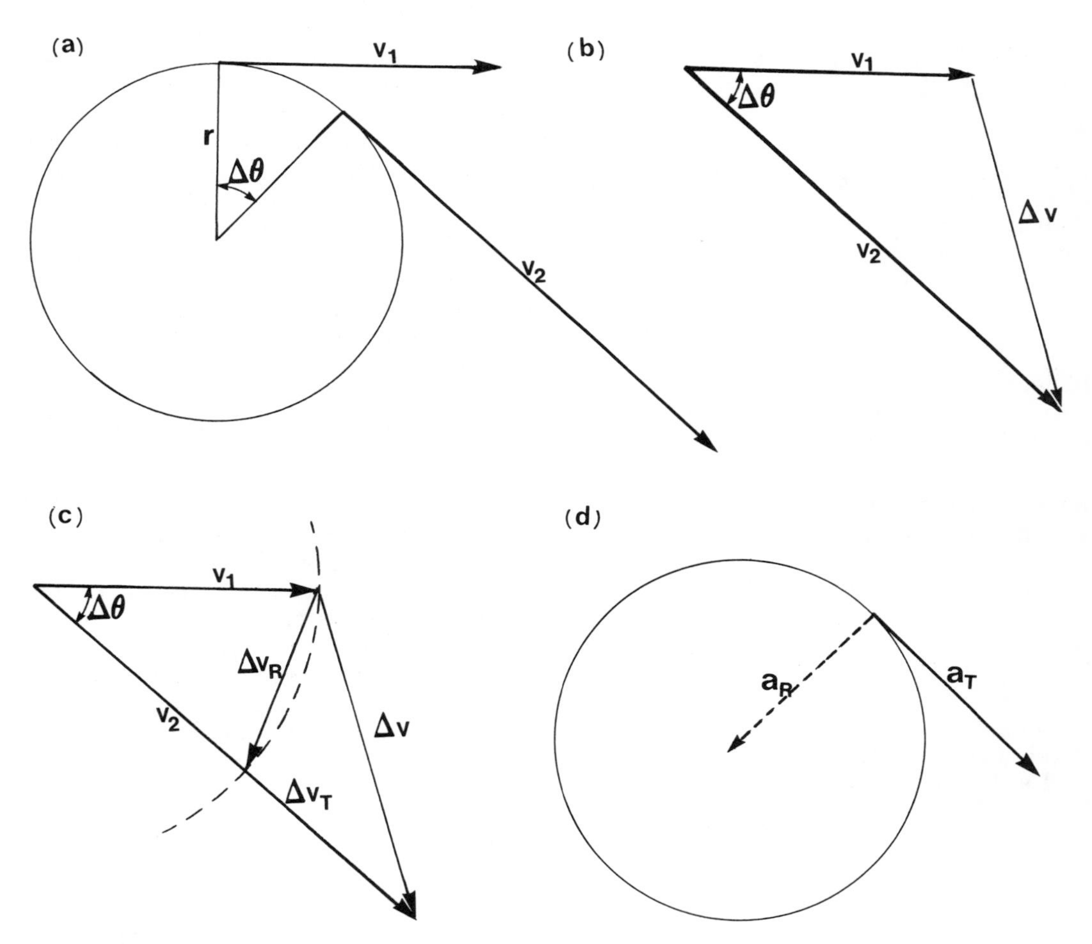

Figure 3.12. Graphical method for deriving instantaneous acceleration; (a) and (b), tangential acceleration; (c) and (d) illustrate the directions of a_R and a_T.

At a time Δt later, a new angular speed ω_2 will be reached because of the average angular acceleration $2\pi/3$ rad/s

$$\omega_2 = 5 + \frac{2\pi}{3} \Delta t$$

$$\omega_2 - \omega_1 = \Delta\omega$$

i.e.,

$$\Delta\omega = \frac{2\pi}{3} \Delta t$$

$$\Delta v_T = \Delta\omega = \frac{2\pi r}{3} \Delta t$$

and

$$a_T = \frac{\Delta v_T}{\Delta t} = \frac{2\pi r}{3} = \frac{2\pi \times 20}{3}$$

$$= 41.95 \text{ cm/s}^2$$

On reviewing the steps to acquire this result, it will be seen that a_T is independent of ω, being constant at 41.95 cm/s² until the steady speed of 300 rpm is reached when it drops to 0.

$$a_R = \omega^2 r = (2\pi n)^2 .20 = 80\pi^2 n \text{ cm/s}^2$$

(where n is the number of rps). For all but very low speeds, less than about $1/20$ rps, a_R is much greater than a_T.

Analogies between Equations for Linear and Rotational Quantities

The reader should note the similarities between the two sets of equations in Table 3.2.

Table 3.2
Analogies between Equations for Linear and Rotational Quantities

Linear	Angular
$a = \frac{\Delta v}{\Delta t}$	$\alpha = \frac{\Delta\omega}{\Delta t}$
$v = v_0 + at$	$\omega = \omega_0 + \alpha t$
$s = s_0 + vt$	$\theta = \theta_0 + \omega t$
$s = v_0 t + \frac{1}{2} at^2$	$\theta = \omega_0 t + \frac{1}{2} \alpha t^2$

SOME NOTES ON CALCULUS

Differential and integral calculus provides a number of mathematical tools of value in kinesiology and biomechanics. In the following discussion the essence of differentiation and integration as useful tools will be provided.

Essentially, differentiation helps to distinguish special features of the variations of one function as a consequence of related variations which occur in a variable which influences the function of concern. For example, a distance-time graph (s versus t) for a runner would show the interrelationship between the parameters distance (s) and time (t). When the runner goes faster, more distance is covered in a selected time interval Δt. "Differentiating" the distance-time characteristic by examining the distances covered in the time interval Δt we can perceive changes in the runner's characteristics. In fact, what we are doing is to create a plot of the speed-time (v versus t) characteristic for a selected time epoch Δt. As Δt is made smaller and smaller for each (s, t) value on the graph, the closer we come to generating the actual instantaneous speed-time characteristic.

Integration is the reverse process of differentiation. Given the speed-time (v-t) relationship for a runner, the distance (s) traveled can be found by integrating (or *adding together*) all of the contributions of distance (s) traveled in each small increment of time Δt, which make up the time span of concern. The distance traveled in each increment is $\Delta s = v \times \Delta t$ (and v will take on a distinct value depending on t). Again, the smaller we make Δt for each (v, t) value on the graph depicting their relationship, the closer we come to generating the actual instantaneous distance-time characteristic.

Differentiation

Figure 3.13(a) shows a plot of s against t for a runner who runs for a total time T. Through point P corresponding to $T/2$ and

$S_{mx}/2$, i.e., $(T/2, S_{mx}/2)$, a straight line (dotted) is drawn to approximate as closely as possible the *s*-*t* curve. Proceeding further away from *P* either to the left or the right results in more deviation of the dotted line from the *s*-*t* curve. Clearly the closer one approaches *P*, the more precisely does the dotted line segment in the region of *P* represent the variation of *s* with *t*. The slope of the curve at *P* can be described by the ratio

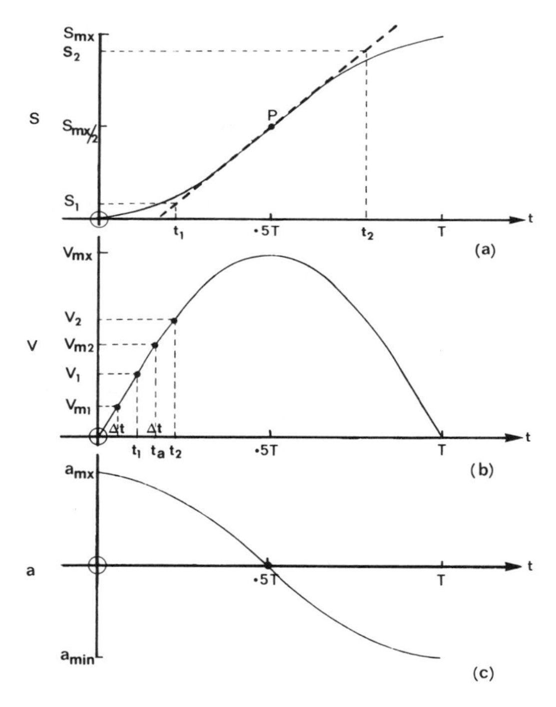

Figure 3.13. Differentiation and integration. Relationships between distance (*s*), speed (*v*), and acceleration (*a*) as functions of time (*t*).

$$(s_2 - s_1)/(t_2 - t_1)$$

Let

$$\Delta s = s_2 - s_1 \quad \text{and} \quad \Delta t = t_2 - t_1$$

then slope $= \Delta s/\Delta t$.

By mathematical definition, as Δt is made to approach zero we obtain the derivative of *s* at point *P*, i.e.,

$$\frac{ds}{dt} = \frac{\Delta s}{\Delta t}, \ \Delta t \to 0$$

is the derivative of *s* with respect to *t* and we have differentiated the function *s*.

It will be recognized that speed

$$v = (s_2 - s_1)/(t_2 - t_1)$$

and thus at any point on the *s*-*t* curve

$$v = \frac{ds}{dt}$$

Thus, Figure 3.13(b) which depicts *v* as a function of *t* can be generated. If we construct the closest approximating slope at each point on the *s*-*t* curve, the corresponding *v* values can be determined and plotted. Note that when the slope of *s*-*t* is 0 at $t = 0$ and $t = T$, so is $v = 0$.

By similar arguments to those just given it can be shown that $a = dv/dt$. Hence the curve of *a* against *t* (Fig. 3.13 (c)) can be derived. Again note that when $v = v_{mx}$ at $t = T/2$, the slope of *v*, i.e., $dv/dt = 0$ and so is $a = 0$.

Integration

If we wish to determine the *s*-*t* curve from a knowledge of the *v*-*t* curve we would use the process of integration. Starting at $t = 0$ on the *v*-*t* curve and realizing that $s = 0$ we know that for $t = 0$, $s = 0$ on the *s*-*t* curve. For a time increment Δt commencing at $t = 0$, i.e., up to a time t_1 as shown in Figure 3.13(b) the speed would have attained a value v_1, but the average speed during Δt would be the value nearly midway between 0 and v_1, i.e., $v_{m1} = v_1/2$.

The distance Δs that would be covered in Δt at an average speed v_{m1} would be $\Delta s_1 = v_{m1} \times \Delta t$. Likewise between t_1 and t_2 the distance $\Delta s_2 = v_{m2}\Delta t$.

Clearly, as *t* values about t_a (corresponding to v_{m2}) are made smaller and smaller, the values such as v_1 and v_2 bounding v_{m2} would more closely approximate v_{m2} until for $\Delta t = 0$ they would all be equal. The contribution to distance in these general circumstances would grow closer to

$$\Delta s = v\Delta t$$

where v is the speed value at a particular value of t. (Note that this can be transposed and written $v = \Delta s/\Delta t$ and as $\Delta t \to 0$, $v = ds/dt$ as before.)

Summing all the Δs contributions as t proceeds we can write

$$\Sigma \Delta s = \Sigma v \cdot \Delta t$$

for $\Delta t \to 0$ we define the mathematical relationship

$$\Sigma \Delta s = s = \int_{t_i}^{t_f} v dt$$

where s is the total distance traveled in the interval t_i (initial time) to t_f (final time) arising from the integrated values of v for all of the time epochs Δt as $t \to 0$. $\int$ is the mathematical sign for integration and the range or limits of integration are often depicted with it as

$$\int_{t_i}^{t_f}$$

From a graphical standpoint, so long as we keep Δt small enough to ensure that $v = (v_1 + v_2)/2$, we can progressively increment s at each step in Δt by an amount $v\Delta t$. In this way, the s-t curve can be generated from the v-t curve. Likewise, the v-t curve can be generated from the a-t curve. Also, it is worth noting that

$$v = \int_{t_i}^{t_f} a \cdot dt$$

General Comments

In general any variables that are mathematically related can be differentiated or integrated with respect to each other, e.g., $y = f(x)$ can be differentiated to give

$$\frac{dy}{dx} = \frac{df(x)}{dx}$$

or integrated according to

$$\int_{x_i}^{x_f} ydx = \int_{x_i}^{x_f} f(x)\, dx$$

For many mathematical functions there are clear-cut relationships for determining differentials or integrals. Often for the purposes of the kinesiologist or biomechanician, graphical methods or numerical techniques implemented on computers, but fundamentally along the lines discussed in this section, are used.

CONCLUSION

The material presented in this chapter should provide the student with a background sufficient to pursue the substance presented in subsequent chapters. Students who find the mathematics too difficult may wish to consult appropriate applied mathematics or physics texts.

CHAPTER 4

Kinetics

INTRODUCTION

While Chapter 3 deals with the kinematics of motion, i.e., a geometric description of movement in terms of its displacement, velocity, and acceleration, it does not account for the sources of motion, i.e., the kinetics. **Kinetics** is concerned with forces that either produce or change the state of rest or motion of a mass, living or inert. This chapter presents the fundamental kinetic concepts needed to understand relevant force, motion, and energy relationships in human movement.

FORCE

Force may be defined as any *action* that causes or tends to cause a change in the motion of an object. It is a vector quantity, i.e., it has both magnitude and direction. Force also has a line of action and a point of application. These are all characteristics of a force.

The action of a force may be direct, such as a push or a pull, or indirect, such as the gravitational attraction between a body and the earth. Force can never be measured directly, but may be estimated by measuring the deflection of a spring or other instrument under the influence of a force (Rodgers and Cavanagh, 1984:84).

Gravitation, the one universal force, was first defined by Sir Isaac Newton in the 17th century.

This **Law of Universal Gravitation** states that:

> ANY TWO BODIES IN THE UNIVERSE HAVE A GRAVITATIONAL ATTRACTION FOR ONE ANOTHER. IF THEIR MASSES ARE m_1 AND m_2 AND THEIR DISTANCE APART, r, IS LARGE COMPARED WITH THE SIZE OF EITHER, THEN THE FORCE ON EITHER BODY POINTS DIRECTLY TOWARD THE OTHER BODY AND HAS THE MAGNITUDE

$$F = G\, m_1 m_2 / r^2 \qquad (4.1)$$

where F is the gravitational force, m_1 and m_2 the respective masses of the two bodies, r the distance between them, and G is a gravitational constant having the same value for all bodies. For an object on or close to a planet such as the earth the resultant force acting between the planetary mass and the object is directed to the center of the planet. Thus, at any place on earth, gravitational force is observed to be directed vertically downward.

The force described by Equation 4.1

depends on the masses m_1 and m_2. **Mass** is defined as the quantity of matter of which an object is composed, and depends upon the volume and density of the object. Mass is related to volume V and density ρ by the relation

$$m = \rho V \tag{4.2}$$

Weight, on the other hand, is defined as a force with which a body is attracted toward the earth (or planetary or lunar type of mass). Therefore, weight is a result of the gravitational attraction which draws people or objects on or near the surface of a planet toward its center.

An examination of the relationship between weight and mass follows logically after a consideration of Newton's Laws of Motion.

Newton's Laws of Motion

The bases for the modern study of motion were laid by Sir Isaac Newton in the 17th century when he formulated his three laws of motion. The early translation from the original Latin is in language which is somewhat archaic, but beautifully expressive:

First Law. "Every body persists in its state of rest or of uniform motion in a straight line unless it is compelled to change that state by forces impressed on it."

Second Law. "The change of motion is proportional to the motive power impressed, and is made in the direction of the right (straight) line in which the force is impressed."

Third Law. "To every action there is always opposed an equal reaction; or, the mutual actions of two bodies upon each other are always equal, and directed to contrary parts" (Resnick and Halliday (1960)).

During the years since Newton's lifetime, these statements have been worked over and reworded to make their meaning clear to each generation of students.

The first law is cited today as follows: A BODY REMAINS AT REST, OR IN A STATE OF UNIFORM MOTION IN A STRAIGHT LINE UNLESS ACTED ON BY AN APPLIED FORCE. Today this has a far more explicit meaning to the average individual than it did a few years ago before space exploration. A rocket remains at rest on its launching pad until ignition of its rocket fuel. It continues to accelerate as long as the "burn" lasts and, when the rocket motors cut off, it has reached escape velocity. After separation, the command capsule continues *in the same straight line* until (1) it is acted on by a short burst of rocket fire for a course correction, and (2) it comes into the pull of the gravitational force of the moon or another planet.

This law is also described as the **law of inertia** because it describes the necessity of requiring a force to change the state of rest or of motion of a body, and it implies as a consequence a resistance to such a change. This resistance to change is termed **inertia**. The mass of a body is a direct measure of its inertia. The concept of inertia is sometimes applied interchangeably with that of mass. A body in space or on the moon will have the same mass or resistance to change in its state of movement or of rest, i.e., *inertia,* as it had on earth. It will take the same amount of force as it did on earth to put it in motion (accelerate it), to stop it moving, or to change its direction. The weightlessness of space does not change this, nor does the one-sixth gravity force on the surface of the moon relative to the earth force at its surface.

The only difference, for a body in space or on the moon, is the absence of or the decrease in the force pulling an object toward the center of a planet or satellite, and that is gravity.

Health professionals often refer to Newton's first law of motion as the law of whiplash because it pertains to neck injuries sustained in motor vehicle accidents by patients who complain of neck pains. During the patient interview, the health care professional elicits information concerning the patient's head position (facing forward, looking in the mirror, or looking over the shoulder) and the relative movement of the vehicle. By applying Newton's first law of motion, it is then possible to hypothesize the injured anatomical structures. Two cases are presented as illustrations.

In the first, subject A is stopped at an intersection, is facing forward, and the car is not equipped with head rests. A's car is subsequently hit from behind by subject B's vehicle. In accordance with Newton's first law of motion, both vehicle A and the driver were at rest; A's vehicle was accelerated by B's vehicle, and, since A's back was resting against the car seat, A's thorax was also accelerated. Given there was no head rest, A's head would continue to remain at rest until it was pulled forward by the neck due to the forward movement of the thorax. The neck is forced into extension, and results in stretching of the anterior structures of the neck and compression of the posterior structures. Since both subject A and A's vehicle were stopped and subsequently accelerated by the second vehicle, any injuries sustained by A are considered as acceleration injuries; i.e., increased acceleration.

The second case considers the events from B's perspective. Subject B (wearing seat belt and shoulder harness) is in motion in a straight line. Subsequently, B's vehicle collides with A's vehicle and stops. When the vehicle stops, B's thorax restricted by the seat belt and shoulder harness also stops, but B's head continues to move forward. This results in stretching of the posterior anatomical structures of B's neck and compression of the anterior anatomical structures. Since both B and the vehicle were initially moving and subsequently stopped by the second vehicle, any injuries sustained by subject B are considered to be deceleration injuries; i.e., decreased acceleration.

Newton's first law leads into the second, which is concerned with force, mass, and acceleration. Actually, this law as Newton stated it used the term motion as the product of mass and velocity, mv (Resnik and Halliday (1960), Tricker and Tricker (1967). Taking this law as Newton stated it, the equation is:

$$F = \frac{mv_2 - mv_1}{\Delta t} \tag{4.3}$$

where m is the body mass, v_1 the velocity at time t_1, v_2 the velocity at time t_2, $\Delta t = t_2 - t_1$, and F is the vector sum of all the forces acting on the body. From Equation 4.3 it will be noted that since v_1 and v_2 are vectors, F must also be a vector. This equation can be treated algebraically to yield:

$$F = m\frac{(v_2 - v_1)}{(t_2 - t_1)} \tag{4.4}$$

It will be recognized that

$$(v_2 - v_1)/(t_2 - t_1) = a$$

the acceleration in the interval Δt.

Hence

$$F = ma \tag{4.5}$$

From this there follows the more modern statement of *Newton's Second Law:* IF A BODY OF MASS, m, HAS AN ACCELERATION, a, THE FORCE ACTING ON IT IS F, DEFINED AS THE PRODUCT OF ITS MASS AND ACCELERATION (F . ma).

Newton's Third Law is very simply expressed as follows: TO EVERY ACTION THERE IS AN EQUAL AND OPPOSITE REACTION. Thus, forces work in pairs. When a foot presses on the ground in walking, the ground pushes back with an equal but opposite force. The ground is acted upon as the foot strikes it and it reacts with an equal force in the opposite direction, causing motion in that direction when possible. (Fig. 4.1) As the running long jumper takes off, the takeoff foot thrusts against the board, the board "pushes back," and the jumper is propelled through the air by a force equal to the thrust of the takeoff foot.

One of the problems encountered in the space program results from this law of action and reaction. When there is no large mass such as the earth or moon to react against the thrust of a foot or a hand, the body itself responds by turning or moving in the opposite direction. This may be illustrated in the laboratory by someone standing on a freely movable turntable with one arm abducted 90°.

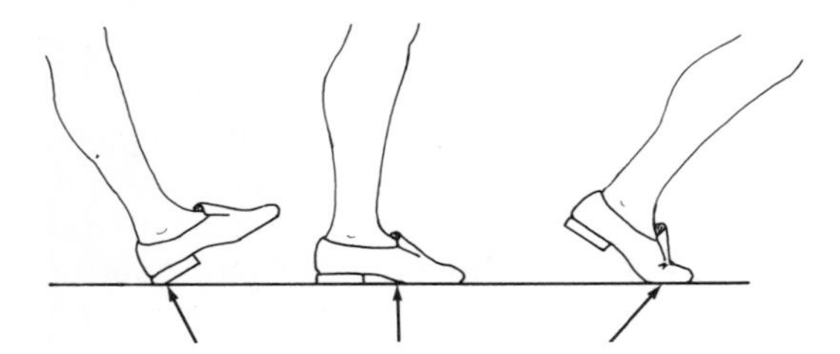

Figure 4.1. Action and reaction. The foot lands and exerts pressure on the ground; the ground reacts with an equal but opposite force; the direction of such force is indicated by the arrow.

Regardless of whether the person swings the arm to the left or right in horizontal flexion or extension, the table reacts by turning the person in the opposite direction.

Acceleration g due to Gravity

Following from Equation 4.5, mass is defined mathematically by

$$m = F/a \qquad (4.6)$$

The force of gravity on earth gives rise to an acceleration g; thus, mass can be derived from a knowledge of F and a with $a = g$ in Equation 4.6.

From an earlier discussion it will be evident that weight

$$W = mg \qquad (4.7)$$

The system of units and conventions used plays roles in describing weight (force) and mass:

If mass is in grams and g is in centimeters per second^2, weight is in dynes, and the system is known as the CGS system, e.g.:

$$W = 10 \text{ grams} \times 980.6 \text{ cm/s}^2 = 9806 \text{ dynes}$$

If mass is in kilograms and g in meters per second^2, weight is in newtons, and the system is known as the MKS system, e.g.:

$$W = 10 \text{ kg} \times 9.806 \text{ m/s}^2 = 98.06 \text{ newtons}$$

Confusion exists within the metric system however, as it is common practice to use the terms gram and kilogram to define both mass and weight. Since weight is the product of mass and the acceleration attributed to gravity, weight is really a force and should be expressed as indicated above. Reference should be made to the international system of units found in Appendix 2 for a review of the unit, formula, and symbol of each metric quantity.

Variations of Gravity

As the amount of gravitational attraction is inversely proportional to the square of the planet's radius (refer to Equation 4.1), the shape of the planet affects the amount of force exerted at any given point on its surface. For example, on earth this gravitational force causes an object to be accelerated 9.806 m/s^2 (SI system) when it is falling at 45° latitude, i.e., halfway between the equator and either pole. However, because the earth is flattened at its poles, making the distance to its center slightly less at these two points, a falling object would be acclerated 9.832 m/s^2 in the polar regions. When falling at the equator, where the distance to the center is greatest, the same object falling at sea level would undergo less acceleration; i.e., 9.78 m/s^2. In other words, the object would be heaviest when weighed at either pole, lighter when weighed at 45° latitude, lighter still at equator sea level, and lightest of all on a mountain peak along the equator, i.e., in Bolivia, Kenya, or Sumatra (Tables 4.1 and 4.2) (Resnick and Halliday (1960)).

Table 4.1
Variation of g with Altitude at 45° Latitude

Altitude	g
m	m/s^2
0	9.806
1,000	9.803
4,000	9.794
8,000	9.782
16,000	9.757
32,000	9.708
100,000	9.598

Table 4.2
Variation of g with Latitude at Sea Level

	Latitude	Meters/s²
Equator	0°	9.78039
	10°	9.78195
	20°	9.78641
	30°	9.79329
	40°	9.80171
	50°	9.81071
	60°	9.81918
	70°	9.82608
	80°	9.83059
Pole	90°	9.83217

Force Systems

The forces acting upon a body are usually summed using vector methods to derive their resultant values. Depending on how forces are applied to given structures that are able to move under their influence, various kinds of motions may result. It is of value to examine the various force systems that exist in order to show fundamental ways in which forces may act.

Linear Force System

When forces act in the same straight line they are said to be colinear in a system called a **linear system of forces.** This is the simplest combination of forces. Figure 4.2 demonstrates a situation in which forces of 10, 3, and 5 N are pushing a body to the right, while at the same time forces of 2 and 4 N are pushing the same body to the left. These forces have direction and magnitude and so are vector forces which may be graphically illustrated by arrows whose length is scaled to represent the amount of force (Fig. 4.2). Vectors pointing to the right or upward are by convention considered as positive, while those pointing in the opposite directions are considered negative. Adding graphically in Figure 4.2(b) we end up with a positive vector 12 units long (Figure 4.2(d)), which is equivalent to 12 N. Mathematically:

$$R = \sum_{i=1}^{n} F_i \qquad (4.8)$$

where R is the resultant of the forces, F_1, $F_2 \cdots F_n$ involved:

$$R = +10-4+5+3-2 = 18-6$$
$$= +12\text{N}$$

Parallel Force System

When the action lines of the forces under consideration are parallel to one another and lie in the same plane, the system may be called a **parallel force system.** As the forces are applied at some distance from each other a different approach is necessary for the analysis and calculations relevant to situations of this kind. If the body upon which parallel forces are acting is to be in equilibrium, the sum of the forces acting in one direction must be nullified by the sum of the forces acting in the opposite direction. Furthermore, cognizance must be taken of the rotational influences that may occur when parallel forces act on a body. This aspect will become more apparent when the subject of levers is discussed later in this chapter. At this

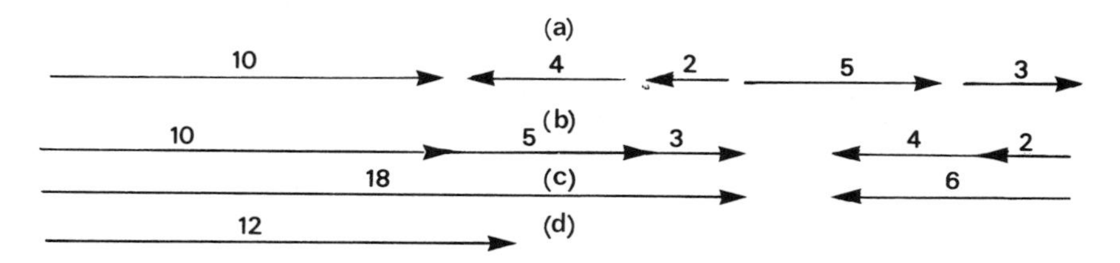

Figure 4.2. Vector diagram of forces described in text. (a) vectors representing each of the different forces (see text); (b), (c), and (d), vectors added graphically.

juncture, it will be sufficient to note that the two aforementioned facets must be considered.

A Special Case: Force Couples

A special case of a parallel force system occurs when there are two forces of equal magnitude acting at a distance from each other and in opposite directions. Under these circumstances they produce a turning action, as in Figure 4.3 in which two individuals cooperate to turn a boat end for end. As long as the force exerted by individual A is equal and opposite to that exerted by individual B, the boat will not go anywhere; there will be no linear displacement or acceleration, and the resultant of the two forces will be zero. But the boat is turned. The forces acting in this situation are known as **force couples** and the *moment* or *torque* is the product of one of the equal and opposite forces multiplied by the distance between them:

$$T_c = Fd \tag{4.9}$$

where the subscript *c* refers to the couple.

In Figure 4.3 each individual is exerting a force of 135 N and they are 2 m apart. Individual A is exerting a clockwise force, thus the resultant of the two forces will be 0:

$$R = \Sigma F = 0$$

$$R = +135 - 135 = 0$$

but the moment or torque:

$$T_c = 135 \text{ N} \times 2 \text{ m}$$

$$T_c = 270 \text{ Nm}$$

In the human body, no true force couples exist. An example commonly

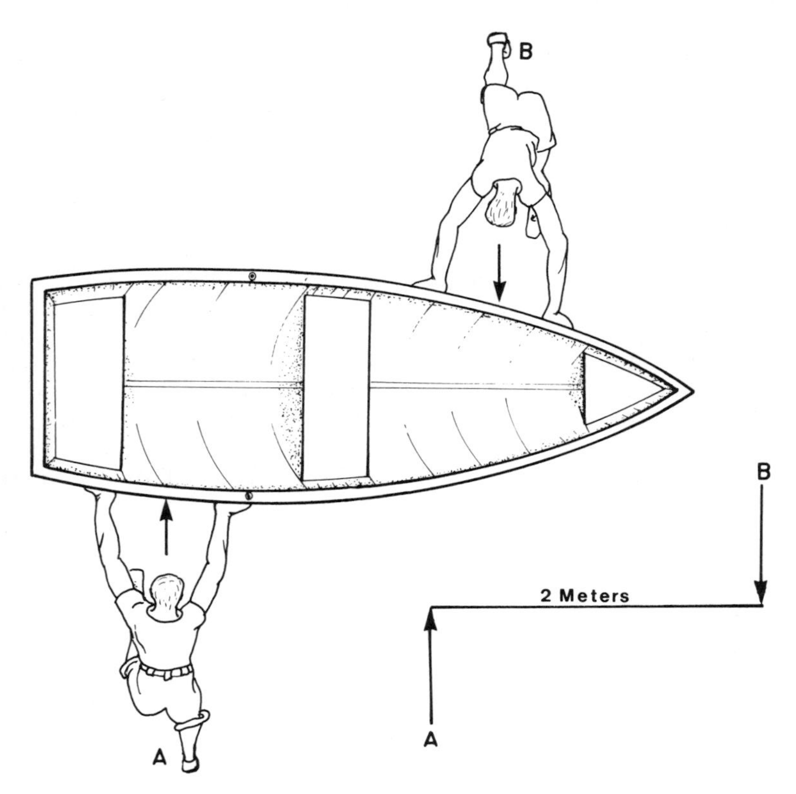

Figure 4.3. Example of a force couple in which equal and opposite forces are applied to each end of the boat. The torque or moment of this couple (T_c) is the product of one force and the distance between them.

cited is the pseudo force couple formed by the upper and lower fibers of the trapezius muscle. These two fiber groups act together to perform several movements, one of which is scapular rotation. Given that the fiber orientation of these muscles is not parallel and that a linear displacement of the scapula occurs when they contract, this is clearly not an example of a force couple.

The human body is, however, frequently called upon to adapt itself to form a force couple relative to an external object. One such example would be the force couple formed by the thumb and index finger in removing a screw lid from a jar. Another example would be the use of a "T"-shaped wrench for removing a tire bolt. Here, the push produced by the right arm would be equal, opposite, and parallel to the pull produced by the left arm.

Concurrent Force System

Forces whose action lines meet at a point are said to operate in a **concurrent force system.** Such forces may be applied to a body from two or more different angles so that projections of their action lines will cross. This intersection need not be inside the body. Figure 4.4(a) presents a simple situation where the action lines intersect within the body, while (b) illustrates an intersection outside the body. Another example occurs in normal quiet standing where the line of gravity falls slightly anterior to the ankle joint so that gravity has a clockwise action on the body and only the counterclockwise tension in the calf muscle maintains erect posture (Figure 4.5).

General Force System

In some situations, the forces acting on a body are not linear, parallel nor concurrent. Any situation in which the forces do not meet the requirements of any of the previous categories is categorized as a **general system of forces.** A general force system always includes at least four forces (LeVeau, 1977).

In Figure 4.6, an example of a general force system can be seen. Forces *A* and *R* are a pair of opposite forces acting in the same line; and forces *X* and *F* are equal and opposite forces acting in another line which is distinctly different from the first. (Action and reaction are equal and opposite.) The four forces are not linear or parallel to each other, nor are they concurrent.

Consider the weight is initially at rest. When the arm force *A* exceeds the rope pull *R* which is equal to the suspended weight (for a frictionless pulling system), the weight will be lifted and the subject's stiffened body will move backwards while pivoting about the feet. When *R* exceeds *A*, the body will tend to move forward.

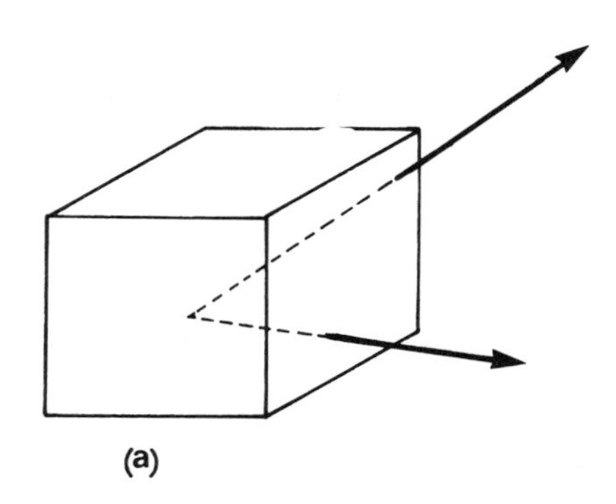

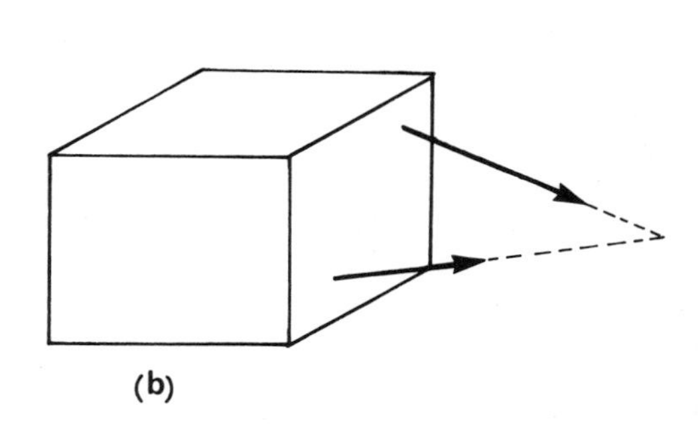

Figure 4.4. Concurrent force systems. (a) lines of force intersect inside the body; (b) lines of force intersect outside the body.

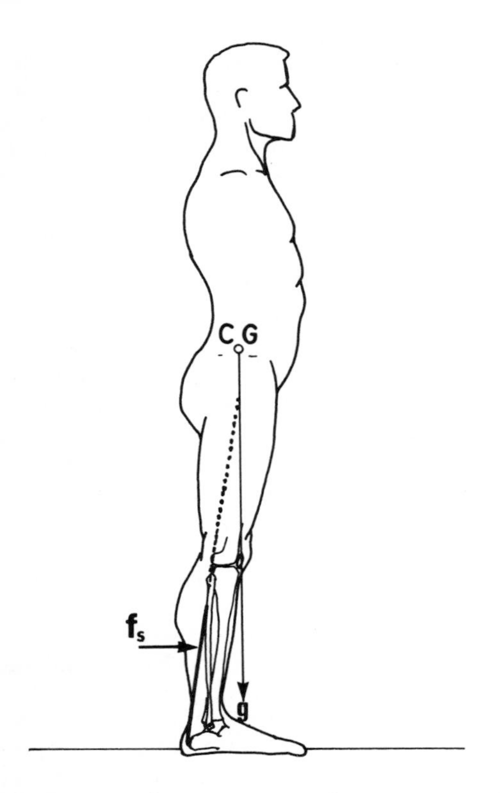

Figure 4.5. The concurrent force system of gravity g and soleus muscle force *fs*, maintaining the upright posture. CG, center of gravity.

Types of Forces

Descriptions of the force systems enables us to better understand the **types of forces** or *stress* usually encountered by the human body. The stress which develops within a structure in response to a load is usually expressed as a force per unit area. Its units are newtons per square meter or pascal. These types of forces are known as either compression or decompression, normal or shear.

Compression forces act in opposite directions within a linear system so that they push material together. If compression forces are great enough, a shortening and widening of the material is effected. **Decompression** or **tension forces**, on the other hand, act in opposite directions within a linear system so that they pull material apart. Depending on the magnitude of the tensile force, a lengthening and narrowing of the material may be accomplished (LeVeau, 1977:223; Rodgers and Cavanagh, 1984:92).

Normal forces work in the same plane so that a force is applied perpendicular to a surface and may have the effect of changing the length in a structure. **Shearing forces** also work in the same plane and cause the surface of one body to slide past the surface of another adjacent to it (LeVeau, 1977:223; Rodgers and Cavanagh, 1984:92). Examples of each of these types of forces will be presented in subsequent material after the concept of levers within the human body has been explained.

With all of these concepts as a basis, it is possible to proceed to solving problems that arise in the study of human movement, in particular those involving kinetics. Two kinds of forces, internal and external, act on the human body. Internal force is exerted primarily by muscle, either in the form of a shortening contraction, or in the presence of an outside force, a lengthening contraction. The major external force is, of course, gravity. As gravitational force pulls all bodies toward the center of the earth, its action line is always vertically downward. Other external forces include those of impact as in catching or striking, falling or contact in sports, water resistance when swimming, the force exerted by a fiberglass pole in vaulting, etc. The action line of these external forces other than gravity depends upon the situation being analyzed. As all of these forces, both internal and external, have direction and magnitude, they are vector quantities and subject to vector analysis.

Composition and Resolution of Forces

Many times when a number of forces are acting on a body it is desirable to find a single force or resultant that will have the same effect on the body as that of the combined forces that it replaces. This process is known as the composition of forces. This problem of finding a resultant of two or more forces has as its concomitant the problem of resolving a single force into two or more components such that the combined action of the two new forces will be equivalent to that of the original force. The most common situation that the student of human movement will encounter is the necessity to resolve a single force into two components.

Graphical Resolution of Forces

Any single force may be replaced by two or more components whose combined action will be the same as that of the original force. The simplest situation of resolving a force into two components involves the construction of a parallelogram of forces such that the original force forms the diagonal (Fig. 4.7). In each of the above cases the pairs of force components of the force F are indicated by P and Q, P' and Q', and P'' and Q''. These forces, acting in unison on a body, will produce the same effect as the single force F. However, it is more desirable to specify the direction that each of the two replacements must take, and the

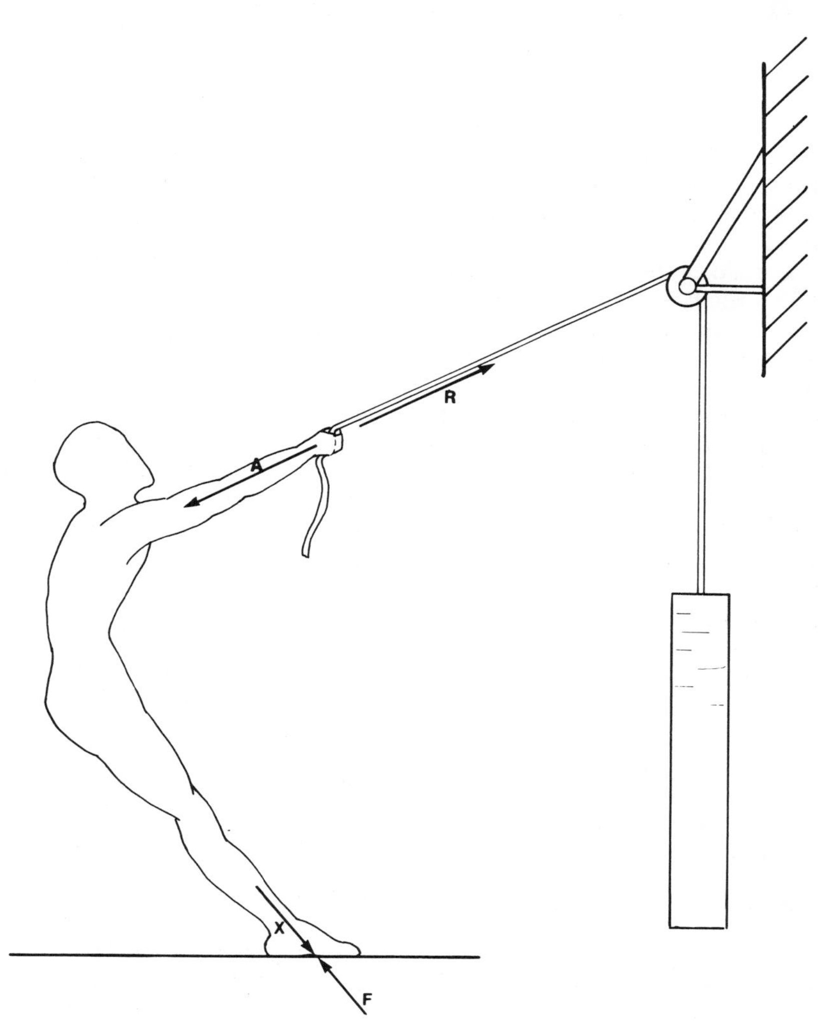

Figure 4.6. General force system; two pairs of forces. *A*, the force exerted by the arms; *R*, the force exerted by the rope; *X*, the force exerted by the feet; *F*, the force exerted by the floor.

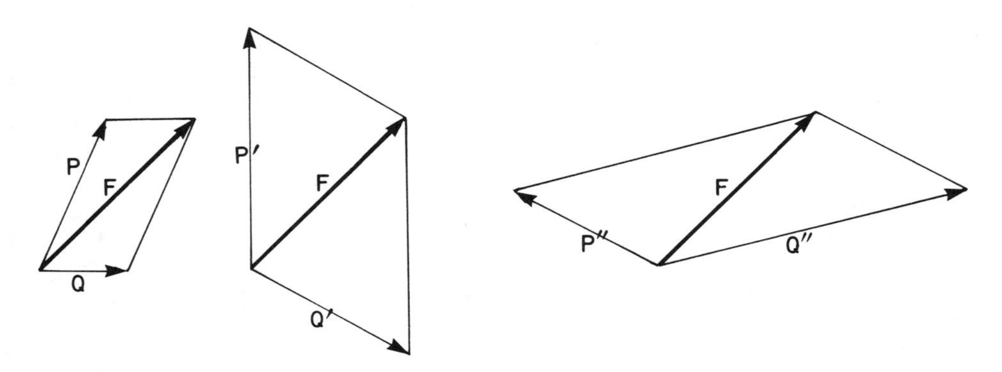

Figure 4.7. Resolution of force F into two components: P and Q, P' and Q', and P'' and Q'' by constructing parallelograms of force around force F.

most frequent procedure is to resolve the force F into two components at right angles to each other. These directions are normally horizontal and vertical, forming an X component on the horizontal or xaxis, and a Y component along the vertical or y axis (Fig. 4.8). The force F is at 45° with the horizontal, and the vector $\overrightarrow{AB}$ represents the force. The x and y axes are drawn through point A, the origin of the force. Perpendiculars from point B to the two axes define the two components, vectors $\overrightarrow{AC}$ (the X component) and $\overrightarrow{AD}$ (the Y component).

The two component force vectors form the two legs of a right angle triangle, with the force as the hypotenuse, so the Pythagorean theorem is applicable: "The square of the hypotenuse is equal to the sum of the squares of the two sides." Consequently,

$$\overrightarrow{AB}^2 = \overrightarrow{AC}^2 + \overrightarrow{AD}^2$$

$$\overrightarrow{AB} = \sqrt{\overrightarrow{AC}^2 + \overrightarrow{AD}^2}$$

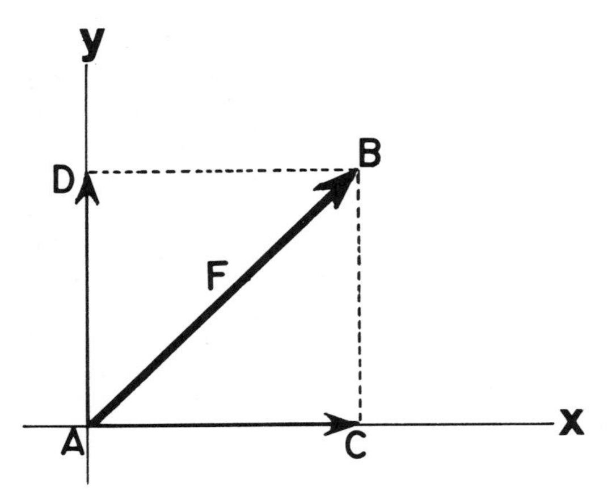

Figure 4.8. Resolution of force F into vertical and horizontal components. See text.

Graphical Composition of Forces

A number of forces acting on a single body can be reduced to a single force whose action will be the equivalent of the combined action of all of the original forces. When it is desirable to manipulate forces in this manner there are two general approaches, graphical and mathematical. Parallel forces can best be composed mathematically, while linear and concurrent forces can be treated by both methods. A graphical solution of a linear force problem is presented above under "Linear Forces." Finding the resultant of a pair of concurrent forces is the reverse of resolving a force into two components as in Figures 4.7 and 4.8. The pair of concurrent forces becomes the sides of a parallelogram, and the single resultant force R is the diagonal of the parallelogram which originates from the junction of the two forces.

In Figure 4.9 two forces, P of 10 N, and Q of 7.5 N, are acting on a single body.

Problem. Find the magnitude and the direction of the resultant (the single force that can replace P and Q).

Solution. Construct a parallelogram of

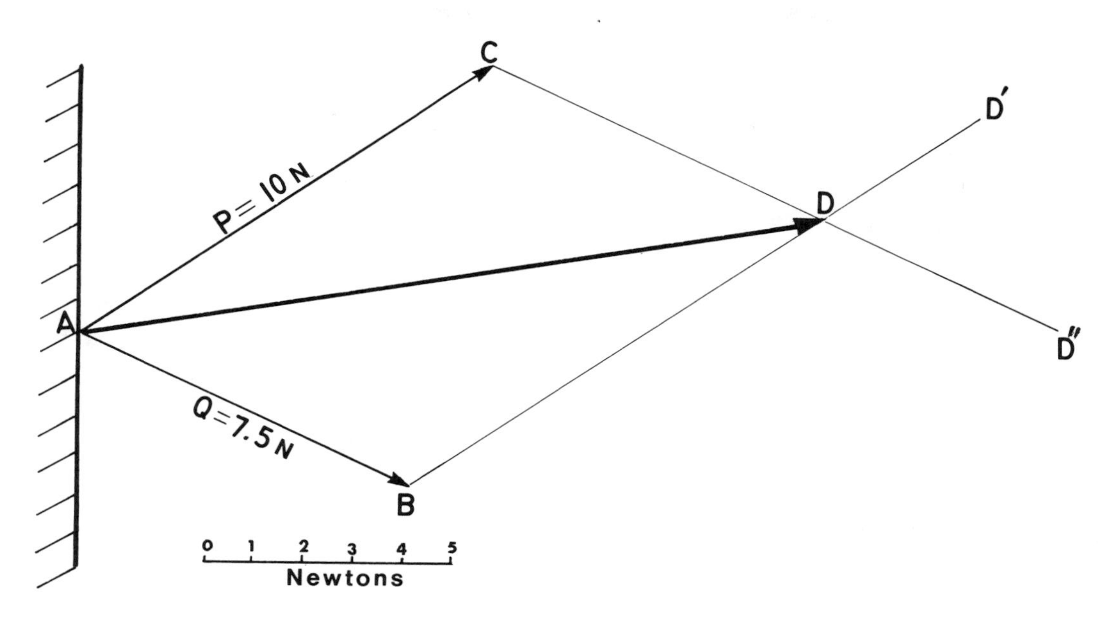

Figure 4.9. Graphical composition of forces P and Q. See text.

forces by drawing line BD' parallel to vector AC, and line CD'' parallel to vector AB. D is the point of intersection of BD' and CD''. The vector AD is the resultant of forces P and Q.

More than Two Forces

With three forces A, B, and C it is possible to construct two parallelograms. The three vectors A, B, and C are drawn from a common origin and appropriately scaled along their lines of action. The resultant D of two components, say A and B, is determined by a parallelogram of forces and then another parallelogram constructed in which C and D are the vectors to be summed. In this way the resultant of the three forces is arrived at. Clearly, this method may be extended to cover as many forces as required. For many vectors it is a laborious method. Another useful graphical method is the solution which derives from the construction of a polygon of forces.

Polygon of Forces

In this construction, vectors are added graphically to each other by drawing them sequentially (and in appropriate lines of action with pertinent scaling) such that the termination of one vector is the starting point for drawing the next one. Figure 4.10 illustrates the situation for four vectors $\vec{A}$, $\vec{B}$, $\vec{C}$, and $\vec{D}$.

The vector R that closes the polygon shown in Figure 4.10(b) is the resultant.

Mathematical Composition and Resolution of Concurrent Forces

This technique has already been illustrated by way of an example in Chapter 3 and it relies on the essential trigonometric functions sin, cos, tan, etc. (See Appendix 3 for a review of these relationships.) Unless the angles that each force makes with the horizontal are known, a force diagram is drawn, reproducing the action lines of the forces in question, and a horizontal x axis is added, intersecting the action lines. (The y axis is not always needed but may be drawn in if desired.) Figure 4.11 is a force diagram of the problem solved graphically in Figure 4.9. Both the x and y axes have been added at the origin of the two forces. The action line of

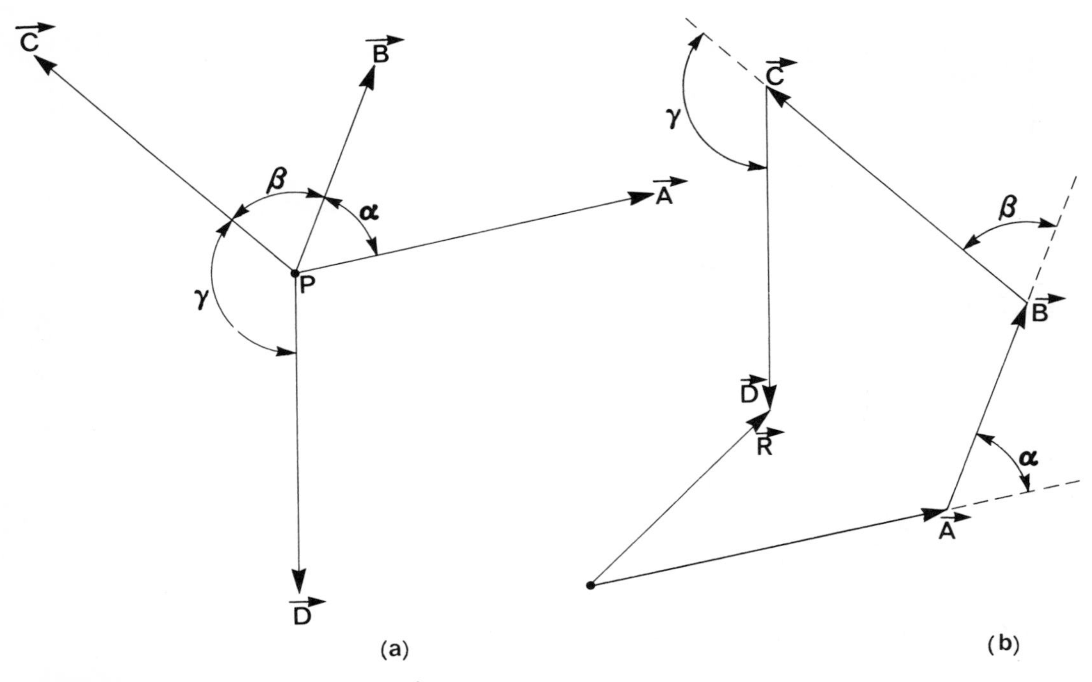

Figure 4.10. (a) forces $\vec{A}$, $\vec{B}$, $\vec{C}$, and $\vec{D}$ at a point *P*. (b) constructing the polygon as shown leads to the resultant force *R* in magnitude and direction.

force P lies in the first quadrant of the x-y system (Fig. 4.12), so both the *X* and *Y* components are positive. On the other hand, the action line of force Q lies in the fourth quadrant and, while the *X* component is positive, the *Y* component is downward and therefore negative. By measuring the angles on the diagram, force P is at 35° with the x axis and force Q forms a downward angle of 25° with the same axis.

The *X* component, R_X, of the resultant

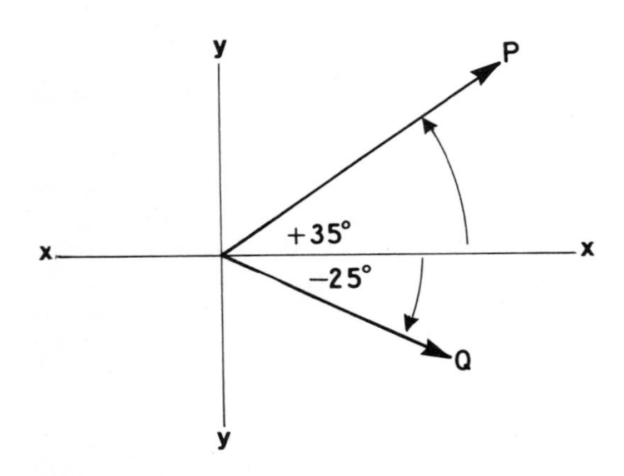

Figure 4.11. Force diagram from problem solved graphically in Figure 4.9 with horizontal x and vertical y axes added. See text.

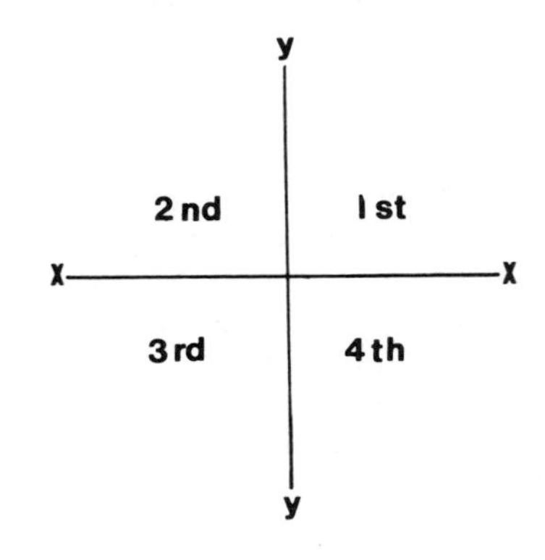

Figure 4.12. x and y axis system and the four quadrants.

force will equal the sum of the X components, ΣF_X, of the forces involved, and similarly the Y component, R_Y, of the resultant will equal the sum of the Y components, ΔF_Y, of the force involved, i.e.

$$R_X = \Sigma F_X$$

and (4.10)

$$R_Y = \Sigma F_Y$$

To find the resultant (in magnitude and direction) of forces P and Q (Fig. 4.11) we proceed as follows, first with a table:

Table of known data

Force	P=10N	Q=7.5N
Angle with horizontal	+35°	−25°
Sine	0.57358	0.42262
Cosine	0.81915	0.90631

The equation:

$$\begin{aligned} R_X &= P \cos 35° + Q \cos 25° \\ &= (10\text{ N} \times 0.81915) \\ &\quad + (7.5\text{ N} \times 0.90631) \\ &= 8.195\text{ N} + 6.7973\text{ N} \\ &= 8.2\text{ N} + 6.8\text{ N} \\ &= 15.0\text{ N} \end{aligned}$$

The horizontal component of forces P and Q is 15 N.

$$\begin{aligned} R_Y &= P \sin 35° - Q \sin 25° \\ &= (10\text{ N} \times 0.57358) \\ &\quad - (7.5\text{ N} \times 0.42262) \\ &= 5.73858\text{ N} - 3.16965\text{ N} \\ &= 5.7\text{ N} - 3.2\text{ N} \\ &= 2.5\text{ N} \end{aligned}$$

The vertical component of forces P and Q is 2.5 N.

The resultant being sought is the hypotenuse of the right triangle whose two sides are represented, respectively, by vectors 15.0 and 2.5 units long.

By using the Pythagorean theorem the magnitude of the resultant may be determined:

$$\begin{aligned} R^2 &= 15^2 + 2.5^2 \\ &= 225 + 6.25 \\ &= 231.25 \\ R &= \sqrt{231.25} \\ &= 15.2\text{ N} \end{aligned}$$

The resultant equals 15.2 N, but the direction has yet to be located, i.e., the angle that the resultant makes with the horizontal. Angle θ is the angle whose sine is Y divided by R. This is expressed as:

$$\begin{aligned} \theta &= \arcsin Y/R \\ \theta &= \arcsin 2.5/15.2 \\ &= \arcsin 0.1645 \\ &= 9° \, 28' \end{aligned}$$

Alternative Solution. The magnitude and direction may also be determined as follows:

$$\begin{aligned} \theta &= \arctan Y/X \\ &= \arctan 2.5/15.0 \\ &= \arctan 0.1666 \\ &= 9° \, 28' \end{aligned}$$

This gives the direction of the resultant but not its magnitude, which is equivalent to the hypotenuse of the right triangle. Cos $\theta = X/R$ and $\sin \theta = Y/R$ so either may be used to find R, the hypotenuse, as $R = Y/\sin \theta$ or $X/\cos \theta$.

$$\begin{aligned} R &= Y/\sin \theta \\ &= 2.50/0.16447 \\ &= 15.2\text{ N} \end{aligned}$$

The resultant is 15.2 N upward at an angle of 9° 28′ (cf. results found by the Pythagorean theorem).

Positive and Negative Components

Whether or not a force component is positive or negative or zero depends upon its direction in relation to the x-y axis system which divides a plane into four quadrants (Fig. 4.12).

A force whose direction upward and to the right places it in the first quadrant will have both X and Y components positive. When the direction is still upward but to the left, the force will be in the second quadrant, and the Y component remains positive but the X component becomes negative. Any force directed downward and to the left will be in the third quadrant and both X and Y components will be negative, while forces directed downward and to the right are in the fourth quadrant where the Y component is still negative but the X component is again positive.

To summarize: All forces directed upward have positive Y components. All forces directed to the right have positive X components. All forces directed downward have negative Y components. All forces directed to the left have negative X components.

Equilibrium Forces

For a body to be in a state of equilibrium it is necessary for the total forces acting upon it to be equal to 0. Mathematically:

$$\Sigma F = 0$$

MOMENTUM

Momentum is the product of the mass of a body and its velocity, and as such it is a vector quantity, a *quantity of motion* possessed by a body: $P = mv$. If mass is in kilograms and velocity in meters per second, P is expressed as kilogram meters per second.

Application of Force and Changes in Momentum

The equation of Newton's original statement of his Second Law

$$F = \frac{mv_2 - mv_1}{\Delta t} \qquad \text{(from 4.3)}$$

involving the time rate of change of momentum may be used for determining the force involved in striking a ball if the necessary data are available: the duration of the contact of the implement with the ball; the velocities of the ball before and after it was struck; and the mass of the ball. It is only in recent years, since ultra-high-speed photography has made the acquisition of such data possible, that the use of this formula has become practical in sports analysis. Unfortunately, as the speed of the struck ball also is influenced by the elastic qualities of the ball and possibly those of the striking implement as well, the estimate of the force involved will not be entirely accurate.

Equation 4.3 may be rearranged as follows:

$$F \times \Delta t = mv_2 - mv_1 \qquad (4.11)$$

where the product $F \times \Delta t$ is referred to as the *impulse* that causes the *change in momentum* ($mv_2 - mv_1$). The impulse consists of a force F applied for a short duration Δt.

It is evident that a small force applied for a long enough period of time may produce momentum changes comparable to a far larger force applied for a briefer interval.

To achieve a large change of momentum in a given period Δt, a large force F is required. Thus, in projectile skills such as throwing, greater body forces are needed to attain higher projectile velocities. This explains why baseball pitchers may hurt their arms when they attempt to achieve maximal velocity of the ball by applying a maximal amount of force over as brief an interval as possible. The resulting tangential and radial accelerations allied with upper extremity movements create forces which are responsible for the strained ligaments, damaged bursae, and possibly torn muscles that pitchers occasionally suffer.

Receiving a Force

When considering Equations 4.3 and 4.11 from another viewpoint, that of catching a fast ball or landing from a high jump, the greater the increase in the amount of time consumed in catching the ball or for completing the landing, the less will be the force felt by the catcher or jumper at any one instant while his body is receiving the force and the less will be the damage to body tissues. In catching a fast ball, even with a catching mitt, the hands are stretched forward to meet the ball and are drawn toward the body as the ball contacts the glove. All of this increases the time during which the momentum of the ball is absorbed by the catcher.

Landing pits for long and high jumps were originally filled with sand, then with sawdust, wood shavings, or tanbark. Today most of them are filled with chunks or sheets of foam rubber. The jumper, if he lands on his feet (this is chiefly in the case of the long jump), dorsiflexes at the ankles, flexes the knees, hips, and spine, and may go into a roll. The high jumper or pole vaulter may land on his shoulders and back or with a roll or both. In any case the force generated by his momentum is spread over as large an area of his body as possible so that any one area is impacted by a comparatively small amount of force. This type of landing also increases the duration of the landing process so that there is less force for the body to absorb from instant to instant. The landing medium also absorbs some of the momentum, as the particles of sand, shavings, tanbark, or foam rubber fly in all directions when some of the jumper's momentum is transferred to the landing medium. The ultimate result is that the force felt by the jumper at any one instant over any one area of the body is decreased to the point where there is little if any damage to the tissues.

The boxer rolls with the punch for the same purpose: to decrease the force with which he is hit by increasing the time of contact, which in turn decreases the change in momentum per unit of time. It is the abrupt change in momentum from a large quantity to little or none that is harmful to the body (e.g., an automobile collision). This corresponds to the condition of a large impulsive force (see Equation 4.11).

The application of the principles embodied in Equations 4.3 and 4.11 is of great importance in the prevention of athletic injuries. The emphasis on fitness for health resulted in the advent of "aerobic dance," "jazzercise," and other similar activities. At first little attention was given to the physical facilities and the conditions under which the exercises were conducted. Often the exercises were performed on nonresilient surfaces with the participants wearing inappropriate shoes or even no shoes. As the number of performers increased, so did the number of patients who visited sports medicine clinics and were diagnosed as suffering from stress fractures of the tibia. Stress fractures occur as a result of impact loading, and the high incidence of such fractures is a direct result of repeated impact loading encountered during many program regimens that include hops, leaps, and bounces. The remedy for the problem is one of prevention.

The injury may be prevented by reducing one of the risk factors, *i.e*, impact loading. The solution may be deduced by referring to Newton's Second Law of Motion and Equation 4.4. The subject's mass, (m) is not likely to change during the exercise period. The velocity changes, ($v_2 - v_1$) also will remain unaltered because the activity regimen must be maintained to achieve cardiovascular benefit from the repeated leaps, bounces, and hops.

Time is the only factor in the equation that is available to alter in order to minimize the magnitude of the impact force. The duration of the impact may be lengthened effectively, (i.e., decreasing the acceleration) by the use of shoes with good absorption properties and/or performing the exercises on gymnastic-type mats or well-sprung resilient floors. By reducing the impact force, the risk of fracture is also reduced.

Conservation of Momentum

The above illustrations all have spoken of momentum as being "absorbed" over as long a period of time as is conveniently possible, or that some of the momentum has been transferred under each of the circumstances. This slow deceleration, as has been pointed out, is accomplished so that the amount of force applied to the body at any given instant is as small as it can be made while the momentum is being altered. These facts imply that momentum is not lost or destroyed, and it is not.

If one observes the interaction of two or more bodies with m_1v_1, m_2v_2, m_3v_3. . . etc., in a closed system in which there are no resultant or outside forces involved and records the sum of the various momenta both before and after the interaction, it will be found that they are identical in value. Mathematically:

$$\sum_{i=1}^{n} m_i V_i = K \quad (4.12)$$

where $\Sigma_{i=1}^{n}$ means the summation of the product following, using all of the values of $i = 1$ through n. K is a constant value. This is the *law of conservation of momentum*.

At the same time it must not be forgotten that the law of conservation of mass also applies to the aforementioned interaction so that:

$$\sum_{i=1}^{n} m_i = M \quad (4.13)$$

where M is a constant value.

Whatever happens, the momentum lost by one object or body is gained by another. The chunks of rubber in the landing pit fly around as the jumper lands, and whatever momentum is not transferred to the rubber is eventually absorbed by the earth, whose mass is so enormous that we cannot measure any change that is caused. Similarly, when the outfielder catches a line drive, the momentum of the ball (whose mass is extremely small) is transferred to his body and then, as with the jumper, to the earth.

WORK

Work is done on an object when a force, F, expressed in N, acts on an object and the object moves through a distance, s, expressed in m, as a result of this force. The units for work in the SI system are Nm or J. To illustrate this definition, consider the examples presented in Figure 4.13, where a force is applied to an object with the objective of moving the object horizontally (from left to right) through a distance of 2.5 m.

In a linear system, the general equation for work, W, is:

$$W = F (\cos \theta) s \quad (4.14)$$

where F is the applied force, θ is the application angle of the applied force relative to the direction of movement of the object and s is the displacement due to force F. By substituting the appropriate values into Equation 4.14, the work may be calculated for each of the examples presented in Figure 4.13. Accordingly, for Figure 4.13(a), the appropriate substitution yields:

$$W = 56 \text{ N} (\cos 0^0) \text{ 2.5 m}$$
$$W = 56 \text{ N} (1) \text{ 2.5 m}$$
$$W = 140.0 \ Nm \text{ or } J$$

For Figure 4.13(b):

$$W = 64.67 \text{ N} (\cos 30^0) \text{ 2.5 m}$$
$$W = 64.67 \text{ N} (0.866) \text{ 2.5 m}$$
$$W = 140.0 \ Nm \text{ or } J$$

For Figure 4.13(c):

$$W = 112 \text{ N} (\cos 60^0) \text{ 2.5 m}$$
$$W = 112 \text{ N} (0.5) \text{ 2.5 m}$$
$$W = 140.0 \ Nm \text{ or } J$$

And for Figure 4.13(d):

$$W = \infty \text{ N} (\cos 90^0) \text{ 2.5 m}$$
$$W = \infty \text{ N} (0) \text{ 2.5 m}$$
$$W = 0 \ Nm \text{ or } J$$

Obviously, all of the 56 N of force is directed horizontally in Figure 4.13(a). In (b) the force had to be increased to almost 65 N to move the intended distance of 2.5 m because some of it is directed onto the supporting surface; i.e., only .866 × 65 or 56 N

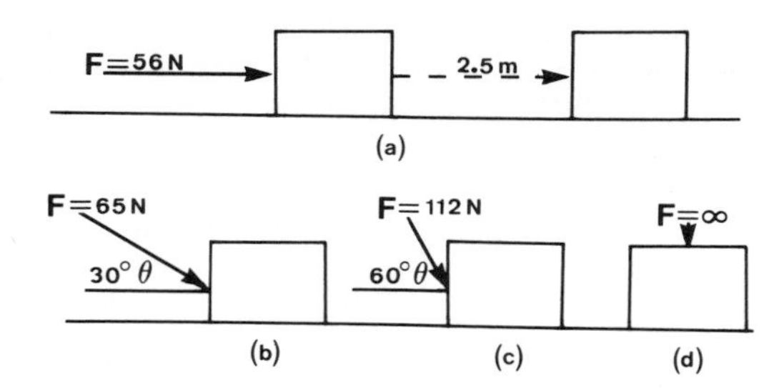

Figure 4.13. Schematic examples of work where F is the applied force, and θ is the angle of application of the force; (a), (b), (c), and (d) represent four different angles of force application. See text.

of force is applied in the direction of the intended movement. Similarly, in (c) the force had to be twice as large as (a) in order to move the 2.5 m distance and a large part of it (.866 × 112) is directed onto the supporting surface. Only half of it is applied in the direction of the intended movement. In the example from Figure 4.13(d), it is obvious that the object will not move as a result of the applied force, F, no matter what its size, because none of the force is directed horizontally. All of the force is directed vertically onto the supporting surface and therefore no work is done.

Work may also be defined in terms of Newton's Second Law of Motion as expressed in Equation 4.3. When considered in these terms, Equation 4.14 becomes:

$$W = ma\,(\cos\theta)\,s \qquad (4.14a)$$

Consider the following example where an object with a 5-kg mass, which is initially at rest, is accelerated due to gravity (9.8 m/s²) and undergoes a displacement of 4.9 m in 1.0 s. By substituting into Equation 4.14a, the quantities become:

$$W = 5\text{ kg }(9.8\text{ m/s}^2)(\cos 0^0)\ 4.9\text{ m}$$

$$W = 5\text{ kg }(9.8\text{ m/s}^2)(1)\ 4.9\text{ m}$$

$$W = 240.10\ Nm\ or\ J$$

Thus, a falling body can also do work.

Similarly, in an angular system, the concept of work may be considered. Let's look at a revolving door at the entrance to a hotel. The doorman has noticed that one of the revolving doors is offering a considerable amount of resistance and is becoming increasingly difficult to turn. He decides to test it and learns that a force of 200 N, applied 1.0 m from the center pivot of the door, is required to keep the revolving door moving. He decides to calculate just how much work is done during one complete revolution. Referring back to Equation 4.14, we can recall that in a linear system

$$W = F\,(\cos\theta)\,s$$

and when the force is applied directly in line with the direction of movement of the object, the cos θ term equals 1, reducing the equation to

$$W = Fs$$

This problem differs from the previous linear examples in the identification of the displacement. In the current angular example, the displacement is equal to the circumference of the circle walked by the doorman. Thus, in this example, Equation 4.14(a) becomes

$$W = Fr\,\theta$$

where θ equals the angular displacement expressed in radians, and r equals the radius of the revolving door. The total work done by the doorman in making one complete revolution can be calculated by substituting the appropriate values as follows:

$$W = 200\text{ N }(1)\,(2\,\pi)$$

$$W = 1256.64\ Nm\ or\ J$$

As can be seen, the product of force and its perpendicular distance from the axis is **torque**. So, rewriting the equation, one may say that work in a rotatory system:

$$W = \mathrm{T}\,\theta$$

The doorman reports his findings and the hotel installs an automatic revolving door such that the door is always in motion at a constant velocity and operates on a frictionless bearing. Since the direction of a person's push is always in the direction of the person's movement and at right angles to the direction of the applied force, i.e., *the centripetal force*, the work done by a person during one revolution of the door will be zero.

The exercise therapist or physical educator will find that the everyday applications of work will often involve a force which changes as the displacement changes. To illustrate this concept, consider the following example. A person is performing a bench press against a 1000-N spring (the characteristics of which are presented in Figure 4.14). During one lift the spring is moved 0.35 m. In spring resistance circuits, the resistance offered by the spring is proportional to the displacement (provided the elastic limit of the spring has not been exceeded). Thus from the graph, 0.35 m is equivalent to a 700-N force. Also, the force may be calculated. The resistance, F_1 at 0.35 m displacement, is equal to

$$F_1/0.35 \text{ m} = 1{,}000 \text{ N}/0.5 \text{ m}$$

$$F_1 = \frac{1{,}000 \text{ N } (0.35 \text{ m})}{0.5 \text{ m}}$$

$$F_1 = 700 \text{ N}$$

The total positive work done during one lift may be determined by calculating the area under the line (cross-hatched area).

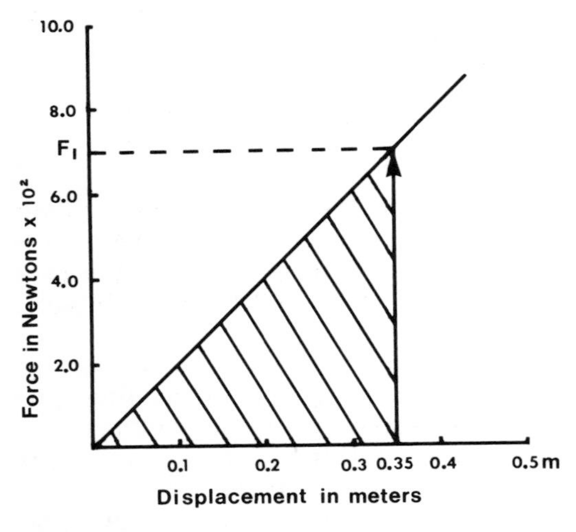

Figure 4.14. Spring characteristics of a 1000-N spring.

The area of this section of the rectangle is calculated (in the same way that the area of a triangle is calculated) by

$$\text{Area} = W = \tfrac{1}{2}\, Fs$$

$$W = \tfrac{1}{2}\,(700 \text{ N})\,(0.35 \text{ m})$$

$$W = 122.50 \; Nm \text{ or } J$$

For those who feel confident with a calculus approach, another way of solving the problem is to identify the equation of the line:

$$F_1 = y = \text{slope } (x) + y \text{ intercept}$$

$$F_1 = y = \frac{1000}{\text{N}}\,(x) + 0$$

$$y = 2000\,(x)$$

Therefore, for $x = 0.35$ m, $y = 700$ N. Work may then be calculated from the integral of the line equation:

$$W = \int F_1 \, dx$$

$$W = \frac{2000 \; N/m}{2}\,(0.35\text{m})^2$$

$$W = 1000 \; N/m \; (0.35 \text{ m})^2$$

$$W = 122.50 \; Nm \text{ or } J$$

Up to now, all discussions of *work* imply that the work done is **positive;** i.e., a force acts in a parallel system in the direction of the movement. Also, if the change in total *energy*, i.e., the sum of all the mechanical energies of a system, is positive, then work is done on the body. However, there are instances in which a force acts parallel to, but in the opposite direction of the movement; e.g., lowering a weight in a controlled manner from a height. In such situations, the change in total energy is negative and is called **negative work**. The reader is urged to consult the section on energy in this chapter for a complete discussion of mechanical energy concepts.

POWER

Power is defined as the time-rate of doing work, or the time-rate of expending energy. The units of measurement are joules per second which is the equivalent of watt in the SI system. Power is computed by dividing the work done by the time over which it occurs. It is important to note that the quotient is *average power*.

$$P \text{ (average)} = W/t \qquad (4.15)$$

If it is desirable to express work as the product of force and the distance moved, then the equation is:

$$P \text{ (average)} = Fs/t$$

Since change in displacement per change in time is the average velocity (recall Chapter 3 on Differentiation), average power also may be calculated as the product of force and average velocity.

$$P \text{ (average)} = Fv \text{ (average)}$$

This equation may be used to express instantaneous power providing the time interval is extremely small.

Power may be expressed in rotatory motion in similar fashion. The angular analogues of force and displacement are *torque*, expressed in *Nm*, and *angular displacement*, expressed in *radians*. So, the power equation for rotatory motion is either:

$$P \text{ (average)} = T\theta/t \text{ or}$$

$$P \text{ (average)} = T\omega$$

ENERGY

Energy is a scalar quantity and is often described as the capacity to do work. It appears in many forms—electrical, chemical, nuclear, mechanical.

In a mechanical context, when a force F is applied to a body so that it moves through a distance s, the work done on the body or energy expended is

$$W = Fs \qquad \text{(from 4.14)}$$

Energy makes use of the same units as work. In the SI system force is expressed in newtons, distance in meters, and energy in joules.

The breakdown of gasoline provides the energy to drive the internal combustion engine; the splitting of the high-energy phosphate bond (~ Ⓟ) of adenosine triphosphate (AT-Ⓟ ~ Ⓟ ~ Ⓟ) provides the energy for muscular contraction. These are examples of chemical energy. Mechanical energy, which is our concern at the moment, involves both potential and kinetic energy.

Potential Energy

Potential energy (*PE*) is also spoken of as gravitational energy or energy of position. Accordingly, it is defined by

$$PE = \text{weight} \times \text{height}$$

(i.e., height relative to reference position) or

$$PE = mgh \qquad (4.16)$$

where mg is the product of mass m and gravitational acceleration g to yield gravitational force or weight, and h is the height of the mass m relative to a chosen reference level (usually taken to be the surface of the earth).

A 90-N boulder balanced on the edge of a 21-m cliff has a potential energy of 1890 Nm relative to the foot of the cliff. It could do considerable damage to a highway at the foot of the cliff.

A gymnast hanging from the stationary rings has a certain amount of potential energy. If his weight (i.e., the product of mass and gravity) is 624 N and his center of gravity is raised 0.15 m when he leaps upward to grasp the rings, he will increase his potential energy by 93.6 *Nm* in relation to the floor.

If the gymnast changes his position to the front uprise causing his center of gravity to rise approximately another 1.2 m, his potential energy will increase further by 748.8 *Nm*.

Thus, the total increase in potential energy relative to the starting position is 842.4 *Nm or J*.

Kinetic Energy

Kinetic energy *(KE)* as its name implies, is the energy of motion. Recently, in the research literature a distinction is made between two types of kinetic energy named kinetic energy of translation and kinetic energy of rotation. The distinctions are appropriate because they stem from the definitions for the two types of motion (cf. beginning section of Chapter 3). The amount of **translational kinetic energy** *(TKE)* in a moving body is determined by:

$$TKE = \frac{1}{2}mv^2 \quad (4.17)$$

i.e., translational kinetic energy equals one-half the mass times the square of the velocity. The angular analogue of this equation defines **rotational kinetic energy** *(RKE)*:

$$RKE = \frac{1}{2} I \omega^2 \quad (4.17a)$$

i.e., rotational kinetic energy equals one-half the moment of inertia times the square of the angular velocity. The total kinetic energy in a system is simply the total of *TKE* and *RKE*.

The quantity of kinetic energy of any given mass depends solely on its linear velocity (and/or angular velocity) while the potential energy of that mass depends solely upon its position. When all three forms of energy are present, the total mechanical energy of the system is the total of *TKE, RKE,* and *PE*.

Conservation of Energy

The *law of conservation of mechanical energy* states that, in a closed system where there are no outside forces present, the sum of the kinetic energy and the potential energy is equal to a constant for that system:

$$PE + KE = \text{a constant} \quad (4.18)$$

so that there is no change in the total amount of energy in the system. Thus, there is no change in the sum of the potential and kinetic energies for that system at any one instant in any given situation.

Suppose the gymnast depicted on the flying rings in Figure 4.15 swings through the arc shown. His overall body motion would be like that of a pendulum. At each high point of his swing he has only potential energy since his velocity momentarily falls to 0. With respect to the lower point that he passes through he has only kinetic energy. In general, $PE = mgy$ where y is the level of the mass m above the reference location. Thus

$$PE + KE = mgy + \frac{1}{2}mv^2$$

At the top of the swing obviously

$$y = h \quad \text{and} \quad v = 0;$$

at the bottom of the swing

$$y = 0 \quad \text{and} \quad v = v_{max}$$

where v_{max} is the maximum velocity attained.

For the top of the swing

$$PE + KE = mgh + 0 = mgh$$

which by the law of conservation of energy must be fixed for the system.

Therefore, for the low point or bottom of the swing:

$$PE + KE = 0 + \frac{1}{2}mv^2 = \frac{1}{2}mv^2$$

i.e.

$$mgh \text{ (at the top)} = \frac{1}{2}mv^2 \text{ (at the bottom)}$$

or

$$h = v^2/2g \quad (4.19)$$

By rearranging and taking the square root of both sides of the equation

$$v = \sqrt{2gh} \quad (4.20)$$

Thus, from a knowledge of h, v may be calculated.

It should be realized that the effects of frictional forces (air friction, ring-on-rope friction, and the like) have been ignored in this treatment. In fact for any system or flight path where friction is negligible the potential energy and the highest point of the path will provide the constant for any

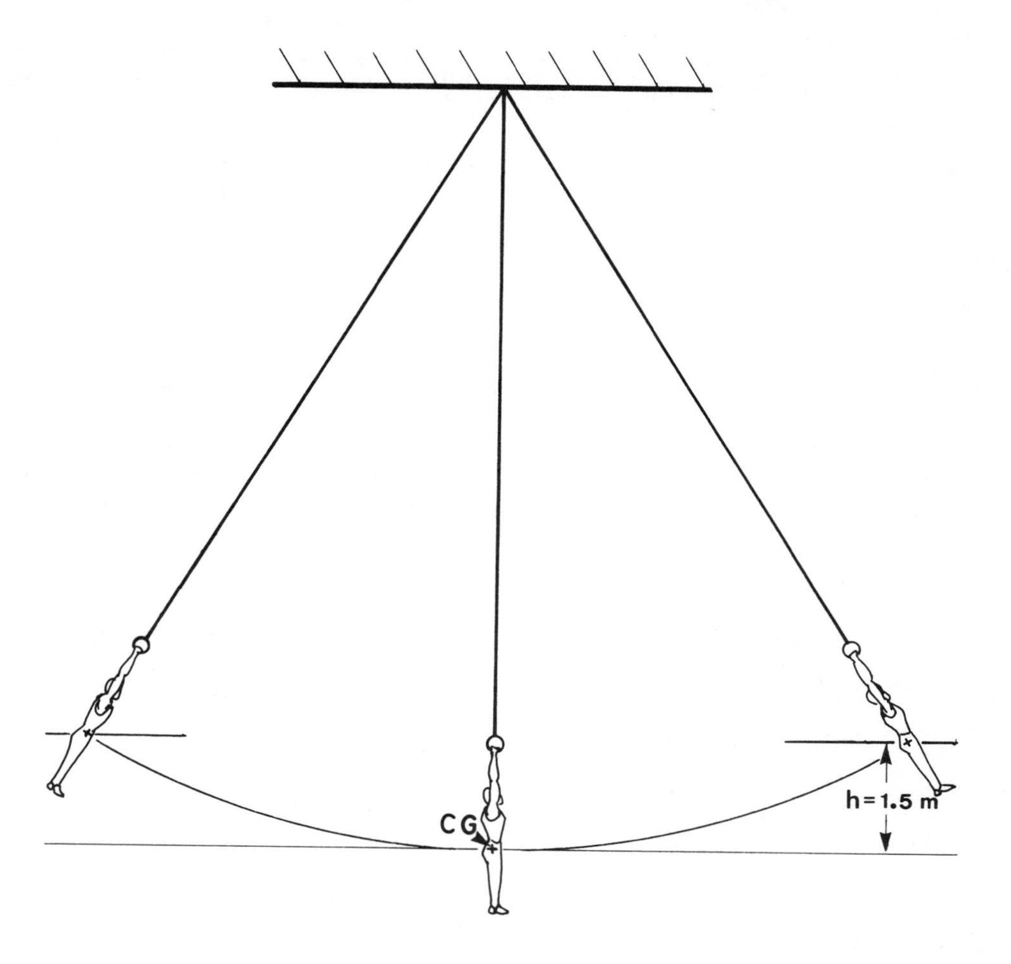

Figure 4.15. Gymnast on flying rings. See text.

other point in the path, and this potential energy will be equal to the maximal kinetic energy of the system.

If the gymnast's center of gravity rises 1.5 m during the swing and his weight is 624 N, the constant is most easily calculated at the top of the swing:

$$PE = (624)\,(1.5) = 936\ Nm$$

As *KE* is 0 at the top of the swing, 936 *J* is the constant energy for the system. As *PE* + *KE* will always equal 936 *J*, it is a simple calculation to determine the kinetic energy for a drop of any given distance.

At the lowest position, velocity *v* follows from Equation 4.20.

$$v = \sqrt{2(9.81)(1.5)}$$

$$= \sqrt{29.43}$$

$$= 5.42\ \text{m/s}$$

If there has been a drop of 0.3 m from the highest position, then the kinetic energy at that position is determined as follows:

$$PE + KE = 936\ \text{Nm}$$

$$624\ (1.5 - 0.3) + KE = 936$$

(The drop is 0.3 m; therefore, the height above the reference position is 1.5 − 0.3 = 1.2 m.) That is, *PE* at this position is

$$624\ (1.2) = 748.8\ \text{Nm}$$

Hence *KE* in this position

$$= 936 - 748.8$$

$$= 187.2\ \text{Nm}$$

The velocity at this position is obtained from

$$\tfrac{1}{2}mv^2 = 187.2$$

$$\tfrac{1}{2}\left(\frac{624}{9.81}\right) v^2 = 187.2$$

$$31.8\ v^2 = 187.2$$

$$v^2 = 5.89$$

$$v = 2.43\ \text{m/s}$$

Work-Energy Principle

The work done on a body is equal to the change in either the potential energy or the kinetic energy or simply "the change in the energy level of the body" (Rodgers and Cavanagh, 1984:88). By using the gymnast on the flying rings as an example, we may illustrate this fact. The *PE* at the top of the swing is 936 *J*. After a drop of 0.3 m, the *PE* is 748.8 *J* or a change of 187.2 *J*. Similarly, the *KE* at the top is 0 *J*; and after a drop of 0.3 m, the *KE* is 187.2 *J*, or a change of 187.2 *J*. Work done on the gymnast may be treated as for a linear system, and can be calculated as the product of force, 624 N, and distance moved, 0.3 m, or 187.2 *J*.

At this juncture, the student may wish to test comprehension of the concepts of energy, work, and momentum with the following example, which is illustrated in Figure 4.16. Solutions and answers are supplied in Appendix 1. In this problem, the mass of the patient-wheelchair complex is 100 kg, the acceleration due to gravity is taken to be 9.81 m/s^2, the height at the start is 1.0 m, and the angle of the ramp with respect to the horizontal is 15°. The problems are:

1. Find *PE* at the top of the ramp.
2. Find *KE* at the bottom of the ramp.
3. Find velocity of the man/chair combination at the bottom of the ramp.
4. Find the linear momentum P at point C.
5. How much work has been done upon arrival at point C?
6. How long (from a position starting at point A) would it take to stop the man/chair combination if a 150-N force is applied in the opposite direction to the velocity?

LEVERAGE

A brief review of the principles of levers perhaps may assist the student in understanding how the laws of leverage apply to human movement. A lever is a device for transmitting force, and it is able to do work when work is done upon it. It is gen-

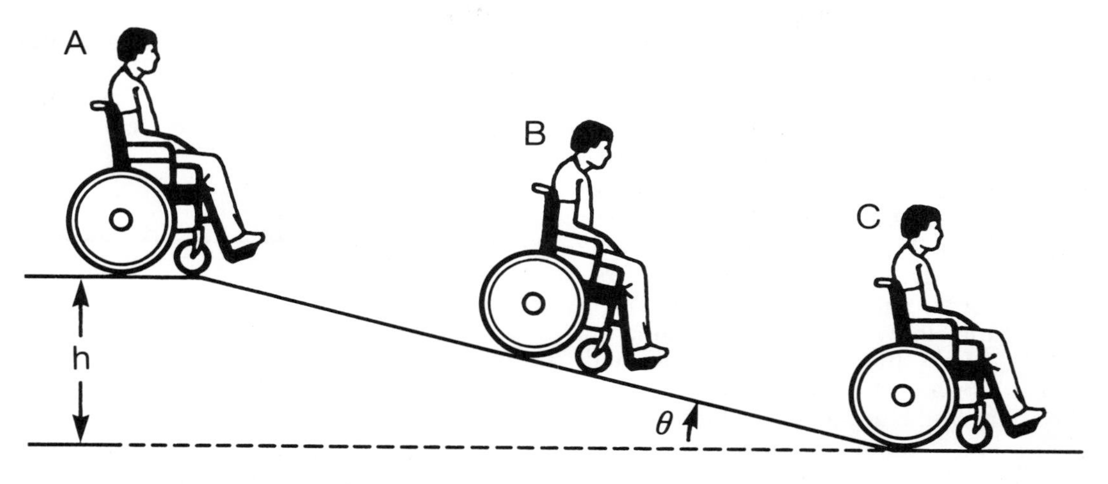

Figure 4.16. Schematic for problem of a wheelchair on a ramp. The wheelchair starts at position *A* at a height *h* of 1.0 m, on a ramp with an angle θ of 15°, and finishes at position C when $h = 0$.

erally a rigid bar or mass (but need be no particular shape), which rotates around a fulcrum (fixed point) on an axis perpendicular to the plane of motion. The rotation is caused by a force applied to the rigid bar or mass. If this force or effort is used to overcome a resistance, it is designated as the effort *E*, and all parts of the lever between the axis and the point where the effort is applied are designated as the lever arm of the effort or simply the effort arm, *EA*. The resisting force or resistance (sometimes termed load) *R* is a second force which tends to rotate the lever in the direction opposite to the effort. The lever arm of the resistance or *RA* consists of all parts of the lever between the axis and the point where the resistance is applied.

In Figure 4.17 the fulcrum is at the point of contact of the hammer head and the board; the effort is applied as indicated in the figure so that the *EA* includes all of the handle of the hammer below the point where the effort is applied, plus that part of the hammer head from the shaft to the fulcrum. The resistance is the force required to withdraw the nail being pulled and is applied to the claws of the hammer by the nail head. The *RA* then is all of the hammer head between the fulcrum and the point where the claws are gripping the nail head.

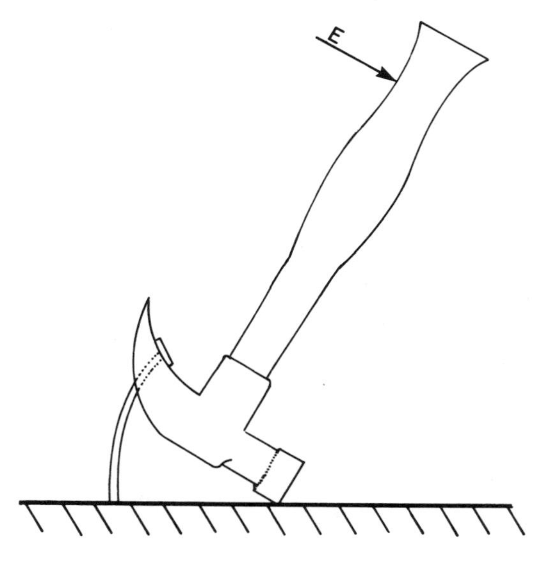

Figure 4.17. The hammer as a lever. *E* is the line of force of the effort used to pull the nail. The effort arm is made up of all parts of the lever between *E* and the fulcrum where the hammer head contacts the board.

Classification of Levers

Levers fall into three categories or classifications depending upon the relationship between the positions of the fulcrum *F*, the effort *E*, and the resistance *R*.

First-Class Levers. When the fulcrum on the axis is located between the effort and the resistance, the *EA* may be equal to, greater than, or less that the *RA* (Fig. 4.18). This type of lever may be used to gain speed if the fulcrum is nearer the point where the effort is applied (Fig. 4.18(b)) or to gain force if the fulcrum is closer to the resistance (Fig. 4.18(c)). Examples of first-class levers in the body include the extension of the elbow by the triceps while the motion of the hand and arm is resisted, and the free foot being plantar-flexed at the ankle while the ball of the foot and toes act upon a resistance such as an accelerator pedal in an automobile.

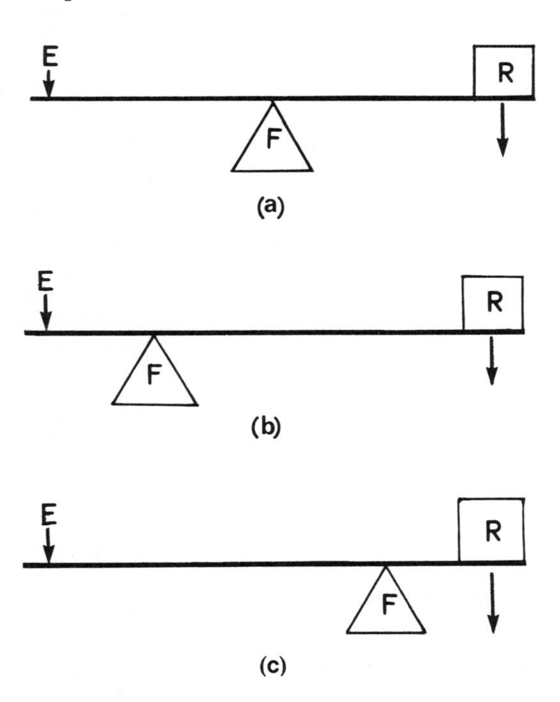

Figure 4.18. First-class levers. *E*, effort; *F*, fulcrum; *R*, resistance, or load. See text.

Second-Class Levers. When the resistance *R* lies between the axis and the effort so that consequently the effort arm length *EA* is always greater than the resistance arm length *RA*, the lever is a second-class lever (Fig. 4.19). A man rolling a wheelbarrow is an example of such a system. The man supplies the effort at the handles, the load is in the barrow, and the fulcrum is in the axis of the wheel. Examples within the body are discussed later in the text.

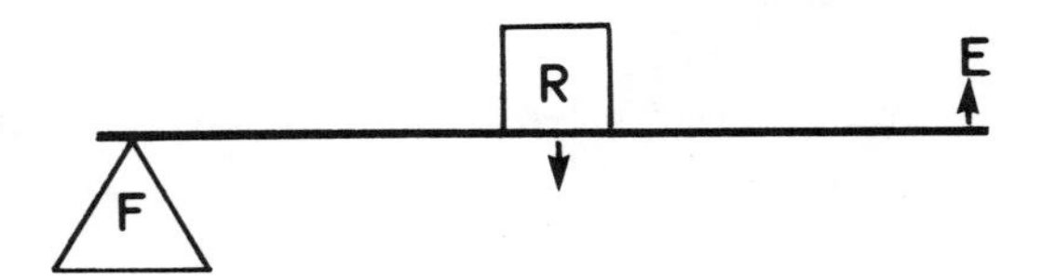

Figure 4.19. Second-class lever. *E*, effort; *F*, fulcrum; *R*, resistance or load.

Third-Class Levers. When the positions of *E* (effort) and *R* (resistance) are reversed, the lever is a third-class lever (Fig. 4.20). *RA* (resistance arm) length is always longer than *EA* (effort arm) length. There are many third-class levers in the human body; among them are the flexion of the forearm about the elbow caused by the biceps, and the flexing of the thigh about the hip joint by the iliopsoas.

Figure 4.20. Third-class lever. *E*, effort; *F*, fulcrum; *R*, resistance or load.

Moment of Force

The moment of force, *M*, allied with a force such as *E* (effort) or *R* (resistance) acting on a lever is defined as the product of the force and the perpendicular distance to its line of action from the axis: the moment for resistive forces is:

$$M_R = R \times RA$$

the moment for effort forces is:

$$M_E = E \times EA$$

It will be realized that for all three classes of levers the resistive moments must be exactly counterbalanced by the effort moments to achieve equilibrium. For equilibrium

$$M_E = M_R$$

$$E \times EA = R \times RA$$

Taking ratios of forces to one side of the equation leads to

$$\frac{R}{E} = \frac{EA}{RA}$$

This ratio is defined as the mechanical advantage (*M.Adv.*) of the lever, i.e.,

$$M.Adv. = \frac{R}{E} = \frac{EA}{RA} \quad (4.21)$$

A large value of *M.Adv.* implies that little effort is required to balance a large resistance.

It should be noted that for a frictionless fulcrum, the energy expended in moving a resistance exactly equals that applied through the effort. Therefore, it follows that the ratio of distances moved by the resistance and effort will be the inverse of *M.Adv.* At this stage it will be helpful to refer to Figure 4.21. This shows a first class lever. If the effort E is increased or decreased about the equilibrium value, resistance and effort extremities will respectively move along the arcs Arc_R and Arc_E

Recalling Equation 3.11:

$$s = r\theta$$

The distances travelled along Arc_R and Arc_E for angle θ shown will be

$$s_R = RA\ \theta, \quad \text{and} \quad s_E = EA\ \theta$$

If we consider that R and E are continually applied at right angles to the lever then work done on R (see Equation 4.14)

$$W_R = R \times RA\theta$$

Similarly, the work done by E

$$W_E = E \times EA\theta$$

From Equation 4.21

$$R \times RA = E \times EA$$

and therefore

$$W_E = W_R$$

as discussed earlier.

Also

$$s_R/s_E = RA/EA = E/R = \frac{1}{M.Adv.}$$

another result that was alluded to earlier.

In the case of a third-class lever *EA* is always less than *RA*. It thus follows from Equation 4.21 and the foregoing discussion that the mechanical advantage is less than 1, i.e., the effort E has to exceed the resistance *R*; but at the same time, when movement occurs, the distance traversed by *R* is greater than that traversed by *E*. Therefore, the velocity of movement of *R* is greater than that of *E*.

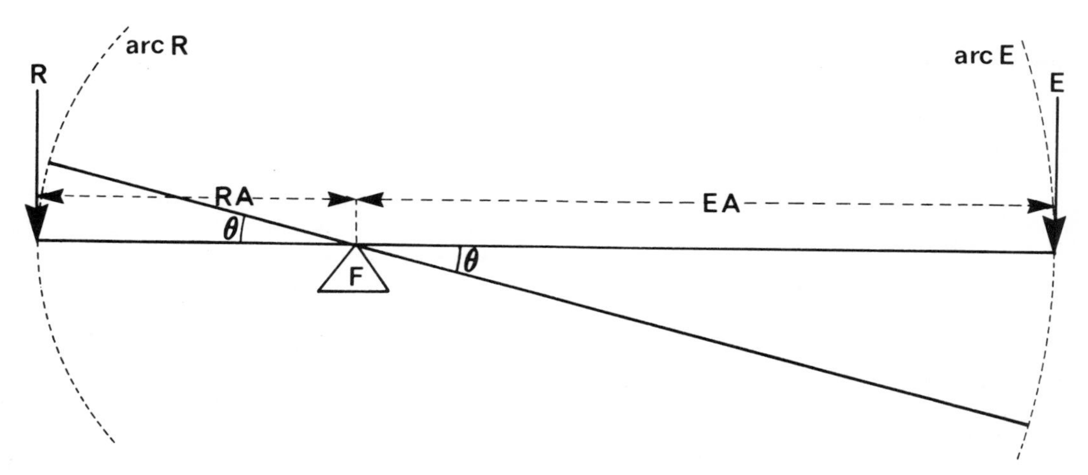

Figure 4.21. Relationships between distances moved for effort and resistance forces.

Force at the Fulcrum

The force acting at the fulcrum depends on the class of the lever and the effort E and resistance R forces. In the case of the first-class lever (Fig. 4.18) the fulcrum must withstand a net force F_1 of:

$$F_1 = E + R$$

i.e., F_1 is a reaction force directed upward.

For the second-class lever (Fig. 4.19) the fulcrum has to oppose a force F_2;

$$F_2 = R - E \quad (R > E)$$

i.e., F_2 is a reaction force directed upward. Turning to the third-class lever (Fig. 4.20) the fulcrum for F_3:

$$F_3 = E - R \quad (E > R)$$

i.e., F_3 is a reaction force directed downward. (Hence the fulcrum arrangement shown.)

Some examples will be helpful at this juncture. Note that the levers themselves are assumed to be weightless.

Example 1. Given a load of 40 N and that *EA* and *RA* are both 2 m. The moment of force of the resistance M_R (resistance multiplied by the perpendicular distance from the line of force of R to the axis through the fulcrum) is:

$$M_R = R \times RA$$

$$= 40\ N \times 2\ m$$

$$= 80\ Nm$$

This requires an equal but opposite moment of force to balance it:

$$E \times EA = R \times RA$$

$$E \times 2\text{ m} = 80\ Nm$$

$$E = 80/2$$

$$E = 40\text{ N}$$

Fulcrum forces are:
Class 1: 80 N upward
Class 2: 0 N
Class 3: 0 N

Example 2. Given that *EA* is equal to 2 m while *RA* is only half of that at 1 m and R is 40 N (this could occur with either a first or second class lever (Fig. 4.22(a) and (b)).

The moment of force of the resistance M_R is 40 *Nm*. Clearly

$$E \times 2\text{ m} = 40\text{ N} \times 1\text{ m}$$

$$E = 40/2$$

$$E = 20\text{ N}$$

Only 20 N of force are needed to balance the resistance, which is only half as much force as required in the first situation. Thus we can say that a lever system with an *EA* twice as long as the *RA* has a mechanical advantage (*M. Adv.*) of 2 (see Equation 4.21).

Fulcrum forces are:
Class 1: 60 N upward
Class 2: 20 N upward

Example 3. Given the same 40-N load as R but an *RA* of 2 m and and an *EA* of 1 m,

$$E \times 1\text{ m} = 40\text{ N} \times 2\text{ m}$$

$$E = 80/1$$

$$E = 80\text{ N}$$

So an effort of 80 N is now needed to support the 40-N load. In this case

$$M.\ Adv. = R/E = \frac{40}{80} = 1/2 \quad \text{and}$$

$$M.\ Adv. = EA/RA = \frac{1}{2} = 1/2$$

Fulcrum forces are:
Class 1: 80 + 40 = 120 N upward
Class 3: 80 − 40 = 40 N downward

Torque

In the simple levers used above to illustrate lever classes, *EA* and *RA* were always perpendicular to the line of force of E and R, respectively, and each ran directly to the axis passing through the fulcrum. Since the perpendicular distance from a line of force to an axis through the fulcrum is the **moment arm** of that force (*MA*), we can write

$$E \times MA_E = R \times MA_R \qquad (4.22)$$

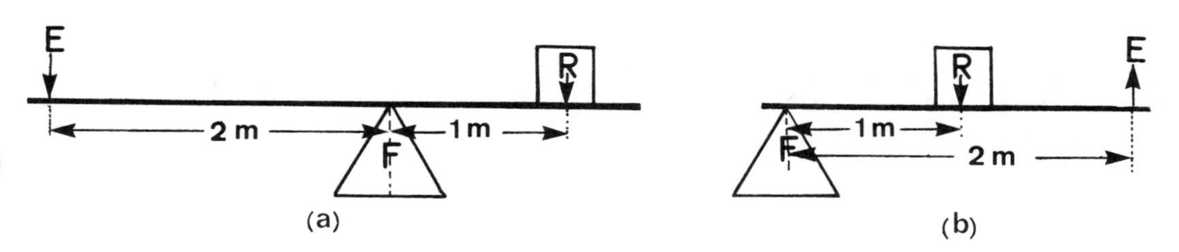

Figure 4.22. Examples of levers with 2-m effort arms and 1-m resistance arms.

where MA_E is the moment arm of the effort and MA_R is the moment arm of the resistance. The product of $MA_R \times R$ (and of $MA_E \times E$) is a moment of force, sometimes called torque, where:

$$T = fd \qquad (4.23)$$

(the Greek letter tau) is the *torque* or *moment of force*, *f* is the force (effort or resistance), and *d* is the moment (torque) arm, i.e., the perpendicular distance from the line of force to the axis. The units of torque are, therefore, force and distance, as newton-meters for example.

Both $E \times MA_E$ and $R \times MA_R$ are moments of force, more simply known as *moments.* As indicated earlier all moments are expressed in double units; e.g., dyne-centimeters, or newton-meters, depending on the system of measures being used.

A **torque**, a **moment of force**, or more simply a **moment**, tends to cause rotation about an axis. Torques causing clockwise rotations are considered positive; when they tend to rotate a body or lever in the opposite direction, counterclockwise, they are considered negative.[a] Thus, in an equilibrium situation the sum of all torques or moments must be equal to 0 and Equation 4.22 above is written:

$$(E \times MA_E) - (R \times MA_R) = 0 \qquad (4.24)$$

or in more general terms:

$$\Sigma M = 0 \quad \text{or} \quad \Sigma T = 0 \qquad (4.25)$$

[a] This is a matter of convention. The opposite convention, i.e., anticlockwise positive, clockwise negative, can be used. Whichever is adopted should be used consistently.

The sum of the moments is equal to 0.

Returning to the earlier example of the hammer as a lever (Fig. 4.17) the nail is exerting a counterclockwise or negative moment, while the effort to pull the nail is clockwise and this moment is positive.

Example 4. The problem is to calculate the amount of force that must be applied at *E* to pull the nail, assuming that the nail is resisting with a force of 67 N.

The solution involves determining the moment arms involved in both the resistance and the effort. One cannot simply draw a line from the nail head to the axis, or from the point of application of the effort *E* to the axis, and measure the distances involved. *A moment arm must be perpendicular to the line of force with which it is associated.* The first step then is to make a force diagram of the situation by adding the lines of force to Figure 4.17 (see Figure 4.23). The line of force of the nail, which is perpendicular to the hammer claws at the point at which the nail head is resting on them, is indicated by the line X—Y. The moment arm of the resistance, MA_R is perpendicular to this line and runs directly to the axis where the hammer head is resting on the board. The line of force of the effort is indicated by line *A - B* (which is perpendicular to the handle). The MA_E runs from the axis and is perpendicular to *A - B* (Fig. 4.23). The MA_E is 0.24 m long and the MA_R is 0.11 m long. Since

$$\Sigma M = 0$$

$$E \times MA_E - R \times MA_R = 0$$

$$(E \times 0.24) - (67 \times 0.11) = 0$$

$$E = 7.37/.24$$

$$E = 30 \text{ N}$$

This 30-N force is just sufficient to balance the nail's resistance. To pull it will need more than this, possibly another 1/5 N or, more likely, another full N of force which would mean about 31 N of force in total. The mechanical advantage of this system then is greater than 2, i.e.,

$$M.\ Adv. = R/E = 67/30 = 2.2$$

Likewise,

$$M.\ Adv. = MA_E/MA_R = 0.24/0.11 = 2.2$$

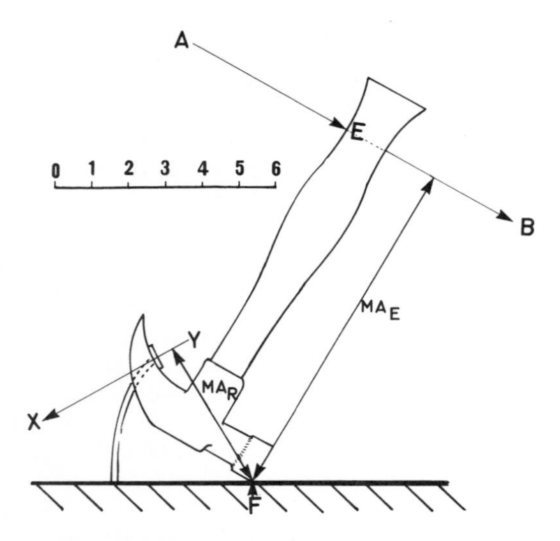

Figure 4.23. Force diagram of hammer. *A-B*, line of force of the effort; *X-Y*, line of force of the resistance (nail); *F*, fulcrum (on the axis); MA_R moment arm of the resistance; MA_E moment arm of the effort.

EQUILIBRIUM

In the solution of equilibrium problems the sum of all the moments must be brought to 0 when an unknown force or moment arm is to be determined. Also, the sum of the forces must be equal to 0:

$$\Sigma M = 0 \qquad \Sigma F = 0 \tag{4.26}$$

and in equilibrium problems involving concurrent forces:

$$\Sigma X = 0 \qquad \Sigma Y = 0 \qquad \Sigma Z = 0 \tag{4.27}$$

where X, Y, and Z are the respective components of the forces involved. The Z component is needed only when the forces and resultants under study are in two or more different planes and a three dimensional approach is needed.

Equilibrium may be either passive, active, or dynamic. While passive equilibrium is always static, static equilibrium need not always be passive. A body holding a pose is static, but the equilibrium being maintained may be completely passive, as occurs when an individual is reclining and completely relaxed or, being relaxed, is wholly supported in some manner. On the other hand, if the pose is being held against gravity (or any outside force) as in Figure 4.24(a) and (b), or even in quiet standing, muscular force is needed to hold the body segments in place. Under these circumstances the equilibrium is now active as muscular tension is creating force moments to balance those generated by gravity as the pose or stance is assumed. Dynamic equilibrium, on the other hand, is that maintained while the individual is actively performing some form of motion or locomotion such that he keeps his center of gravity over a constantly changing base of support, as in Figure 4.25.

Equilibrium under Static Conditions

A body is in **stable equilibrium** when its *potential energy is at a minimum* and work must be done on it to cause a change in position. A living body is in stable equilibrium, then, only when it is lying down on a horizontal surface adequate to support it. In order to change its position it must exert internal (muscular) force or have an outside force act upon it. When a given posture is maintained, the center of gravity is held over the base of support and the line of gravity (LG) falls within that base (Fig. 4.24(a) and (b)). If the individual is erect or nearly so, such that the

Figure 4.24. Static equilibrium. (a) and (b) show two positions called scales or arabesques.

Figure 4.25. Illustration of dynamic equilibrium (photography by W. Patriquin for the Boston *Herald-Traveler*).

center of gravity (CG) is some distance above the base, the individual has potential energy relative to the base. The larger the base, the greater is the amount of force that it takes to move the LG outside that base. Under this and other circumstances in which the *potential energy remains constant* (i.e., there is very little or no change in the height of the CG),the body is in **neutral equilibrium**. The individual is in **unstable equilibrium** when the *potential energy is at a maximum* and the base of support is extremely small so that it takes very little force to move the LG outside the base. This situation may change very rapidly to a dynamic equilibrium or relapse into a neutral or even a stable equilibrium, depending on the individual's degree of motor control.

Equilibrium under Dynamic Conditions

A baby experiences only stable equilibrium conditions during the first months of life; when movement on hands and knees commences, a small amount of potential energy is gained but an almost neutral equilibrium is maintained. When toddling commences, problems increase: a neutral equilibrium can only be maintained through the aid of a parent's hand, or holding on to furniture. However, the baby soon develops dynamic equilibrium which varies between the neutral and the unstable as learning to walk, run, and jump occur.

Over a Minimal Base. During all forms of locomotion any living body strives to maintain its balance, and all of its postural reflexes are geared to this end.

This equilibrium may be unstable at times, as when a ballerina dances on her toes or a figure skater performs as in Figure 4.25, or neutral as soon as she increases the size of her base and resorts to a less specialized form of locomotion.

A child learning a new motor skill or a new form of locomotion such as riding a bicycle must learn to maintain as near neutral equilibrium as possible on a narrow, moving base. The inertia resulting from the linear motion contributes to the child's dynamic equilibrium and as long as the combined CG of body and machine continue to travel in the desired direction and their line of gravity falls on the line tangent to the front and rear wheels, balance is maintained. At the same time the rider must be prepared to fight this inertia when making a turn. Then an inward force must be created by steering into the turn and by an inward lean.

Over a Changing Base. When a gymnast or tumbler performs rolls, headsprings, and handsprings, cartwheels, etc., the base of support changes from the feet to other parts of the body. If the LG does not continue to travel in the direction of the next point of contact, dynamic equilibrium is lost and the performer collapses to the mat (stable equilibrium).

Physical therapists employ the principles of stable equilibrium when selecting walking aids for patients with unstable gait patterns. Gait training usually begins in the parallel bars and, as the patient's stability improves, the base of support is gradually reduced. The progression may be first to a walker, subsequently to two crutches, then a cane, and ultimately to full weight bearing with no walking aids.

TORQUE AND ROTATIONAL MOTION

Torque or Moment of Force

In dealing with rotational movements or motion, the moment or torque fulfills the same function that force performs for motion in a straight line: it is the size of the moment that increases or decreases the angular velocity of a body and so produces acceleration or deceleration of the rotational movement.

Moment of Inertia

In discussing Newton's First Law earlier in this chapter we mention that the terms mass and inertia may be used interchangeably. In rotational movement the moment of inertia I is the analogue of mass in linear motion. If a body is divided into a large number of very small parts, and a typical particle has a mass m which is at a perpendicular distance r from the axis of rotation, then its contribution to the moment of inertia I is mr^2. Under these conditions the moment of inertia is equal to the sum of all such contributions from all portions of the body (say n altogether) and:

$$I = \sum_{i=1}^{n} m_i r_i^2 \qquad (4.28)$$

where m_i is the mass of the i th portion and r_i its distance from the center of rotation.

The distance r for any particular portion is constant only for one given axis. As a body or body segment may rotate about different axes, each r_i will change with each change of axis. Thus, the moment of inertia is not a fixed constant for a body but is dependent on the distribution of the mass about the axis of rotation or the perpendicular distances. I may also be written as follows:

$$I = MK^2 \qquad (4.29)$$

The radius of gyration K is defined as the distance from the axis of rotation of a point at which the total mass m of a body might be concentrated without changing the moment of inertia of the body.

$$\left(M = \sum_{i=1}^{n} m_i \right)$$

The moment of inertia is the measure of the resistance of a body at rest to rotatory motion, or, if rotating, to change its state of rotation. As torque exerts a turning force on a body, it is analogous to the force in the equation $F = ma$. Therefore it may be stated that:

$$T = I\alpha \qquad (4.30)$$

Torque is equal to the product of the moment of inertia and the angular acceleration and so is analogous to force F as I is analogous to mass m and α to acceleration a in linear motion.

Angular Momentum

Linear momentum P equals the product of mass and velocity, so angular momentum L is the product of the moment of inertia and the angular velocity:

$$L = I\omega \qquad (4.31)$$

IF THE SYSTEM IS CLOSED AND THERE IS NO EXTERNAL TORQUE ACTING ON THE SYSTEM, THE TOTAL ANGULAR MOMENTUM REMAINS UNCHANGED, EVEN THOUGH THE MOMENT OF INERTIA MAY BE ALTERED. This is the law of conservation of angular momentum. Note that while the quantity of mass remains unchanged, r or K (and so r^2 or K^2) may be changed.

An example involving the use of a "frictionless turntable" (Fig. 4.26) will serve to illustrate the use of the above relationships.

Imagine a man standing on a frictionless turntable holding a 60-N barbell in each hand; his arms, with extended elbows, are abducted 90°. At the start of the experiment the barbells are held at 0.90 m from the rotational axis. He starts spinning at a rate of 1 rps (revolution per second, 360°/s). While he is spinning at this rate he pulls the barbells into his shoulders so that they are only 0.15 m from his axis of rotation. The problem is to determine the angular velocity with the barbells drawn in. Under the circumstances, any change in the moment of inertia of his body will be so extremely small it may be ignored. Therefore as

$$I = mr^2$$

$$I = (60\ \text{N/g})(0.90\ \text{m})^2$$

$$I = 48.6\ \text{Nm}^2/\text{g} \qquad \text{(from 4.28)}$$

$$L = I\omega$$

$$L = \frac{48.6}{g}\,\omega_1 \qquad \text{(from 4.31)}$$

where ω_1 corresponds to 1 rps and $\omega_1 = 2\Pi \times 1$ radians/s.

When the hands draw the weights close

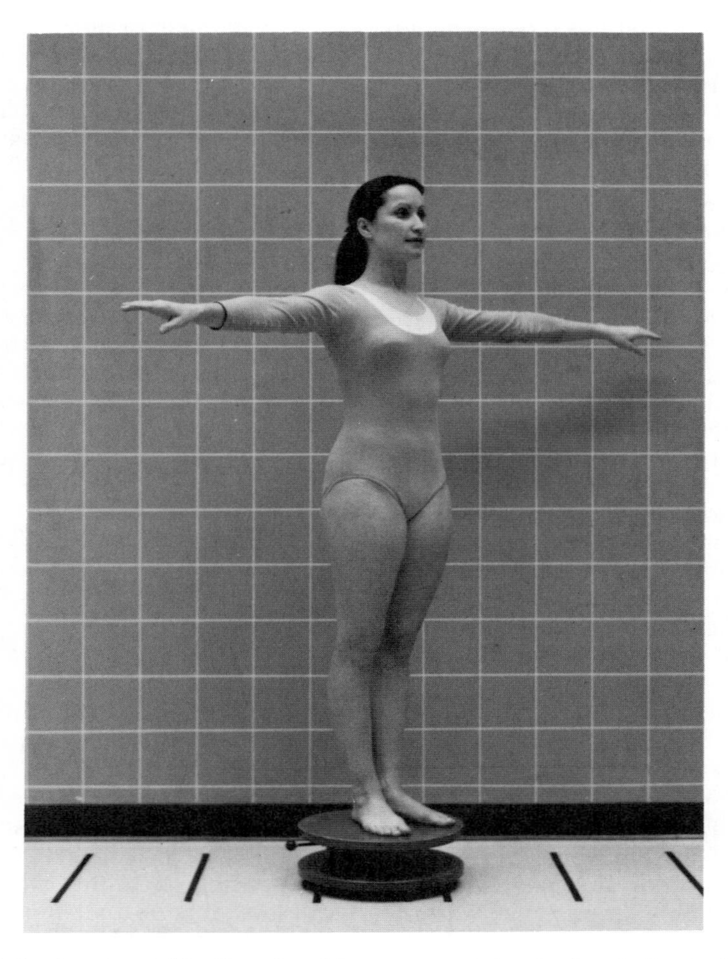

Figure 4.26. Frictionless turntable. Female subject can rotate her body relative to the floor using a "frictionless turntable" that has a mechanical bearing with minimal friction.

to the body, their mass is unchanged but r has shrunk to 0.15 m. Then

$$L = \frac{60\ N}{g}\ (0.15\ m)^2(\omega_2)$$

where ω_2 corresponds to the new angular velocity.

As L is unchanged (law of conservation of angular momentum), it follows that:

$$\frac{48.6}{g}\ \omega_1 = \frac{1.35}{g}\ \omega_2$$

$$\omega_2 = \left(\frac{48.6}{g}\right)\left(\frac{g}{1.35}\right)\omega_1$$

$$\omega_2 = 36\ \omega_1$$

Hence, the new spinning rate = 36 rps.

The greater the mass at a greater distance from the axis of rotation, the slower the man will rotate. When the mass is drawn closer to the axis, r or K is decreased so that the moment of inertia becomes smaller and, as the angular momentum remains unchanged, the increase in the angular velocity must balance the equation.

Use of these principles forms the basis of control in many skills, from figure skating to gymnastics to rebound tumbling to more elaborate competitive dives. In the air, the diver's (or tumbler's) body rotates about its center of gravity and the performer can regulate the speed of rotation by the posture or postures assumed. In a tuck position, the radius of gyration mentioned earlier will be quite short, making the moment of inertia comparatively small, and the diver will spin rapidly about a transverse axis in the frontal plane. Thus, a good diver may complete one and a half somersaults from the low (1 m) diving board. If the performer feels that the spin is too fast, the angular velocity may be decreased by opening the tuck slightly, increasing the radius of gyration and thus the moment of inertia. On the other hand, if the diver feels that the number of somersaults planned will not be completed at the present speed of rotation, the rate of spin may be increased by tightening up the tuck (or pike).

In twisting dives or leaps or skating spins, the performer rotates about an axis running lengthwise through the body, and controls the moment of inertia, and hence the angular velocity, by the position of the arms during the twist or spin, just as the man did on the frictionless turntable. By these techniques, many a diver has "saved" a dive that started to go wrong.

Analogies between Linear and Rotational Motion

Throughout the preceding section the analogies occurring between rotational and linear motion are indicated. Table 4.3 summarizes these analogies and serves as a reminder to the reader of the many important concepts and laws presented.

FRICTION

Frictional forces arise whenever one body moves over the surface of another. **Friction** is defined as "the tangential force acting between two bodies in contact that opposes motion or impending motion. If the two bodies are at rest, then the frictional forces are called **static friction**" (Rodgers and Cavanagh, 1984:85). Thus, when human locomotion takes place (Fig. 4.1), friction plays a role in many ways. For effective propulsion when each foot in turn reacts with the floor it is important that there be no slip between the floor surface and the contacting body member. In this case large frictional forces exist between the two interacting surfaces. With an oily floor surface, the friction is reduced and both propulsion and stepping are hindered. Different body movement skills are required for effective balance and progression.

Joints in the human body have members which in general articulate by rolling and sliding over each other. To minimize frictional forces appropriate surface properties and lubrication are required. This accounts for the anatomical structure of joints which are bathed in synovial fluid. Frictional forces exist when tendons slide within guiding synovial sheaths.

Table 4.3
Analogies between Linear and Angular Motions

Linear Motion	Angular Motion
Distance (s)	Angle (θ)
Velocity (v)	Angular velocity (ω)
$v = \Delta s/\Delta t$	$\omega = \Delta\theta/\Delta t$
Acceleration (a)	Angular acceleration (α)
$a = \Delta v/\Delta t$	$\alpha = \Delta\omega/\Delta t$
Mass (m)	Moment of inertia (I)
	$I = \Sigma mr^2 = MK^2$
Force (F)	Torque (T)
$F = ma$	$T = I\alpha$
Impulse	Angular impulse
$F\Delta t = m\Delta v$	$T\Delta t = I\Delta\omega$
Momentum (P)	Angular momentum (L)
$P = mv$	$L = I\omega$
Kinetic energy (TKE)	Kinetic energy (RKE)
$KE + 1/2\ mv^2$	$KE = 1/2\ I\omega^2$
Potential energy (PE)	Potential energy (PE)
$PE = mgh$	$PE = mgh$
Work done (W)	Work done (W)
$W = Fs$	$W = T\theta$
Power (Fs/t)	Power ($T\theta/t$)

The foregoing discussion suggests that frictional forces depend upon surface properties, lubrication, the mode of movement between surfaces, and the nature of the forces holding them together. An understanding of the ways in which such factors may be controlled is advantageous in appreciating aspects of human movement as well as designing devices of use to man.

Figure 4.27 depicts a block of material A, having weight W and resting upon a surface B. A horizontal force P is exerted in attempts to move A over B. W represents the weight of A. R is the ground reaction force of the surface B. It exactly counterbalances W. The force P required to move the block must exceed slightly the force of friction F between objects A and B. By experimentally exploring the relationship between P and W for movement to occur it is found that

$$P = F_{max} = \mu W = R$$

i.e.,

$$F_{max} = \mu R \tag{4.32}$$

where μ is the coefficient of static friction, and is a factor which relates to the surface properties of A and B, and to the reaction force R between the two surfaces. Static friction may be thought of as "starting friction" or "inertial friction" because it is the friction which resists the start of motion.

Examples of static friction are numerous since it would be impossible to walk or achieve any form of locomotion without static friction. Imagine trying to take the first step and walk on an icy surface as compared to that on rough concrete. The difficulty in walking on an icy surface is directly related to the reduced friction force between the soles of the feet and the icy surface.

From Equation 4.32 it is evident that F_{max} is independent of the surface area between A and B, and is proportional to the normal force; i.e., the ground reaction force of B on A.

"If there is relative motion between the two bodies, then the forces acting between the surfaces are called **kinetic friction**" (Rodgers and Cavanagh, 1984:85). When sliding between the surfaces occurs, which is one type of kinetic friction, a **coefficient of sliding friction** μ_s which is always lower than μ, pertains. It is important to note that the force of sliding friction is in line with, but opposes, the direction of movement and is perpendicular to the force holding the two objects together. As with static friction, the force

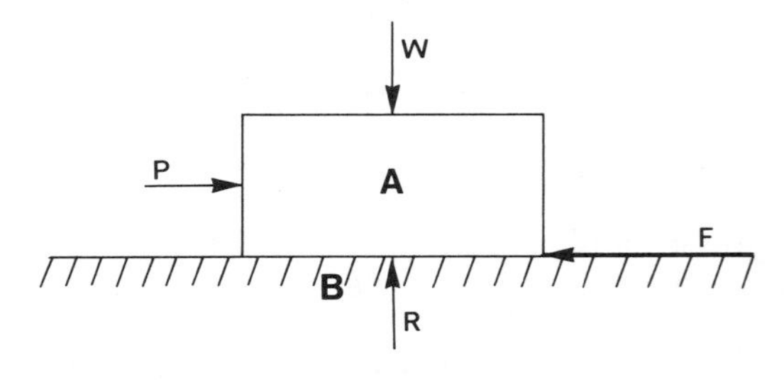

Figure 4.27. Diagrammatic representation of friction force.

of sliding friction is proportional to the force holding them together and is totally dependent on the nature of the surfaces. Examples of sports in which slidng friction plays a major role are the sports of curling or shuffleboard. Thus, the force for sliding friction

$$F_s = \mu_s R$$

and (4.33)

$$F_s < F_{max}$$

Coefficients of friction μ and μ_s are normally within a range from 0.1 to 1.0. For example, for wood on wood it is approximately 0.4; for metal on metal, 0.2; for greased metal on metal, 0.05. These coefficients of friction provide examples of the effects of different materials, the roughness of the contact surfaces, and the influence of lubricants.

When work is done by moving against frictional forces, such work generates heat between the surfaces, especially if the movement is very fast.

A second type of kinetic friction is known as rolling friction. **Rolling friction** results as a consequence of the deformations of both surfaces as one rolls over the other. Rolling friction is generally less than μ or μ_s by 100 to 1000 times, and depends on the reaction forces and the radius of curvature of the rolling member. Examples of sports in which rolling friction plays a major role are golf, bowling, and billiards.

IMPACT AND REBOUND

If a moving object collides with another object, probably it will rebound from that object. The extent of rebound is dependent on many factors, which include the masses and velocities of the two objects, the elasticity of the two, the friction between the two surfaces, the energy lost, and the angle with which the two objects collide (Broer and Zernicke, 1979:125; Luttgens and Wells, 1982:330).

Objects that impact with each other are subject to the law of conservation of momentum. The importance of this statement may be recognized in the equation:

$$m_1v_1 + m_2v_2 \text{ (before collision)}$$
$$= m_1v_1 + m_2v_2 \text{ (after collision).}$$

If two billiard balls collide on colinear paths, the momenta of the two after collision will change based on this principle. Rolling friction will play an important part; some energy will be dissipated in the form of heat and sound of impact; but the major influence will be the elasticity of the two balls.

A measure of an object's elasticity is expressed as its **coefficient of restitution**. When an object collides with another, the objects will deform. The extent to which an object may regain its shape following impact is manifested in its extent of rebound. Thus, if a basketball is dropped onto a well-sprung gymnasium floor, the height of its rebound is dependent on the ability of the basketball to gain its former shape **and** the ability of the floor to restore its surface also. The coefficient of restitution may be determined by dropping the basketball onto a standardized surface and inserting its height of bounce and height from which it was dropped into the following formula:

$$e = \sqrt{\frac{\textit{bounce height}}{\textit{drop height}}}$$

A coefficient less than 1.0 is obtained because mechanical energy is lost in the impact. Restitution coefficients are dependent on composition of the ball and the surface it contacts, temperature, and velocity of impact. As an example, the coefficient of a new tennis ball might be about 0.67, but when it is well worn, the coefficient increases to 0.71. A tennis match played on concrete, whose coefficient is 0.74, is a very different match from one played on grass with a coefficient of only 0.43 (Hay, 1978:80-81). It is obvious that a tennis player who enjoys playing when the ball bounces quite high is going to opt for the concrete surface and the well-worn tennis ball. Similarly, the squash player seeking greatest bounce will warm the squash ball before starting the match.

So far, the discussion has focussed on direct impacts, i.e., the angle of impact with the surface is 90°, and the ball is compressed evenly. For impacts at other angles, consideration must be given not only to the coefficient of restitution but also to the frictional force. The measure of a ball's elasticity will affect the vertical component of the rebound, while the parameters of friction will affect the horizontal component.

A ball approaching a flat surface at an angle θ (measured with respect to the horizontal) will rebound at an angle less than θ. The angles before and after impact are usually described with respect to a line placed at the point of contact and perpendicular to the surface; thus, the **angle of incidence** is that angle bounded by the path of the ball just before impact and the perpendicular line. The **angle of reflection** is described in similar terms as the angle between the perpendicular line and the path the ball takes immediately after impact. To illustrate, the angle of reflection of a bounce pass in basketball will be larger than the angle of incidence because of coefficients of restitution of ball and floor, friction, and no spin. Interestingly, following impact, a basketball which approaches with no spin will gain top spin in the rebound. This happens because as the ball impacts with the floor, the "back" of the ball will compress further than the "front," which is "free" to roll forward. As the ball leaves the surface, the roll commences and the ball has top spin.

FLUID MECHANICS

The fluid medium in which the human body functions generally has an influence on performance. Most human activity takes place in the fluid medium of air and for some purposes, the characteristic influences of this medium may be neglected. Air resistance is usually not of consequence at the low speeds of movement involved in unaided human terrestrial activity.

However, a sky diver relies on air-resistance forces and his interactions with them (e.g., by changing the profile he presents to the air flow). An Alpine skier takes advantage of the "egg-shape" of his body as does a speed skater, all in an effort to reduce air resistance to forward motion.

Water is the other medium in which humans perform. The activities range from swimming, scuba diving and sailing to the activity of taking a bath. In these activities, the water resistance forces are of major consequence to the activities. A scuba diver contends almost exclusively with water resistance forces, while a swimmer must contend with forces of both water and air as the activity is at the interface of the two fluids. Similarly, a sailor must contend with the air resistance forces on the boat's sail and the water resistance forces on the hull.

When a body moves through a medium, the resistance to that movement stems from two forces designated *lift* and *drag*.The force which opposes forward motion is called **drag** and is dependent on the extent to which the medium is disturbed. Drag takes many forms and it is appropriate to identify three kinds of drag as *surface*, *form*, and *wave*.

Surface drag results when the boundary layer surrounding a body becomes unstable. This may be caused by a variety of conditions, such as the velocity of the flow relative to the body itself, the amount of surface area of the body, its smoothness of surface, and the type of fluid involved.

Form drag depends largely on the shape of a body, particularly its cross-sectional area perpendicular to the direction of fluid flow. It results when the difference is taken between a force acting on the "front" of a body and another force acting on the "rear" of the same body. Normally, when a body is progressing through a medium, the force on the front of the body is a high pressure force resulting from a laminar flow. The force on the rear of the body is a low pressure force resulting from a turbulent flow. The laminar flow is always larger than the turbulent flow (i.e., the high pressure is always greater than the low pressure) so the difference becomes the force known as form drag. As with any force, these values are vector quantities.

Wave drag is unique to situations in which a body is moving between two media (e.g., air and water). The movement of the body between the two media causes waves in the denser medium.

The other force which resists movement through a medium is **lift**, so-called because it provides a "lifting" action as the body moves through the medium. Lift works at right angles to drag (Fig. 4.28) and both are vector quantities. The amount of lift depends on the orientation of the body moving through the fluid, i.e., the object's **angle of attack**.

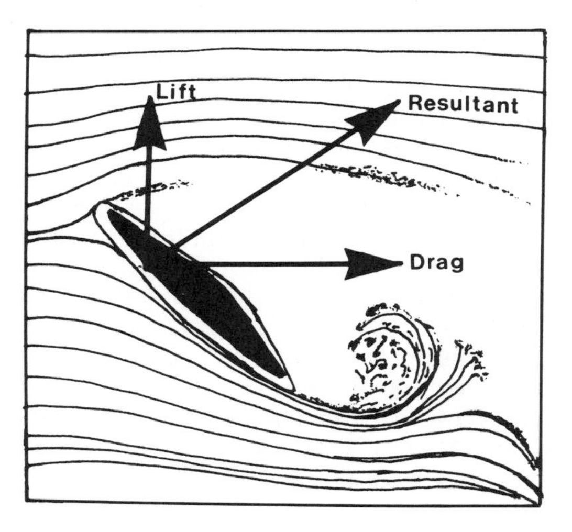

Figure 4.28. Lift and drag forces acting on a discus mounted in a wind tunnel. The lines indicating the path followed by the air are obtained by injecting smoke through small jets positioned upstream (drawn after Hay, J.G., 1978. *The Biomechanics of Sports Techniques.* Englewood Cliffs, N.J.: Prentice-Hall, Inc., p. 172, from a photograph by Richard V. Ganslen).

From the relation between drag and lift, it is obvious that there cannot be any lift without some drag, which explains why the angle of attack must be greater than zero degrees. Usually, the objective is to maximize lift and minimize drag. The compromise between lift and drag is sometimes referred to as the *lift-drag ratio*.

Aerodynamics

Lift is a component of an aerodynamic force that opposes the force of gravity and is dependent on the relative flow of air past a body. In aeronautics, lift is dependent on the relative flow of air past the wings of a plane (i.e., the design of the wings determines the lift factor). Car racers take advantage of this fact, but for a different reason. Foils on race cars are mounted in reverse to those of planes (e.g., upside-down) so that the "lift" component helps to hold the car close to the ground.

In athletics, the discus is probably one of the best examples to illustrate lift and drag principles (Fig. 4.28). A discus thrower seeks an angle of attack of the discus that optimizes the lift-drag ratio and maximizes range.

Objects that are spherical and travel through the medium of air are subject to the Magnus effect (so-named for the German scientist who first described it). A spherical body such as a ball may rotate as it moves through air. It will carry with it a boundary layer of air which rotates with it (Fig. 4.29). The ball will seek the path of least resistance, which will be on the side of the ball that is rotating in the same direction as the direction of the oncoming air. The effect is to cause the ball to curve in the direction of this low pressure of air. Examples of sports in which the Magnus effect plays a major role include tennis, golf, and football. Balls which spin around a horizontal axis have either topspin or backward spin, a feature that tennis players use to advantage. Spin about a vertical axis is known as either right or left spin, and is sometimes called hook or slice by the golfer.

Finally, spin about an axis that lies along the line of flight is called gyroscopic action and acts to stabilize the object as it moves through the air medium, a fact that the football quarterback uses to advantage in throwing the football.

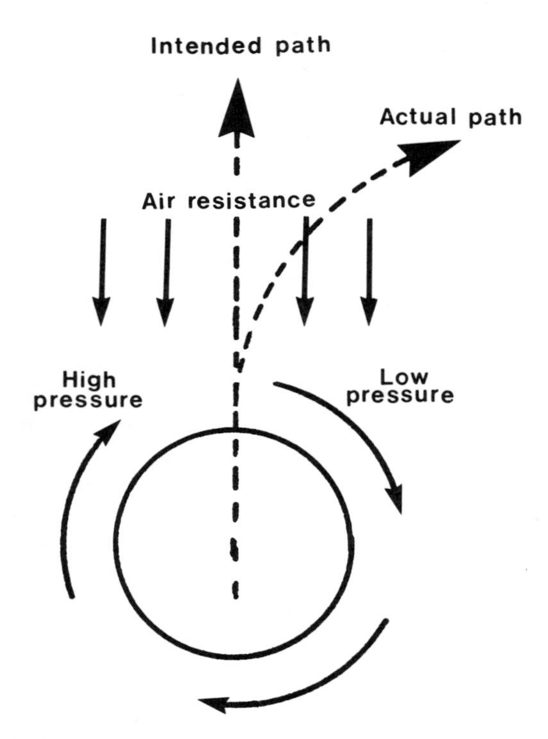

Figure 4.29. The Magnus effect; the spin of a ball will cause an imbalance in the resistance forces and the ball will seek the path of least resistance, namely, low pressure.

Rebound and Spin

A tennis ball that hits the court surface with topspin will bounce lower than if it had no spin. This is provided that the angle of incidence is the same in both cases. The rebound angle is the resultant of two forces: friction force caused by the spin and rebound force without spin. The friction force for a topspin ball is forward because the spin force on the bottom of the ball against the court surface is backward (Broer and Zernicke, 1979:129). Upon contact, the ball is projected forward and therefore does not bounce as high as the ball with no spin.

Tennis coaches and players alike usually are unable to comprehend this fact because their experience is that topspin balls bounce high and arch in flight to the backcourt. The essential fact that is overlooked is that the topspin ball travels a curved path as a result of the Magnus effect imposed by the spin imparted by the racquet. As the ball approaches the court, the curve steepens and at contact the angle of incidence approaches 90°. Thus, the reflected (bounce) angle is also large. The net effect gives the impression that a topspin ball bounces higher than one with no spin. In fact, if a no-spin ball were to contact the court at the same angle as the topspin ball, it would bounce higher. It should be noted that a ball with no spin will pick up some topspin as a result of contact with the court surface.

Similarly, a tennis ball with backspin will bounce higher than one without spin. Again, the impression is that the ball bounces lower than a ball with no spin. Other factors cause the "lower bounce." The friction force is backward and markedly affects the horizontal component, causing the ball to "skid" upon contact, with very little bounce.

A ball with side spin left or right "kicks" left or right after impact. Once again, the resultant (the actual rebound) is the vector sum of the friction force and the rebound force without spin.

Descriptions of the mechanics involved when balls hit implements are no different from what has already been described. In the case of the tennis ball impacting with the tennis racket, it is customary that the principles governing both spin and rebound are involved. Add to these the fact that the tennis racket is also moving. A good example of this interaction is to discuss the method of imparting spin to a tennis ball. Spin is produced by sliding the face of the racket across the ball thus imparting a friction force to a ball that is also impacting at some angle. For topspin, the player swings the racket from a low to a high position, while for backspin the racket moves from high to low. A golfer who has problems with a slice off the tee probably swings the club head from the "outside" to the "inside" of his golf swing arc, thus sliding the face of the club head across the golf ball. And the table tennis player slides the racket across the bottom of the ball, which will cause it to spin and "kick" off the table after bouncing. Examples are numerous and especially well-known by the practitioners of the sports.

Hydrodynamics

As the density of the fluid medium increases, fluid resistance and other properties have a very important effect on the mechanics of human performance. Water is a fluid medium in which most humans function by floating and swimming. Walking in water up to the neckline is an energy-consuming endeavor relative to performing the same act in air. The fluid frictional forces resisting motion in water are far more substantial than those in air. Thus, other methods of propulsion with greater energy economy are employed. Aspects pertinent to body equilibrium in fluids such as water include buoyancy and the body's center of buoyancy.

Buoyancy and Center of Buoyancy

Buoyancy is the upward force that any fluid exerts on an object or body that is partially or entirely immersed in it. Its force is equal to the weight of the volume of fluid which that object or body displaces. This is derived from Archimedes' principle that a body immersed in a fluid experiences a buoyant force equal to the weight of the fluid it displaces. As gravity is considered as acting at the center of gravity (CG) of a body, so buoyancy may be considered as acting at the center of buoyancy (CB) of a body. If an object is of uniform density, CG and CB will coincide. However, neither boats, floats, fish, nor animals including humans, have bodies of uniform density, so in them CB does not coincide with CG.

The center of buoyancy is defined as the center of gravity of the volume of the displaced fluid *before its displacement*. If one imagines a volume of water the same shape as the swimmer, the location of the CG of this volume corresponds exactly to that of the volume of a swimmer taken to have a uniform body density. The swimmer's CB, then, is located in the region which displaces the largest volume of water. Normally this is the thoracic region, although a very obese individual with large hips and thighs will have the CB nearer to the pelvic region.

Static Equilibrium in the Water

Static equilibrium in the water occurs only when the CB and CG are in the same vertical line. When a swimmer tries to float motionless the body will rotate until this condition is met (Fig. 4.30). A vertical floater (Fig. 4.30(c)) may, by altering the position of the extremities, move the CG cephalad in the body and thus change the floating angle (Fig. 4.30(d)). If an individual normally floats at a large angle with the horizontal, the float should commence from the vertical position so that any rotation will result from the buoyant force of the water. If a swimmer who normally floats with the body close to the vertical commences the float from a horizontal position, the legs will drop with a gravitational acceleration that may pull the head below water if no effort is made to check the downward rotation.

Dynamic Equilibrium in the Water

As water offers support to a swimmer who makes progress by pushing against, or pulling the body along, the surface of the water, dynamic equilibrium is not a problem. This holds as long as the usual horizontal or near-horizontal mode of progress is maintained. However, when a swimmer attempts various skills or stunts used in synchronized swimming, it may be found difficult to keep oriented to the stunt being performed. Spears (1966) speaks of this as space orientation and points out that a synchronized swimmer must have complete body awareness which takes place through kinesthesia, i.e., through all parts of the proprioceptive system. A swimmer is continually moving

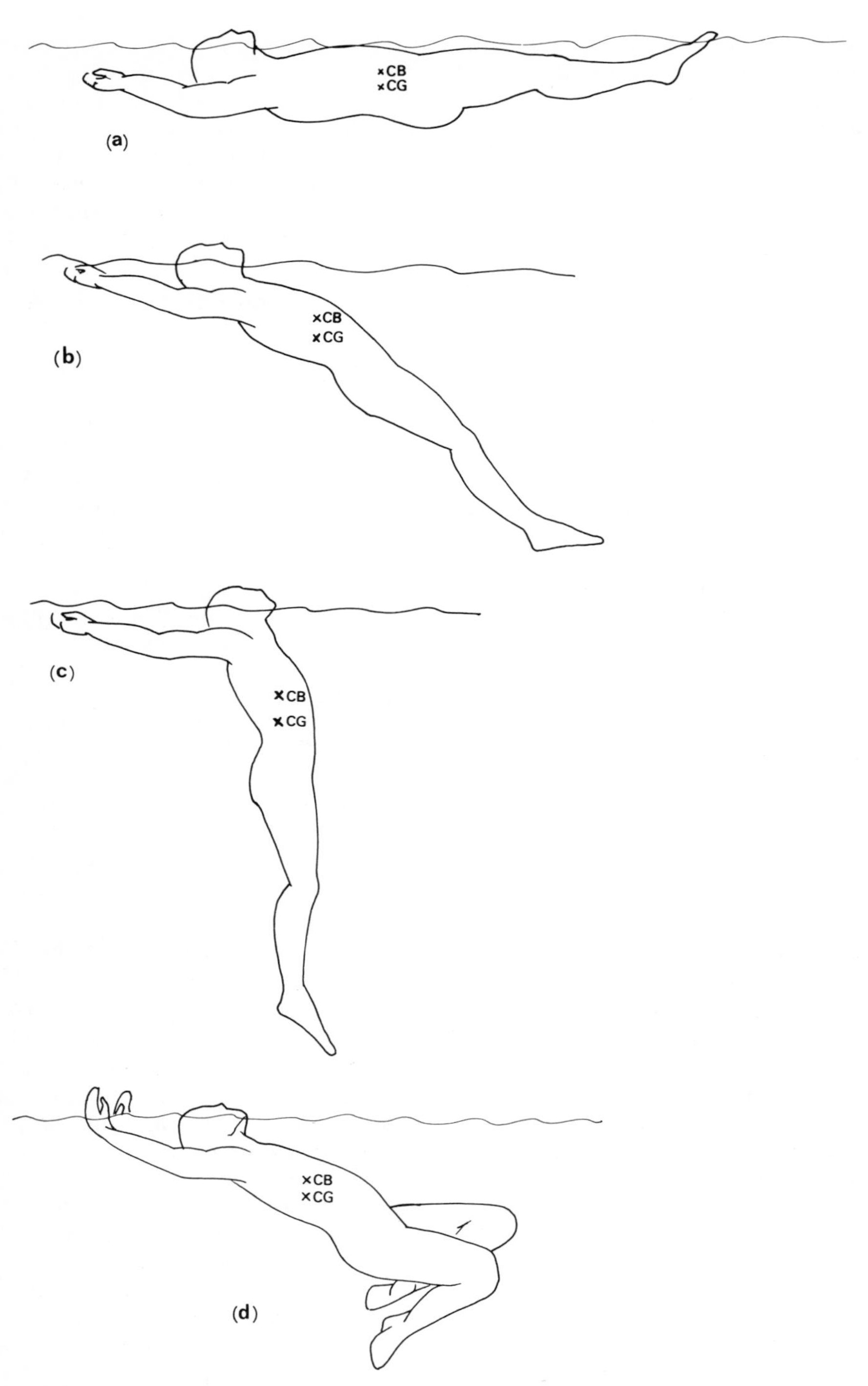

Figure 4.30. Floating positions dependent on the relationship between the swimmer's centers of buoyancy and gravity. (a) horizontal floater; (b) average floater; (c) vertical floater; (d) change of floating angle consequent on shift of body position relative to (c).

through three dimensions largely unfettered by gravity and, until she has been well-schooled to this condition, disorientation may easily result.

Many stunts require a performer to keep the movement in a given plane and for the head to emerge either at the point of submersion (dolphins, somersaults, and related skills) or to emerge some distance away in a given direction, as in a porpoise. Sometimes a stunt involves a full or a half-twist so the swimmer should emerge facing in a different direction. A performer should not depend solely on the inner ear mechanisms to maintain orientation to the skill, because they may be insufficient. This is particularly true in the early stages of learning synchronized skills; thus, the beginning swimmer orients herself by visual cues, checking the direction in relation to markings on the pool bottom or sides or in relation to nearby swimmers. In open, murky water this is, of course, more difficult. As she becomes more familiar with the performance, the synchronized swimmer learns to depend on other proprioceptive cues, such as joint position and muscle tension, and she eventually achieves the total body awareness that enables her to perform an expressive movement.

Physical therapists make use of hydrodynamic principles when they prescribe exercise in water. This is commonly referred to as hydrotherapy. A male patient is used to illustrate a common rehabilitation exercise progression. The patient has a limited range of motion of hip abduction and his hip abductor muscle strength is weak. On land, he is unable to reproduce a normal gait pattern without the use of a cane (used in the hand opposite to the involved hip). Without a cane, the patient demonstrates an exaggerated shifting of his weight towards the involved hip while weight bearing on this side. This gait deviation effectively reduces the resistance moment arm, and as a result, less effort is required by the involved hip abductors in order to maintain equilibrium while walking.

Increased Range of Motion

The first therapy is directed toward increasing the range of motion. The exercise is carried out in a hydrotherapy pool with the patient in the horizontal floating position (Fig. 4.30(a)). External buoyancy support devices are placed under the head, mid-back, and legs to ensure that he maintains this position effortlessly. To encourage increased active hip abduction, the patient is asked to slowly abduct the hip. Since all resistance forces at the hip joint have been minimized, he will be able to obtain an increased amount of active hip abduction compared with performing the exercise while standing on land.

Exercise Progression for Strengthening Muscles

The same exercise that is described above is used in the initial strengthening program. The next stage is to increase the speed with which the patient performs the exercise. Fluid mechanics comes into play since the faster-moving limb causes increased turbulence in the water, which in turn increases the resistance to movement (i.e., increased drag).

An alternate exercise progression technique is performed with the patient standing on the normal, contralateral, lower limb, with the water level at mid-thorax. The patient then actively abducts the involved lower limb. Due to buoyancy, the resistance is less than that when a patient performs this exercise on land; however, it is a more difficult exercise than the same exercise performed in the horizontal floating position.

Finally, a functional approach to the strengthening of the patient's hip abductor muscle group may be accomplished by having him practice walking in the hydrotherapy pool, first with the water at shoulder level and then gradually at lower and lower water levels.

Drag Forces in Water Medium

It is difficult to account for all the dynamic forces acting upon all of the body parts in an activity as complex as swimming. Some of the forces are the same as those in aerodynamics.

Surface drag, form drag, and wave drag all provide resistance forces to a swimmer. A good example to describe wave drag is the situation when a swimmer, who is continually plunging arms and legs into and out of the water (and the air), makes waves. The total resistance to forward motion of a swimmer is the total of the three drag forces.

Once again, as with aerodynamics, a lift-drag ratio that will minimize the drag and maximize lift is sought by the swimmer. As a result, competitive swimming styles have changed to achieve this principle. Good examples are (1) changing breast stroke style from wedge kick to whip kick, and even dolphin kick, for the purpose of minimizing surface drag and form drag; (2) changing front crawl free style from flutter kick to limited kick, or even no kick, for the purpose of minimizing surface drag, form drag, and wave drag.

FREE-BODY DIAGRAMS

Chapters 3 and 4 include a variety of topics concerned with kinematics and kinetics under conditions of both static and dynamic situations. Before attempting to apply these concepts to solving problems in the human body in Chapter 5, it is important to properly "dissect" each problem and identify all of the forces acting on a body. This may be done by isolating the body from its surroundings and then meticulously accounting for each force and its effect on the isolated body. The method involves the construction of a diagram named a **free-body diagram** (Fig. 4.31(b)). The free-body diagram provides a clear and uncluttered presentation of the various vector forces acting on a body. The direction and magnitude of each force and torque are shown by an arrowhead and length of arrow, respectively. The point of application of a force is located at the tail of the arrow (Luttgens and Wells, 1982:339). Free-body diagrams are oriented with respect to conventional *x-y* coordinate systems. A table of forces and their direction and point of application (Table 4.4) should be consulted when first constructing a free-body diagram.

Table 4.4
Direction and Point of Application of External Forces[a]

Force	Direction of Force	Point of Application
Weight (*W*)	Downward (toward center of earth)	Center of gravity of body
Normal reaction (*R*)	Perpendicular to contact surface	Contact
Friction (*F*)	Along contact surface (perpendicular to normal reaction)	Contact
Buoyancy (*B*)	Upward	Center of buoyancy of body
Drag (*D*)	Opposite the direction of oncoming fluid flow	Center of gravity of body
Lift (*L*)	Perpendicular to drag	Center of gravity of body

[a]From Luttgens, K., and Wells, K. F., 1982. *Kinesiology: Scientific Basis of Human Motion.* Philadelphia: Saunders College Publishing, p. 339.

The analysis should then be carried out by analyzing each directional parameter and writing equations of motion. The equations of motion for the free-body diagram in Figure 4.31 are shown below:

$$\text{Vertical: } R_y - W = \frac{W}{g}\ddot{y}$$

$$\text{Horizontal: } R_x = \frac{W}{g}\ddot{x}$$

$$\text{Rotational: } T - R_x r \cos\theta - R_y r \sin\theta - I_{cg}\ddot{\theta} = 0$$

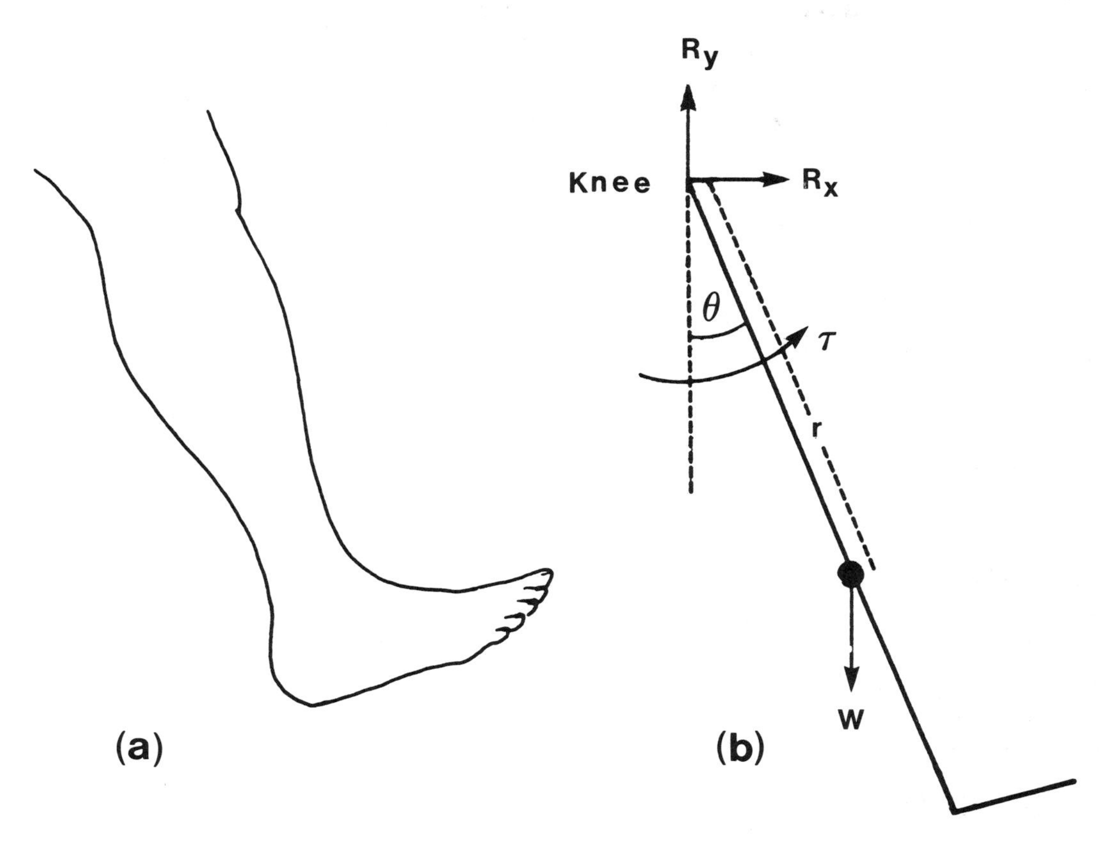

Figure 4.31. (a) Space diagram and (b) free-body diagram of a lower leg during the swing phase of gait. $\vec{R}_x$ and $\vec{R}_y$ = joint reaction forces; W = segment weight; θ= segment angle; τ = joint moment of force; r = distance to segment center of gravity. (From Rodgers, M.M., and Cavanaugh, P.R., 1984. Glossary of biomechanical terms, concepts and units. *Physical Therapy*, Vol. 64, No. 12:90.)

where $\ddot{x}$ and $\ddot{y}$ = components of acceleration in X and Y directions; I_{cg} = moment of inertia of segment about center of gravity; θ = segment angle; r = distance to segment center of gravity; $\ddot{\theta}$ = angular acceleration; R_x and R_y = joint reaction forces; W = segment weight; T = joint moment (Rodgers and Cavanagh, 1984:90).

At the start, to test comprehension, the reader may wish to construct a free-body diagram of Figure 4.30(b) by drawing the appropriate arrows representing the forces of that static problem. Careful attention should be paid to the direction of the arrowhead, the length of the arrow, and the location of the arrowtail. The skill of constructing free-body diagrams is acquired slowly and may be tested in the ensuing chapter when not only static problems but also dynamic problems are posed.

CONCLUSION

Having dealt with a number of important fundamental aspects of applied mechanics, we may begin now to consider biomechanical examples that rely on the fundamental principles embodied in this and the previous chapter.

CHAPTER 5

Application of Mechanical Principles to the Human Body

INTRODUCTION

With the background provided by the foregoing chapters, particularly Chapters 3 and 4, it is now possible to examine a variety of examples relating both kinematic and kinetic principles to the human body. Some examples deal mainly with static equilibrium situations. In these examples, references are made to specific muscles or muscle groups, and the assumption is that the reader has some knowledge of gross muscle anatomy. (Muscle structure and function as well as some "micro" muscle anatomy are found in Chapter 6.) When dynamic aspects are presented, the examples are specific for force analyses of the whole body. Also, a number of techniques are presented for determining the center of gravity of segments of the body as well as of the total body.

MOMENT ARMS IN THE BODY

In the complex levers which are characteristic of the musculoskeletal system in the human body, the moment arm (*MA*) of a muscle is the perpendicular distance from the line of force or action line of the muscle to the axis of the joint involved (Fig. 5.1). This perpendicular distance, the *MA*, must be used in all calculations of muscle force, and must not be confused with the length of the lever arm (i.e., the distance from the point of attachment of the muscle to the joint axis).

Because a muscle always has one attachment close to the joint at which it acts, a muscle lever arm and an even shorter *MA* are typical of all joints and body levers. When movement is produced by a muscle exerting tension as it shortens, the presence of any second-class levers in the musculoskeletal system is precluded, because the length of the resistance moment arm of the body lever is always greater than that of the effort moment arm.

Classes of Levers in the Body

In normal quiet standing, the center of gravity is anterior to the promontory of the sacrum, and the line of gravity falls anterior to the ankle joint passing through the metatarsals (Fig. 5.2). Electromyograms of quiet standing show slight but constant activity in at least one soleus muscle under these conditions (Joseph (1960); O'Connell (1958)). This tension counteracts the downward-forward moment caused by gravity and so prevents the body from falling forward to the ground.

Quiet standing then is a static, balanced situation, in which muscular effort balances the pull of gravity, and posture is maintained. Also, these forces (soleus tension and gravity) operate a first-class lever system; the axis through the fulcrum in the ankle joint lies between the soleus, acting behind, and gravity, acting in front of the ankle axis.

To rise on the toes the center of gravity must be shifted forward until the line of gravity falls over, or even momentarily slightly in front of, the toes themselves. At the same time, the heels come off the floor as metatarsophalangeal extension takes place. This action may simply be caused by forward sway (i.e., gravity) or possibly by contraction of the long and short toe extensors pulling on their proximal attachments, thus causing toe extension while the toes

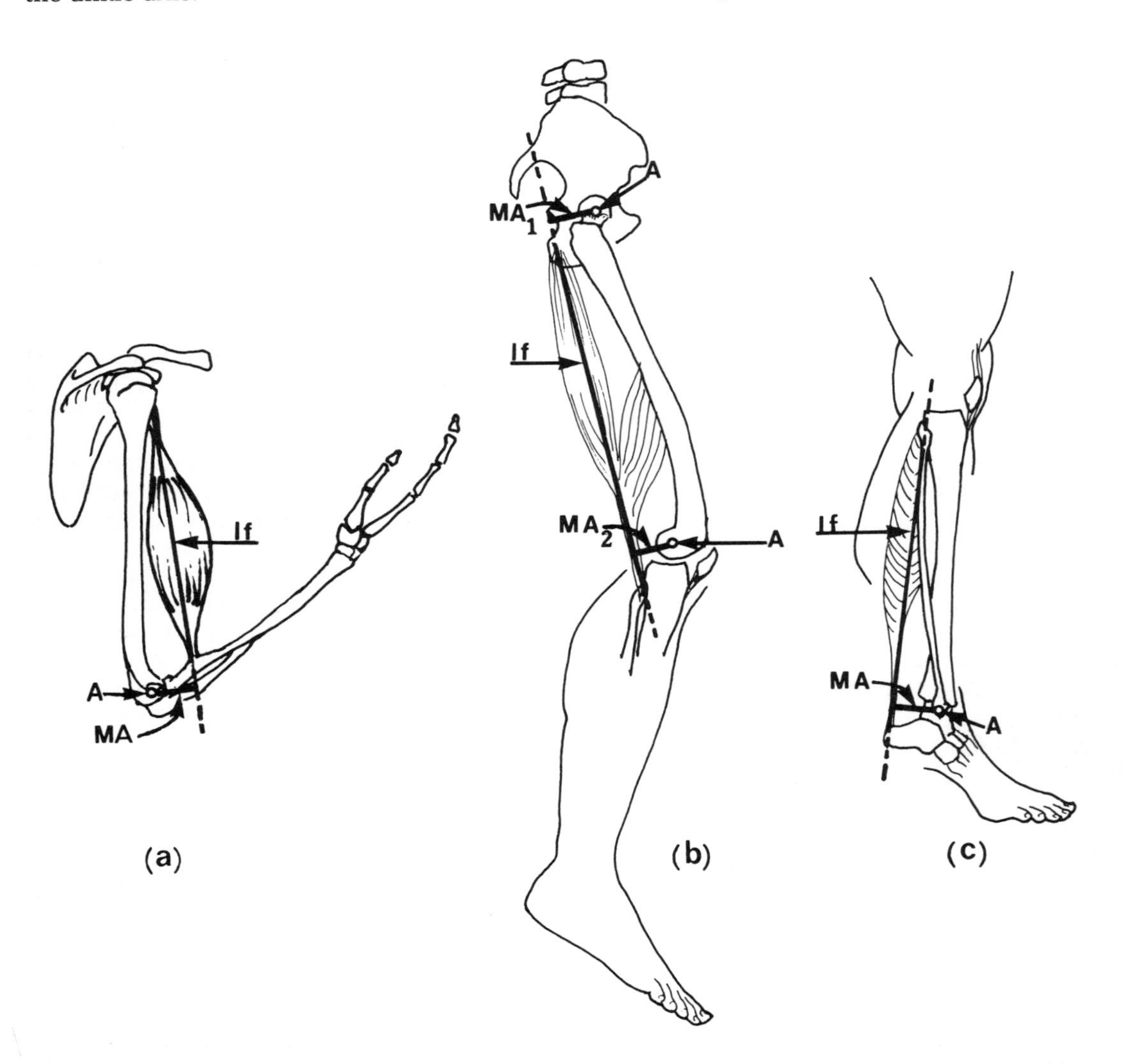

Figure 5.1. Muscle moment arms. (a) biceps brachii (elbow joint); (b) biceps femoris (knee and hip joints); (c) soleus (ankle joint). *lf*, line of force (action line) of the muscle; *A*, joint axis; *MA*, moment arm of the muscle.

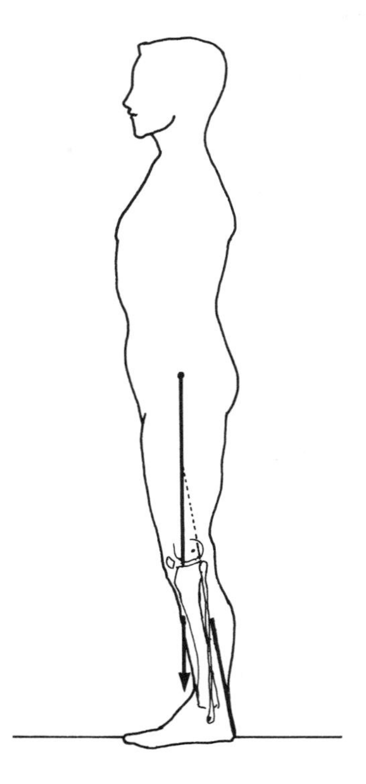

Figure 5.2. Line of gravity passing through the dorsum of the foot.

are still weight bearing. (Whether the toe extension is caused or contributed to by one, both, or neither might be determined electromyographically.) The forward sway immediately elicits increased activity in the calf muscles (Basmajian (1974); Joseph (1960); O'Connell (1958)), which pull back on the tibia and lower femur to move the body backward at the talotibial joint. The line of gravity continues to pass through the decreased base (the heads of the metatarsals and the toes) and, under the circumstances, produces extreme plantar flexion, characteristic of the posture when a subject has risen on the toes.

Thus, gravity and the toe extensors offer two possibilities for the effort of toe extension. In the first instance (Fig. 5.3(a)), a first-class lever system is in operation whether the axis is considered to be at the ankle or at the metatarsophalangeal joints; i.e., both joints are situated between effort (gravity) and resistance (tension in the plantar- and toe-flexing muscles). In the second instance (Fig. 5.3(b)), the axis is at the metatarsophalangeal joints, and the effort is contraction of the toe extensors, which have an *MA* shorter than that of the resistance due to gravity. Therefore, the leverage is third-class.

On the other hand, some kinesiologists consider that the brachioradialis, acting on the forearm to flex it at the elbow joint, is the effort in a second-class lever that supports a load halfway between the radial attachment and the elbow (Fig. 5.4). Their rationale makes use of the concept of an effort arm (all parts of the lever between the axis and the point where the effort is applied), *EA* in Figure 5.4(a). Then, they postulate a weight hung midway on the forearm, *R* in the same figure. Superficially, this seems to be an example of a second-class lever, with the resistance located between the effort and the axis and the effort arm longer than the resistance arm. However, one cannot use the length of such an effort arm to determine the amount of force that the brachioradialis must exert to support the load hanging from the forearm. Rather, the moment arms indicated in Figure 5.4(b) must be used in the calculations. Thus, the equivalence with a third-class lever is shown in the figure.

However, if we consider the brachioradialis, or any elbow flexor, as a *resistance* supporting the forearm against gravity, while a heavy object is lowered (Fig. 5.5), this lever system becomes second class. The muscle moment arms at the elbow are always shorter than the distance from the joint axis to the point where any load can be applied, and so, when the weight of the load is the force causing the movement, the system becomes a second class lever. This will be true in any situation where muscle shortening is the effort of a *third-class* lever. It is obvious that, as soon as this muscular tension becomes a resistance to a reversal of joint action caused by an outside force, the lever system is reversed also and becomes a second-class lever. Under these circumstances the muscle is contracting eccentrically and performing negative rather than positive work.

Advantages and Disadvantages of Short Muscle Moment Arms and Longer Lever Arms

Any lever system with a short effort moment arm requires a force greater than the load if the lever is to move that load. This is one of the seeming disadvantages of the human body, in which all muscular moment arms are short in proportion to the levers they move. However, there is in fact an advantage to this situation: a very small movement of the short end of the lever is magnified in direct proportion to the length of the lever being moved (Fig. 5.6). The pectoralis major has shortened only 2 cm and has moved the upper extremity through an angle of 83°. Since, in the same amount of time, the distal end of the lever moves a much larger distance than does the point of muscular attachment, the distal end is moving proportionately faster. This is also illustrated in Figure 5.7. The entire arm, forearm, hand, and tennis racket are moving as a unit through the same angular displacement in the same amount of time, but the distance, i.e., the length of the arc, traversed by the elbow joint is obviously shorter than that of the hand, and both are shorter than that of the racket. Since it is being moved

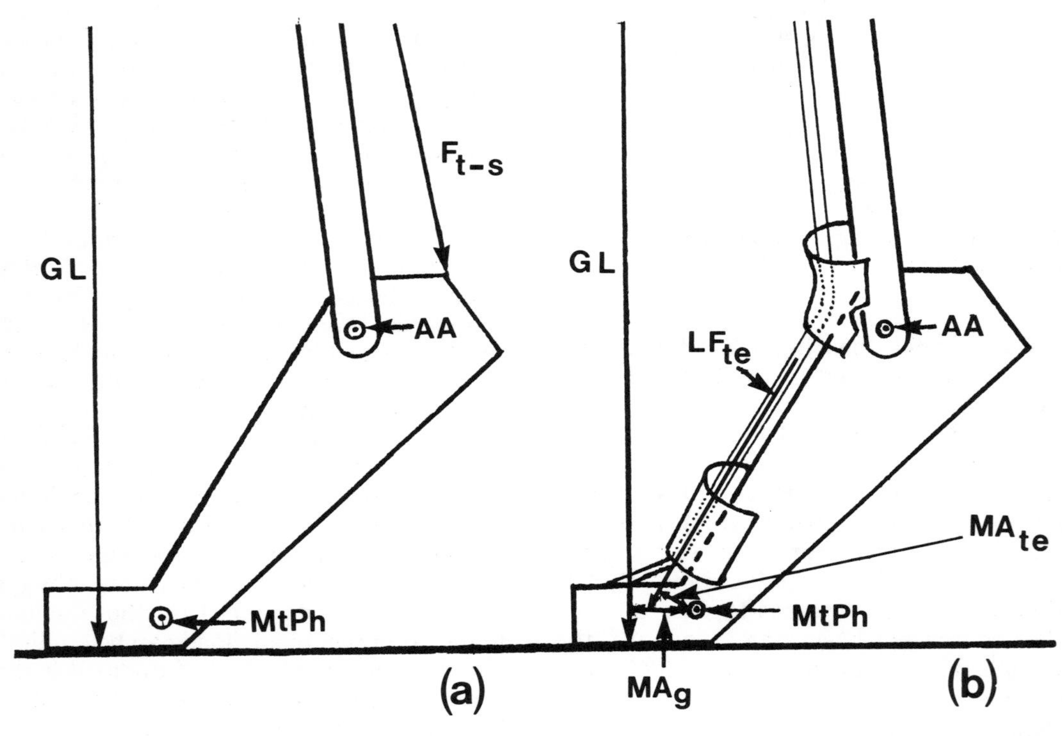

Figure 5.3. Standing on the toes. (a) force diagram as a first-class lever (gravity effort). *AA*, ankle axis; *MtPh*, metatarsophalangeal axis; $F_{t\text{-}s}$, line of force of triceps surae (gastrocnemius-soleus-plantaris); *GL*, gravity line. This is a first-class lever system as the joint axis is between the line of gravity (the force causing the forward motion of the body which moves the center of gravity over and briefly anterior to the toes) and the resisting force (the tension exerted by the plantar flexors). (b) force diagram as a third-class lever (toe extensors are effort). *AA*, ankle axis; *MtPh*, metatarsophalangeal axis; LF_{te}, line of force of the toe extensors; MA_{te}, moment arm of toe extensors; *GL*, gravity line; MA_g, moment arm of gravity. MA_g is longer than MA_{te} and both lie on the same side of the metatarsophalangeal axis.

through the same angle that the elbow has traveled, the racket head has moved curvilinearly almost 3½ times as far in the same amount of time, and so is moving at 3½ times the speed of the elbow.

Lengthening a body lever by an implement to gain speed is made use of in any sport which uses a club, bat, or racket to project an object. The longer the lever, the greater will be the speed of the impact surface of the implement, but size, length, weight, and shape are limited not only by the rules of the particular sport involved but by anatomical considerations as well. A tennis racket suitable for Ivan Lendl or Boris Becker would be a handicap to most women and certainly to a child or most adolescents. While the overall length of a tennis racket is limited by the rules, the weight and grip size are varied to suit a variety of players. The same is generally true of other striking instruments from baseball bats to polo mallets. In the latter instance, not only the arm length of the player but the size of the polo pony must be considered.

A boy who tries to hit a tennis ball with a racket that is too heavy will either "choke up" on the grip, effectively shortening the racket, try to hit the ball while pulling the implement closer to the body, or he will swing the racket more slowly. In the first and second instances, he is shortening the overall length of his lever so that less effort is needed for the work of moving the load and more of the muscle force is available for speed of movement (Halverson (1966)). In the third instance, he is compensating for insufficient muscle force by using his limited strength for work rather than power.

Based on the observations of children compensating for large implements, heavy balls, and standard-size court sizes and field dimensions, several investigators have

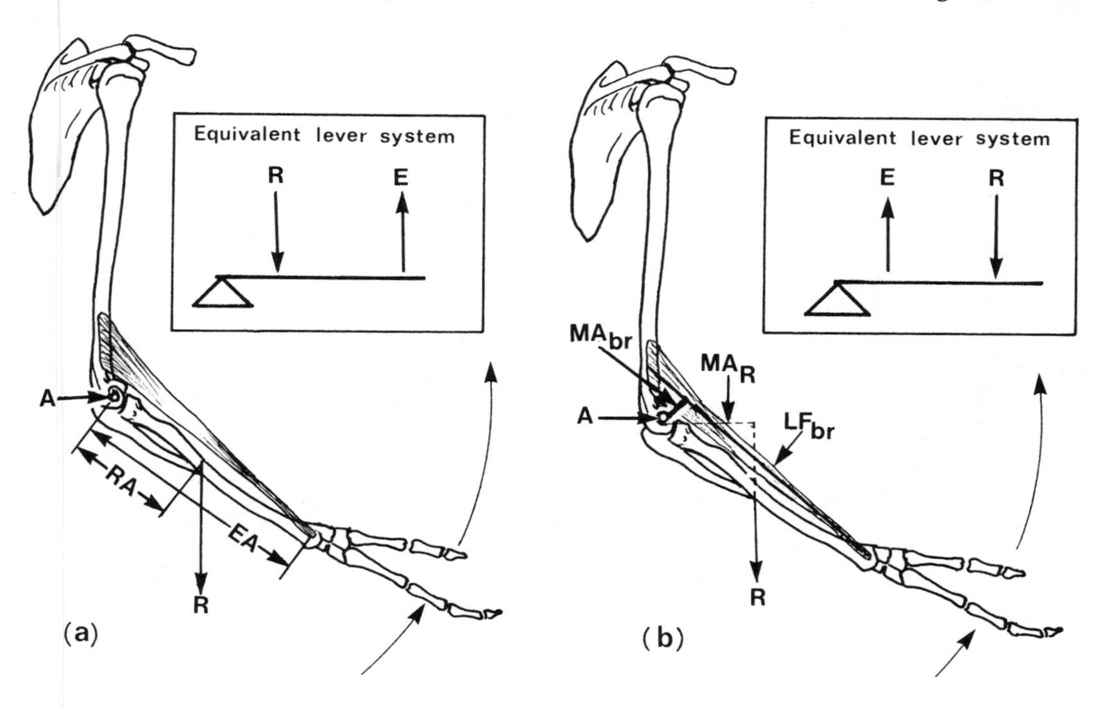

Figure 5.4. Brachioradialis as the effort in elbow flexion: a second- or third-class lever? (a) erroneously postulated second-class lever. *A*, axis containing fulcrum of elbow joint; *R*, line of force of resistance (load); *RA*, resistance arm; *EA*, effort arm. (b) third-class lever. *A*, axis of elbow joint; *R*, line of force of resistance (load); *MA*, moment arm of the resistance; LF_{hr}, line of force of brachioradialis; MA_{br}, moment arm of brachioradialis.

made recommendations for sports in which children compete (Orlick (1975 and 1984); Ward and Groppel (1980); Bearpark (1981); Pooley (1984); and Waddell (1986)). It would appear that some overall questions need answers from researchers in biomechanics, such as, for each anthropometric measure, how large and heavy a ball is needed for such sports as soccer and basketball, how long and heavy a racket and how large a racket head are needed for all of the racket sports, and how large a court or field is needed for all sports played by youngsters.

The compensatory tactics children use illustrate the fact that to conserve force for extra speed, the moment arms of body levers may be shortened by flexing distal joints and so decreasing the force necessary at the proximal joint to move the limb. A runner makes use of this principle by flexing the swing leg sharply at the knee as it is brought forward. This moves the center of gravity of the lower extremity closer to the hip axis and so decreases the moment arm of the lower extremity, thereby decreasing the amount of force (and the moment) necessary to move the thigh, leg, and foot forward. This extreme knee flexion is particularly obvious in sprint running (and in Standardbred trotting horses). Because of this sharp knee flexion the hip flexors can shorten rapidly and the speed of the run is maintained.

Elementary body mechanics necessitates that loads should be carried as close to the body as possible, thus decreasing the moment arm of the load in relation to the joints involved. This decreases the strain put on lifting muscles, joint capsules, and ligaments.

Although identifying the various classes of levers as they occur in the body gives insight into the forces necessary to move them, such identification is not vital to movement analysis *per se*. What is vital is to remember that the body is a system of levers and so obeys the laws of levers and mechanics. It has effort arms and resis-

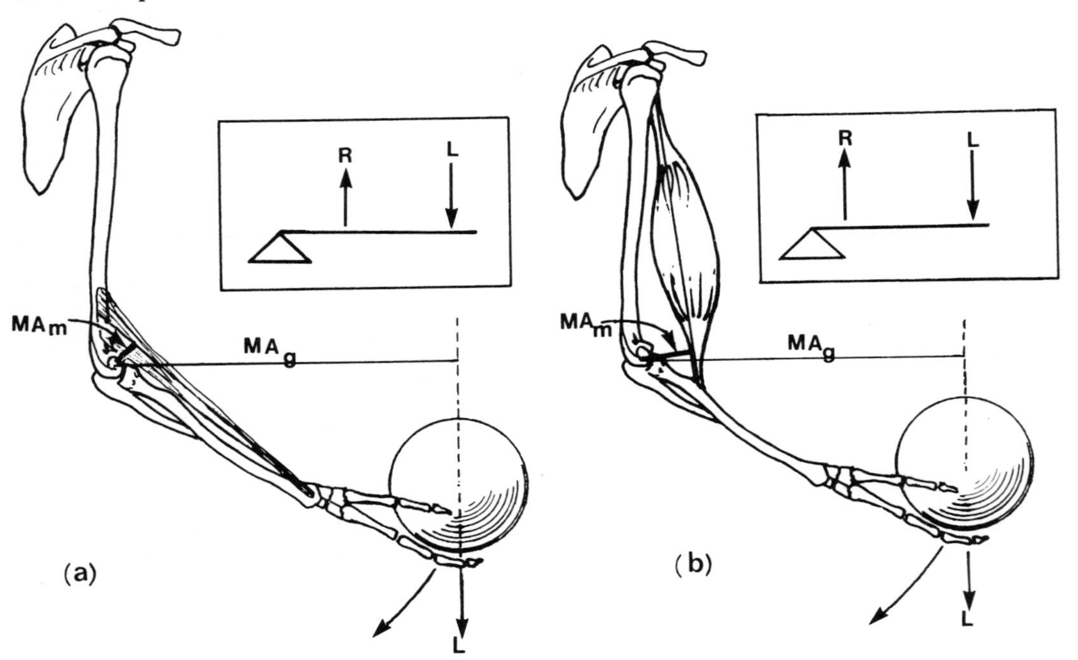

Figure 5.5. Elbow flexors acting as resistance: second-class levers. (a) brachioradialis as resistance; (b) biceps as resistance. *L*, load or gravity as the moving force or effort; MA_g, moment arm of gravity; MA_m, moment arm of resisting muscle. The insets show equivalent lever systems of the second-class.

tance arms, moment arms of the effort (MA_E) and moment arms of the resistance (MA_R). Muscles may be moving forces in a motor skill or they may resist the action of an outside force such as gravity by a lengthening tension, which controls and slows the movement. On the other hand, if a very fast movement is desired, muscles may work with gravity to accelerate the movement, as when an axe is swung down hard to split a log.

Skilled motor activity comes from making optimal use of levers, decreasing or increasing their length as occasion demands, and timing the muscular control to act on them with an optimal amount of force.

A number of problems in statics and dynamics follow in applying some of the mechanical principles to the human body.

EXAMPLE 1—Resultant of Forces of Two Gastrocnemius Heads

It is assumed that the forces acting at each of the two heads of the gastrocnemius are represented by vectors $\vec{AB}$ and $\vec{AC}$ in the plane of the paper as shown in Figure 5.8. Given that the forces exerted by the two heads are equal, it is required to determine the magnitude and direction of the resultant force.

By graphically constructing the parallelogram *ABDC*, with the length *AB* equal to the length *AC*, vector $\vec{AD}$ is obtained. This represents the resultant force whose magnitude (established by measurement) is 1.82 times *AB* or *AC*. Its direction is upward along the line which bisects the angle *BAC*.

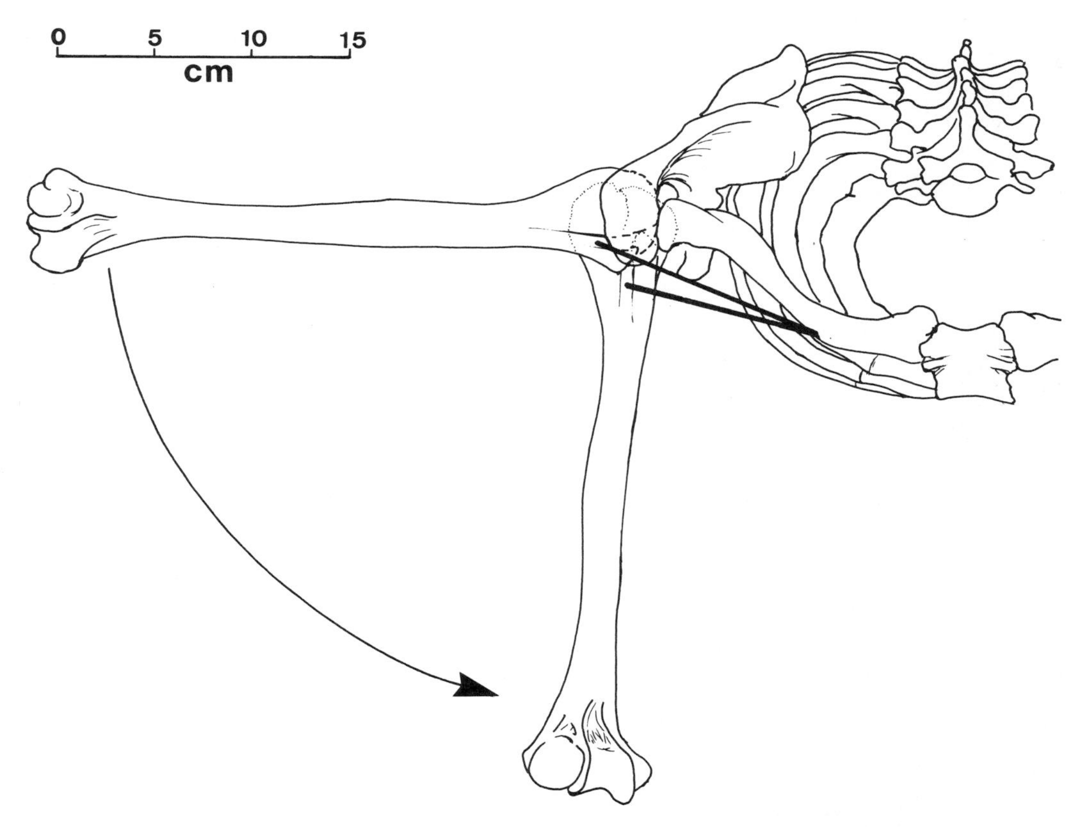

Figure 5.6. Effect of small amount of muscle shortening on range of movement. Clavicular head of pectoralis major has shortened 2.0 cm and the humerus has moved through an 83° arc.

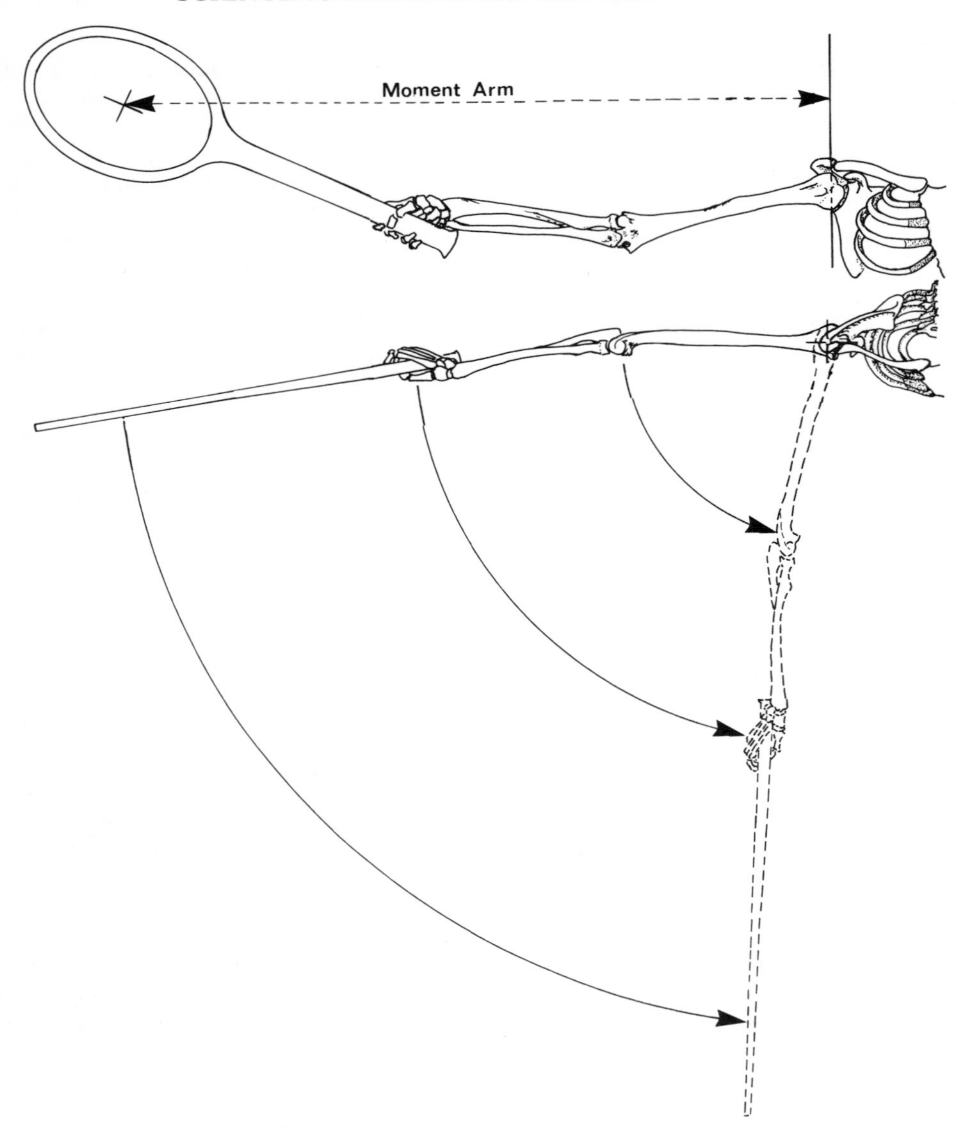

Figure 5.7. Forward swing of a tennis forehand drive.

If some values for magnitude and direction are supplied for each of the two gastrocnemius heads in Figure 5.8, a trigonometric solution may also be used. For example, if the angle vector $\vec{AB}$ makes with the right horizontal is 100° and the angle vector $\vec{AC}$ makes with the horizontal is 70°, what is the magnitude and direction of the resultant (i.e., the angle the resultant makes with the horizontal) when: (a) $\vec{AB} = 1000$ N and $\vec{AC} = 1000$ N; (b) $\vec{AB} = 1200$ N and $\vec{AC} = 1200$ N; and (c) $\vec{AB} = 1000$ N and $\vec{AC} = 1200$ N? Answers to each of these questions may be found in Appendix 1.

EXAMPLE 2—Resultant of Forces of Two Heads of Pectoralis Major

Figure 5.9 shows a graphical composition of the forces exerted by the clavicular and sternal heads of the pectoralis major.

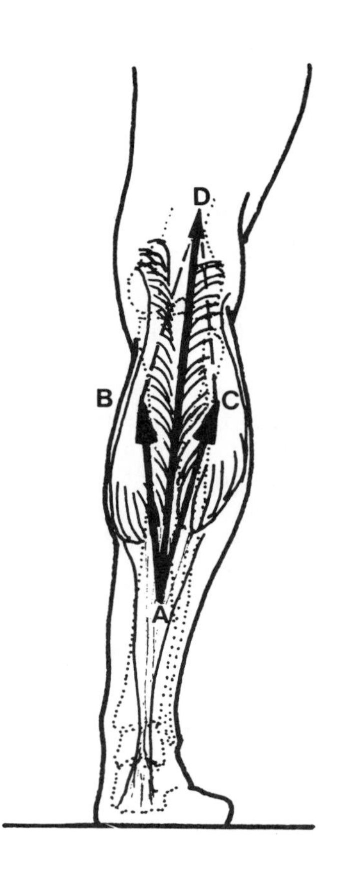

Figure 5.8. Graphical composition of the forces exerted by the two heads of the gastrocnemius.

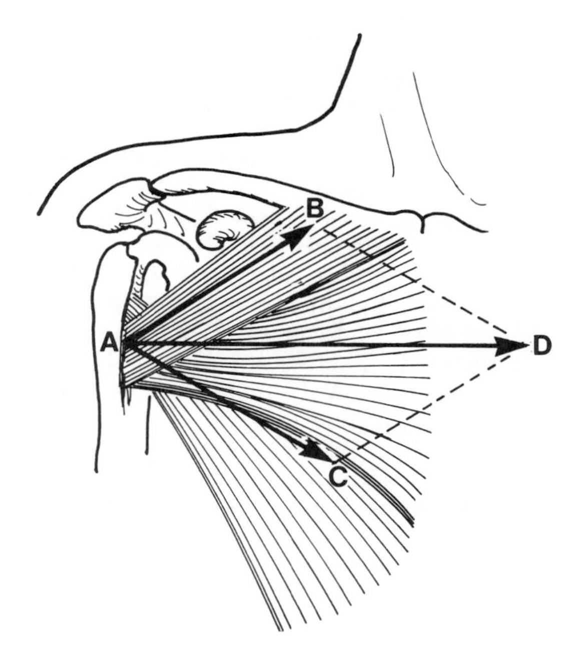

Figure 5.9. Graphical composition of the forces exerted by sternal and clavicular heads of the pectoralis major.

$\vec{AB}$ is the clavicular component of force and $\vec{AC}$ the sternal one. The vector $\vec{AD}$ is the resultant force of the entire muscle acting at point A.

The reader might wish to substitute some real numbers into the problem and attempt a trigonometric solution as with the previous problem. If so, first try working the problem so that the relative magnitude and direction are the same as presented in Figure 5.9. Second, show a muscle imbalance in the opposite direction so that vector $\vec{AC}$ is smaller than $\vec{AB}$, for example, signalling that the sternal head of the pectoralis major has decreased strength due to mild paralysis.

EXAMPLE 3—Three Abductor Forces Acting at the Hip Joint

Once again, a statics problem shown in the plane of the paper is posed. Consider the three abductors acting about the hip joint: gluteus medius, gluteus minimus, and tensor fascia lata. Each of these three muscles has a different shape and mass and a slightly different angle of pull, although they all attach on or near the greater trochanter of the femur.

To evaluate the resultant force in magnitude and direction necessitates a knowledge of the magnitudes and lines of action of the individual muscle forces. Inman (1947) determined the action lines on the pelvis. By postulating that their separate contributions to abduction were proportional to their individual masses, he established the following relative force values:

Tensor fascia lata	1
Gluteus minimus	2
Gluteus medius	4

On this basis and knowing the directions of the action lines, the relative resultant force may be determined. Figure 5.10 indicates the action lines from the trochanter *A*. That of the tensor fascia lata, *AB*, is shown as attaching at a point that is very close to its junction with the iliotibial band. The action line of the gluteus medius is represented by *AC* and that of the gluteus minimus by line *AD*. Thus, $\vec{a}$, $\vec{b}$ and $\vec{c}$ are the respective vectors acting on the pelvis. The resultant force of these three muscles may be determined graphically in two different ways: (1) by the parallelogram method illustrated above and (2) by constructing a polygon of forces.

Solution by Parallelograms of Force. With three forces it is necessary to construct two parallelograms. The three vectors $\vec{a}$, $\vec{b}$, and $\vec{c}$ in Figure 5.10(b) are drawn from a common origin and at the same angles as in Figure 5.10(a). In this case, for reasons of clarity, the first parallelogram is drawn between vectors $\vec{a}$ and $\vec{c}$ with a resultant vector $\vec{R'}$. The second parallelogram is drawn between vectors $\vec{R'}$ and $\vec{b}$ to find the final resultant vector $\vec{R}$.

Solution by Polygon of Forces. Referring to Figure 5.10(c), it will be seen that the first vector $\vec{a}$ is drawn at the same angle to the horizontal as in Figure 5.10(a). The second vector $\vec{b}$ is added to the end of $\vec{a}$ and the third one $\vec{c}$ is similarly added to $\vec{b}$. The vector that closes the polygon is the resultant, $\vec{R}$, of the three forces.

In both solutions, the resultant relative force is found to be just less than 7 units. This value is close to the arithmetic sum of the relative forces since they all act in approximately the same line.

Solution by Resolution into Horizontal and Vertical Components (Algebraic Approach). Given the proportional forces

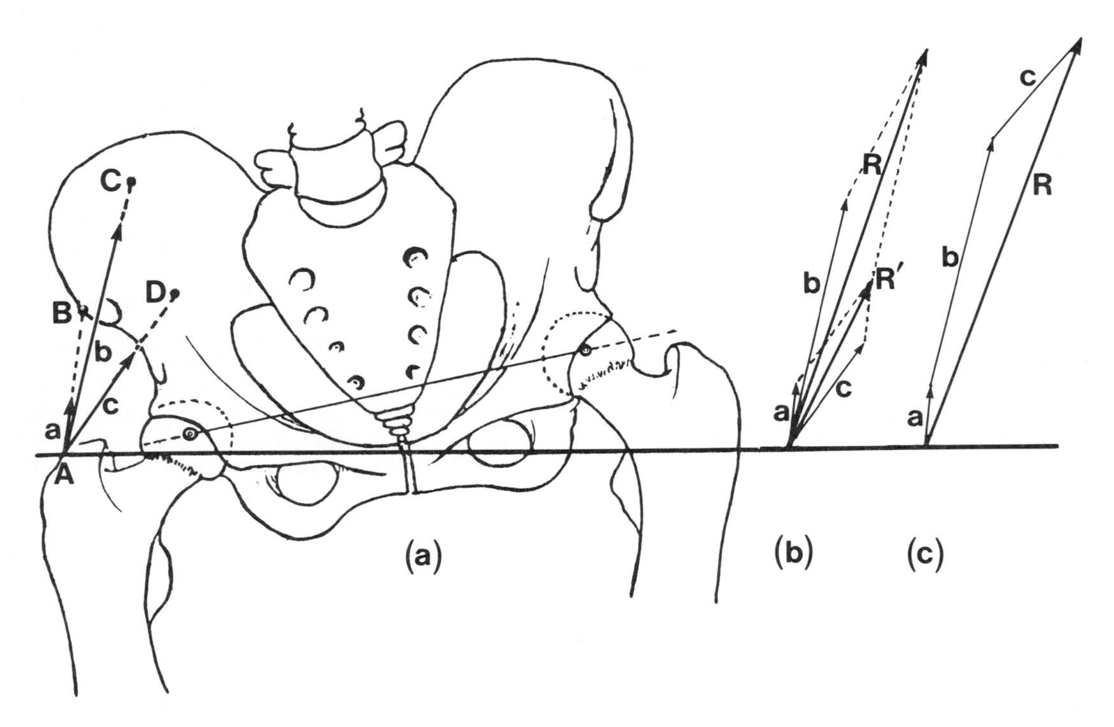

Figure 5.10. Abductor forces acting at the hip joint. (a) pelvis and hip joint with abductor action lines (redrawn after Inman (1947)); (b) graphical solution of resultant by parallelograms; (c) graphical solution of resultant by a polygon of forces.

exerted by each of the muscles in the group, it is convenient to convert the proportions to percentages of the total force exerted by the three muscles. Thus, gluteus medius contributes four-sevenths (approximately 57%) of the force, gluteus minimus two-sevenths (approximately 29%) and tensor fascia lata one-seventh (approximately 14%). The relative amount of force in the resultant can be obtained by using the percentage of the total force available from the group as a whole. For the purpose of this problem we will assume the total abductor force to be 100% and that each muscle is exerting its proportional amount of force as shown in Table 5.1.

Table 5.1
Known Data

Muscle	Gluteus medius	Gluteus minimus	Tensor fascia lata
Force	57%	29%	14%
Angle with horizontal	76°	55°	81°
Sine	0.97030	0.81915	0.98769
Cosine	0.24192	0.57358	0.15643

Horizontal component:

$$R_x = \Sigma F_x = 57 \cos 76° + 29 \cos 55° + 14 \cos 81°$$
$$= 57 \times 0.24192 + 29 \times 0.57358 + 14 \times 0.15643$$
$$= 13.79 + 15.63 + 2.19$$
$$= 31.61$$

Vertical component:

$$R_Y = \Sigma F_Y = 57 \sin 76° + 29 \sin 55° + 14 \sin 81°$$
$$= 57 \times 0.97030 + 29 \times 0.81915 + 14 \times 0.98769$$
$$= 55.31 + 23.75 + 13.83$$
$$= 92.89$$

The angle θ that the resultant makes with the horizontal is given by

$$\theta = \arctan R_Y/R_X$$

$$\theta = \arctan +92.89/+31.61$$

$$\theta = +71.2°$$

The resultant force at the greater trochanter is directed upward and to the right, making an angle of 71.2° with the horizontal. Correspondingly, the force acting on the pelvis is directed downward and to the left.

EXAMPLE 4—Momentum

Moving to a dynamics problem, consider a football running back of mass m_1 running toward the goal line with velocity v_1 and a linebacker of mass m_2 in the end zone, running directly toward the running back with velocity $-v_2$. Both players have to leap over a pile of linemen and they collide in midair above the goal line. Note that they are airborne and have colinear trajectories when they collide. Figure 5.11 provides a diagrammatic representation of the situation. The linebacker holds onto the running back after collision and they move as a combined unit. If the running back has a velocity of 9 m/s and a mass of 72 kg while the linebacker has a velocity of 6 m/s and a mass of 100 kg, what is the final velocity of the merged players, and in what direction do they move?

Figure 5.11. Schematic diagram of events before collision on the football field. RB running back; LB, linebacker.

$$m_1v_1 + m_2v_2 = (m_1 + m_2)\, v_f$$
$$(72)(9) + (100)(-6) = (72 + 100)\, v_f$$
$$648 \text{ kg m/s} - 600 \text{ kg m/s} = 172\, v_f$$
$$v_f = 0.28 \text{ m/s}$$

forward over the goal line.

Thus, the two players move in the direction of the running back's movement before collision, and a touchdown is scored.

If, on the other hand, the velocity of the linebacker is 8 m/s, what is the result?

$$m_1v_1 + m_2v_2 = (m_1 + m_2)\, v_f$$

$$648 \text{ kg m/s} - 800 \text{ kg m/s} = 172\, v_f$$

$$v_f = -0.88 \text{ m/s}$$ backward away from the goal line.

In this instance, no touchdown is scored.

EXAMPLE 5—Impulsive force

Once again, treating the human body as one unit, an example is used from the sport of ice hockey. A collision occurs between hockey players with the same masses as the previous problem, i.e., 72 and 100 kg. The two men slide as one mass almost without friction. They would travel at an initial velocity $v_i = 6.5$ m/s until they slam into the boards, where their combined final velocity v_f becomes 0 m/s. If, as they hit the boards, their bodies are compressed approximately 5 cm by the force of impact, the force with which they hit the boards may be determined. In order to determine the amount of this force, the impulse equation (Equation 4.4) is used:

$$F\Delta t = mv_2 - mv_1$$

$$F\Delta t = mv_f - mv_i$$

The length of time of the impact is also unknown but may be determined from the available data. The amount of compression (5 cm) is the distance that the bodies travel during the impact. As their velocity changes from 6.5 to 0 m/s during this time, the average velocity v_{av} is used in the final determination by substitution in the formula $v = s/t$ or $t = s/v$.

$$v_{av} = \frac{v_f + v_i}{2} = \frac{0 + 6.5}{2} = 3.25 \text{ m/s}$$

By substitution

$$\Delta t = \frac{5 \text{ cm}}{3.25 \text{ m/s}} = \frac{.05 \text{ m}}{3.25 \text{ m/s}} = 0.015 \text{ s}$$

$$F = (mv_f - mv_i)/\Delta t = \frac{(171)(0) - (172)(6.5)}{0.015} = \frac{-1118 \text{ kg m/s}}{0.015 \text{ s}} = -74{,}533.33 \text{ kg m/s}^2 = -74.53 \times 10^3\text{N}$$

F is an impulsive force of 74.53×10^3 N imparted by the boards to the impacting bodies.

EXAMPLE 6—Effect on Impulsive Force of Increasing Time of Impact

The effect of increasing the time of impact can be graphically illustrated if a comparison is made between forces generated (1) when a 45-kg boy drops 1.0 m from a high bar and lands "hard" with little or no flexing in his joints, so that the total cushioning of his body (shoes, feet, spine, etc.) lowers his center of gravity approximately 2.5 cm; and (2) when he lands softly on the balls of his feet and his joints all flex so that the total drop of his center of gravity *after* his feet touch the floor is approximately 25 cm.

The time for a 1.0-m drop under the influence of gravity follows from

$$s = \tfrac{1}{2}gt^2$$

i.e.,

$$t_{\text{drop}} = \sqrt{2s/g} = \sqrt{(2)(1)/9.81} = 0.45 \text{ s}$$

The velocity at the end of the drop

$$v_f = v_i + gt = 0 + 9.81(.45) = 4.41 \text{ m/s}$$

The average velocity during impact is

$$v_{av} = 0 + 4.41/2 = 2.205 \text{ m/s}$$

"Hard" Impact. Since the cushioning distance during impact is 2.5 cm, the time of impact

$$\Delta t_i = \frac{0.025}{2.205}$$

$$= 0.011 \text{ s}$$

and as

$$F = \frac{mv_f - mv_i}{\Delta t}$$

$$= \frac{(45)(0) - (45)(4.41)}{.011}$$

$$= 18{,}040.91 \text{ N}$$

$$= 18.04 \times 10^3 \text{ N}$$

"Soft" Impact. In the second instance, the soft landing, the time of drop, and the velocities are the same, but the distance has increased 10 times, so now

$$\Delta t_i = \frac{0.25}{2.205}$$

$$= 0.113 \text{ s}$$

Hence

$$F = \frac{mv_f - mv_i}{\Delta t}$$

$$= \frac{(45)(0) - (45)(4.41)}{.113}$$

$$= 1756.19\text{N}$$

$$= 17.56 \times 10^2 \text{ N}$$

On the other hand, if the boy uses a greater amount of eccentric, lengthening contraction sufficient to slow his joint actions further, the time would be increased beyond this point and the force on his feet would be decreased. Suppose that the time lapse from toe touch to deepest knee bend was 0.36 sec; then

$$F = 198.45/0.36$$

$$= 551.25 \text{ N}$$

This is considerably less force than in either of the two previous examples; i.e., 3.1% of that of the hard landing and only 31% of that of the first soft landing. These calculations serve to illustrate some of the typical forces to which the human body may be subjected without sustaining serious injury, and they provide objective evidence for using controlled "cushioning" when landing from a height.

It should be noted that as the time for impact Δt increases, the resulting force decreases. Since the force cannot be less than 441.45 N (the weight of the boy), the impulse equation must be used with caution as the time period Δt grows larger. (In essence, the longer the time, the less the impact, i.e., the force F undergoes variations from instant to instant during impact).

EXAMPLE 7—Muscle and Joint Reaction Forces

Returning to statics problems, it is desired to calculate the total muscle force (TMF) of the biceps and the force on the distal end of the humerus when the outstretched hand of a person is supporting a 45-N shot as shown in Figure 5.12.

The combined weight of the forearm and hand is 15 N and the center of gravity is 16 cm from the elbow axis; the line of gravity of the shot falls 34 cm from the elbow axis. (Note that all forces, including the action line of the biceps, are vertical and perpendicular to the long axis of the forearm.)

The moment arm of the biceps is not known, nor can it be accurately measured *in vivo*. It may be estimated on a living subject who flexes his elbow to the desired angle by measuring the perpendicular distance from the biceps tendon to the humeral epicondyle; it can be measured with a fair degree of accuracy in a cadaver or on a skeleton.

In this instance the moment arm of the biceps about the elbow will be taken as 5.6 cm.

The line diagram in Figure 5.12(b) is a simplified version of Figure 5.12(a), and so it does not show the reaction force at the elbow joint. Apart from this, there are two downward forces and a third upward one,

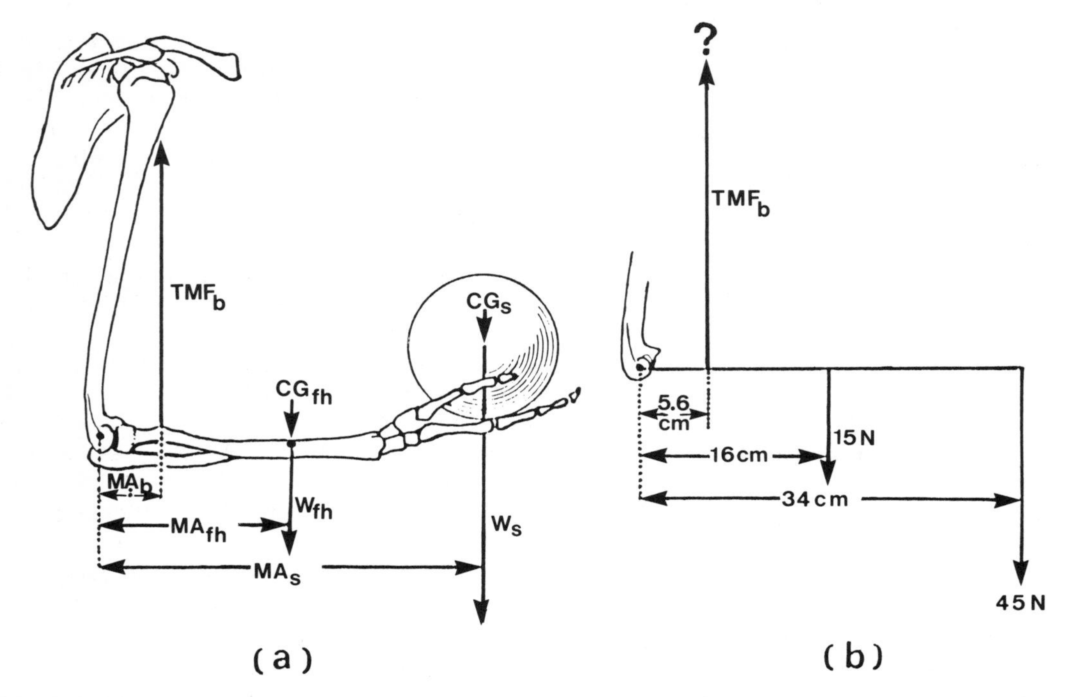

Figure 5.12. (a) drawing; (b) force diagram. MA_b, moment arm, biceps; MA_{fh}, moment arm, forearm plus hand; MA_s, moment arm of the shot; TMF_b, total muscle force of biceps; W_{fh}, weight of forearm and hand; W_s, weight of the shot; CG_T, center of gravity of forearm and hand; CG_s, center of gravity of the shot.

all parallel and each with its own moment arm. As this is an equilibrium situation, the sum of the moments is equal to 0:

$$\Sigma M = 0 \quad \text{or} \quad \Sigma T = 0$$

The weight of the body segments and of the shot in the hand tend to rotate the forearm clockwise (CW) about the elbow joint and so are considered positive, while the biceps force tends to rotate the forearm in the opposite or counterclockwise (CCW) direction and so is considered negative:

$$\Sigma M = (W_{fh} \times MA_{fh}) + (W_s \times MA_s) - (\text{TMF}_b \times MA_b) = 0$$

$$\Sigma M = (15\text{N} \times 16 \text{ cm}) + (45 \text{ N} \times 34 \text{ cm}) - (\text{TMF}_b \times 5.6 \text{ cm}) = 0$$

$$TMF_b \times 5.6 \text{ cm} = 2.4 \text{ Nm} + 15.3 \text{ Nm}$$

$$TMF_b = 17.7 \text{ Nm}/.056 \text{ m}$$

$$TMF_b = 316 \text{ N}$$

It is noteworthy that the biceps must exert about 7 times as much force as the weight of the load that it supports.

This same biceps force also pulls the forearm against the articular surfaces of the humerus so that there is pressure against them. Let H equal the unknown force on the humerus:

$$\Sigma F = 0$$

$$\Sigma F = +15 \text{ N} + 45 \text{ N} - 316 \text{ N} - \text{H} = 0$$

$$H = 60 \text{ N} - 316 \text{ N}$$

$$H = -256 \text{ N}$$

an upward force. There is a vertical force on the distal end of the humerus of 256 N as a result of the biceps supporting the 45-N shot in the hand. This force is distributed on the articular surfaces of the humerus and the mean pressure is the force divided by the surface area on which the force bears.

EXAMPLE 8—Resolution of Total Muscle Force of Biceps into Rotatory and Secondary Components

Assuming a constant TMF of 450 N, this force is to be resolved into two components for each of several angles at the elbow, commencing with total extension and proceeding progressively until full flexion. The two mutually perpendicular components to be identified are the rotatory component which acts at right angles to the long axis of the bone that it moves (in this case the forearm); and the secondary component which acts along the long axis. The nature of the latter component is to be discussed.

Figure 5.13 shows the requisite resolution into the two components for a series of different angles. In each case, the action line of the rotatory component originates at point *A*, the intersection of the component originates at point *A*, the intersection of the biceps action line with the long axis of the forearm, *x*-*x*, and is perpendicular to this axis. The action line of the secondary component is coincident with the long axis, and its direction may easily be determined by constructing the parallelogram of forces around a given TMF of the biceps. Vector $\overrightarrow{AB}$ is drawn along the biceps action line and is 450 units long.

In Figure 5.13(a) the forearm is extended and the action line of the biceps makes an angle of approximately 8° with the long axis of the forearm. With a TMF of 450 N the rotatory component, vector $\overrightarrow{AC}$, is only 63 N, while the component represented by vector $\overrightarrow{AD}$ is 446 N. As the elbow is flexed the angle of pull of the biceps action line increases, as does the rotatory component. Maximal rotatory force is achieved at (c) when the action line of the biceps is at 90° with the forearm axis and all of the 450 N of contractile force is available for supporting or moving the forearm. With continuing elbow flexion the rotatory component decreases as the secondary component increases but its direction is now away from the joint. In (d) the muscle is pulling the forearm away from the joint with a 411-N force (vector $\overrightarrow{AD}$), and the rotatory force is 183 N (vector $\overrightarrow{AC}$).

It should be noted that when the elbow is in extension beyond the value reflected in Figure 5.13(c), i.e., Figure 5.13(a) and (b), the secondary force $\overrightarrow{AD}$ is directed toward the joint and is increased with more extension. Being directed toward the joint, this force is of a stabilizing-compressing nature, tending to hold the joint together. As the joint goes into flexion beyond the position shown in Figure 5.13(c), i.e., Figure 5.13(d) and (e), the direction of $\overrightarrow{AD}$ is now away from the joint; the secondary force becomes one of dislocation-decompression, increasing with greater flexion.

LOCATING THE CENTER OF GRAVITY

Much time and energy have been expended in efforts to pinpoint the location of the center of gravity in the human body. The earliest recorded effort is that made by Borelli in 1679, who balanced the body over a prismatic wedge in three different planes (Dawson (1935)). There the matter rested until the 19th century, when a number of different studies were made, notably by Mosso (1884) and by Braune and Fischer in 1889 (Duggar (1966)). Much work has been done in this field since then, and by the end of the first third of this century the method now commonly used in kinesiology laboratories was coming into general use.

With Gravity Board and Scale

The method now used in college kinesiology laboratories was, according to Cooper and Glassow (1968), devised in 1909 by Lovett and Reynolds (Fig. 5.14). This method is based on the fact that when a body is in equilibrium the sum of the gravitational moments acting on the body is equal to zero. The subject is weighed (weight = W_t) and then steps onto the gravity board. The board is tared [a] by setting the scale to zero so that the weight registered on the scale includes only the gravitational force acting on the body parts.

Taking moments about the line of action of the body force (which line of action must pass through the board), the clockwise moment CWM is the product of the force recorded on the scale W_s and the perpendicular distance from the line of action to the knife edge resting on the scale. If the distance between the knife edges is D, then the counterclockwise moment CCWM about the body is the product of the weight supported by the distal knife edge (total weight minus weight registered on the scale, i.e., $W_t - W_s$) and the perpendicular distance d_2 from the distal knife edge and the plane containing the line of gravity ($d_2 = D - d_1$). By convention CW forces are considered positive and CCW forces are considered negative.

For equilibrium

$$\Sigma M = 0$$

i.e.,

$$\Sigma M = \text{CWM} - \text{CCWM} = 0$$

$$\text{CWM} = W_s d_1$$

and

$$\text{CCWM} = (W_t - W_s)d_2$$

$$(W_t - W_s)d_2 - W_s(D - d_2) = 0$$

$$(W_t - W_s)d_2 = W_s(D - d_2)$$

$$W_t d_2 - W_s d_2 = W_s D - W_s d_2$$

$$W_t d_2 = W_s D$$

$$d_2 = W_s D / W_t$$

The line or plane of gravity is d_2 units from the *knife edge on the floor.* If the subject stands in his footprints drawn on a piece of paper which is placed with one edge at a given marker, say 100 cm from the distal knife edge of the board, the plane of gravity can be easily recorded on the footprints.

This method is useful in obtaining the location of the line of gravity in relation to the base of support at a specific instant, i.e., the time when the scale was read. When a subject stands on a gravity board, the pointer on a dial, or the beam of the balance, of a supporting scale is never completely still as there is a continuous slight swaying of the body on the heads of the tali, as was first reported by Hellebrandt *et al.* (1937).

When an attempt is made to locate the primary transverse plane by this method (the transverse plane passing through the center of gravity), only an approximation can be made, for several reasons. When the subject is supine there is a change in the distribution not only of the blood throughout the circulatory system but of the abdominal viscera as well. Also, Mosso (1884) and Cotton (1932) discovered that the locus of the center of gravity of a subject in a recumbent position shifted slightly with each deep inspiration (Duggar (1966)). This effect is undoubtedly a result of the abdominal viscera being forced caudad by the contraction of the diaphragm. Such a deep inspiration might well give a center of gravity a locus closer to that when the body is erect.

The use of the gravity board requires that the subject assume a static position, and some fairly elaborate apparatus has been set up to determine the locus of the line of gravity simultaneously in two planes. One technique which has become popular in kinesiology laboratories makes use of three scales, with the gravity board supported by three adjustable pointed bolts, one resting on each scale (Fig. 5.15). When the board has been leveled it is ready for use. Rasch and Burke (1967) use a large rectangular board so that they can place a subject in various side-lying positions corresponding to different diving or gymnastic conformations. Waterland and Shambes (1970) use a similar arrangement but with a smaller board, and they recommend one in the shape of a right triangle. Two of the scales are so placed that their

[a] If this is not possible, the weight of the board registered on the scale must be subtracted from the weight registered when the subject steps on the board.

dials face in the same direction and at 90° to the face of the third scale. The subject is photographed simultaneously from the side and front by synchronized cameras, one picture including the face of scale A and the other the faces of scales B and C (Fig. 5.15).

Calculations for locating the primary

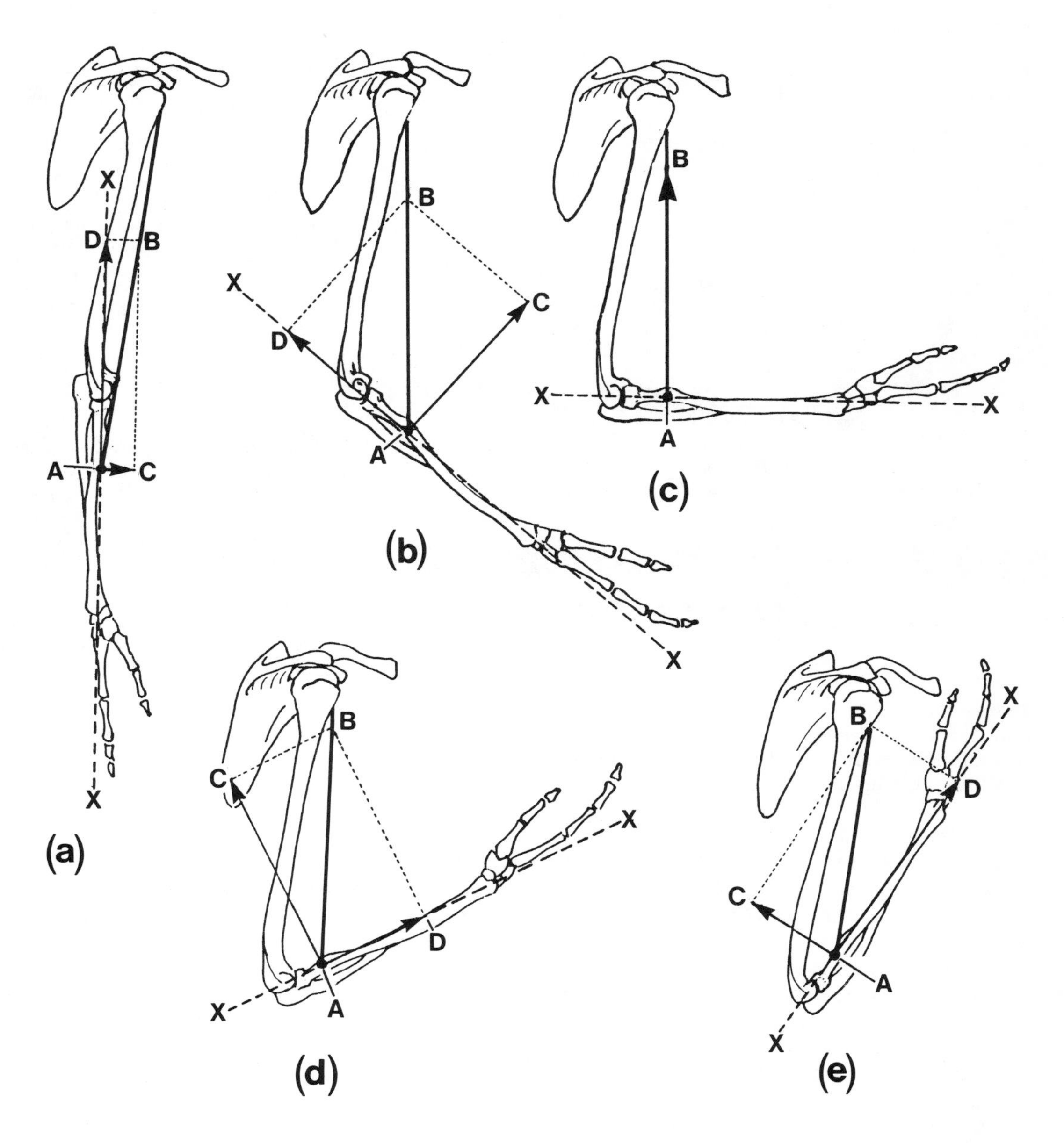

Figure 5.13. Resolution of total muscle force of biceps (TMF_b) into rotatory and secondary components. *AB* representing TMF_b is maintained at 450 N. *X* − *X* , long axis of forearm; *AC*, rotatory component; *AD*, secondary component. Note the change in direction of the secondary component from that in (a) and (b) after going past (c) as in (d) and (e). See text.

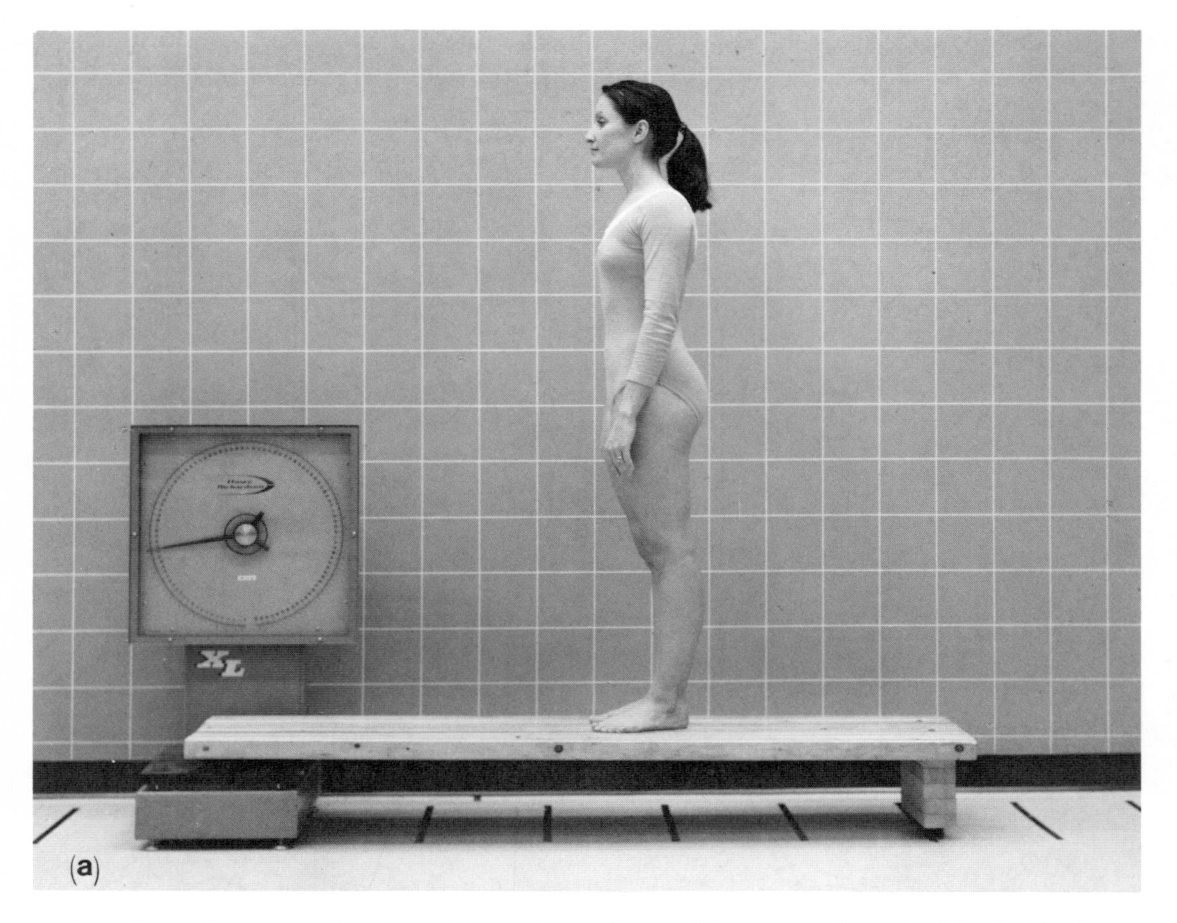

Figure 5.14. Equipment for determining primary planes of the center of gravity. Note knife edges supporting the gravity board. It is of interest to note the increased reading on the scale for (b) as compared with (a) when the subject rises on her toes, thus moving her center of gravity ahead of the position it occupied in (a).

planes which pass through the center of gravity are based on the same principles presented earlier, and procedures are similar.

Taking moments about the dotted line (in Fig. 5.15) which represents one plane of gravitational forces on the body:

$$\text{Clockwise moments: } CWM = W_A d_1$$

where W_A is the weight registered by scale A and d is the perpendicular distance from the plane of gravity to bolt a. Counterclockwise moments: $CCWM = (W_B + W_C)$ d_2 where W_B and W_C are the respective indications of scales B and C, and d_2 is the perpendicular distance from the line connecting bolts b and c to the plane of gravity.

$$\Sigma M = CWM - CCWM = 0$$

leads to

$$W_A d_1 - (WB + WC)d_2 = 0$$

Recognizing that $d_1 + d^2 = D$, $d_1 = D - d_2$ and substituting in the moment equation gives

$$W_A D - W_A d_2 - W_B d_2 - W_C d_2 = 0$$

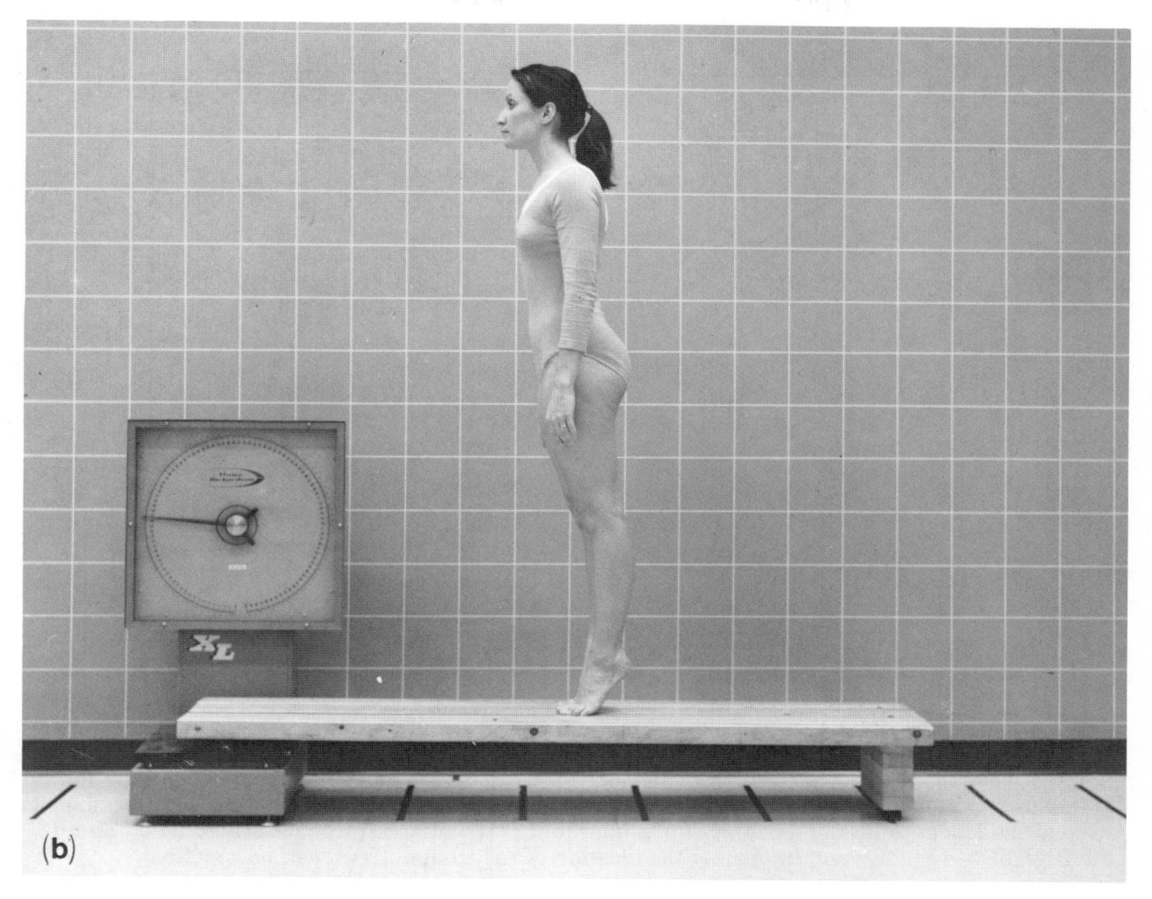

Figure 5.14(b).

i.e.,

$$W_A D - (W_A + W_B + W_C)d_2 = 0$$

Since the weight of the subject

$$W_t = W_A + W_B + W_C$$

it follows that

$$W_A D - W_t d_2 = 0$$

thus

$$d_2 = W_A D / W_t$$

To find the plane of the center of gravity parallel to the line between bolts a and b, the same procedure is followed, i.e.,

$$\Sigma M = W_C d_1' - (W_A + W_B)d_2' = 0$$

where $d_1' + d_2' = D'$ and d_1' is the perpendicular distance from C to the line representing the plane of gravity perpendicular to line BC, i.e., $d_2' = W_C D'/W_t$.

Swearingen (1949) used equipment consisting of five platforms mounted one above the other. The top platform supported the subject in an adjustable seat, the second and third platforms were horizontally adjustable, and the fourth platform was separated from the fifth, or base, by a ball-and-socket joint in the center and electrical contact points at each corner. Measurement of the CG "center of gravity" location for the upright seated posture, for example, was made by balancing the system with the subject tipped hori-

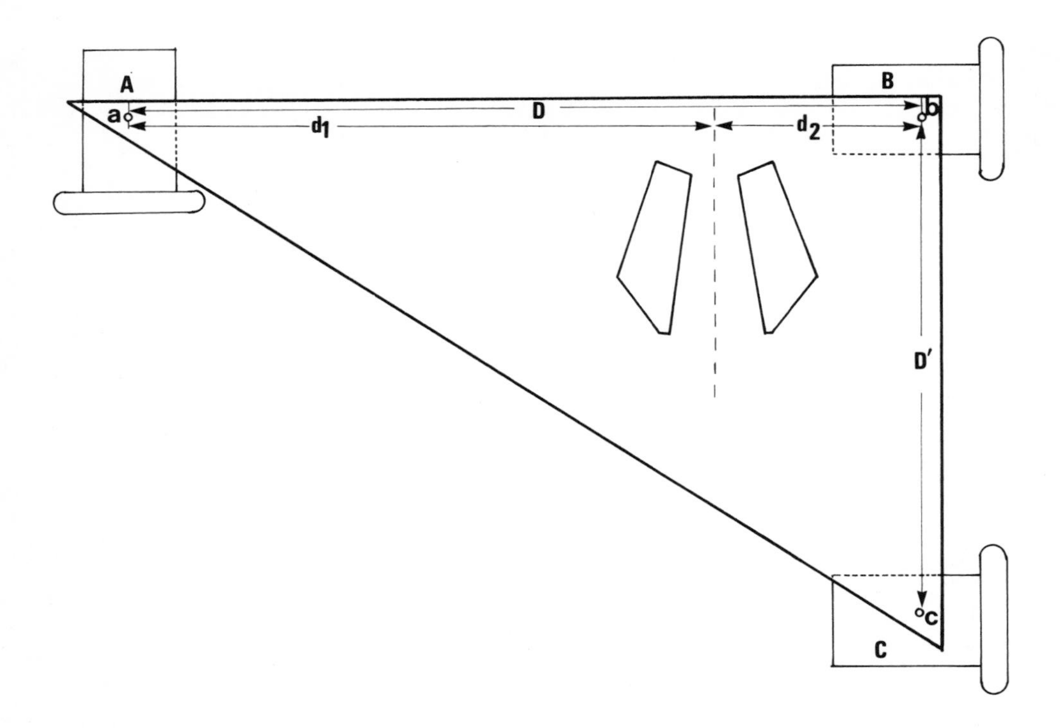

Figure 5.15. Schematic diagram of type of gravity board currently in use. A, B, and C, scales supporting the board; a, b, and c, pointed bolts supporting the board on the corresponding scale; *D*, the distance between bolts a and b; *D′*, the distance between bolts b and c. While the diagram is annotated for determining the location of the dotted line via d_1 and d_2 thus defining the sagittal plane, it should be evident that the location of a similar (but perpendicular) line described by d_1' and d_2', where $D_1' = d_1' + d_2'$, and indicating the location of the frontal plane, can be specified.

zontally, then rebalancing with the seated subject tipped approximately 20 degrees from the horizontal. Two planes passing through the CG were thereby established, and the exact location could be measured with respect to any suitable reference point (Duggar (1966)).

By Segmental Method

All of these methods are moderately accurate subject to the limitations mentioned earlier, as well as to the accuracy of the scales used in the study. However, none of these methods can be used in dynamic situations. A solution to this problem was suggested by Dawson as early as 1935. He suggested that the subject be "appropriately fotograft & the foto transfered to coordinate paper. The subj. is then weighed & measured & compared with the subj. described in the tables (Braune and Fischer's measures, Table A4.1, Appendix 4). One of the 4 cadavers is selected as being most like the subj. in bild & the data from this cadaver ar therefore used in the following calculations. The wts. of the parts of the subj. (wt.) are computed on the assumption that the weight of the part of the subj. =

(wt. part cadaver × wt. whole subj.)/wt. of whole cadaver

"Having determined the wt. of the various parts of the subj. & noted upon the coordinate paper the position of the cc. g. (centers of gravity) of the parts, it is a simple mathematical calculation to determine the position of the c. g. of the body as a whole. By the use of a system of coordinates the calculations wer stil further simplified so that it became necessary merely to insert the values ascertained in an appropriate equation which could be solved."[b]

This approach has become greatly simplified with the advent of minicomputers coupled with digitizing tablets and stop-motion film projectors. Also, it is no longer necessary to rely solely on Braune and Fischer's data. Dempster (1955) collected similar material on eight cadavers and went even further. After determining segmental weights and centers of gravity he measured the specific gravity of each segment. All of these data are presented in Appendix 4, Tables A4.1 through A4.4. Dempster's cadavers averaged approximately the same height as those used by Braune and Fischer but averaged about 4 kg less in "weight" and were considerably older.

Dempster also measured the volumetric displacement of limb segments of 38 young men, 6 rotund, 11 muscular, 10 thin, and 11 of medium physique (see below for description of this technique). Table A4.5, Appendix 4, presents fractions of body weight calculated from these data by taking the product of each segmental displacement (weighed to the nearest gram) and the specific gravity of the segment. The average was then taken for each segment in each group. The availability of these data makes it possible to calculate the moment of force exerted by the weight of each segment of the body. As the center of gravity (CG) is located at the point where the sum of all moments is equal to zero, it becomes a simple matter to locate this center by the mathematical composition of parallel forces.

When using any table of fractions of body weight, it is not necessary to calculate the moment of each segment: all that is necessary is to take the product of the fraction of body weight given in the table and the moment arm of that segment. The sum of these figures presents the length of the moment arm (*MA*) of the CG from the selected axis.

It should be remembered that all of these data are based entirely on measurements made from men and so do not yield as accurate results when applied to women. The authors have found that there may be a discrepancy of several centimeters between the location of the primary transverse plane (CG) as calculated by the segmental method when compared with that determined from the supine position on the gravity board.

Immersion Technique for Determining Segmental Weights. Greater accuracy may be obtained by the limb immersion technique. Figure 5.16 illustrates Dempster's immersion tanks, which are made of stainless steel. Versions less expensive but just as accurate can be made of plastics by any good plastics workshop (Fig. 5.17). Table A4.3, Appendix 4, presents Dempster's landmarks and directions for the immersion technique. These joint centers, detailed in Table A4.3, may be marked on the subject with a skin pencil or felt point

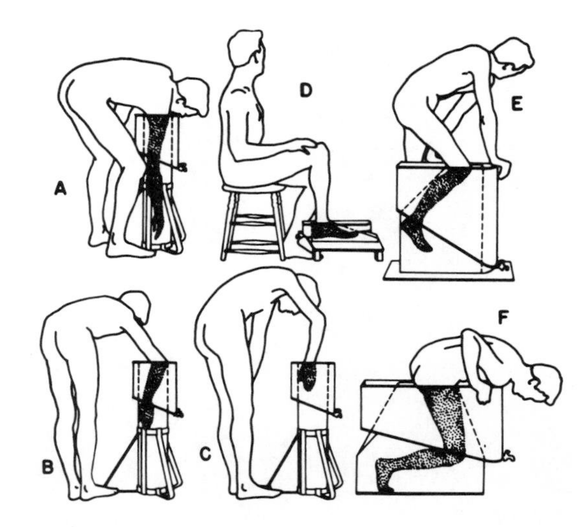

Figure 5.16. Immersion tanks used by Dempster (1955).

[b] The above quotation is an accurate reproduction from *The Physiology of Physical Education* by Percy Dawson, M.D. Dr. Dawson was a member of a group known as the Simplified Spelling Board and insisted on this format for what was, at the time, an excellent physiology text.

pen so that when the limb is immersed it can be lowered only as far as the desired mark. The displaced water is then drained from the jacket into a tared vessel and weighed to the nearest gram or ounce. This weight of displaced water is then multiplied by the specific gravity of the segment measured (Table A4.4, Appendix 4) to obtain the segmental weight. There will, of course, still be some slight inaccuracy as the specific gravities were obtained from elderly male cadavers, but this is not considered significant at this time. Hopefully some day similar specific gravity measures for women will become available. However, comparison of the CG location for women (as well as men) supine on the gravity board with that determined by the above technique on the same subject and in a similar posture shows a closer agreement than with the other methods.

Mathematical Determination of Segmental Weights. A reasonably accurate mathematical derivation of segmental weight has been devised by Barter (1957) of the United States Air Force. He used data from both Braune and Fischer's and Dempster's reports on cadaver measures and developed the series of regression equations presented in Table A4.6, Appendix 4. This material is also available in Duggar (1966), who states that "The standard deviations of the residuals do not incorporate the uncertainty of the regression value itself based on only 12 cadavers, so that the total error of prediction will be greater for body weights differing from the average cadaver weight, 59.4 kg. Extrapolation of these regression equations to include persons weighing more than the heaviest cadaver, 75.3 kg, is unreliable."

Complete accuracy in locating the CG is not possible at this time. ". . . The most exacting measurements cannot locate the CG of a particular man to within 0.3 cm" (Duggar (1966)). However, a fair estimate of the accuracy of the segmental calculations can always be made by comparing the results of such calculations made from a standing posture with those obtained with the subject supine on the gravity board. "Calculations based on appropriate segment data and applied to particular subjects whose (link) length and posture can be measured should be accurate to within 1 cm" (Duggar (1966)).

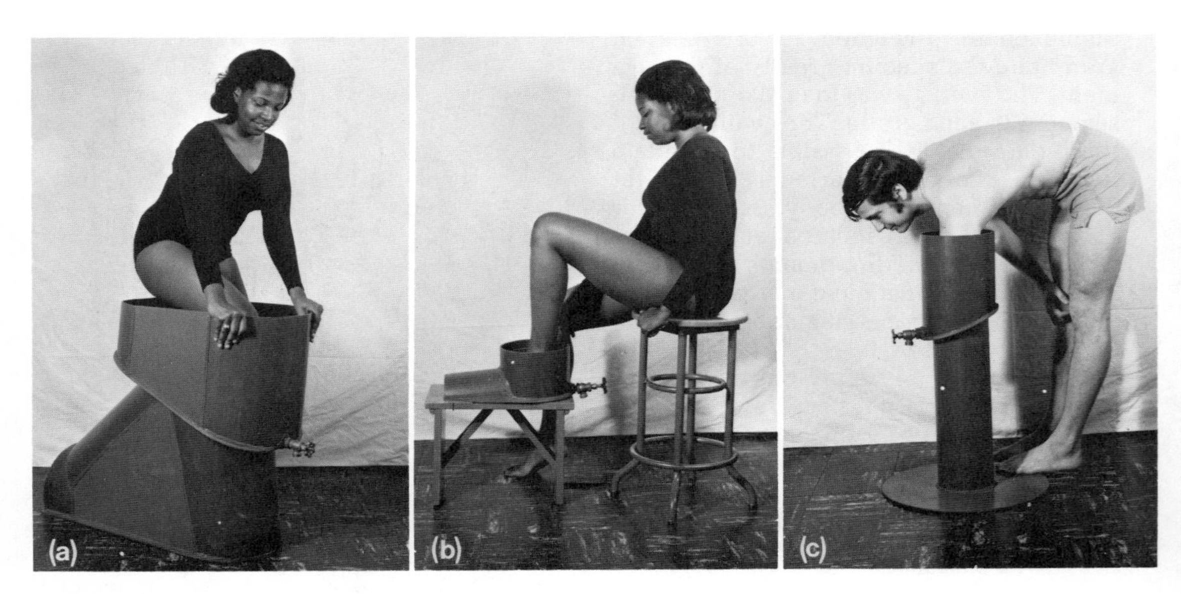

Figure 5.17. Immersion tanks made of plastic.

EXAMPLE 9—Calculation of Center of Gravity of Body Segments

The mass of a man's hand is 0.4 kg and the hand measures 18 cm from tip to wrist when extended. The mass of the forearm is 0.9 kg and is 26 cm long from the wrist to the elbow joint. It is desired to determine the location of the center of gravity of the combined forearm and hand.

Referring to the tables in Appendix 4 it is possible to establish the diagram in Figure 5.18. Table A4.3 (Appendix 4) indicates that the center of gravity of the forearm (CG_a) is located 43% of the forearm length measured from the elbow axis or 11.18 cm; the center of gravity of the hand (CG_h) is at 50.6% of the length of the extended hand from the wrist axis, which places the CG of the hand approximately 26 cm + 9.11 cm = 35.11 cm from the elbow axis.

The total downward force W_T of the hand and forearm is

$$W_T = W_a + W_h$$

where W_a = weight of the arm (forearm) alone and W_h = weight of the hand, i.e.,

$$W_T = 8.83 \text{ N} + 3.92 \text{ N} = 12.75 \text{ N}$$

which acts at CG_T.

Invoking the principle of moments,

$$W_T \times MA_{W_T} = \Sigma M$$

i.e., the moment of force due to the combined weights of the arm and hand must equal the sum of the moments of the individual weights.

The left side of the equation becomes

$$12.75 \text{ N} \times X \text{ where } X = MA_{W_T}$$

The right side is

$$(8.83)(11.18) + (3.92)(35.11)$$

Therefore:

$$12.75\ X = 98.72 + 137.63$$

$$12.75\ X = 236.35 \text{ N cm}$$

$$X = 18.54 \text{ cm}$$

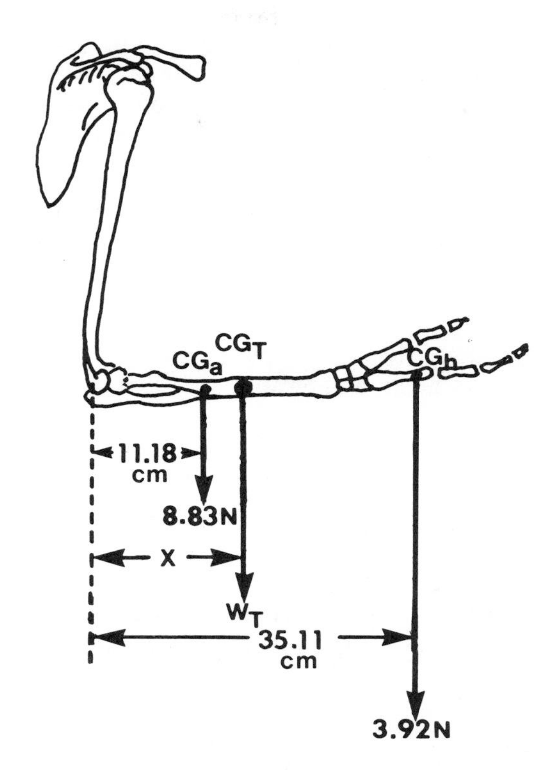

Figure 5.18. Illustration for Example 9.

The center of gravity of the combined forearm and hand lies 18.54 cm distal to the axis through the elbow.

It should be recognized that the calculation just performed will, at best, produce a good estimate of the location of the center of gravity of the combined body segments. This is so since *in vivo* measurements on a particular living subject are not easy to perform, and reliance has to be placed on estimates arising from data accumulated by workers such as Dempster (1955) and along the lines reflected in Appendix 4.

EXAMPLE 10—Shift in Center of Gravity with Relative Body Movement

If the hand of a man in Example 9 is flexed to 90° at the wrist, by how much would the position of the combined arm and hand CG shift in the X direction?

In this case the same mathematical relationships as indicated in Example 9 will obtain, but the CG of the hand will be located at 26 cm (approximately) from the elbow. Let the position of the combined CG relative to the elbow be X'. Then

$$12.75\ (X') = (8.83)(11.18) + (3.92)(26)$$
$$12.75\ (X') = (98.72) + (101.92)$$
$$X' = 15.74 \text{ cm}$$

The shift in center of gravity is

$$\Delta X = X - X'$$
$$\Delta X = 18.54 - 15.74$$
$$\Delta X = 2.8\text{cm}$$

i.e., there is a shift of 2.8 cm toward the elbow.

EXAMPLE 11—Location of Line of Gravity by Segmental Method

Figure 5.19 shows tracings made from film strips synchronized when using two mutually perpendicular cameras. The particular pose is a segment of a gymnastic stunt, the scale, and known in ballet as an arabesque.

It is desired to determine the positioning along the X axis (in Fig. 5.19(a)) of the gravity line for the portion of the body which is supported by the right hip joint, using the segmental method alluded to earlier.

To begin, Figure 5.20, a link diagram of Figure 5.19(a), is constructed to scale using the performer's measurements. Next, the horizonal or X moment arms in the sagittal plane for each segment are measured and the moments about the origin (intersection between X and Z axes located at the right hip) are computed. Those moments occurring to the right of the supporting hip axis will cause a clockwise (CW) rotation of the trunk on the right femur and are by convention considered as positive. Those moments occurring to the left, on the other hand, will cause counterclockwise (CCW) moments and are considered as negative; i.e.,

$$\begin{aligned}\Sigma M = {} & (W_{ut} \times MA_{ut}) + (W_{lt} \times MA_{lt} \\ & + (W_h \times MA_h) - (W_{lth} \times MA_{lth}) \\ & - (W_{ll} \times MA_{ll}) - (W_{lf} \times MA_{lf}) \\ & + (W_{ra} \times MA_{ra}) + (W_{rfa} \times MA_{rfa}) \\ & + (W_{rha} \times MA_{rha}) + (W_{la} \times MA_{la}) \\ & - (W_{lfa} \times MA_{lfa}) + (W_{lha} \times MA_{lha})\end{aligned}$$

where W is weight of the segment, MA is the moment arm of the segment, and the lower case subscripts identify the segment: ut, upper trunk; lt, lower trunk; h, head; lth, left thigh; ll, left leg; lf, left foot; ra, right arm; rfa, right forearm; rha, right hand; la, left arm, lfa, left forearm, lha, left hand.

Table 5.2 summarizes the data and facilitates the required calculations.

In any parallel force system the product of the resultant and its moment arm is equal to the sum of the moments; i.e., $R \times MA_R = \Sigma M$. From Table 5.2, R is approximately 493.8 N; then

$$493.8 \text{ N} \times MA_R = 2302.8 \text{ Ncm}$$
$$MA_R = 4.66 \text{ cm}$$

Thus, the gravity line passes through a point 4.66 cm anterior to the right hip (origin).

$$\Sigma M = 2302.8 \text{ Ncm}$$

Should it be necessary to locate the center of gravity of the portion of the body supported about the right hip joint, the same process would be followed, calculating the Z or vertical moments and then the Z position of the center of gravity. Also, we have information about the center of gravity only in the sagittal plane. Frontal plane information can be derived from the tracing in Figure 5.19(b) and by carefully describing mathematically body components in that plane; the Y component can be calculated in a similar vein. For repetitive calculations involving a variety of body poses, computer-aided approaches are indicated.

EXAMPLE 12—Shift in Line of Gravity with Position

Figure 5.21 shows an altered scale position in link diagram form. Note that

Figure 5.19. Tracings of a scale from film. (a) from the side; (b) from the front.

Table 5.2
Segmental Data: X Moments (see Fig. 5.20)

Segment	Weight	Plus Moment Arms	Plus Force Moments	Minus Moment Arms	Minus Force Moments
	N	*cm*	*Ncm*	*cm*	*Ncm*
Upper trunk	105.5	18.5	1951.8		
Lower trunk	132.6	1.7	225.4		
Head	49.4	27.9	1378.3		
L thigh	89.2			25.4	509.3
L leg	31.6			61.9	1956.0
L foot	8.5			88.9	755.7
R arm	23.4	46.9	1097.5		
R forearm	11.3	70.6	797.8		
R hand	3.8	91.4	347.3		
L arm	23.4	4.5	105.3		
L forearm	11.3			19.5	220.4
L hand	3.8			41.9	159.2
Total	493.8		5903.4		3600.6
					$\Sigma = 2302.8$

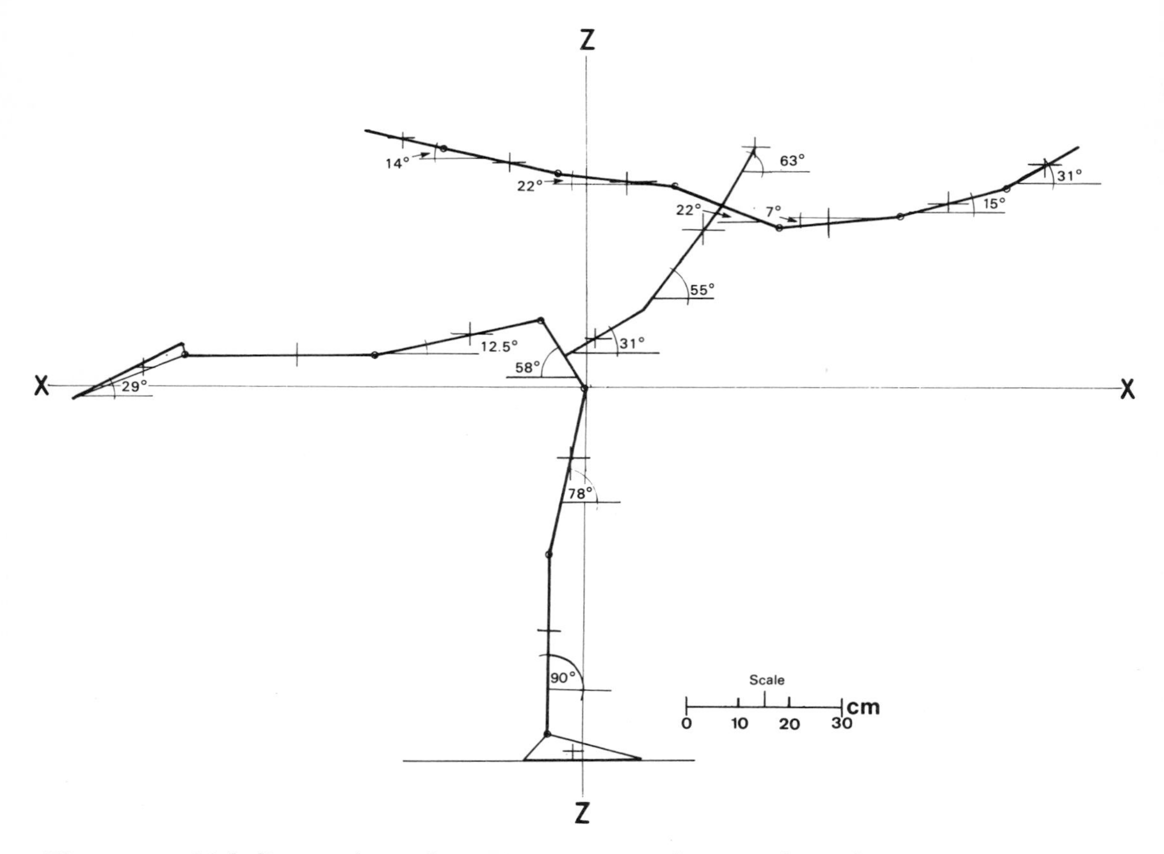

Figure 5.20. Link diagram drawn from Figure 5.19 to scale, using the performer's measurements.

both arms are flexed at the shoulder in this case. It is desired to determine the shift in the gravity line from the situation shown in Figure 5.20.

Table 5.3 shows the tabulated calculations in a manner similar to Table 5.2. i.e.,

$$R \times MA_R = \Sigma M$$

$$MA_R = (\Sigma M)/R$$

$$MA_R = 8412.6/493.8$$

$$MA_R = 17.04 \text{ cm}$$

Thus, the line of gravity is now 17.04 cm anterior to the hip axis. There has, therefore, been a comparative anterior shift relative to the conditions depicted in Figure 5.20 of 17.04 − 4.66 or 12.38 cm.

$$\Sigma M = 8412.6 \text{ Ncm}$$

EXAMPLE 13—Estimation of Total Muscle Force Exerted by Hamstrings

For the subject depicted in Figures 5.19 and 5.20 it is desired to determine the muscle force exerted by the hamstrings which must play a major role in maintaining the subject's pose. Next, the effect on the hamstring, force required for the modified position shown in Figure 5.21 is to be determined.

It should be realized that it is impossible to measure moment arms of muscles or to indicate a muscular line of pull with complete accuracy *in vivo*. If a skeleton or a cadaver could be positioned exactly as the subject, measures of moment arms and angles could be made and used. The next best method is to draw the skeletal struc-

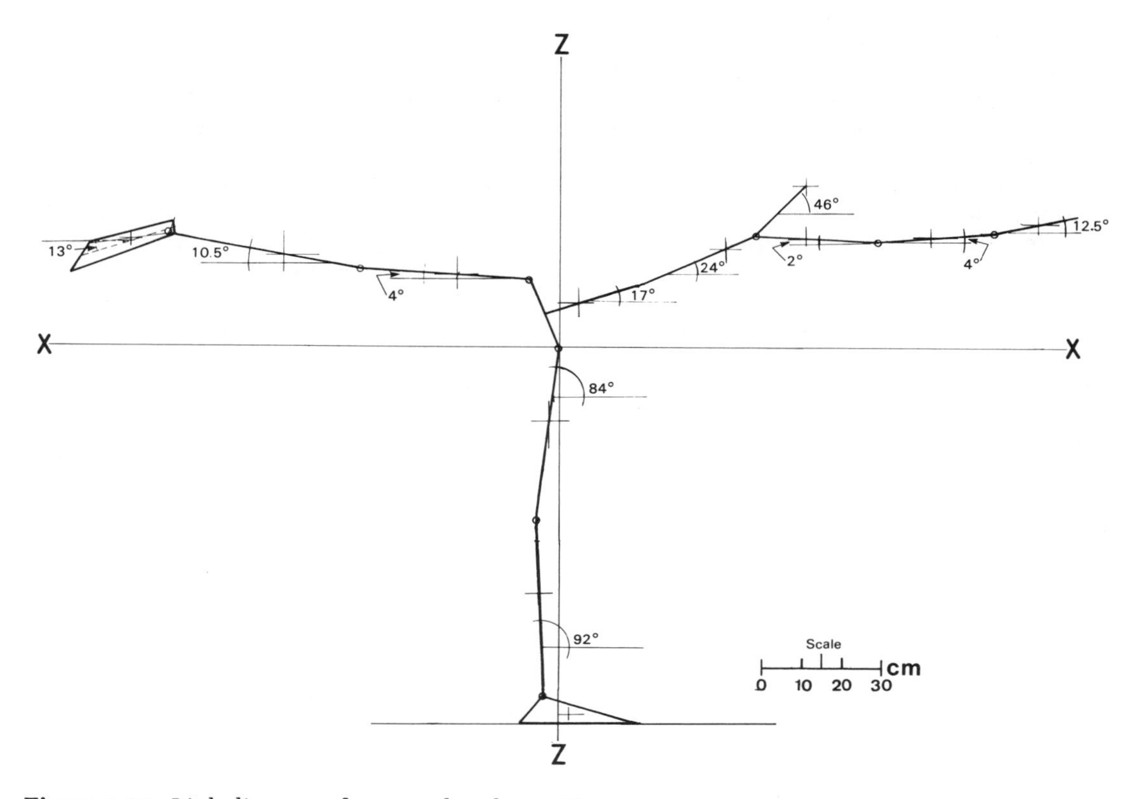

Figure 5.21. Link diagram of a second scale position.

Table 5.3
Segmental Data: X Moments (see Fig. 5.21)

Segment	Weight	Plus Moment Arms	Plus Force Moments	Minus Moment Arms	Minus Force Moments
	N	*cm*	*Ncm*	*cm*	*Ncm*
Upper trunk	105.5	34.3	3618.7		
Lower trunk	132.6	5.7	755.8		
Head	49.4	61.6	3043.0		
L thigh	89.2			30.5	2720.6
L leg	31.6			69.9	2208.8
L foot	8.5			107.9	917.2
Both arms	46.7	80.6	3764.0		
Both forearms	22.7	93.9	2131.5		
Both hands	7.6	124.5	946.2		
Total	493.8		14259.2		5846.6
					$\Sigma = 8412.6$

tures within the body outline as accurately as possible and to work from these. This has been done in Figures 5.22(a) and 5.23(a), which provide a means whereby appropriate approximations and estimates can be made. The essence of the approach to be made is in the proper construction of a force diagram which represents the situation of concern.

To properly construct the force diagram we must know the horizontal distance of the line of pull of the hamstrings from the hip axis as well as the similar distance to the line of gravity. A horizontal line is drawn through the hip axis of Figures 5.22(a) and 5.23(a) and another through the hamstrings from the femoral condyle (which alters the line of pull somewhat from attachment to attachment) to the attachment on the ischial tuberosity. The intersection of this line with the horizontal gives the distance from the hip axis as well as the angle the line of pull makes with the horizontal. As we have already located the line of gravity (the X or sagittal component of the center of gravity) in each case of concern (Examples 11 and 12), we can proceed with the force diagrams as shown in Figures 5.22(b) and 5.23(b).

For Figure 5.22 the line of gravity falls 4.66 cm anterior to the hip axis. This is shown in Figure 5.22(b) as the line R located the perpendicular distance MA_R from the hip joint axis A. By measurements

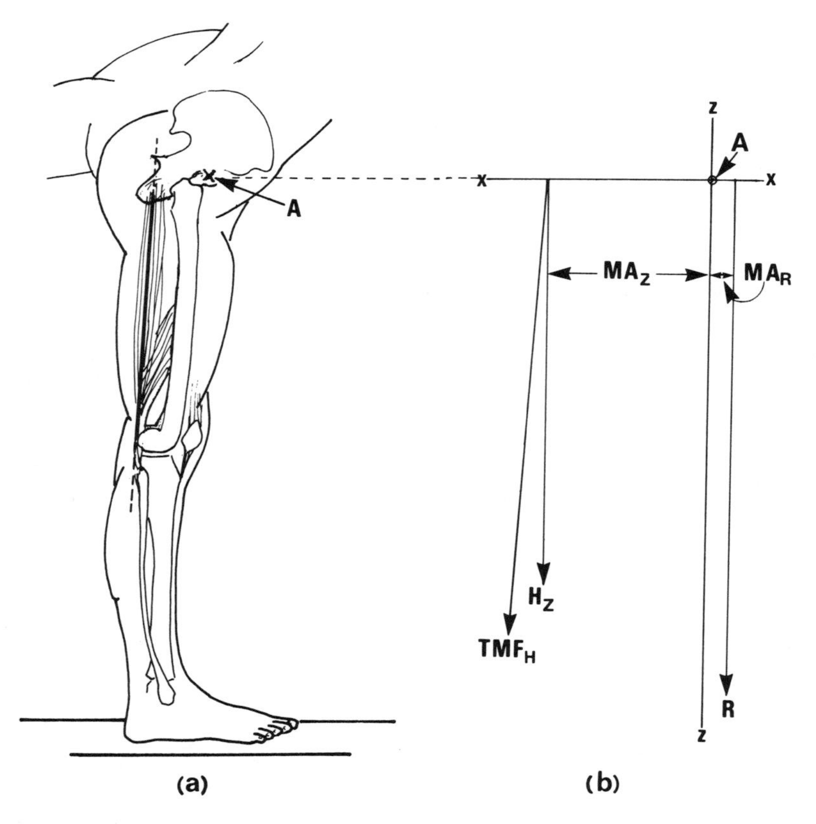

Figure 5.22. Supporting limb.(a) details pertinent to Figure 5.19 with bony structures and hamstring action line. (b) force diagram for determining hamstring TMF. x — x, horizontal coordinate through hip axis A; z — z, vertical coordinate through hip axis; MA_R, moment arm of the resultant weight of the supported body segments, 4.66 cm; R, line of force (action line) of the resultant weight; TMF_H, line of force of hamstrings; MA_Z, moment arm of Z component of hamstring force; H_Z, Z component of hamstring force.

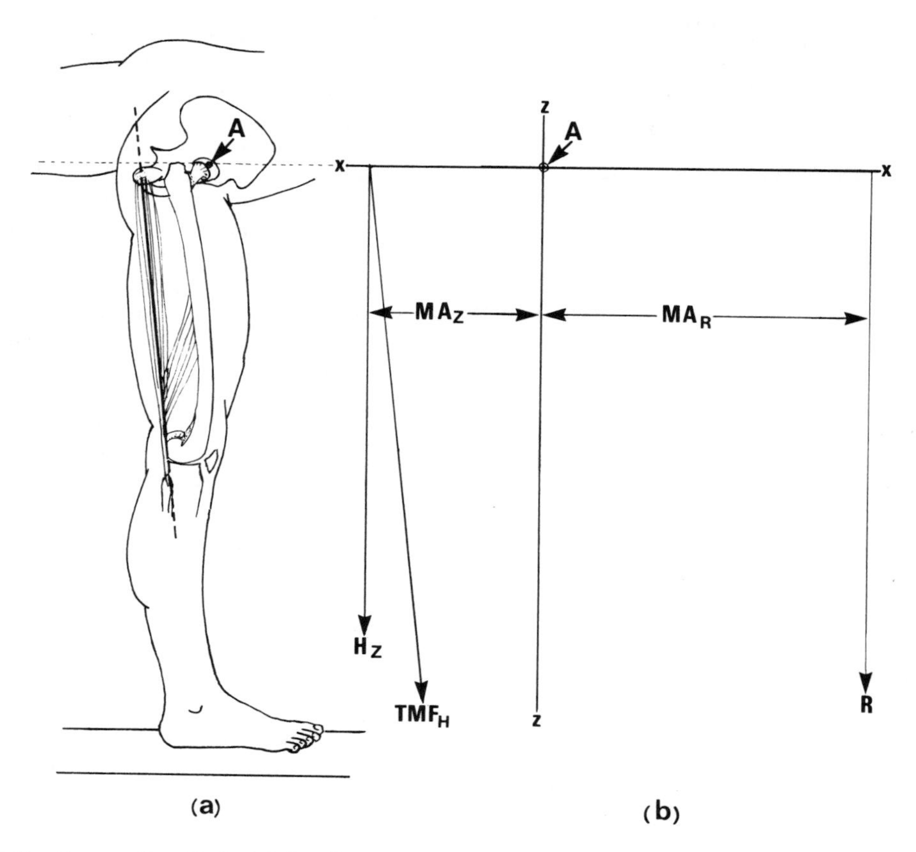

Figure 5.23. Supporting limb. (a) details pertinent to Figure 5.21 with bony structures and hamstring action line. (b) force diagram for determining hamstring *TMF*, x − x, horizontal coordinate through hip axis *A*; z − z, vertical coordinate through hip axis; MA_R, moment arm of the resultant weight of the supported body segments, 17.04 cm; *R*, line of force (action line) of resultant weight; TMF_H, action line of hamstrings; MA_Z, moment arm of *Z* component of hamstring *TMF*; H_Z, *Z* component of TMF_Z.

made upon Figure 5.22(a) we find that the line of pull of the hamstrings crosses the horizontal 8.6 cm posteriad and forms an angle of 81° with the horizontal.

Thus, TMF_H, which represents the total muscle force of the hamstrings can be located as shown, its distance from *A* on the *X* axis being MA_Z.

The gravitational forces on all of the body segments supported by the right hip joint exert a clockwise moment about that joint while the supporting muscles (hamstrings) exert a counterclockwise moment.

Since this is an equilibrium situation where $\Sigma M = 0$, we can solve for the unknown force (TMF_H). The vertical component of TMF_H is depicted in Figure 5.22 by H_Z. For equilibrium then, it follows that

$$R \cdot MA_R - H_Z MA_Z = 0$$

i.e.,

$$H_Z = R \cdot \frac{MA_R}{MA_Z}$$

substituting the values of these elements

$$H_Z = 493.8 \times 4.66 / 8.60$$

$$H_Z = 267.57 \text{ N}$$

From trigonometric considerations, it follows that

$$\text{TMF}_H = \text{H}_Z/\sin 81°$$

$$\text{TMF}_H = 267.57/0.9877$$

$$\text{TMF}_H = 270.90 \text{ N}$$

Proceeding in a similar fashion for the situation shown in Figure 5.23, the line of gravity falls 17.04 cm anterior to the hip axis, and by measurement upon Figure 5.23(a) the hamstring line of force crosses the horizontal through the hip axis 9.1 cm posteriad and forms an angle of 84° with it.

Thus,

$$H_Z = 493.8 \times 17.04 / 9.1$$

$$H_Z = 924.65 \text{ N}$$

and

$$\text{TMF}_H = \text{H}_Z/\sin 84°$$

$$\text{TMF}_H = 924.65/0.9945$$

$$\text{TMF}_H = 929.76 \text{ N}$$

It is interesting to note that this is about 3.5 times the muscular effort required in the first position shown in Figure 5.22.

It should be realized, however, that these figures are derived by making a number of assumptions, foremost of which is that the gluteus maximus is not participating; second, that the moment arm of the biceps femoris as measured from the drawing is also that of the semimembranosus and semitendinosus muscles (i.e., the entire hamstring group); and third, that the drawings used were at least moderately accurate for their purpose. We have also reduced a problem to one in which only two dimensions are considered. For more exactness and specificity it would be necessary to calculate all of the force components along the mutually perpendicular three axes *X*, *Y*, and *Z*.

EXAMPLE 14—Estimation of Muscle Force Exerted by the Abductors: Gluteus Medius, Gluteus Minimus, and Tensor Fascia Lata

It is required to analyze and determine the muscle forces exerted b the abductor group acting about the hip joint.

There are three abducting muscles: gluteus medius, gluteus minimus, and tensor fascia lata. Their lines of pull all converge at or near the greater trochanter (Fig. 5.3). Rather than selecting an arbitrary resultant action line, we can compose these three lines of force into a single resultant, basing our solution on the work done by Inman (1947) (see Example 3). Figure 5.24 is also estimated from Inman's (1947) description, and the angles components a, b, and c form with the horizontal are measured. With these items in mind, we can build a vector diagram of the three muscles and, by following any one of the three different methods discussed in Example 3 determine the resultant line of abduction force.

The algebraic approach for determining this force seems to be the most appropriate. The first step of this solution involves resolution of each force vector into its horizontal or *Y* component (which lies in the frontal plane) and its vertical or *Z* component. The *Y* component is always equal to the cosine of the angle that the vector makes with the horizontal multiplied by its magnitude, while the *Z* component is equal to the sine of the same angle times its magnitude. As the *Y* and *Z* ordinates are drawn through the point of convergence of the three vectors (Fig. 5.24), we see that two of the vectors fall in the first quadrant and so will have positive *Y* and *Z* values. The third vector, however, falls in the second quadrant and so will have a negative *Y* value while the *Z* value remains positive.

Starting with the horizontal or *Y* component:

$$R_Y = \Sigma f_Y = \Sigma f_Y \cos \theta = 14 \cos 55° + 29 \cos 73° - 57 \cos 91°$$

$$R_Y = 14\ (0.5736) + 29\ (0.2924) - 57\ (0.0175)$$

$$R_Y = 8.03 + 8.48 - 0.99$$

$$R_Y = 15.52$$

Similarly:

$$R_Z = \Sigma f_Z = \Sigma f \sin \theta$$

$$R_Z = 14 \sin 55° + 29 \sin 73° + 57 \sin 89°$$

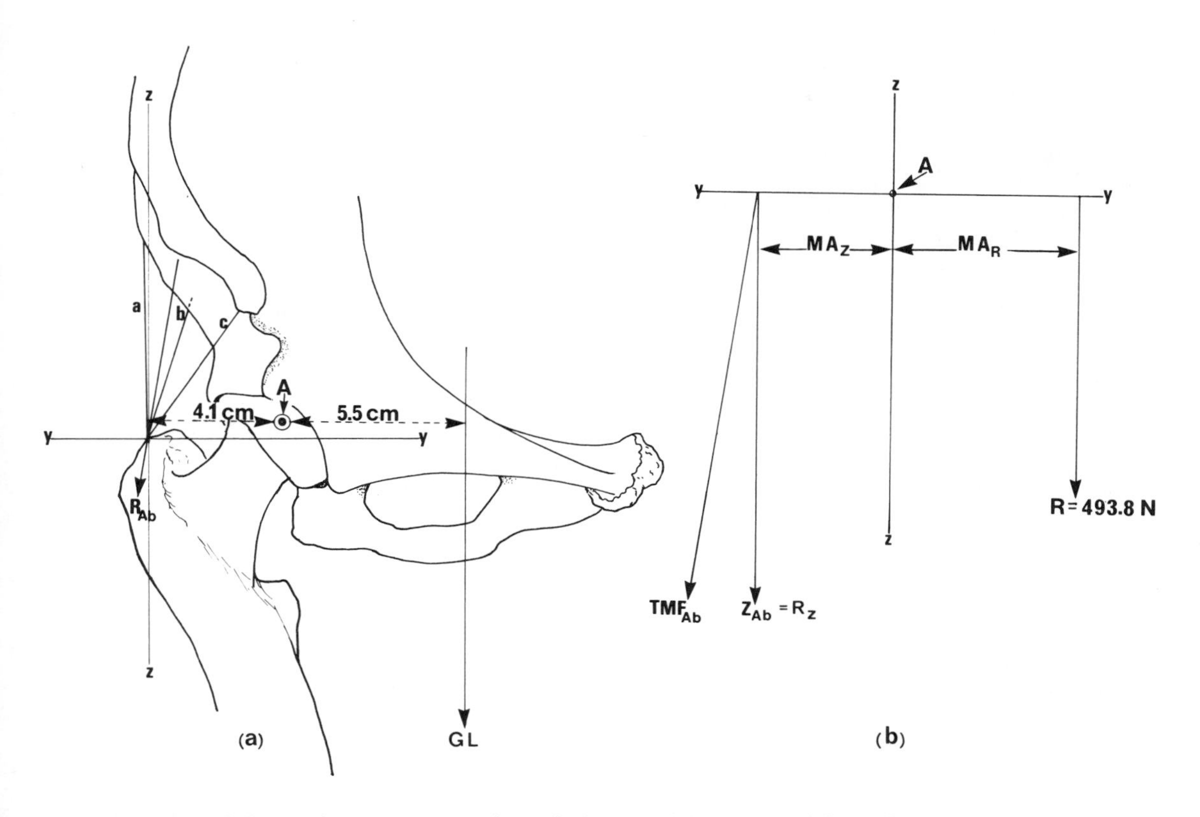

Figure 5.24. Abductor forces acting at the right hip. (a) enlargment of the right pelvis and proximal femur from Figure 5.19.(b). a, action line of the gluteus medius; b, action line of the gluteus minimus; c, action line of the tensor fascia lata; *GL*, gravity line of the lever; R_{ab}, total muscle force of abductors. *Y* and *Z* coordinates drawn through the point of convergence of the three action lines. (b) force diagram of the forces acting at the hip in the frontal plane, *Y* and *Z* coordinates through the hip axis *A*; *R*, the action line of the resultant weight of the lever supported by the hip; MA_R, moment arm of the resultant; TMF_{ab}, total muscle force exerted by the three abductors; Z_{ab}, the vertical or *Z* component of the abductor force; MA_z, moment arm of the *Z* component.

$$R_Z = 14\ (0.8192) + 29\ (0.9563) + 57\ (0.9999)$$

$$R_Z = 11.47 + 27.72 + 56.99$$

$$R_Z = 96.19$$

θ, the angle that the abducting resultant makes with the horizontal is

$$\theta = \arctan R_Z/R_Y$$

$$\theta = \arctan 96.19/15.52$$

$$\theta = 80.83°$$

Assuming that each of the three abductor muscles contributes proportionately to maintaining the pose, we may now build our force diagram and calculate the total muscle force (TMF) exerted by the abductors. As before, the weight of the body supported by the right hip is about 493.8 N. Examination of Figure 5.19(b) shows the pelvis with an upward tilt at the right supporting hip. It is not necessary to sum the moments of each segment as we know that the gravity line must fall within the base of support, the right foot. Experience has shown us that if a subject takes this pose on a gravity board to determine the primary sagittal plane, the scale needle (or beam) is constantly vibrating, indicating lateral sways of the body over the foot. Thus, we can safely assume a vertical line passing through the foot as a line of gravity (Fig. 5.25). On Figure 5.24(a) the horizontal distance mediad from the hip axis is 5.5 cm, while that from the abductor resultant is 4.1 cm laterad from the axis. With this information we can draw the force diagram in Figure 5.24(b) and proceed with the problem.

For equilibrium about the joint $\Sigma M = 0$ and therefore the vertical muscle component

$$R_Z = 493.8 \times 5.5 / 4.1 = 662.4 \text{ N}$$

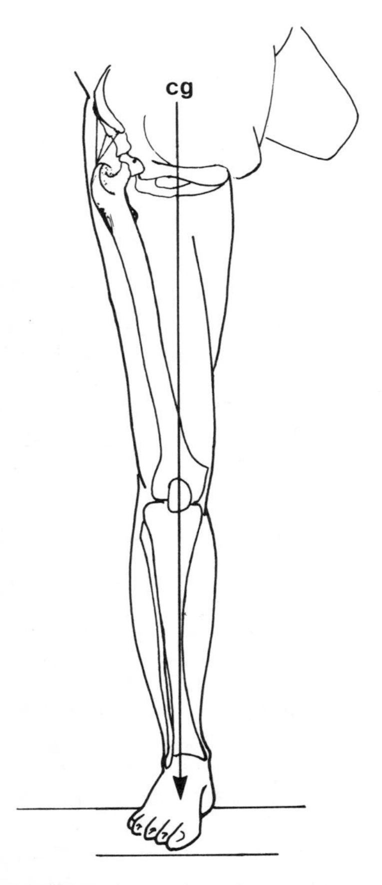

Figure 5.25. Supporting limb, (see Fig. 5.19(b)) showing right pelvis and femur.

and the total muscle force of the abductors

$$\text{TMF}_{ab} = R_Z / \sin 80.83°$$

i.e.,

$$\text{TMF}_{ab} = 662.4 / 0.9864$$

$$\text{TMF}_{ab} = 671.5 \text{ N}$$

EXAMPLE 15—Compression Forces at the Acetabulum

It is required to determine the compression forces acting on the acetabulum for the static subject depicted in Figures 5.19 and 5.20. In effect, the amount of force with which the head of the femur is pressing against the joint is to be calculated. Such force is a result of the combination of body weight and muscle actions. The muscular actions to be considered are the hamstring forces and the abductor forces.

At the outset, the resultant of the hamstrings force and the effective body weight borne at the hip joint will be determined.

From Figure 5.22(b) and the pertinent calculations of TMF_H depicted in Example 13, it is evident that the resultant force acting at the hip joint, A, must be the vector sum of TMF_H and R, the resultant weight.

Resolving the forces into horizontal and vertical components, there results:

Horizontal components:

$$\text{Horiz. comp.} = TMF_H (\cos 81°)$$
$$= -270.90 (.1564)$$
$$= -42.4 \text{ N}$$

N.B. With reference to Fig. 5.22, the direction is negative.

Vertical components:

$$\text{Vert. comp.} = \text{body weight} + TMF_H (\sin 81°)$$
$$= -493.8 + (-270.9)(.9877)$$
$$= -761.4 \text{ N}$$

N.B. With reference to Fig. 5.22, the direction is negative.

Resultant force:

$$R = \sqrt{(\text{Horiz. comp.})^2 + (\text{Vert. comp.})^2}$$

$$= \sqrt{(-42.4)^2 + (-761.4)^2}$$

$$= 762.6 \text{ N}$$

$$\theta = \arctan \frac{\text{Vert. comp.}}{\text{Horiz. comp.}}$$

$$= \arctan \frac{-761.4}{-42.4}$$

$$= 86.81°$$

i.e., the resultant of the hamstrings and effective body weight is 762.6 N directed at 180° + 86.81° relative to the horizontal axis. The reaction force due to this at the acetabulum is directed in the opposite direction (i.e., 180° away). Thus, the acetabular force is 652.6 N at 86.81° to the horizontal (X) axis in the sagittal ($X - Z$) plane (see Fig. 5.22).

TMF_{ab} was found to be 671.5 N at 180° + 80.83° in the $Y - Z$ plane; the acetabular component will be 671.5 N at 80.83°.

If it is argued that this force occurs at the same time as the previously determined acetabular force, the resultant force acting on the acetabulum will be the vector sum of these two forces in three-dimensional space.

Resolving into X, Y, and Z components as follows:

X: $(R_{\text{horiz.+vert.}})(\cos 86.81°)$
$(762.6)(.0556) = 42.4$ N

Y: (TMF_{ab}) $(\cos 80.83°)$
$(671.5)(.1594) = 107.0$ N

Z: $(R_{h+v})(\sin 86.81°) + (TMF_{ab})$
$(\sin 80.83°)$

$(762.6)(.9985) + (671.5)(.9864)$
$761.4 + 662.3 = 1423.7$ N

The resultant force

$$R^2 = X^2 + Y^2 + Z^2$$

i.e.,

$$R = \sqrt{(42.4)^2 + (107.0)^2 + (1423.7)^2}$$

$$= \sqrt{2{,}040{,}929.7}$$

$$= 1428.3 \text{ N}$$

The resultant direction in the $X - Y$ plane is arctan 107.0/42.4 with respect to the X axis, i.e., 68.38°.

The resultant direction in the $Y - Z$ plane with respect to the Y axis is arctan 1423.7/107.0, i.e., 85.70°.

EXAMPLE 16—Projectile Problem

In shooting a basketball free throw, obviously the goal is to put the ball through the hoop. "If it doesn't go through the hoop, then we need to know why. Should we change the amount of force (change velocity), should we change the angle of release, or should we change both?" (Widule (1974)). Of course, there are no answers to these questions until we know some other facts. For example, the distance from the free throw line to the basket is 3.96 m (13 ft) and the basket rim is 3.05 m (10 ft) high. And a 2-m (6.6 ft) tall basketball player will release the ball at a height of about 2.8 m while a 1.22-m (4 ft) tall child will release it below the waist at, let's say, 0.8 m. So, how do we determine what velocity and what angle of release should be used by each of the basketball players?

First, we may start with some trigonometric identities that provide the basis for plotting the path of the projectile. If we use the origin of coordinates in Figure 5.26 as the point of release of the ball from the hands, we may establish a right triangle so that (1) the angle of release (with respect to the horizontal is θ; (2) the range (in this case, 3.96 m) is the adjacent side of the right triangle; (3) the product of initial velocity and time is the hypotenuse; and (4) the actual height with respect to the starting point plus the distance attributed to gravity is the opposite side. If we know the height of the starting point (S.P.), the height of the contact point (the basket or C.P.), and the time of flight, we may calculate the angle of release and the velocity of ball release.

Consider the following identities:

$$\cos \theta = x \,/\, V_0t \text{ or } x = V_0t \cos \theta$$

$$\sin \theta = y + \tfrac{1}{2}gt^2 \,/\, V_0t \text{ or}$$

$$y = V_0t \sin \theta - \tfrac{1}{2}gt^2$$

By substitution and rearrangement:

$$Tan\ \theta = \frac{y + \frac{1}{2}gt^2}{x}$$

and

$$V_0 = \frac{x}{t \cos \theta}$$

The time of flight for each of the players is 1.0 s. The appropriate values are substituted into the first formula as follows:

$$\text{Adult: } Tan\ \theta = (3.05 - 2.80) + \tfrac{1}{2}(9.81)\ (1.0)^2$$

$$\theta = 52.49°$$

$$\text{Child: } Tan\ \theta = (3.05 - 0.80) + \tfrac{1}{2}(9.81)\ (1.0)^2$$

$$\theta = 61.05°$$

Then, to find velocity, the second formula is used:

$$\text{Adult: } V_0 = \frac{3.96}{1(.6089)}$$

$$V_0 = 6.5 \text{ m/s}$$

$$\text{Child: } V_0 = \frac{3.96}{1(.4840)}$$

$$V_0 = 8.18 \text{ m/s}$$

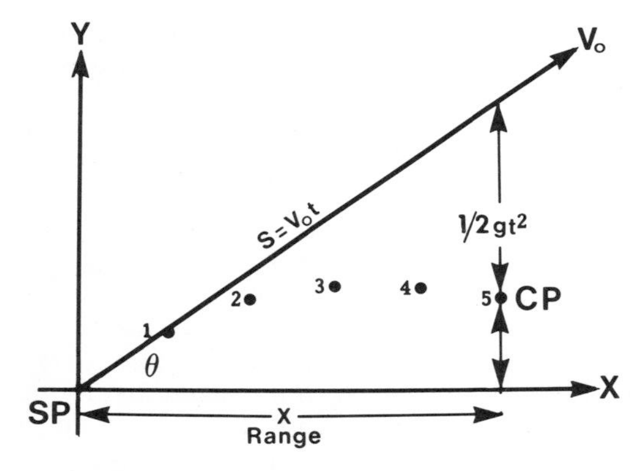

Figure 5.26. Projectile problem: basketball shot for goal. S.P. = starting point; C.P. = contact point; x = range; y = the difference in height between starting point and contact point; θ = angle of projection; V_0 = initial velocity; $\frac{1}{2}gt^2$ = displacement attributed to gravitational force. (Adapted from Widule, C.J., 1974. *Analysis of Human Motion.* West Lafayette, Ind.: Balt Publishers.)

In other words, the adult player selecting an angle of release of about 52.5° and a velocity of 6.5 m/s probably will make a goal. However, the child must release the ball at a much lower height and, therefore, the ball must travel a greater distance. Thus, the child must select a steeper angle of 61° and a higher velocity of 8.18 m/s in order to make a goal. The irony of this arrangement is that the child who is at a height disadvantage is also at a "muscle disadvantage" since the strength of the child probably is less than a fifth that of the adult.

It is obvious that a large range of release angles matched with their initial velocities may be calculated for several release heights. For the coach or player, information concerning angle of release becomes an important tool for improving performance. In addition, other important components of the basketball shot may be obtained. For example, if the angle and velocity of release are known, the horizontal and vertical components of the velocity may be determined and the time to reach the high point of flight and the high point's actual location may be determined. This type of information may be obtained for the adult as follows:

The horizontal component of the velocity:

$$V_h = V_0\ (\cos \theta)$$

$$V_h = 6.5\ (.6089) = 3.96 \text{ m/s}$$

The vertical component of the velocity:

$$V_v = V_0\ (\sin \theta)$$

$$V_v = 6.5\ (.7932) = 5.16 \text{ m/s}$$

The time to the high point may be calculated by recognizing that the high point is reached when vertical velocity minus the product of gravitational acceleration and time is zero. Therefore,

$$t_{HP} = \frac{V_v}{9.81}$$

$$t_{HP} = \frac{5.16}{9.81} = 0.53 \text{ s}$$

The location of the high point may be calculated both horizontally and vertically by using the product of velocity and time to obtain a distance. For horizontal location:

$$HP_{horiz} = V_h\ (t_{HP})$$

$$HP_{horiz} = 3.96\ (0.53) = 2.1\ \text{m}$$

The vertical location is found similarly, except that the acceleration due to gravity must also be taken into account. For vertical location:

$$HP_{vert} = [V_v\ (t_{HP})] - \tfrac{1}{2}gt^2$$

(N.B. The gravity value is calculated for t_{HP}.)

$$HP_{vert} = [5.16\ (0.53)] - 1.38 = 1.4\ \text{m}$$

That is, the high point is located 2.1 m horizontally and 1.4 m vertically from its starting point.

The manipulation of equations to solve problems with projectiles, even when the projectile is human, is a useful tool for the physical educator. Equations of uniformly accelerated motion form the basis for the several formulae and were first introduced in Chapter 3. They are recounted here as follows:

$$V_f = V_0 + at$$

$$S = V_0t + \tfrac{1}{2}at^2$$

$$(V_f)^2 = (V_0)^2 + 2as$$

PROBLEMS

Having been exposed to a number of examples that apply mechanical principles to the human body under conditions of either static equilibrium or whole body analyses, the reader may wish to test comprehension of this material with some sample problems. As with the other problems, the solutions are provided in Appendix 1, and the reader is urged to compare solutions and answers only after making a sincere attempt to answer each problem.

Problem 1. The first example in this chapter (Fig. 5.8), poses a problem concerned with concurrent forces having a common origin. Two heads of the gastrocnemius are analyzed in terms of their contributions to plantar flexion under different conditions of force contributions. This problem adds a third concurrent force, that of the contribution of another plantar flexor, the soleus. If the soleus muscle was added to the model (vector $\overrightarrow{AE}$) so that its angle with respect to the right horizontal was 85°, and the two heads of gastrocnemius are at angles of 100° and 70° as before, what is the magnitude and direction of the resultant (identified with respect to the horizontal) under the following conditions? (a) when the magnitude of each of the three heads of the triceps surae is 1000 N; (b) when the magnitude of $\overrightarrow{AB}$ is 1000 N, $\overrightarrow{AC}$ is 1200 N, and $\overrightarrow{AE}$ is 1000 N.

Problem 2. In the early 1970s, hip prostheses were designed with short femur necks to reduce the torque experienced about the vertex of the prosthesis. Figure 5.27 illustrates the comparison of the normal and shorter femur necks. In Figure 5.27(a), the torque of the hip abductors (force F multiplied by its moment arm d_1) must equal the torque of the body weight (force of the structures supported by the hip W and the respective moment arm d_2), in order to have equilibrium. Obviously, the force of the hip abductors F' must be greater in (b) because of the shorter moment arm d'_1 if equilibrium is to occur. Therefore, the hip abductors have to produce abnormally large forces to maintain rotational equilibrium.

Compare the forces required based on the following data. For the normal length of femur neck (Fig. 5.27(a)), $W = 700$ N, $d_1 = 6$ cm, and $d_2 = 12$ cm. For the shortened femur neck (Fig. 5.27(b)), the body weight and its moment arm are the same, but d'_1 is only 3 cm long. What class of lever is involved and what is the magnitude and direction of the force on the head of the femur in each example?

Problem 3. You have been asked to analyze isometric elbow flexion, when the elbow is at 90° of flexion, in a subject with a complete lesion of the musculocutaneous nerve (Fig. 5.28). There are two resistances, that of the weight in the hand

(66.72 N) and that of the weight of the forearm and hand (18.24 N). (a) Identify the class of lever operating during isometric elbow flexion and give at least two reasons for your choice. (b) Calculate the resistance torque. (c) Calculate the effort force. (d) Calculate the resultant joint force and its angle of application.

Problem 4. In subjects with patella-femoral joint pain, a quadriceps resistance force is applied to assay the amount of discomfort and determine if pain is correlated with the angle of the knee joint (Fig. 5.29). (a) Determine the patella-femoral compression force at each of 30° and 90° of knee flexion by using the graphical

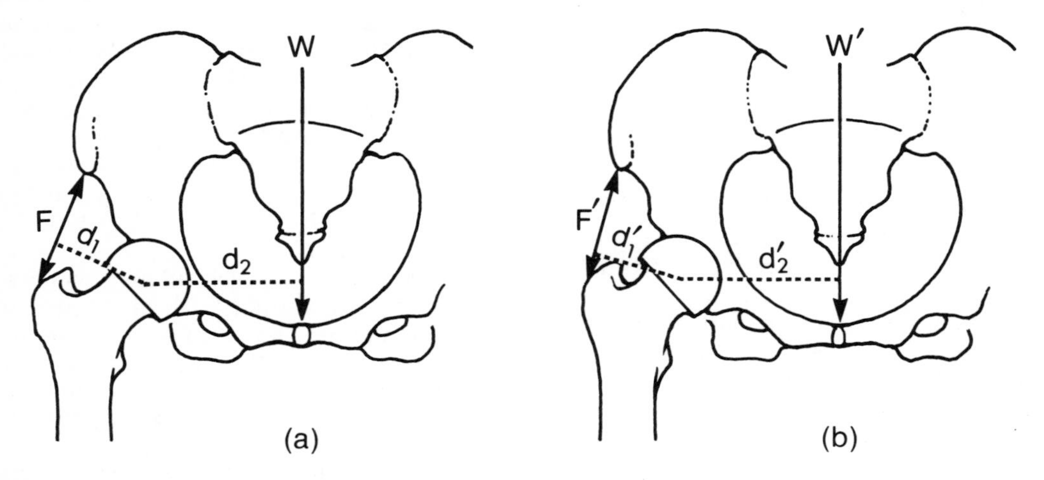

Figure 5.27. Schematic of hip joint. (a) normal; (b) hip prosthesis with neck of femur shortened. See text, problem 2.

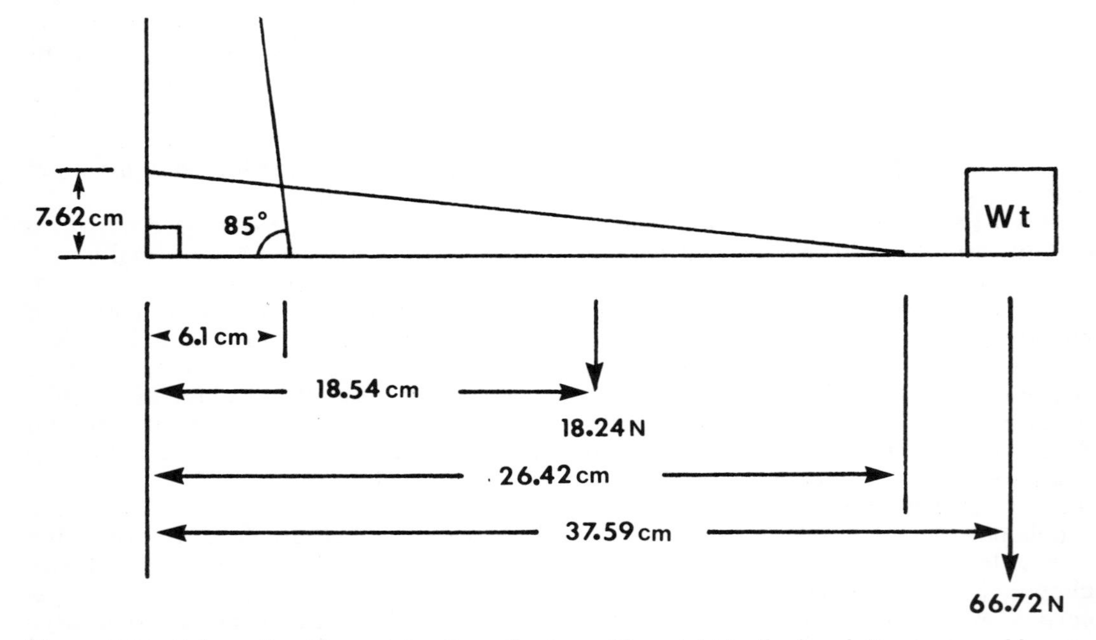

Figure 5.28. Schematic of isometric elbow flexion with weight in the hand. See text, problem 3.

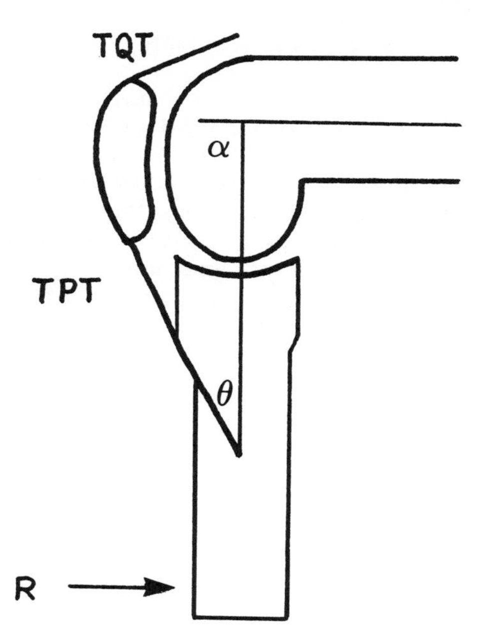

Figure 5.29. Schematic of knee joint at 90° flexion. See text, problems 4 and 5.

method of determining resultant forces, and by directly calculating the force based on the following formula:

$$PFCF = TQT \times \cos\frac{(180 - \alpha) - \theta}{2} + TPT \times \cos\frac{(180 - \alpha) - \theta}{2}$$

where $PFCF$ = patella-femoral compression force, TQT = tension in the quadriceps tendon, TPT = tension in the patella tendon. Assume that the tension in TQT = the tension in TPT and that for each joint angle, the tension force in the patella tendon equals 406.74 N. Note that when the knee joint angle, α, = 30°, the patella tendon angle, θ, = 18°; when $\alpha = 90°, \theta = 1°$. (b) Identify the joint angle where you would expect the subject to have the least discomfort. (N.B.: The values given are adapted from the work of Smidt, (1973).)

Problem 5. Again, refer to Figure 5.29. The anterior cruciate ligament (ACL) helps to reduce anterior shear of the tibia beneath the femur. When strengthening the quadriceps femoris muscle in a person with a weak ACL, the therapist is always concerned with anterior shear. (a) For the two RMAs provided in Table 5.4, calculate the shear force at each of 30° and 90° of knee flexion during an isometric quadriceps contraction. Assume that the shear force is always parallel to the tibial plateau and that the subject is sitting during the exercise. Assume that the effort torque is constant at 135 Nm for each of the cases. Ignore the effect of weight of the leg and foot. Employ the figures from Table 5.4 in your calculations.

Table 5.4

Knee Joint Angle α	RMA	EMA	Patella Tendon Angle θ
30°	30.48 cm	4.88 cm	18°
90°	30.48 cm	3.96 cm	1°
30°	15.24 cm	4.88 cm	18°
90°	15.24 cm	3.96 cm	1°

(b) Based on your calculations, select which of the four isometric conditions is optimal. Also identify whether each of the four conditions will place stress on the anterior or posterior stabilizers of the knee. (N.B.: The values given are adapted from the work of Smidt, (1973).)

Problem 6. In Figure 5.30, a therapist is applying cervical traction to a patient. The traction force is developed by the therapist

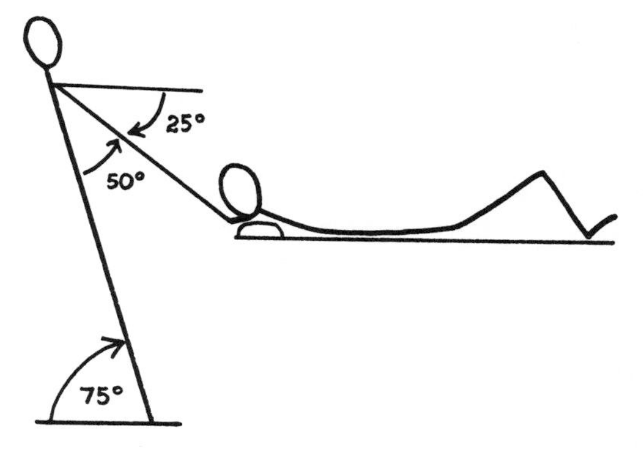

Figure 5.30. Schematic of therapist administering cervical traction to a patient. See text, problem 6.

leaning backwards while maintaining straight arms, and the system is in static equilibrium. The therapist's height is 1.6 m, the mass of the therapist is 50 kg, and the therapist's body proportions from the feet are as follows: level of glenohumeral joint = 85.6% of total body height; level of center of mass = 58.8% of total body height; and level of hip joint = 52.5% of total body height. Find the amount of traction force.

CONCLUSION

At this juncture the student should be familiar with a number of applied mechanics techniques as they relate to determining *musculoskeletal* forces in the human body. Chapter 2 provides the detail of the skeletal components, and Chapter 6 provides detail of the structure and function of the muscular components.

CHAPTER 6

The Skeletal Muscle System

INTRODUCTION

The motor system of the animal kingdom (including humans) comprises three interrelated anatomical systems: the skeletal system, which provides the bony levers that actually generate motion; the muscle system, which supplies the power to move the levers; and the nervous system, which directs and regulates the activity of the muscles. In this chapter we consider skeletal muscle.

Man has about 640 skeletal muscles of many shapes and sizes, from the tiny stapedius muscle of the middle ear to the massive hip extensor, the gluteus maximus. Muscles are situated across joints and are attached at two or more points to bony levers. Movement is produced by a shortening and broadening of the muscle which brings the lever ends into closer approximation.

Muscles differ in shape according to their functions. Some are long and slender for speed and range of movement, such as the biceps brachii; others are sheetlike to form supporting walls, such as the oblique abdominals; and some are multiple-headed to distribute and vary movement, such as the deltoid.

PROPERTIES OF SKELETAL MUSCLE

Muscle has four well-developed characteristic properties: irritability, contractility, distensibility, and elasticity. **Irritability** is the ability of muscle tissue to respond to stimulation. Muscle is our second most highly irritable tissue, being exceeded in this capacity only by nerve tissue.

The most distinguishing characteristic of muscle is **contractility.** By contractility, reference is made to the capacity of muscle to produce tension between its ends: to exert a pull. **Relaxation** is the opposite of contraction. It is entirely passive; the giving up of tension. Both relaxation and contraction progress from zero to maximal values over a finite time. Neither is instantaneous.

Muscles have a third property which is important in their function. They are **distensible:** they can be lengthened or stretched by a force outside the muscle itself. The stretching force can be the pull of an antagonistic muscle, of gravity, or of a force exerted by an opponent. Distensibility is a reversible process and the muscle suffers no harm so long as it is not stretched in excess of its physiological limits.

Finally, a muscle is **elastic:** unless it has been overstretched, it will recoil from a distended length. Distensibility and elasticity are separate and antagonistic properties whose coexistence in the muscle's connective tissues contributes significantly to muscle function. Although essentially each opposes the other, together they assure that contractions will be smooth and that the muscle will not be injured by a sudden strong change in either stretch or contraction.

STRUCTURE OF SKELETAL MUSCLE

It is necessary to consider the structure of skeletal muscle before discussing its functions because structure and function are interdependent and inseparable. We consider, first, muscle structure in its gross aspects, i.e., the muscle as a discrete organ; second, the histology of the component fibers which make up the muscle organ; and finally, the ultrastructure of the contractile machinery of the fibers, the myofibrils.

Gross Structure

Muscles are discrete organs readily recognizable in any animal dissection, being attached in a manner characteristic of each species to bones or shells (with some exceptions) and crossing joints in very specific ways. Each muscle exerts a pull across the joint, and the action line of the muscle in relation to the joint determines the movement which will be produced within the limitations of joint structure.

The Muscle as an Organ

A skeletal muscle is composed of two types of structural components: active contractile elements and inert compliant materials. The contractile elements are contained within the **muscle fibers.** Each muscle is composed of many muscle fibers, a medium-sized muscle containing approximately 1 million. Fibers vary in length from a few millimeters in the stapedius to greater than 34 cm in the sartorius, and they vary in width from 10 to 150 μ. Fibers may run with their long axes parallel to the length of the muscle, as in the sartorius or rectus abdominis. These are known as **parallel** muscles (Fig. 6.1(a)). Muscles in which the fibers are arranged in the form of a spindle, as in the biceps brachii, are classed as **fusiform** muscles (Fig. 6.1(b)). There are also **fan-shaped** muscles (Fig. 6.1(c)), whose fibers fan out from a narrow area of attachment to a broad one. Examples include the pectoralis major, the anterior portion of the internal oblique of the abdominal wall, and the glutei, medius and minimus. Last, there are **pennate** muscles, whose short fibers are arranged in a feather-like pattern

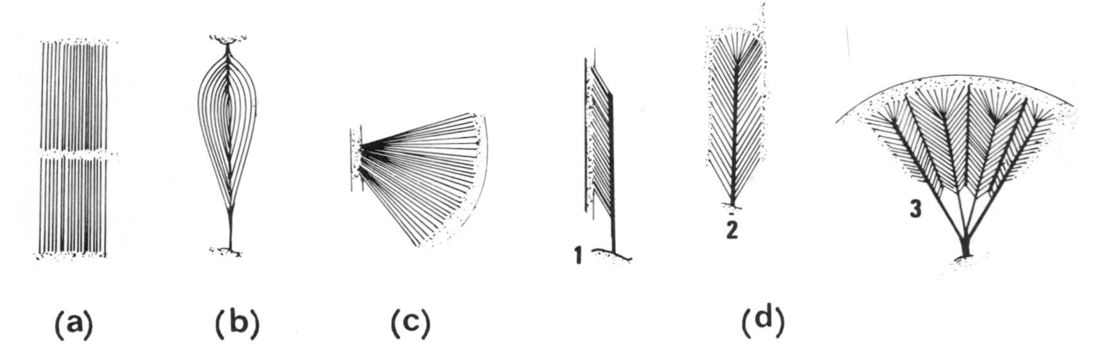

Figure 6.1. Arrangement of fibers in muscles. (a) a parallel muscle; (b) a fusiform muscle; (c) a fan-shaped muscle; (d) pennate muscles: 1, single pennate; 2, double pennate; 3, multipennate.

(Fig. 6.1(d)). This type of muscle may be **single** or **double pennate** as in the forearm muscles, or the fibers may be arranged between multiple tendons as in the deltoid, the gluteus maximus, and the infraspinatus. These are described as **multipennate**. The arrangement of the fibers is related to the function of the muscle concerned. Fast-acting muscles generally have parallel fibers (biceps), while those designed for strength (gastrocnemius) are more often pennate. In general, individual muscle fibers run from tendon to tendon of the muscle; they are occasionally arranged serially, but in humans and other vertebrates they seldom anastomose with neighboring fibers.

About 85% of a muscle's mass consists of the muscle fibers themselves, the rest being composed largely of connective tissues which contain variable proportions of collagen, reticular, and elastic fibers. It is their distensibility and elasticity which assures that the muscle's tension will be transmitted smoothly to the load, and that an elongated muscle will recover its original length after being stretched. The connective tissues provide a complex arrangement of simple, essentially spring-like elements which are the **elastic components** of the muscle and which occur both in series and in parallel with the contractile elements.[a]

A connective tissue sheath, the **epimysium**, surrounds the muscle and sends septa (the **perimysia**) into the muscle to envelop bundles (fascicles or fasciculi) of muscle fibers. Larger bundles may be subdivided into several smaller bundles. From the perimysia delicate strands of fine connective tissue (the **endomysia**) pass inward to invest individual fibers (Fig. 6.2).

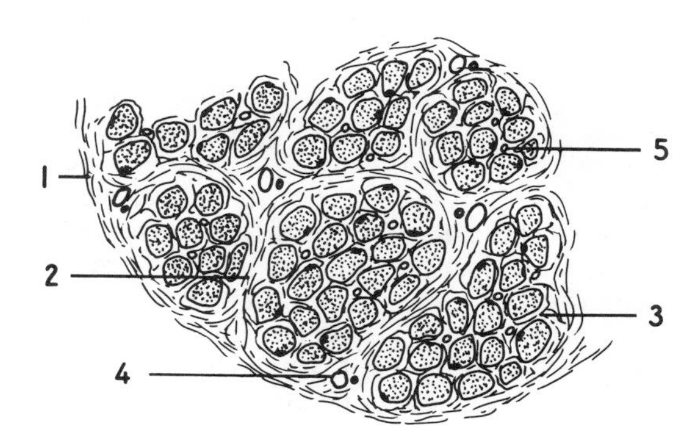

Figure 6.2. Cross-section of muscle to show its fiber bundles and the organization of connective tissues enclosing them. The epimysium (1) surrounds the entire muscle. Septa passing inward from it form the perimysia (2), which encompass bundles of muscle fibers. Note that smaller primary bundles, each with its own perimysium, are enclosed together as a secondary bundle. Delicate strands from the perimysium penetrate the bundles to form the endomysia (3), which invest the muscle fibers themselves. Arteries and veins (4) travel along the strands of the perimysia while capillaries (5) lie in the endomysia.

The total number of fibers, and hence the cross-section area of a muscle, are related to its strength requirements, but bundle size reflects the general function of the muscle. Muscles whose function it is to produce small movement increments, such as those required in manipulation, are composed of small bundles, whereas those concerned with powerful gross movements contain larger fasciculi. As a result the proportion of connective tissue is greater in the muscles which are capable of finely graded movements.

The connective tissues of the muscle blend with the collagen bundles of the tendon, forming a strong and intimate union, the **myotendinous junction.** Connective tissues of muscle and tendon are continuous. They act together as a buffer system against the possibility of too rapid development of contractile force in the muscle.

---

[a] Elastic components are found both in parallel and in series with the contractile elements of the muscle. Because those in parallel contribute only negligibly to passive elastic tension during stretch and become slack during contraction, they are of little interest to us. Those situated in series, however, are important factors in both passive elastic tension and active contractile tension. Therefore, throughout this section "elastic components" will be understood to refer to the series elastic components unless otherwise specified.

Without their distensibility the muscle would be in danger of rupturing its fibers or tearing its attachments by a sudden contraction. Fascia and tendons also act to harness the pull of the muscle fibers to the bony levers. When a relaxed muscle is passively stretched or when it actively contracts, the initial tension developed is due to the elasticity of the connective tissues. In order to do work on a load, a muscle must first stretch out the elastic components until their tension is appropriate to the load before any shortening of the muscle becomes apparent. Until then, there is effectively no load on the contractile elements. Once muscle tension and load are in equilibrium, further expenditure of energy may be used to lift the load and perform external work.

Circulatory and Nerve Supply

Circulation

Muscle obtains a rich blood supply from branches of neighboring arteries. The arteries and veins travel in the epimysium, while arterioles and venules course in the perimysia and capillaries run longitudinally in the endomysia between individual muscle fibers (Fig. 6.2). An abundant circulatory network is provided by frequent transverse linkages between capillaries of adjacent fibers. Capillary anastomoses are especially well developed in the neighborhood of motor endplates. In some instances, dilated cross-connecting vessels appear and are thought to act as reservoirs from which the muscle fibers may draw oxygen during sustained contraction, at which time capillary flow may be significantly reduced by compression of the supply vessels.

Nerve Supply

Nerves enter the muscle near the main arterial branch and divide to distribute both motor and sensory fibers to the muscle bundles. Motor fibers fall into two categories: large fibers (alpha subdivision of Group A of the Erlanger-Gasser classification) and smaller fibers (gamma subdivision of Group A). Each large **alpha motor neuron**, with its cell body lying in the ventral horn of the spinal cord, supplies a number of muscle fibers by successive bifurcation of its axis cylinder. One motor neuron and all of the muscle fibers which its axon terminals innervate constitute a **motor unit**. The number of muscle fibers per motor unit varies considerably with both the size of the muscle and the type of its function. Small muscles and muscles concerned with fine gradations of contraction have necessarily smaller motor units than do larger bulky muscles whose job is the maintenance of strong contraction. For example, motor units in the extraocular muscles of the eye consist of about five muscle fibers per motor unit, while those of the gastrocnemius may have as many as 1900 fibers activated by one motor neuron. It should be mentioned that the size of the motor unit is related to the size of the muscle fascicles, but this does not mean that a single muscle bundle is also a single motor unit. Rather, muscle fibers of a specific motor unit tend to be distributed among several fascicles in a limited area of the muscle.

The smaller **gamma motor neurons** innervate the muscle spindles, providing a means of central regulation of muscle contraction over an indirect pathway known as the gamma loop, which is discussed in detail in Chapter 10 under "Role of Spindle Innervation in Voluntary Movement."

Sensory nerve fibers are also of two general sizes: large Group I neurons whose sensory terminals lie in the muscle receptors (spindles and tendon organs) and smaller Group III fibers which probably subserve the sense of muscle pain.

Microscopic Structure

Histology of the Muscle Fiber

The muscle fiber is a syncytial mass of sarcoplasm, cylindrical in form with a bluntly tapered end and surrounded by a specialized membrane, the **sarcolemma**. The sarcolemma is a unit membrane about

100 Å thick. A number of nuclei lie peripherally just under the sarcolemma, and numerous **myofibrils**, each about 1 μin diameter, lie longitudinally embedded in the sarcoplasm. The myofibrils are the contractile elements of the muscle. There are about one thousand in each muscle fiber.

Viewed under the ordinary light microscope the muscle fiber has a distinct cross-striated appearance of alternate dark and light areas appearing in a regular and repeating pattern. High magnification shows the striping to be a property of the myofibrils, which are oriented in register with dark regions adjacent to dark regions and light to light so that the pattern appears to extend across the whole fiber (Fig. 6.3). Dark areas are known as **A bands** because of their anisotropic (doubly refracting) effect upon polarized light. Each appears to be somewhat lighter in its mid-region, the **H zone**, which is crossed by a darker line, the **M line**. The light regions are known as **I bands** because of their more nearly isotropic effect. Each I band is clearly bisected by a dark line, the **Z disc** or **line**. The region from one Z disc to the next constitutes a **sarcomere**, and this appears to be both the structural and functional unit of the myofibril.

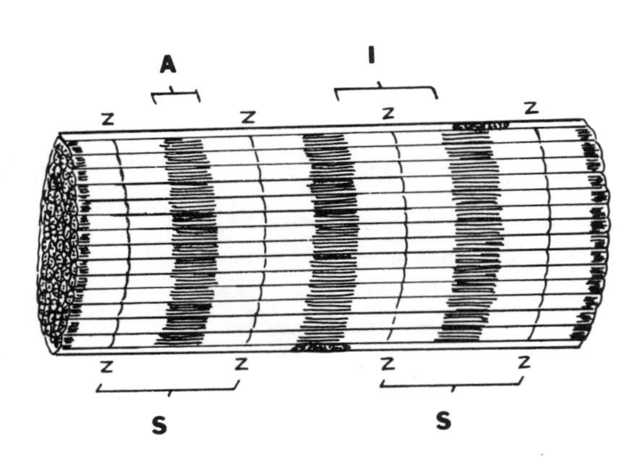

Figure 6.3. Diagram of a portion of a single muscle fiber. The drawing shows that the striations are a property of the myofibrils which lie in the fiber with light and dark bands in register. S, sarcomere; A, A band; I, I band; Z, Z line.

Neuromuscular Junction

As a terminal axon approaches a muscle fiber, its myelin sheath narrows and finally ceases a short distance before the end-plate. It has been firmly established that there is no continuity of nerve and muscle protoplasms. The nerve axon does not penetrate the sarcolemma; its arborizations lie on the surface of the muscle fiber. Axon terminals may lie under the endomysium but do not penetrate the sarcolemma. The neuromuscular junction is formed by axolemma and differentiated sarcolemma in apposition to each other. The structure of the muscle fiber in the junctional region is highly specialized. The sarcolemma is profusely folded into troughs and grooves, presenting a spiny or lamellated appearance (Fig. 6.4). The muscle portion of the neuromuscular junction has been called the **subneural apparatus**. Many muscle fiber nuclei are seen in the vicinity of the junction, accompanied by an abundance of mitochondria. Most authorities believe that the gap between the nerve terminus and the sarcoplasm is a specialized barrier across which excitation must be transmitted by chemical

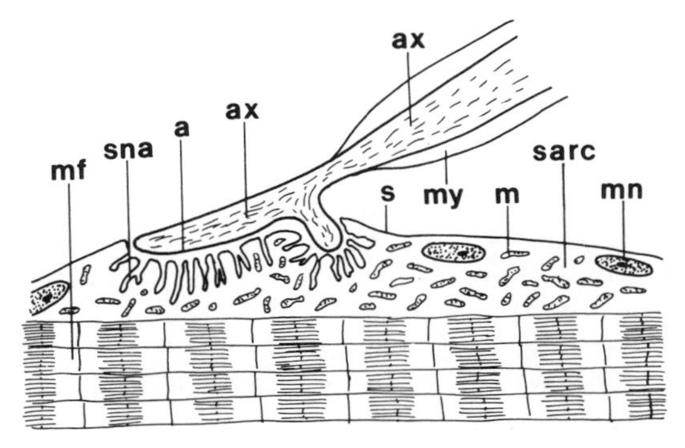

Figure 6.4. Schematic drawing of the neuromuscular junction. a, axolemma; m, mitochondrion; s, sarcolemma, ax, axoplasm; mf, myofibril; mn, muscle nucleus; my, myelin sheath; sarc, sarcoplasm; sna, folds of the subneural apparatus. (Adapted from Couteaux, R., 1960. Motor end plate, structure. In *Structure and Function of Muscle*, Vol. 1, edited by G. H. Bourne. New York: Academic Press, p. 337.)

means (discussed in more detail in Chapter 7 under "Synaptic Transmission").

Chemical Composition of the Muscle Fiber

The skeletal muscle fiber is composed of a semifluid sarcoplasm of water, salts, and other substances in which are suspended the nuclei, myofibrils, a reticular system of tubules, sacs and cisterns, mitochondria, glycogen granules, and lipid droplets. Water constitutes 75%, protein 20%, and other materials 5% of its mass.

Muscle proteins may be grouped into four major factions: proteins of the sarcoplasm, of the stroma, of granules, and of the myofibrils. The sarcoplasmic proteins, which occupy the space between the myofibrils, are readily extracted with water or neutral salt solutions. They include myoglobin, a pigment with a high affinity for oxygen which functions in the transfer of oxygen from the capillaries to the sites of oxidation (the mitochondria), and enzymes concerned with the glycolytic aspects of muscle metabolism.

Stroma proteins are retained in the muscle residue after extraction with strong salts and are difficult to isolate. For this reason, knowledge regarding them is limited. Some are of a collagenous nature and contribute to the structure of the sarcolemma.

Differential centrifugation is used to separate the proteinaceous granules from homogenized muscle. These granules include the nuclei, mitochondria and microsomes. In the intact fiber the mitochondria, and microsomes are located among the myofibrils. Mitochondrial proteins include the enzymes of oxidative metabolism.

Myofibrillar protein consists almost entirely of actin and myosin. Small amounts of tropomyosin and troponin are also present. Actin and myosin compose about half of the total protein content of the fiber and account for most of the contractile material itself.

Glycogen granules account for about 0.5 to 1.0% of the fiber. Lipids occur in small amounts. The principal salts are those of potassium, sodium, calcium, magnesium, and chloride. Nonprotein extractives include creatine, creatine phosphate (CP), adenosine triphosphate (ATP), adenosine diphosphate (ADP), and lactic acid.

Ultramicroscopic Structure of Muscle

The Myofibril

Electron micrographs have revealed that each myofibril is composed of two types of filaments: one thicker and shorter (in the rabbit psoas, about 100 Å in diameter by 1.5 μ in length) and another thinner and longer (rabbit psoas, about 50 Å by 2 μ). These are seen to lie longitudinally in a very definite parallel orientation. Each **thick filament** extends the length of the A band, while a **thin filament** passes from each Z line through the I band and into the adjacent A band as far as the edge of the H zone. Therefore, the denser outer portions of the A band are produced by the overlapping of thick and thin filaments for part of their lengths. The central H zone region is less dense because it contains only thick and no thin filaments. The I band is least dense because it contains only thin filaments. The arrangement of the filaments is shown diagrammatically in Figure 6.5.

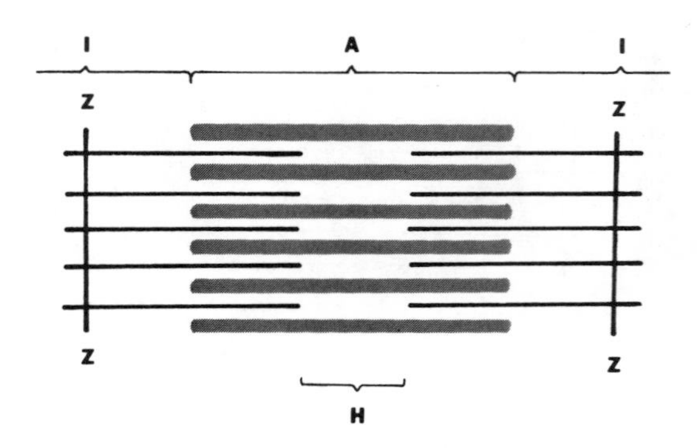

Figure 6.5. Diagrammatic representation of filament arrangement in a single sarcomere of a fibril. A, A band; I, I band; H, H zone; Z, Z line. See discussion in the text.

The bands, zones, and lines may be readily identified in the electron micrographs of Figures 6.7 and 6.10(a) and (b).

The H zone is not homogeneous but shows variations in its density. Across its center is the dense region of the M line, on either side of which there appears a narrow region whose density is lower than that of the rest of the H zone. The lightest portion, known as the **pseudo-H zone**, maintains a constant width regardless of stretch or contraction, indicating that it is a structural feature of the thick filaments and not just another reflection of overlap. Significance of the pseudo-H zone is discussed later in this chapter.

The relation of filaments to fiber banding is clearly demonstrated by cross-sections through these areas (Fig. 6.6). If sections are taken through the denser part of the A band, the thin filaments are found surrounded by thick filaments in an orderly hexagonal array. If sectioned through the H zone of the A band, only thick filaments are present, the thin ones being absent. If sectioned through the I band, only thin filaments are present. Thus, the cross-striations of the muscle fiber seen with the light microscope are found to be due to a repeating pattern of varying filament densities along each myofibril with the patterns of adjacent fibrils in register.

Chemical Structure of Myofibrils. Chemical analysis of the fibrils shows them to be composed of about 20% protein, the rest being a watery suspension of salts and other metabolically important substances. About 80% of the protein consists of **actin** and **myosin** in a ratio of about 1:3. Actin is a low viscosity protein, molecular weight 70,000 to 76,000, while myosin is a more viscous molecule, molecular weight 1,000,000 to 1,500,000. During contraction these two proteins combine to form a complex, **actomyosin**, which is the contractile material per se of the muscle fiber. Smaller amounts of two other proteins, **tropomyosin** and **troponin**, are also present.

Extraction methods specific for myosin remove the thick filaments and result in an electron micrograph lacking the A band. If only actin is extracted, the treated tissue shows loss of much but not all of the content of its I bands, while its A bands are unchanged. Thus, the thick filaments are composed of myosin molecules and the thin filaments are composed of actin plus other substances.

It is reasonable to assume that the association of actin and myosin, long recognized as a first step in muscular

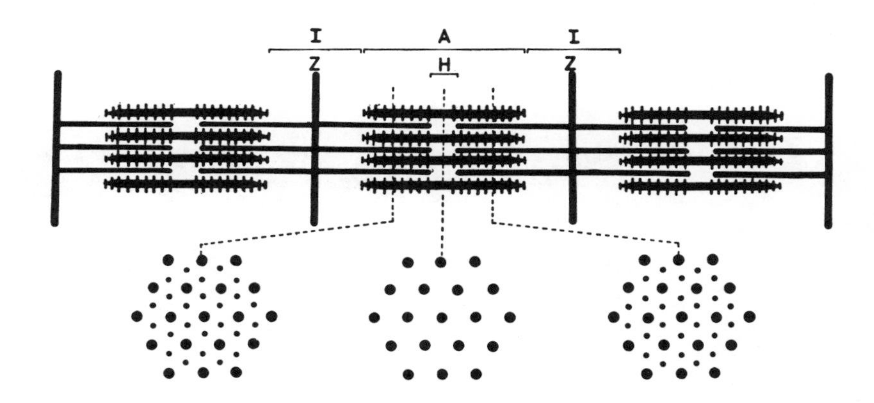

Figure 6.6. Diagrammatic representation of the structure of striated muscle. Shown are overlapping arrays of actin- and myosin-containing filaments, the latter with projecting cross-bridges on them. For convenience the figure is drawn with considerable longitudinal foreshortening. (From Huxley, H. E., 1969. The mechanism of muscular contraction. *Science 164:* 1356–1366, Fig. 1, June, 20, 1969. Copyright© 1969 by the American Association for the Advancement of Science.)

contraction, must involve some form of bonding between the two types of filaments. Electron micrographs show projections extending from the thick filaments in the dense parts of the A band (Fig. 6.7). It has been well documented that these represent an orderly system of cross-bridges which attach the thick filaments to each of the surrounding thin filaments. On any single thick filament, six projections form a helical pattern, each set about 60° farther around the filament and about 60 to 70 Å apart. Thus, the period of repetition is about 400 Å.

Fragmentation of myosin by trypsin digestion yields two subunits, the heavier of which has a molecular weight of about 230,000 and has been designated as **heavy meromyosin** (HMM), and the lighter, with a molecular weight of about 96,000, as **light meromyosin** (LMM). In terms of these molecular subunits, Huxley (1965) found that the myosin molecules of the thick filaments consist of globular heads of HMM and longer tails of LMM, the latter appearing as linear strands with a strong attraction for one another and for HMM. Each thick filament is composed of about 200 to 400 myosin molecules in an overlapping array with their HMM heads projecting outward (probably as the cross-bridges) and their LMM tails overlapping in parallel to form the backbone of the filament. The molecules are oppositely oriented in each half of the sarcomere with their tails directed toward the center (Fig. 6.8(a)). As a result there is a region in the center which is devoid of HMM bridges. This is consistent with the pseudo-H zone identifiable in the striation pattern. The dense M line may result from the crossing of the LMM strands in the middle of the filament.

Actin has been shown to exist in both a globular (**G-actin**) form and a fibrous (**F-actin**) form. The actin filaments consist of a double helix of G-actin molecules with a 360 Å period, probably surrounding an F-actin core (Fig. 6.8(b)). The alternate high and low points of the helix are the active sites for attachment of the HMM bridges. Electron micrographs of disrupted fibrils have shown that actin filaments are firmly attached to the Z discs.

Later studies by electron microscope and X-ray diffraction have indicated that the HMM bridge is further fractionable into a globular subfragment (HMM-S_1) and a linear subfragment (HMM-S_2). Huxley (1969) was the first to suggest that the linear HMM-S_2 is connected at one end to HMM-S_1 and at the other end to LMM by flexible junctions (Fig. 6.8(c)). With such an arrangement, the bridges have a rather wide range of movement and are able to adjust to the variable distances between filaments which have been observed to exist at different muscle lengths. Interfibrillar distances are known to decrease when muscle is stretched and to increase as the muscle is returned to its rest length. In such a system the regularity of bridge

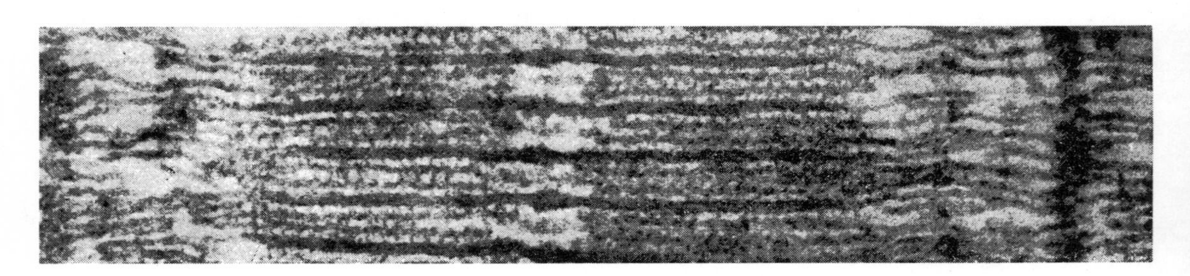

Figure 6.7. Electron micrograph of a longitudinal section through an array of fibrils in which two thin filaments lie between two thick ones. Note the projections from the thick myosin filaments connecting with the thin actin filaments. These are presumably the cross-bridges. (From Huxley, H. E., 1958. The contraction of muscle. *Sci. Amer. 199:* Fig. 1, p. 71. Copyright© 1958 by Scientific American, Inc. All rights reserved.)

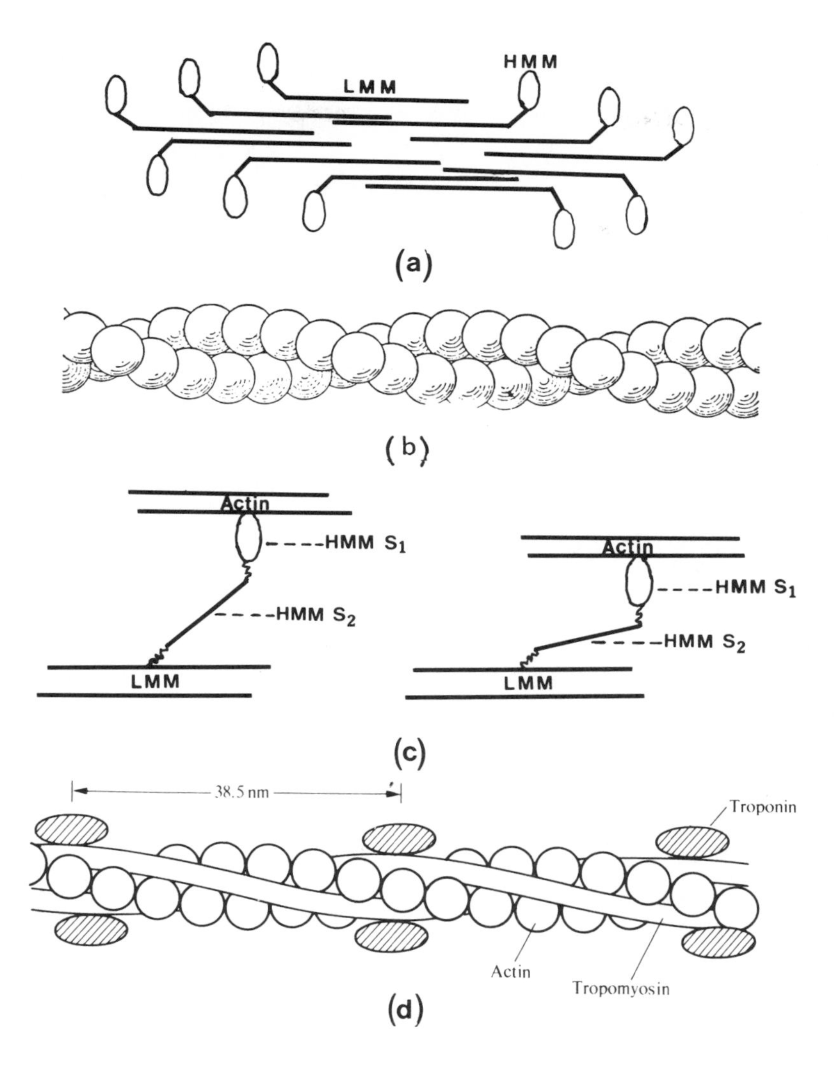

Figure 6.8. Proposed ultrastructure of myosin and actin. (a) aggregation of light meromyosin (LMM) and heavy meromyosin (HMM) into the myosin filament. LMM forms the backbone of the filament while HMM projects as the cross-bridge. (b) actin filament formed by a double helical organization of G-acting molecules. Alternate high and low points may contain the active sites for attachment of HMM cross-bridges. (c) detail of HMM subfragments: globular fragment (HMM-S_1) and linear fragment (HMM-S_2). HMM-S_1 is joined to S_2 by one flexible juncture, and S_2 joins the LMM backbone by another. The HMM bridges are thus capable of considerable movement to adjust, without any changes in their orientation on the myosin filament, to the variable distances between actin and myosin filaments which occur during the contraction cycle. The figure at left shows the position of the bridge when the distance is large, as during contraction; the figure at the right shows the position when the separation is small, as during stretch or relaxation ((a), (b), and (c) adapted from Huxley, H. E., 1969. The mechanism of muscular contraction. *Science 164:* 1356–1366, Fig. 1, June 20, 1969. Copyright© 1969 by the American Association for the Advancement of Science.) (d) the likely arrangement of actin, tropomyosin, and troponin in the thin filaments of muscle. Tropomyosin appears to be arranged along the double helical organization of the actin filament. ((d) from Tribe, M. A., and M. R. Eraut, 1977. *Nerves and Muscle* (Basic Biology Course Book 10), p. 163. Cambridge, England: Cambridge University Press.)

attachments between actin and myosin would be determined by the constant location of the active sites on actin, rather than by exact and invariable positioning of the bridges on myosin.

Sarcoplasmic Reticulum and T System

The electron microscope has revealed another interesting and significant detail of muscle subcellular structure. The endoplasmic reticulum of the muscle fiber displays certain distinctive characteristics which have merited the use of the term **sarcoplasmic reticulum** in reference to it. First, instead of being distributed throughout the cytoplasm, the membrane-limited tubules and cisterns form lace-like sleeves around the myofibrils in a distinct pattern which is repeated in a definite phase relationship to the striation bands of each sarcomere. Second, there is a special arrangement of reticular elements in characteristic groups of three, known as **triads**, consisting of two cisterns separated by a transversely oriented tubule, the **T tubule** (Fig. 6.9). The triads are found at the same site in muscle tissue of any given species. In fishes and amphibians they occur at the Z lines; in reptiles, birds, and mammals, they occur at the A-I junctions (Fig. 6.10). Third, ribonucleic acid (RNA) granules, although present in the sarcoplasm, are absent from the reticulum itself. This suggests that the function of the reticulum may be other than protein synthesis.

The T tubules are now known to be distinct from the sarcoplasmic reticulum, although they are functionally associated with it. The **T system**, as it is called, communicates with the sarcolemma or perhaps represents invaginations of it. The content of the tubules is continuous with the extracellular fluids and contains sodium in significant amounts. The cisterns of the sarcoplasmic reticulum contain a high concentration of calcium. The probable roles of the sarcoplasmic reticulum and the T system in the coupling of excitation and contraction are discussed below under "Excitation-Contraction Coupling."

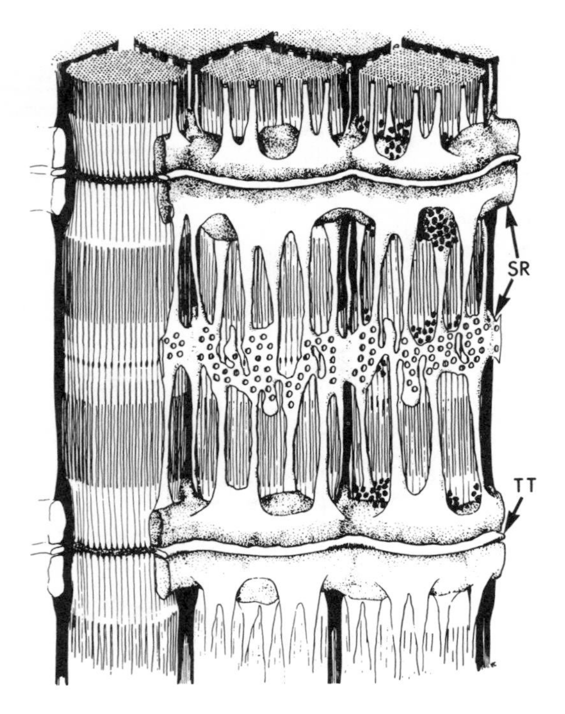

Figure 6.9. Reconstruction from electron micrographs of a portion of an adult frog skeletal muscle fiber. The sarcoplasmic reticulum (SR) is shown surrounding several myofibrils for a length of slightly more than one sarcomere. Two transverse tubules (TT) are seen extending across the figure at the centers of triads located next to the Z lines of the myofibrils. (From Peachey, L. D., 1965. The sarcoplasmic reticulum and transverse tubules of frog's sartorius. *J. Cell Biol.* 25: 209.)

THE NATURE OF CONTRACTION

Excitation

Muscle may be excited and caused to contract by natural or by artificial means. Normally, excitation is accomplished only by the nervous system: nerve impulses arriving at the neuromuscular junction cause the release of a transmitter substance which diffuses across the junction and chemically excites the muscle fiber. However, whether induced naturally or artificially, excitation is evidenced by the

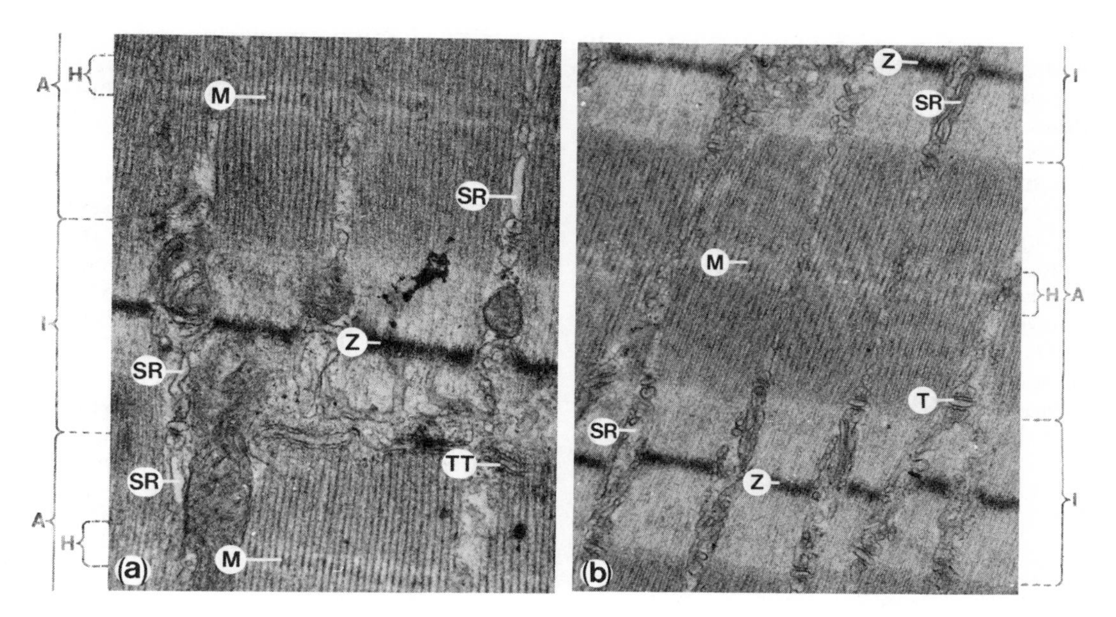

Figure 6.10. Longitudinal section of rat muscle. Two electron microscope photographs of myofibrils from a mammalian (rat) muscle. Original magnification × 46,000. (a) longitudinal elements of the sarcoplasmic reticulum (SR) lie between the myofibrils. Transverse tubules (TT) of the T system are seen at the A-I junctions. M and Z lines are conspicuous. (b) another section shows triads (T) in profile at the A-I junctions on both sides of the dark Z lines especially well in the lower half of the photograph. Longitudinal elements of the sarcoplasmic reticulum extend across the Z lines in both the upper and lower halves of the photograph. The myofibrils, lying as they do with their dark and light bands in register, show the basis for the striated appearance of the muscle fiber. (Courtesy of Dr. G. Harrison and Dr. D. Philpot, NASA Ames Research Center, Moffett Field, California.

generation and conduction of action potentials in the sarcolemma of the muscle fiber.[b] Action potentials travel along the fiber membrane at a speed of 1 to 3 m/sec and initiate the events which lead to shortening of the contractile elements of the myofibrils and the consequent production of tension in the muscle.

Contraction

Muscular contraction requires the expenditure of energy obtained from chemical reactions coupled to a contractile mechanism which uses the energy to generate tension and produce external work. Some facts regarding the processes involved are well established; others still evade understanding and are as yet only theory. The details of the chemistry of contraction and relaxation fall outside the realm of this textbook. The reader is referred to appropriate physiology texts for a detailed account of the chemistry of contraction and relaxation.

In brief, the source of energy for muscular contraction is the high-energy-producing molecule, ATP. In addition, other chemical substances are essential for muscular contraction. Among them are Ca^{++} and Mg^{++} and actin and myosin combined to form actomyosin. The ATP supply is generally maintained through the metabolism of glucose that is available in the blood stream. Glucose is supplied from storage in the liver in the form of glycogen. There is

[b] The generation and conduction of action potentials in cell membranes are discussed in detail in Chapter 7.

also an on-the-spot supply of glycogen in muscle cells. Glucose metabolism is accomplished in two ways, anaerobically by glycolysis in the sarcoplasm and aerobically in the mitochondria via the Krebs cycle and electron transport system.

A supplementary and fast means of supplying ATP in muscle involves CP. When vigorous muscular activity persists, wherein the oxygen supply is insufficient to meet the aerobic needs, glycolysis takes over and a condition of oxygen debt is reached and may be tolerated for a short period of time. Eventually the system must be balanced through the oxidative process.

Heat Production in Muscle

Only about 40% of the energy resident in glucose is captured as ATP-stored energy, and of the chemical energy released from the ATP during contraction only about 30% can be converted into external work. This metabolically useful energy may be called the **available energy**. The rest appears as heat and represents energy wasted as a result of the inefficiency of the chemical and physical processes.

For almost three decades the work of A. V. Hill dominated thinking with regard to the heat production associated with muscular activity. Although some of the recent literature has raised questions concerning some points, it still is reasonable to divide the heat released by an active muscle into two major portions: the **initial heat** which appears during the contraction and the **recovery heat** which appears after relaxation. The ratio of initial to recovery heat is the same for twitch and tetanus, whether isotonic or isometric.

Initial Heat. The initial heat may be further divided into activation heat, shortening heat, and relaxation heat.

Activation Heat. Activation heat is produced upon stimulation and is associated with the appearance of the active state. It is probably related to the breakdown of ATP to initiate contraction and perhaps also to the thermal effects of the release and movement of calcium. It is the basal heat production, appearing whether the muscle shortens or not. It is independent of muscle length and tension. In tetanic contraction activation heat is sometimes called the **maintenance heat**.

Shortening Heat. Shortening heat is the extra heat produced when the muscle is permitted to shorten. It is absent in isometric contraction. Authorities seem to agree that the amount of shortening heat is proportional to the distance shortened and that the rate of its production is a linear function of the velocity of shortening. There is some disagreement as to whether or not it is also load-dependent.

Relaxation Heat. As tension subsides a portion of the initial heat can be identified as relaxation heat. It may reflect the release of the energy which was stored in the elastic components during the development of tension.

Recovery Heat. Recovery heat constitutes a larger fraction of the total heat than does the initial heat. It is produced more slowly and over a relatively long period of time following contraction. It represents heat loss during the reconstitution of ATP by interaction with creatine phosphate and by cellular respiration, and during the resynthesis of glucose and glycogen from lactic acid. It can be subdivided into an anaerobic portion related to glycolysis and a larger aerobic portion which varies with the amount of energy expended (work done) and which reflects the oxidative reactions of the Krebs cycle and electron transport system.

Regulation of Muscle Metabolism

Muscle metabolism is self-regulated. When oxygen is adequately available, glycolysis is depressed, but during vigorous muscular activity the accumulation of reduced coenzyme and other end products accelerates glycolysis. An increase in cellular levels of ADP acts as a potent stimulator for metabolism. Therefore, the rate and amount of ATP used automatically determine the rate and amount of its resynthesis.

Excitation-Contraction Coupling

For many years a gap existed in our knowledge of the means by which the electrical potentials traveling along the surface of the muscle fiber could excite the myofibrils, some of which were deep in the fiber as far as 50 μ away. The time lapse between the spike potential in the sarcolemma and contraction of the sarcomeres is not sufficient to permit a substance to diffuse such a distance. For some time it was suspected that there must exist some form of rapid transmission to conduct excitation inward to the myofibrils. Because the triads were oriented in so definite a relationship to the cross-striation pattern, it was reasonable to think that they might be involved. Huxley and Taylor (1958) showed that differences among species regarding the location of triads correlated with the trigger points for artificially producing local shortening of sarcomeres. For example, current passed through microelectrodes produced local shortening within a frog muscle fiber only when the stimulus was applied at the region of the Z lines, and the triads were most abundant here. In crab and lizard muscle, however, contraction was induced only when stimulation was applied at the A-I junction, and this was the area of triad concentration for muscles of these species. Other species exhibited a similar agreement between local activation of the sarcomere and the distribution of triads.

Abundant evidence indicates that the T system is separate from the sarcoplasmic reticulum and continuous with the sarcolemma and that excitation is conducted inward by this channel. The presence of sodium within the T tubules indicates that typical action potentials may be the manner of transmission.

To link excitation and contraction, excitation must trigger the events of fiber activation which lead to tension development. Calcium, which plays an essential role in initiating the process of contraction, accumulates in the sarcoplasmic reticulum at the site of the triads, having passed along the T tubules from the extracellular fLuids and across the triadic junctions between tubules and sarcoplasmic reticular cisterns. Action potentials traveling along the T tubules are thought to cause the release of Ca^{++} from the cisterns into the sarcoplasm surrounding the myofibrils. The free Ca^{++} then catalyzes the formation of actomyosin; myosin ATPase is thereby activated and splits ATP. The chemical energy thus derived is converted into mechanical energy and the sarcomere contracts. As long as stimulation continues, Ca^{++} continues to be released and contraction is maintained.

Sliding Filament Theory of Muscular Contraction

When electron micrographs are taken of a **stretched** muscle, its sarcomeres are seen to be longer than in resting muscle but their A bands are unchanged. Therefore, stretching has not lengthened the thick myosin filaments. The length of the thin actin filaments also is unchanged, as shown by the constancy of the distance from the H zone of one sarcomere through the Z line to the H zone of the next sarcomere. The H zones, however, have increased and the I bands have lengthened. Apparently the extent of overlap of the two types of filaments has decreased while each filament has maintained its own integrity (Fig. 6.11, compare (a) and (b)).

In a moderately **contracted** muscle, filaments and hence A bands still retain their original lengths but changes are found in other parts of the striation pattern. Sarcomeres are shorter, and H zones and I bands have diminished. The area of overlapping of thick and thin filaments has increased (Fig. 6.11, compare (b) and (c)).

Electron microscopic evidence has led to the theory that the band pattern changes described above are due to the sliding of the filaments past one another, and that contraction is produced by the creeping of the thin actin filaments along the thick myosin filaments, the motion being mediated by chemical interactions between the

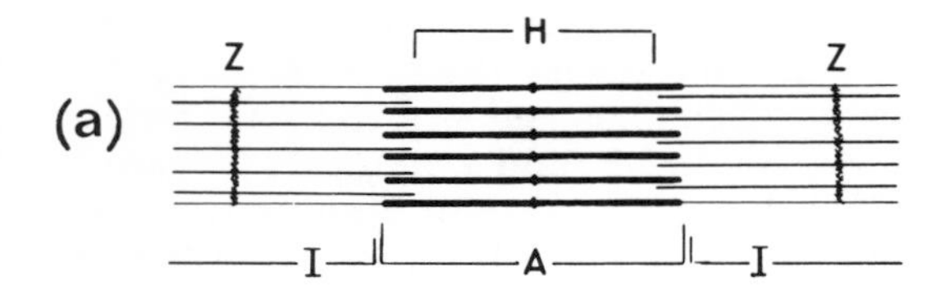

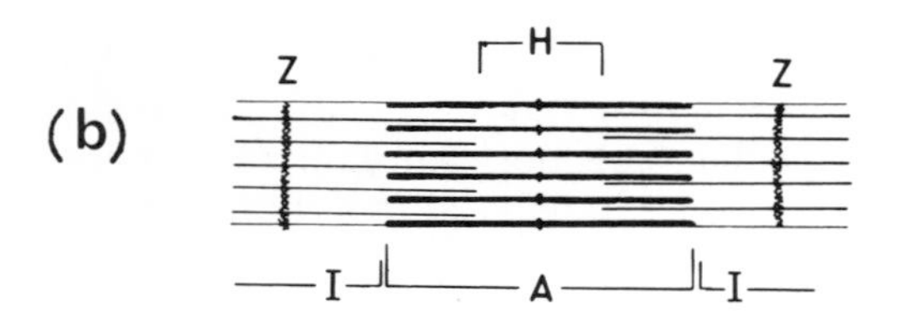

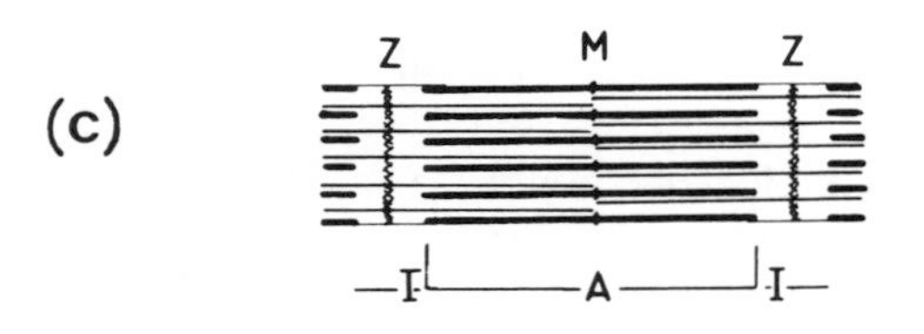

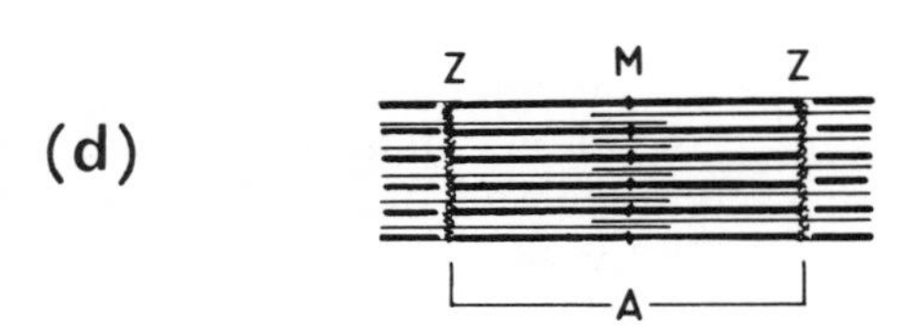

Figure 6.11. Diagrammatic representation of filament relationships in a sarcomere under various conditions. (a) stretched; (b) relaxed; (c) moderately contracted; (d) strongly contracted. A, A band; H, H zone; M, M line; I, I band; Z, Z disc.

filaments. This is the **sliding filament theory**, suggested practically simultaneously by two different groups of investigators, the first composed of A. F. Huxley and R. Niedergerke (1954) and the second composed of H. E. Huxley and J. Hanson (1954). The theory, which has been supported by experimental evidence of many kinds, is now accepted by most physiologists.

Interactions of actin and myosin occur "ratchet style." The myosin bridges oscillate back and forth, shifting their attachments from site to site along the actin filaments. The orientation of HMM molecules in the myosin filaments suggests that bridge action exerts forces which are directionally oriented toward the center of the sarcomere. Therefore, their oscillations will draw the thin actin filaments farther into the A bands, shortening the sarcomere and exerting tension.

X-ray diffraction studies indicate that there is substantial movement of the HMM bridges during contraction, and many theories have been devised to account for bridge movement. Huxley (1969) was the first to suggest that displacement between the two halves of the HMM-S_1 may generate the mechanical force to move the actin filament a finite distance. The bridge then swings back, attaches to another site, and draws the filament farther into the A band. The cycle, repeated a specific number of times by many bridges, results in a shortening of the contractile material and the development of tension at the ends of the muscle fiber.

As the actin filaments advance farther into the A bands, the H zones will be reduced in size. The maximal number of bridges will be able to attach when the tips of the actin filaments have reached the outer borders of the pseudo-H zones. As further contraction advances the filaments first to the M line and then beyond it, actin filaments from the two opposite sides of the sarcomere will meet and then pass by, producing a **double overlap** of thin with thick filaments (Fig. 6.11(d)). In strong contractions extreme double overlapping will cause the H zone to be replaced by an area which is denser than the outer areas of the A bands, where only single overlap occurs. Electron micrographs of cross-sections taken through the sarcomere mid-region under such conditions show twice as many thin filaments as are otherwise present. Even in maximally shortened muscle, however, the pseudo-H zones are still distinguishable, remaining constant in location and dimensions and lighter than adjacent portions of the double overlap regions. This is consistent with the absence of HMM bridges in the pseudo-H zone, where the tails of the oppositely oriented myosin molecules compose the A band center. In supercontraction Z lines appear to butt against the ends of the thick filaments and the thin filaments are in maximal double overlap within the A bands.

When in strong contraction the actin filaments are pulled into the sarcomeres far enough to enter the bridge areas of the opposite halves of the A bands, it is reasonable to suppose that their orientation with respect to these bridges will be abnormal. Any interaction which might occur between the filaments could not be expected to contribute to tension development and would probably oppose it. If the bridges attach, their action will be oppositely directed to that for tension production. At the least one would expect the intrusion of such "wrong way" filaments to interfere physically and/or chemically with the normal interaction of bridges with their own filaments. Double overlap with its probable consequences has been suggested as a significant factor in the observed reduction of tension capacity in a shortened muscle.

The shortening of sarcomeres adds up to produce shortening of myofibrils, which in turn adds up to produce shortening of the muscle fiber. Contractile tension of the muscle, therefore, represents a summation of the short-range forces acting at multiple bridges between the myosin and actin filaments. Each cycle of a bridge contributes its small part to the production or maintenance of tension.

Action of the bridges cannot be synchronous. While some are pulling, others must be just attaching and still others detaching to shift to a new site of attachment. The peak tension at any moment reflects the average number of bridges which are active at that moment.

Relaxation

Physical Changes in Relaxation

When muscle fibers no longer receive impulses from their motor neurons, they relax. Muscular relaxation is completely passive. It is basically a cessation of tension production and may or may not be associated with lengthening of the previously shortened muscle. Muscle fibers are incapable of lengthening themselves actively. If gross lengthening occurs, it is brought about by a force outside the muscle itself, such as gravity, contraction of antagonists, or assumption of a load. However, as the bridges detach at relaxation, the internal elastic force which was built up within the fibrils during contraction is released. Recoil of the elastic components then restores the fibrils to their uncontracted lengths. The passive lengthening of the fibrils will be concomitant with the slipping of the actin filaments away from the centers of the sarcomeres.

Magnitude of Contractile Tension

The tension developed by a contracting muscle is influenced by a number of factors such as the characteristics of the stimulus (normally nerve signals transmitted to the muscle), the length of the muscle both at the time of stimulation and during the contraction, and the speed at which the muscle is required or made to contract.

The Stimulus

Most of what has been learned about muscle has been derived from studies using stimulation by electrical pulses. Although it is an artificial stimulus, electricity has distinct advantages for experimental purposes because it can be precisely controlled. The intensity, form (time course of rise to and duration of peak intensity), and frequency of pulses can be arbitrarily selected and varied as desired. Measurable responses of the muscle can be correlated with the quantitated stimulus characteristics. Interestingly, new therapeutic locomotor aids are available using controlled electrical stimuli applied to the motor nerve or muscle when regular nerve function is disrupted.

A curarized muscle may be stimulated directly by pulses applied to the muscle tissue or indirectly by pulses applied to its motor nerve fibers. The response of the whole muscle, of a single motor unit, or of one muscle fiber may be studied under controlled conditions.

The Single Pulse

Response of Muscle to a Single Pulse. If a single electrical pulse of adequate intensity is applied directly to a muscle fiber, the fiber will respond in an **all-or-none** fashion. Increasing the intensity of the pulse will not increase the magnitude of the fiber's response. It is important to mention here that the all-or-none response of the muscle fiber is determined by the all-or-none character of its excitation and not by any all-or-none limitations inherent in the contractile mechanism itself.

When a single adequate pulse is applied to a whole muscle, the muscle will respond with a quick contraction, followed immediately by relaxation. Such a response is called a **twitch.** Its magnitude will vary with the number of muscle fibers which respond to the stimulus and this will vary directly with the intensity of the pulse up to a finite maximal intensity.

The twitch is an indication of force development by the muscle. After a short **latent period** tension becomes evident and rises in a hyperbolic manner to a peak (the **contraction period**). It then declines over a slightly longer time course to zero (the **relaxation period**) (Fig. 6.12(a)).

The time course of the development of overt tension in the twitch is influenced by the interaction of the contractile components of the muscle fibrils with the elastic components of the muscle. Figure 6.12(b) illustrates the sequence of events and their influence on the shape of the twitch curve.

1. The active state, which indicates an increased resistance to stretch and a rise in heat production as compared to resting muscle, is evident even before tension appears. It reaches full intensity abruptly, is maintained for about half of the contraction period, and then progressively declines during the rest of the contraction period. It is reflected diagrammatically in Figure 6.12(b).
2. The contractile components begin to undergo activation during the latter half of the latent period. As they shorten, the elastic components of the muscle are stretched and begin to exert passive elastic tension. Elastic tension is low at first. During this time the contractile elements are able to shorten rapidly.
3. When the active state is at full intensity, about halfway through the contractile period, the elastic tension is rising rapidly.
4. As the active state begins to decline in the latter half of the contractile period, its intensity is still sufficient to continue to stretch the elastic components, and tension continues to mount but at a decreasing rate. The twitch curve begins to round off.
5. At the peak of the twitch curve, tension in the contractile and elastic elements is in equilibrium.
6. Beyond the peak, as the active state continues its decay, developed tension falls below elastic tension and the elastic components recoil, stretching out the contractile components. Overall tension falls.
7. Decay of the active state is completed before tension returns to zero. The fact that tension outlasts the active state is partially explained on the assumption that the breaking of cross-bridges within muscle elements requires more time than their formation. Therefore, the recoil of the elastic components lengthens the contractile material less rapidly than the rate of decay of the active state.

The time course and intensity of the active state are studied by the techniques of **quick stretch** and **quick release.** Because of the elastic components and the viscosity of muscle tissue, the externally measured force exerted in a twitch is less than the full capability of the contractile material, that is, less than the intensity of the active state. The viscoelastic effect may be

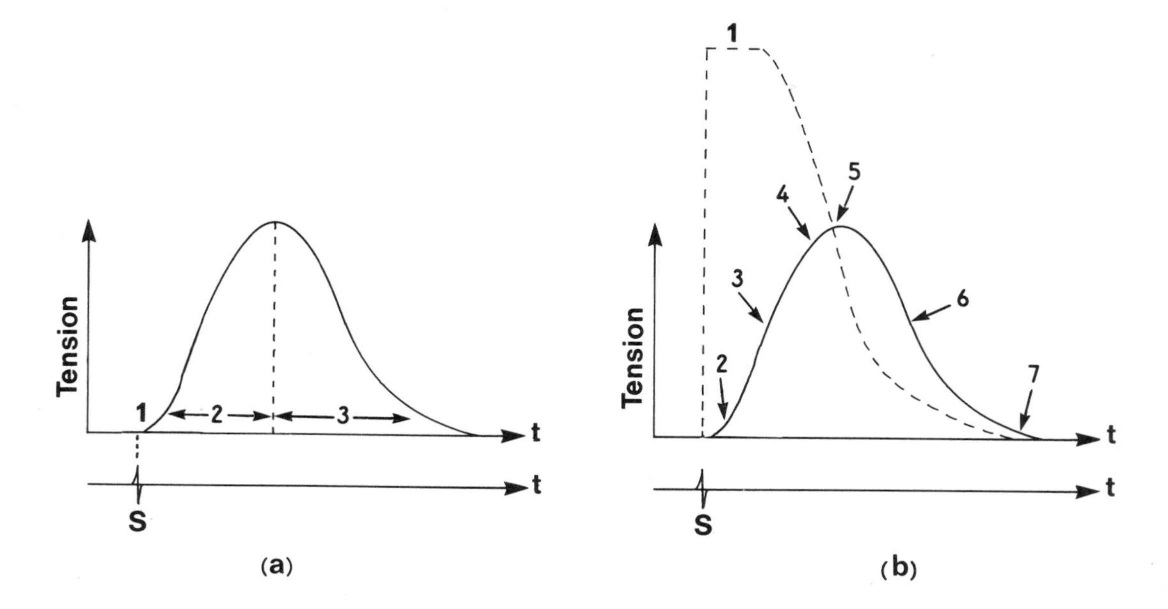

Figure 6.12. (a) tension development in a muscle twitch as a function of time. S, stimulus; 1, latent period; 2, contraction period; 3, relaxation period. (b) relationship of active state and tension development. Active state (broken line) is superimposed on the twitch tension curve (solid line). 1, peak of active state; 2, elastic components begin to exert tension; 3, tension rises rapidly; 4, as active state begins to decline, tension continues to rise but at a decreasing rate; 5, at the peak of the twitch curve, contractile and elastic tensions are in equilibrium; 6, as the active state continues to decay, recoil of the elastic components stretches out the contractile elements and tension falls; 7, active state decay is complete before tension has returned to zero. See text for further discussion.

counteracted and the full tension characteristics of the contractile elements registered by employing quick stretch or quick release. If, coincident with stimulation, the muscle is given a short, quick stretch which pulls out the elastic elements just slightly beyond what their effective excursion would be, the muscle is relieved of the necessity of stretching out the elastic components and its full tension is revealed. By this means the onset, rise time, and duration of the peak intensity of the active state can be determined.

The time course of the decay of the active state is studied by the method of quick release, in which the fiber is stimulated to contract isometrically until its full active state has been developed. Then it is suddenly released to a slightly shorter length. Tension falls immediately but is quickly redeveloped, at a rate exceeding that in a normal twitch. The peak level, however, is lower. By varying the time of release and plotting redeveloped tension against time, a curve reflecting the decline of the active state is obtained (Fig. 6.12(b))

Characteristics of the Single Pulse and Their Influence on the Muscle Twitch. An adequate stimulus may be defined as any environmental change, external or internal, which arouses in the contractile material an active state of sufficient magnitude to produce measurable tension. Whether natural or artificial, the environmental change must meet certain minimal requirements with regard to its basic characteristics: the magnitude or intensity of the change; its abruptness or rate of rise; and the duration of its application. Within physiological limits, increase above minimum in any of these will induce an increased response in the muscle.[c]

Intensity. A single electrical pulse must have a certain minimal intensity to be effective. This minimal level is an inverse measure of the irritability of the tissue; the smaller the minimal intensity, the greater the irritability. The minimal effective intensity is designated the **threshold** or **liminal stimulus.** These terms refer to the weakest stimulus which will evoke a barely perceptible response. **Subthreshold** and **subliminal** refer to a stimulus of inadequate intensity. As the intensity of the single pulse is increased above the minimum, contractile tension in the muscle increases progressively as a result of the activation of more and more muscle fibers. Finally, an intensity is reached which evokes the maximal response of which the muscle is capable. Presumably all fibers are then active. Further increase in intensity will not be accompanied by further increase in contraction. The weakest stimulus intensity which will evoke maximal contraction of a muscle is called the **maximal stimulus.**

Abruptness or Rate of Rise. A weak but adequate pulse with a rapid rate of rise from zero to its preset intensity will evoke a stronger contraction than will a pulse of the same intensity with a slower rise. A minimal rate is required even for an intense stimulus. If intensity rises too gradually, there will be no response at all; the stimulus is then ineffectual. For any stimulus of adequate intensity, the more abruptly it is applied the greater will be the response it evokes, within the limits of the muscle's capacity. The greater the intensity the less rapidly it need rise to produce a given level of response.

A common experience illustrates the principle. If the hand is plunged abruptly into hot water of about 44°C, the response (sensation of heat) resulting from the abruptness of the change in skin temperature from about 34 to near 44°C will be greater than if the change is made gradually by first immersing the hand in water at skin temperature and then slowly raising the temperature to 44°C. If the rate of temperature change is too slow, the change will be imperceptible.

Duration. For a stimulus of adequate intensity and rise rate, the duration of its peak intensity will influence its effectiveness. Within limits, the longer its duration the greater will be the muscle's response. Exclusive limits are found at both extremes: the duration can be so short that no response will occur in spite of the fact that the same intensity and abruptness would be sufficient with longer duration, or the duration can be so long that the response decreases until it ceases altogether. The latter is a common experience in the laboratory when direct current is used to stimulate tissue. The muscle responds at the closing of the circuit but ceases to respond as current flow continues at the constant (peak) level. The duration of the peak intensity has exceeded the response capabilities of the tissue.

The relationship of intensity and duration of single current pulses in the production of a barely perceptible contraction is

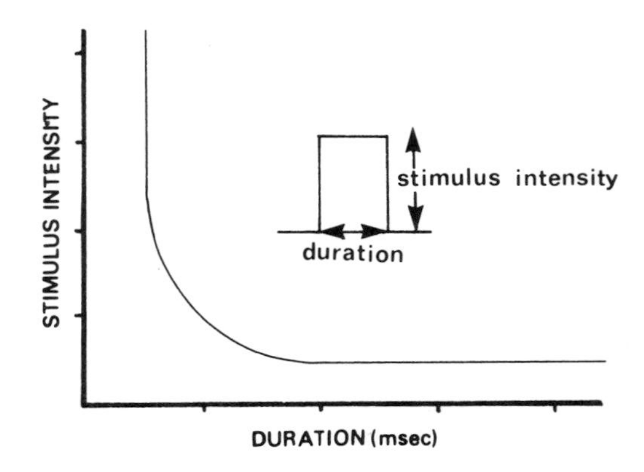

Figure 6.13. Intensity-duration curve. The upper limb of the curve indicates that a very strong stimulus must be applied for at least a minimal duration in order to be effective. The lower limb shows that below a certain minimal intensity a stimulus will not induce a response regardless of its duration. Between two limbs intensity and duration are inversely related.

[c] In a single fiber, only an increase in the frequency of stimuli will produce an increase in its all-or-none response.

presented in the intensity-duration curve shown in Figure 6.13. Note that both the upper and lower ends of the curve become straight lines, one vertical and the other horizontal, neither meeting the axes. The upper end indicates that even a very strong stimulus must be applied for at least a minimal duration to be effective. The lower end shows that below a certain minimal intensity a stimulus will not induce a response regardless of its duration. Between these limits, the greater the intensity, the less duration is required to produce a response.

For a stimulus of constant intensity and rate of rise, the longer its duration, the greater will be the response up to a finite limit (Fig. 6.14(a)). The duration required for a given stimulus to evoke a perceptible response is its **excitation time** and is, within the limits discussed above, inversely related to the intensity.

If a stimulus of constant intensity and duration is applied at various rates of rise, effectiveness will be directly related to the rate. The more abruptly the stimulus is applied, the greater will be the muscle's response. As the rate decreases, the response will diminish until ultimately, regardless of intensity, the stimulus becomes ineffectual (Fig. 6.14(b)).

The decreased effectiveness of a constant stimulus intensity at long duration and/or low rate of rise is designated **adaptation** or **accommodation.** Many tissues besides muscle adapt to a gradual or persistent stimulus. The physiological changes which are induced by the stimulus are apparently reversed at a rate which is faster than their development under the

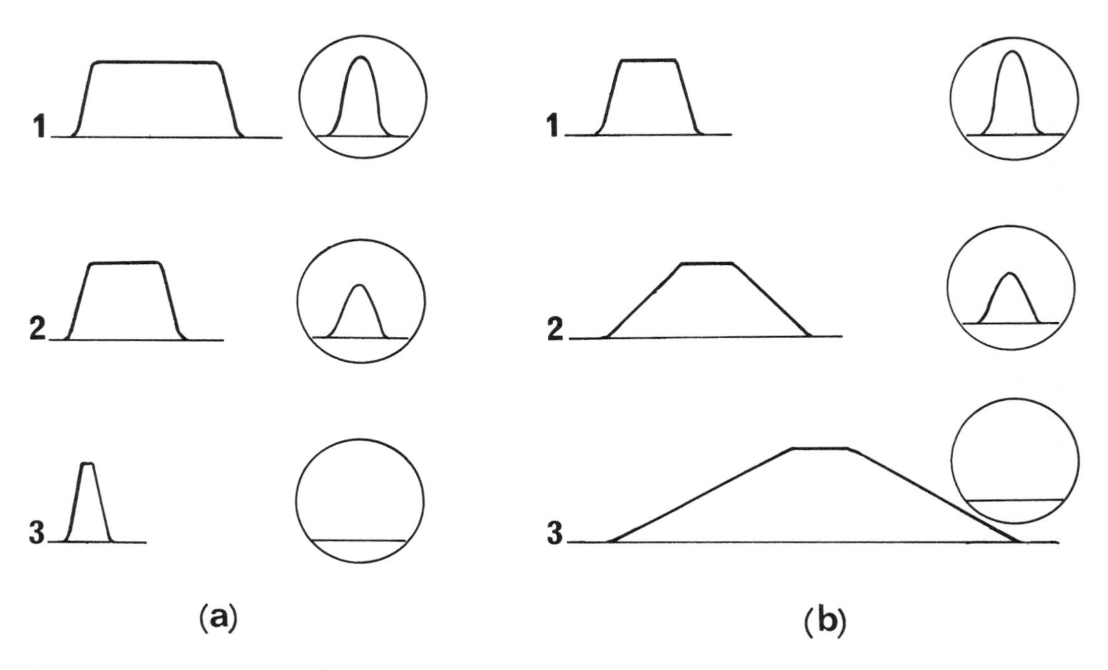

Figure 6.14. Comparison of single pulses with rate of rise and duration of peak intensity. Hypothetical responses are shown in circle insets. (Note: the time scale of the pulses is greatly exaggerated as compared with that of the responses.) (a) duration of pulse. Three pulses of identical intensity and rate of rise but with different durations are shown: 1, moderate duration: probably the most effective; 2, short duration: less effective; 3, duration too short: ineffectual. (b) rate of rise. Three pulses of identical intensity and duration but with different rise times are shown: 1, the most rapid rise: the most effective stimulus; 2, less rapid rise: less effective; 3, least rapid: least effective (response much reduced or absent).

existing conditions. In the case of muscle tissue, excitatory processes may be inadequate to activate the tissue or, if activated, the magnitude or persistence of the active state may be insufficient to stretch out the elastic components enough to produce overt tension.

The rate of rise and duration of electric pulses may be varied as required for the principle under study. For most studies of concern to us, a pulse of rapid rise and short duration is used, with variations in its intensity appropriate to experimental objectives.

Repetitive Stimulation

Response to Repetitive Pulses. If an adequate stimulus is applied to a muscle fiber repeatedly at a rate rapid enough so that each succeeding stimulus reactivates the contractile elements before the previous tension has completely subsided, successive responses summate, each building upon the previous until a maximal level is achieved. If stimulation is continued, the contraction peak is maintained at this level. Such a response is known as **tetanus** or **tetanic contraction.** Ultimately, fatigue will cause the peak level to decline progressively. When stimulation ceases, contraction terminates and the fiber relaxes, tension subsiding quickly to zero. If, however, the repetitive stimulation is too prolonged, **contracture** will result and relaxation will be very much slowed as compared with normal. Unlike rigor, contracture is reversible.

Effect of Frequency of Pulses upon Response. The frequency of stimulation, usually expressed as cycles per second or **hertz**, determines both the shape and the magnitude of a tetanic contraction traced on a myograph by an excised muscle. When pulses are delivered with a time span (period) which places successive stimuli during the relaxation phase of the preceding response, the contraction approaches a tremor and a scalloped tracing results. This is **incomplete tetanus.** With a time span (period) which is short enough to allow for restimulation during the contraction phase, the tracing is smooth. This is **complete tetanus** (Fig. 6.15). Within physiological limits, the shorter the period (i.e., the greater the frequency) the smoother the curve and the greater the tension development will be. If, however, the period is shortened beyond a certain point, the refractory period will be encountered. The **absolute refractory period** is a short space of time immediately following stimulation during which the muscle cannot be reexcited regardless of stimulus intensity. This is followed by a longer period, the **relative refractory period,** during which irritability is gradually regained and the tissue will respond to a stimulus which is appropriately greater than threshold. The earlier the pulse falls in the relative refractory period, the greater its intensity must be to be effective. Although both portions of the refractory period last for a finite time, in muscle both have been completed before tension begins

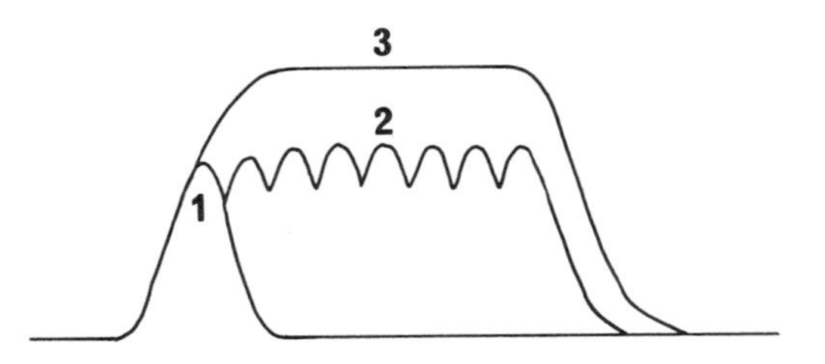

Figure 6.15. Response to repetitive stimulation. Curve 1, single twitch in response to a single stimulus; curve 2, incomplete tetanus in response to low-frequency repetition of the stimulus; curve 3, complete tetanus in response to higher-frequency repetition of the stimulus. Same stimulus magnitude used in all three.

Tetanus-Twitch Ratio. The tension developed in response to repetitive pulses is greater than that evoked by a single pulse of the same magnitude. The **tetanus-twitch ratio** varies with different muscles and may be as great as 5 (rat gastrocnemius). To explain the greater tension developed in a tetanic contraction, it has been postulated that in a twitch the short duration of the active state allows too few bridge movements to permit the contractile material to shorten enough to fully stretch out the elastic components before the active state begins to subside. Hence the full capacity for tension production cannot be realized. Repetitive stimulation, however, by maintaining the active state, permits continuation of bridge activity. The stretching of the elastic components is completed and full tension is developed.

Post-Tetanic Potentiation in Muscle. In many muscles, especially when curarized, if twitch responses to single pulses are recorded before and immediately after a period of tetanic stimulation, the post-tetanic twitch shows an increase in magnitude and a steeper rise of tension than the pre-tetanic control. This phenomenon is known as **post-tetanic potentiation.** The effect occurs whether the muscle is stimulated directly or indirectly by its motor nerve.

Potentiation is maximal shortly after the repetitive stimulation and then decays exponentially at a rate which is dependent on both the frequency of pulses and the number delivered in the train. Short trains produce potentiation without any alteration of the twitch duration, but longer trains result in lengthening of the contraction time and of the half-relaxation time (the time required for tension to drop to 50% of its peak value).

Conclusion

To be adequate, stimulation must consist of an appropriate combination of intensity, rate of rise, duration, and frequency to excite muscle fibers and to activate their contractile material sufficiently to produce measurable tension. Within the limits discussed, the adequacy of the stimulus is determined by the interaction of these mutually interdependent characteristics.

In the living body an adequate environmental change results in the generation and conduction of a train of action potentials in motor neurons which becomes responsible for excitation of the muscle fibers. The frequency of impulses reflects the effectiveness of the stimulation and determines the magnitude of the muscle tension developed. Because the magnitude and form of the nerve impulses are constant for any given set of body conditions, the frequency of these impulses is the most significant characteristic in determining the muscle's response.

Muscle Length

The most obvious property of muscle is its capacity to develop tension against resistance. The length of the muscle at the time of activation markedly affects its ability to develop tension and to perform external work. Muscle tension may be measured in terms of the greatest load which can just be lifted or as the maximal tension readout on a strain gauge or tensiometer.

When a muscle contracts, the contractile material itself shortens, but whether the whole muscle shortens or not depends on the relation of the internal force developed by the muscle to the external force exerted by the resistance or load. The terms "force" and "tension" are often used erroneously as synonyms. Tension is a scalar quantity having magnitude only, while force is a vector quantity having both magnitude and direction. The term **tension** is used in this discussion to refer to the magnitude of the pull of the muscle as it would be registered on a strain gauge arranged in line with the muscle axis. **Internal force** is used to refer to the tension magnitude acting in the direction of the action line of the muscles under given conditions, and **external force**refers to the resistance opposing the muscle.

Types of Muscle Contraction

There are three identifiable types of muscle contraction designated according to the length change, if any, induced by the relationship of internal and external forces. The three types are isometric, isotonic, and eccentric contraction.

Isometric Contraction. If the internal force generated by the contractile components does not exceed the external force of the resistance and if no change of muscle length occurs during the contraction, the contraction is **isometric.** The available energy expended by the muscle and the tension produced against the resistance may be considered to be in equilibrium.

No contraction in the body is purely isometric because at the fibril level the contractile components do shorten. Their shortening, as discussed above under active state, is offset by stretching of the elastic components. By current usage, an isometric contraction is one in which the *external* length of the muscle remains unchanged. (Isometric contraction is sometimes called static contraction.)

Concentric Contraction. If the constant internal force produced by the muscle exceeds the external force of the resistance and the muscle shortens, producing movement, the contraction is **concentric.** (This is sometimes called a shortening contraction.) Energy utilization is greater than that required to produce tension which will balance the load, and the extra energy is used to shorten the muscle. During concentric contraction work is done by the muscle on the load. This is regarded as *positive work*. Recalling Equation 4.14, work (W) is the product of the force (F) and the distance (x) it is moved through

$$W = Fx$$

A muscle can develop greater tension in isometric than in concentric contraction because none of the available energy is expended in shortening. In concentric contraction the greatest load that the muscle can lift is about 80% of its maximal isometric tension.

Eccentric Contraction. If to an already shortened muscle an external force greater than the internal force is added and the muscle is allowed to lengthen while continuing to maintain tension, the contraction is called **eccentric**. (The term lengthening contraction is sometimes used.) The energy expended by the muscle is less than the tension exerted on the load, but the muscle acts as a brake controlling the movement of the load. In eccentric contraction a muscle can sustain greater tension than it can develop in isometric contraction at any given equivalent static length. During an eccentric contraction work is done by the load on the muscle. This is logically taken as negative work. Naturally, negative work is measured in the same units as positive work.

In lifting and then lowering a given weight W through a vertical distance h the amounts of positive and negative work for the musculature involved (say for biceps) would be identically equal to Wh. While chemical energy is expended by the muscle in both instances, the energy cost of the negative work is considerably less than for the positive work. The difference is indicated by a lower oxygen uptake during the negative work, being about one-tenth as much in human subjects. Other estimates have placed the cost at one-third to one-thirteenth of that required for the equivalent amount of positive work.

Eccentric contractions are very common. Every movement in the direction of gravity is controlled by an eccentric contraction. Examples include sitting, squatting or lying down, bending forward or sideward, going down stairs, stooping, placing any object down onto a surface, etc. In eccentric contractions the active muscles are those which are the antagonists of the same movement when it is made against gravity. Sitting or squatting is controlled by leg extensors, not flexors; lying down, by hip flexors, not extensors; lowering a load, by shoulder flexors, not extensors. Electromyograms (which reflect the electrical activity of muscle and hence provide insights regarding muscle contractions) show not only that anatomically antagonistic muscles are actively controlling the eccentric movement but also that the electrical activity in these muscles is less than when the same muscles are contracting concentrically to do the same amount of positive work with the same load over the same distance and at the same speed.

Comparison of Tension Curves. Singh and Karpovich (1966) were among the first to demonstrate that the force developed by elbow flexor and extensor muscles in eccentric contraction exceeds that in both concentric and isometric contraction at most muscle lengths. Using an instrument designed by Singh, they measured the manifestations of muscle force through the entire range of motion at the elbow joint, simultaneously recording the angle through which the forearm was moving. With the elbow flexors, eccentric force was consistently the greatest of the three over the entire range of motion and concentric was least (Fig. 6.16(a)). With the elbow extensors, although eccentric force was lowest of the three at the start (shortened position, elbow at 140°), it had exceeded concentric force when the angle reached 120°, and by 100° and lower it had surpassed isometric force as well (Fig. 6.16(b)).

Daily activities involve a continual shifting from one to another type of contraction and of combinations of the three types, as required. During movements the changing lengths of lever arms and of angles of pull, for both muscle and load, introduce complexities which require complicated processes of neuromuscular integration to properly adjust the number and activity of motor units to the task.

Relationship of Muscle Tension to Length

The **initial length** of a muscle, i.e., its length at the time of stimulation, influences the magnitude of its contractile response to a given stimulus. A stretched muscle contracts more forcefully than when it is unstretched at the time of activation. This is true whether the contraction is isometric, isotonic, or eccentric. Within physiological limits, the greater the initial length, the greater will be the muscle's tension capability. Parallel-fibered muscles exert maximal total tension at lengths only slightly greater than rest length. Muscles with other fiber arrangements have maxima at somewhat greater relative stretch. In general, optimal length is close to the muscle's maximal body length, i.e., the greatest length that the muscle can attain in the normal living body. This is about 1.2 to 1.3 times the muscle's rest length. Tension capability is less at shorter and longer lengths. Therefore, a muscle can exert the greatest tension or sustain the heaviest load when the body position is such as to bring it to its optimal length. In isotonic contractions the increased tension and longer length permit greater shortening; hence, more work can be done or, alternatively, the same work can be done at lower energy cost. The diminished energy cost of eccentric contraction is in part due to this **stretch response,**[d] but other factors are also involved, as evidenced by the capacity to produce greater tension than with either isometric or isotonic contractions at most equivalent lengths.

The relationship of tension to muscle length may be presented graphically in the form of a **tension-length curve** in which tensions in an isolated muscle are plotted against a series of muscle lengths from less than to greater than the resting length (Fig. 6.17). Both the passive elastic tension (curve 1) exerted by the elastic components in the passively stretched muscle and the total tension (curve 2) exerted by the actively contracting muscle are plotted. Since total tension (T_T) represents the sum of elastic tension (T_E) plus the developed tension (T_D) of the contractile elements, the latter may be found by subtraction and is represented by the difference $T_T - T_E$ between the curves for total and elastic tensions. Values for developed tension are shown as curve 3. Note the following facts regarding developed tension.

1. At less than 50% of rest length the muscle cannot develop contractile tension.
2. At normal (intact) rest length the muscle is already in slight passive elastic tension. At this length the muscle produces its greatest *developed* tension ($T_D = T_T - T_E$).

---

[d] The stretch response should not be confused with the stretch reflex. The latter is a response mediated by the nervous system, while the former is a property of muscle tissue independent of nerve.

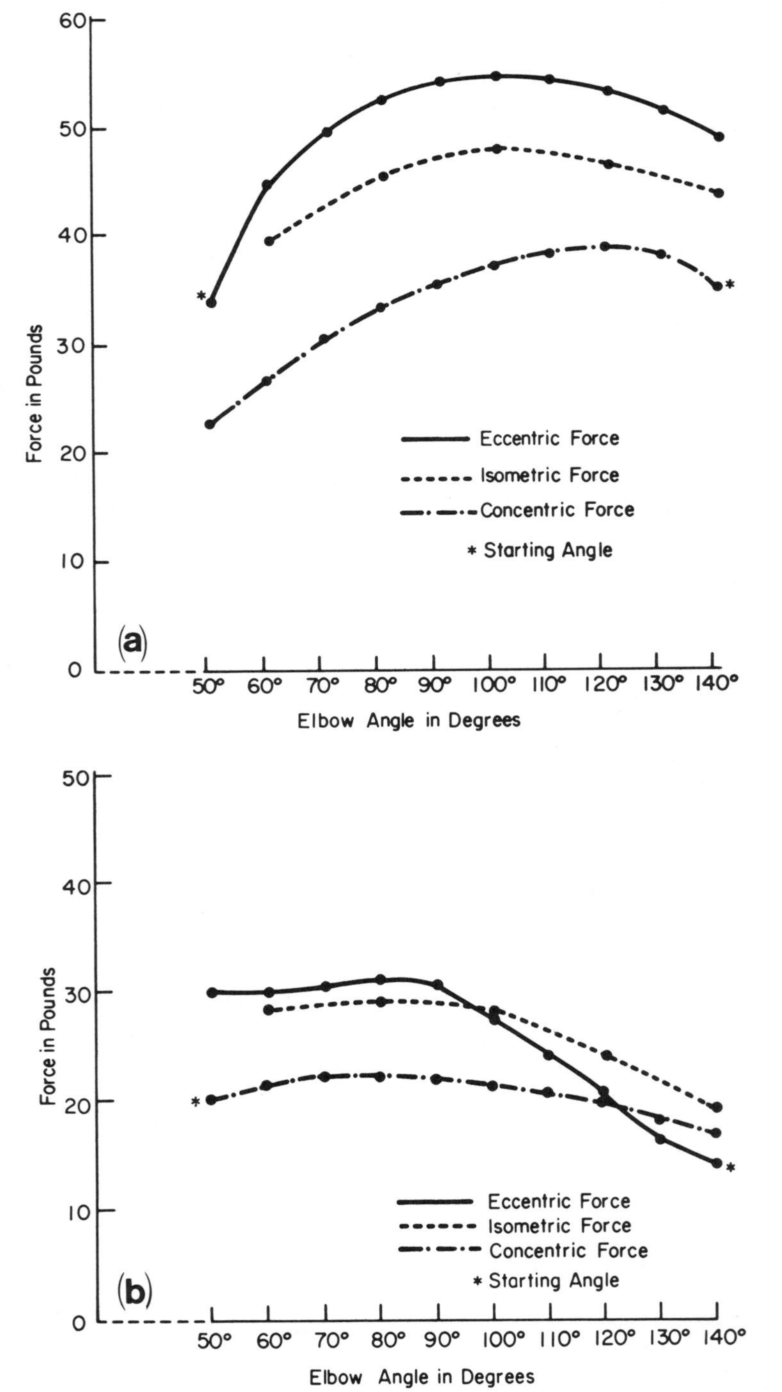

Figure 6.16. Concentric, isometric, and eccentric tension curves. Curves of maximal concentric, isometric, and eccentric tension of (a) forearm flexor muscles and (b) forearm extensor muscles. (From Singh, M., and Karpovich, P.V., 1966. Isotonic and isometric forces of forearm flexors and extensors. *J. Appl. Physiol.* *21:* 1435.)

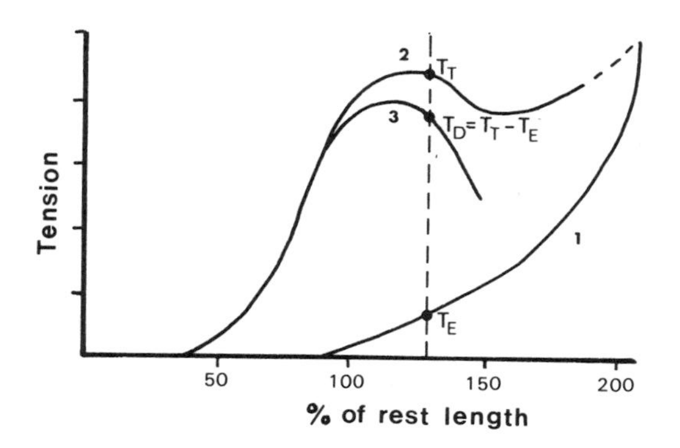

Figure 6.17. Tension-length curves for isolated muscle. Curve 1, passive elastic tension T_E in a muscle passively stretched to increasing lengths; curve 2, total tension T_T exerted by muscle contracting actively from increasingly greater initial lengths; curve 3, developed tension calculated by subtracting elastic tension values on curve 1 from total tension values at equivalent lengths on curve 2, i.e., $T_D = T_T - T_E$.

3. When contraction is initiated at a length longer than rest length, although *total* tension is greater than at rest length, *developed* tension has already diminished and declines progressively at all greater lengths.
4. At extreme lengths (far right end of the curves) total tension would ultimately become equal to elastic tension, developed tension being zero.

Maximal contractile tension is assumed to be developed when sarcomere lengths are such that maximal single overlap of actin and myosin filaments exists. At greater lengths the number of cross-links diminishes as overlap decreases, and at shorter lengths double overlap results in reduced tension as a result of the antagonistic action of bridges. Gordon et al. (1966) investigated the tension-length relationship in frog skeletal muscle fibers at various sarcomere lengths. Their results are plotted in Figure 6.18. In these fibers the mean sarcomere length was 2.5 μ and the mean filament lengths were 1.5 μ for myosin and 1.0 μ for actin. Maximal tension was developed at sarcomere lengths of 2.0 to 2.25 μ. At greater lengths tension decreased linearly, becoming zero at about 3.65 μ. At shorter lengths tension declined gradually with decreasing length until about 1.7 μ and then dropped abruptly to zero at about 1.27 μ.

Drawing upon these data and the electron microscope evidence of filament relationships in contracted muscle, we may postulate the stages of the sliding filament process most probably associated with significant lengths on the tension-length curve presented above. Figure 6.19(a) through (d) presents four of these stages and demonstrates the concept of a quantitative relationship between tension and the number of bridges linking actin and myosin filaments.

Speed of Contraction

Most isolated nonloaded muscles normally shorten by about 50% or less of their rest length. The absolute amount by which any muscle can shorten depends upon the length and arrangement of its fibers, the greatest shortening occurring in the long parallel-fibered muscles such as the biceps and sartorius. In intact muscle, shortening is further limited by the structure of joints, the resistance of antagonists, and any load which opposes the muscle.

Intrinsic Speed of Shortening

The intrinsic shortening speed of a muscle reflects the rate of shortening at the sarcomere level. It is limited by the rate at which bridges can attach, move, and detach and by the rates of the chemical reactions involved. With muscle attachments severed, shortening speed of the contractile material is maximal but no tension is developed. A muscle can produce tension only when shortening against resistance, and the amount of tension developed is equal to the load.

When shortening against resistance, speed varies inversely with the load. Therefore in isotonic contraction the less the resistance, the more nearly maximal is the rate of shortening. This may be ex-

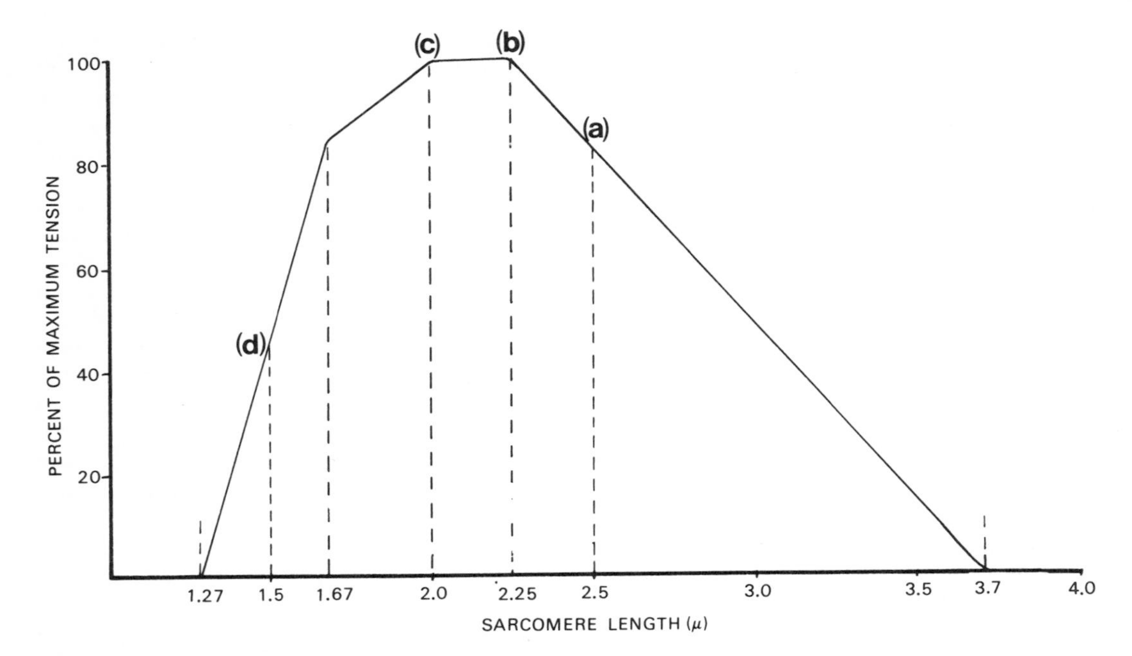

Figure 6.18. Tension-length curve for frog muscle at various sarcomere lengths. The letters on the tension curve and the broken vertical lines relate tension to significant sarcomere lengths. Note that tension is 0 both at the shortened length of 1.27 μ and at the extended length of about 3.7 μ, is maximal over lengths 2.00 μ and 2.25 μ, and declines rapidly below 1.67 μ and above 2.25 μ. (After Gordon, A. M., Huxley, A. F., and Julian, F. J., 1966. Variation in isometric tension with sarcomere length in vertebrate muscle fibers. *J. Physiol. (London) 184:* Fig. 12, p. 185.) See Figure 4.8 for diagrams of probable filament relations in myofibrils at some of the lengths.

plained as follows: the active state arises abruptly upon stimulation and persists for a relatively fixed period of time; the less resistance which is met by the contractile material, the more readily the bridges function and the greater the distance of shortening accomplished during the persistence of the active state.

When a muscle is required to shorten more rapidly against the same load, less tension is produced than when it shortens more slowly. This may be due to the fact that fewer links are formed between actin and myosin in the shorter time available and that the bridges which do form are detached more quickly. Consequently at higher speeds fewer bridges will be attached at any given moment and less tension is produced.

Hill (1965) pointed out that the load determines the rate of the chemical reactions associated with contraction and that the magnitude of the velocity depends on the difference between the actual load being lifted and the maximal magnitude of force of which the muscle is capable. Bárány's work (1967) supports the correlation of ATPase activity and speed of contraction in 14 different muscles of mammals, lower vertebrates, and invertebrates.

In isometric contraction, different levels of tension are achieved at the same rate. Since no shortening is involved beyond that needed to stretch out the elastic components, the rate of tension development is constant, determined by the active state. Hence the time to reach any given tension will be proportional to the tension: lower tensions will be achieved sooner than higher tensions.

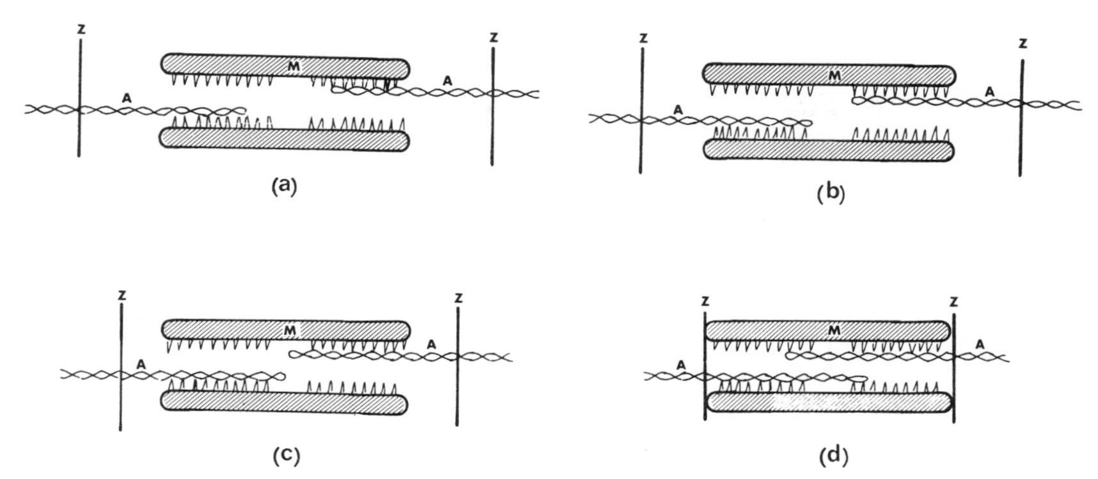

Figure 6.19. Schematic drawings of filament relationships at various stages of the sliding filament process associated with significant lengths on the tension-length curve presented in Figure 6.18. M, myosin; A, actin; Z, Z disc. (a) sarcomere length 2.5 μ. Actin filaments are partially overlapping the myosin filaments in the A band. Not all bridges can attach. Tension capability about 85% maximum. (b) sarcomere length 2.25 μ. Actin filaments are in maximal single overlap with the bridge-containing regions of the myosin filaments. Maximal tension capability. (c) sarcomere length 2.0 μ. Actin filaments have reached the center of the A band. Still maximal tension capability. (d) sarcomere length 1.5 μ. Z discs have collided with the ends of the myosin filaments. Ends of the actin filaments have passed into the bridge area of the opposite half of the sarcomere. Forty-five percent maximal tension capability.

Force-Velocity Relation

Speed is a scalar quantity, lacking the component of direction, while velocity is a vector quantity having both magnitude and direction. Therefore, the term *speed* has been used in the discussion of the rate of intrinsic shortening of the contractile material and the rate of tension development within the muscle, for direction was not significant. The term *velocity* is used to discuss the rate of muscle shortening against external resistance, i.e., the rate of movement, for in such considerations direction is an influential factor.

The velocity at which a muscle shortens is influenced by the force that it must produce to move the load. In concentric contraction the relationship is evidenced by the decrease in velocity as the load is increased (Fig. 6.20, solid line curve). Shortening velocity is maximal with zero load and reflects the intrinsic shortening speed of the contractile material. Velocity reaches zero with a load just too great for the muscle to lift; contraction is then isometric and maximal force can be produced.[e] When more muscle fibers are activated than are needed to overcome the load, the excess force is converted into increasing velocity and therefore greater distance of movement. A commonly experienced example is the exaggerated movement which occurs when one lifts a light object anticipated to be much heavier.

In eccentric contraction, values for shortening velocity become negative and the muscle's ability to sustain tension increases with increased speed of lengthening, but not to the extent which might be expected from extrapolation of the short-

[e] As previously stated, greater forces can be *sustained* in eccentric contraction than can be *produced* in either isometric or isotonic contraction.

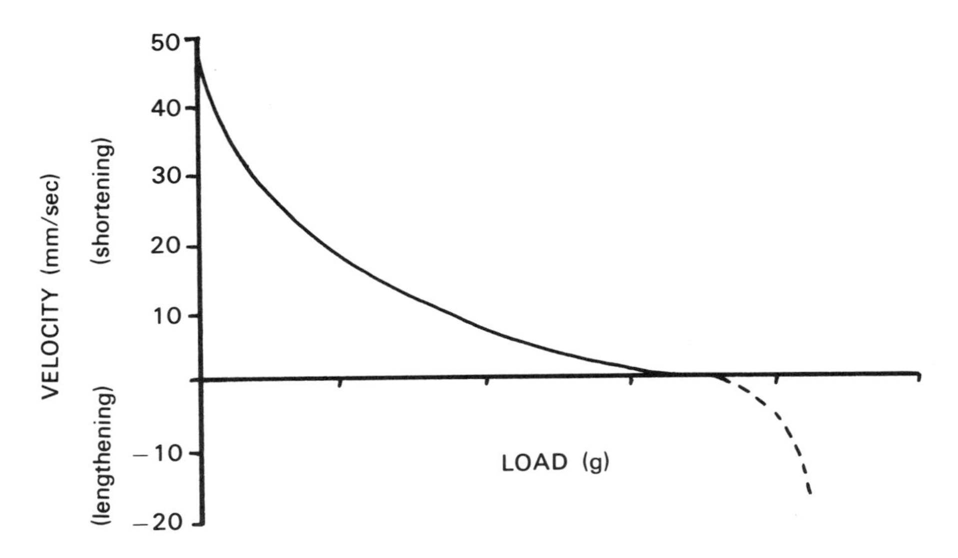

Figure 6.20. Relationship of velocity of shortening to tension in isotonic contraction (———). As the load is increased, velocity of shortening decreases, reaching 0 with a load just too great for the muscle to lift. In eccentric contraction, shortening velocities become negative and tension increases with increased speed of lengthening (– – –).

ening curve (see broken line in Fig. 6.20, extending the curve from the hyperbola of the force-velocity curve below the abscissa into the area of lengthening velocity).

In isotonic contractions the length and tension of the elastic components do not change once they are sufficiently stretched to permit the load to be raised. Therefore, isotonic twitch myograms reflect the velocity as well as the extent of shortening. When an excised frog gastrocnemius muscle records twitch responses with different loads, the lighter the load, the higher is the twitch curve and the steeper its rising slope (Fig. 6.21). In other words, the lighter the load, the greater the amount of shortening per unit of time. With a light load (Fig. 6.21(a)) the muscle's maximal velocity as measured over the steepest part of the contraction is 17 mm in 0.01 s or 1.7 m/s. With a moderate load (Fig. 6.21(b)) velocity is 12 mm in 0.01 s or 1.2 m/s, and with a heavy load (Fig. 6.21(c)) velocity is 5 mm in 0.01 s or 0.5 m/s.

Winter (1979) developed an interesting three-dimensional plot to demonstrate the relationship among force, length, and velocity (Fig. 6.22). If force is a function of both length and velocity, "the resultant curve is actually a surface, which represents only the maximal contraction condition." The more usual contractions would be at fractions of this maximum and surface plots would be required for each level of contraction; i.e., at 75%, 50%, or 25% of maximum (Winter (1979)). It is important to notice that Winter's clever schematic presents no new material, but offers the convenience of combining information from two other figures representing length-tension (Fig. 6.18) and force-velocity (Fig. 6.20) relationships in muscle. If a point on the surface depicted at (a) was analyzed in Winter's plot (Fig. 6.22), it would be concluded that this point represented isometric contraction, at resting length, with velocity zero and with force, which is dependent on all of these variables, at maximum. Moving from (a) to (b), the contraction has changed to a concentric one, the length is greater than resting, i.e., 1.1 × that of resting length, the velocity of shortening is relatively low, and the

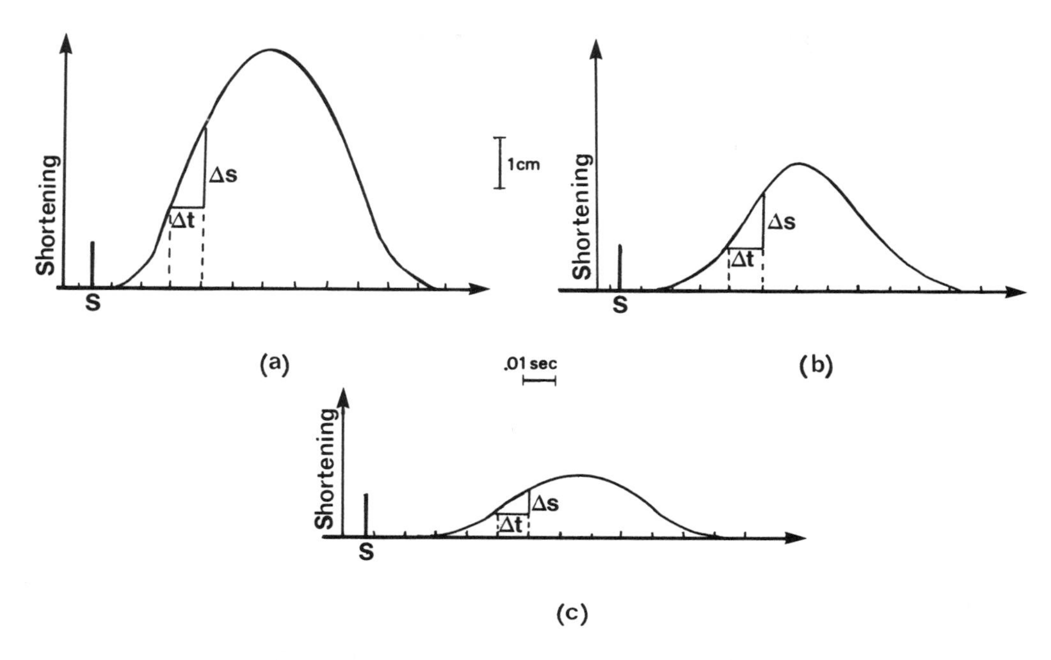

Figure 6.21. Shortening velocity of excised frog gastrocnemius with three different loads. Stimulus, S; Δt, 0.01 s taken at steepest part of contraction; Δs, amount of shortening in millimeters during that time unit. (a) with a light load, maximal rate of shortening (as taken over the steepest part of the contraction) is 17 mm/0.01 s or 1.7 m/s. (b) with a moderate load, velocity is 12 mm/0.01 s or 1.2 m/s. (c) with a heavy load, velocity is 5 mm/0.01 s or 0.5 m/s.

tension is considerable (0.8 × maximum isometric tension) but less than that at (a).

Optimal Velocity

The force-velocity relationship may also be stated conversely in terms of the influence of velocity upon force: a rapidly contracting muscle generates less force than does one contracting more slowly. From the standpoint of efficiency, however, energy expenditure is least when work is done at moderate velocity. With any given load, if the velocity of shortening is gradually increased, the work output of the muscle rises at first, reaches a peak, and then declines. Therefore, for any load the optimal velocity lies somewhere intermediate to the slowest and fastest shortening rates. Furthermore, the greater the load, the lower is the optimal velocity with which it may be moved.

Velocities below optimum are uneconomical because force must be maintained over a longer time; hence, more energy is expended to achieve the same amount of shortening. Velocities above optimum waste energy because of the need to employ a greater number of muscle fibers to achieve the same force. A plausible explanation is as follows: (1) the tension developed will be proportional to the number of bridges attached at any moment; (2) chemical reaction rates dictate that a finite and constant time is required for a bridge to attach; (3) the greater the speed at which the active sites on the actin filament move past the myosin bridges, the fewer the bridges that can attach and the less will be the tension; (4) as a result, more muscle fibers must be recruited to achieve the necessary force. Optimal speed probably reflects the greatest speed which will still allow a sufficient number of bridges to attach to provide the required tension.

Most individuals will unconsciously perform at optimal velocity if allowed to

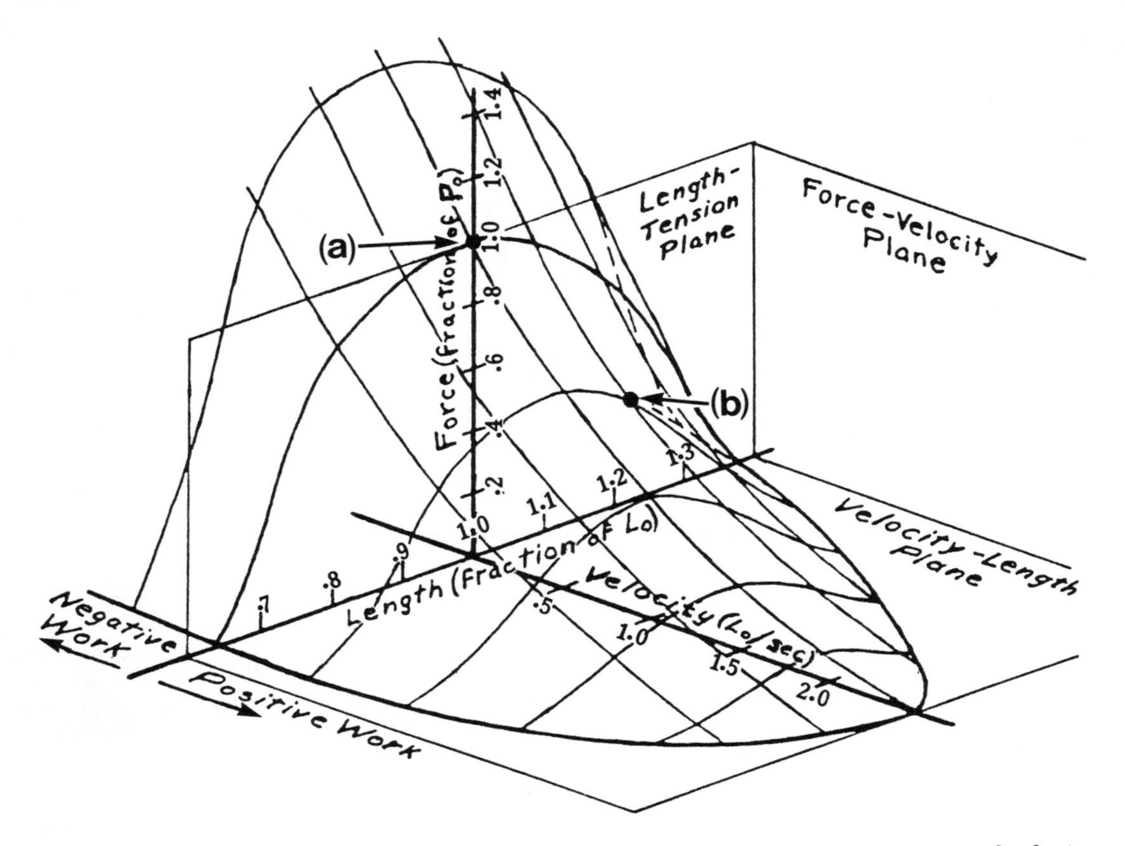

Figure 6.22. Three-dimensional plot showing change in muscle tension as a function of velocity and length. See text page 170. (After Winter, D.A., 1979. *Biomechanics of Human Movement*. New York: John Wiley and Sons, p. 120.)

do so. Optima vary for the same individual with different loads and in different types of activity and among individuals for the same load or activity. In athletic performance efficiency is often sacrificed for other objectives, and rates above optimum are deliberately adopted.

In isolated vertebrate muscle the Q_{10} of velocity is 2.5.[f] Although studies have not been made directly in humans, empiric evidence suggests that the velocity of muscle shortening is improved by "warming up." Experiments on the velocity effects of local warming or increased core temperature might indicate that artificial prewarming of athletes to a safe degree would improve their performance records. Physical educators have differing opinions on the value of "warm-ups."

[f] Q_{10} refers to the extent to which the rate of a chemical reaction is increased by a 10°C rise in temperature. It is usually measured within the range 15 -35°C. A Q_{10} of 2 indicates that the rate is doubled. For most biochemical reactions Q_{10} lies between 2.5 and 3.0.

Slow and Fast Muscle Fibers

Although the previous discussion has considered striated muscle in general, there is abundant evidence that there are two types of skeletal muscle, distinguishable by speed of contraction and endurance. Over 100 years ago Ranvier observed that some muscles of the rabbit were redder in color and that those muscles contracted in a slower and more sustained manner than did the paler muscles of the same animal.

Since then the designations of red and white muscles have become synonymous with slow and fast contraction, respectively. In addition to a slower contraction-relaxation cycle, red muscles have lower thresholds, tetanize at lower frequencies, fatigue less rapidly, and are more sensitive to stretch than the faster white muscles.

As might be expected, individual muscle fibers reflect these differences in contractile behavior. Investigations by a number of workers have revealed histological and biochemical differences which distinguish the two types of muscle fibers and which correlate with the physiological differences between fast (white) and slow (red) muscles. These are listed in Table 6.1.

Examination of the table suggests obvious relationships between the several categories which are consistent with differences in speed and endurance. The larger size, greater density of fibrils, and lower viscosity of the white fibers should contribute to greater speed. Their abundant sarcoplasmic reticulum and T tubules located at the A-I junctions should also favor fast response, by providing for the transport of glycolytic enzymes and for large scale release of Ca^{++} ions in the vicinity of the cross-bridges. Conversely, in the red fibers, reduction in the sarcoplasmic reticulum and sparsity of T tubules are appropriate to slower response. The discrete endplates and multiple folds in the subneural sarcolemma of the fast fibers may be expected to provide for increased activation. Where multiple innervation exists in slow fibers, it is characteristic to find small graded action potentials, which distinguish them from fast fibers, which have rapidly rising, larger potentials.

The large glycogen stores and high ATPase activity in the white fibers will favor speed of response but quicker fatigue is to be expected because of their lesser blood supply and predominantly glycolytic metabolism. The increased endurance of the red fibers is consistent with their rich blood supply and abundance of mitochondria which support an essentially oxidative metabolism. Furthermore, their small diameters provide a greater surface for exchange of gases, ions, and metabolites than is provided by an equivalent mass of larger white fibers. Rapid K^+ depletion progressively alters the ionic gradients and ultimately limits the ability of the fast fibers to perform work. In the slow fibers presumably a steady state is reached in which K^+ gain resulting from recovery processes is in equilibrium with the loss incurred during contraction. As a result, endurance is enhanced. The greater elasticity and slower initial decay of the active state which is characteristic of the red fibers can account for their mechanical fusion at lower frequencies and for their prolonged twitch times.

Most human striated muscles contain both types of fibers but in differing proportions which determine the color of each muscle. Some show a characteristic arrangement or zonation of the fiber types within the muscle; in others the two types are randomly distributed. In such muscles as the gastrocnemius, tibialis anterior, and flexor digitorum longus, fast fibers predominate, although slow fibers may also be present. In many mammals the soleus muscle appears to consist entirely of slow fibers. The preponderantly slow-fibered muscles are the antigravity muscles, adapted for continuous body support. Their sensitivity to stretch results in a continuous mild (tonic) activity even at rest. The predominantly fast-fibered muscles are phasic muscles which produce quick postural changes and fine skilled movements. At rest they are electrically silent.

Recent studies indicate that in mammals the slow red fibers should be further subdivided on the basis of differences in enzymatic activity, especially succinic dehydrogenase and ATPase activity, into two subcategories. As a result, muscle fibers may be classified into three types, which have been designated as A, B, and C. Type A fibers represent the classic fast, pale fibers, whereas types B and C represent two types of slow, red fibers. Stein and Padykula (1962) found all three types

Table 6.1
Comparison of Fast and Slow Muscle Fibers: Histological, Physiological, and Biochemical

	Fast	Slow
Histological differences		
Size and color	Fibers large and pale	Fibers smaller and redder because of greater myoglobin content
Sarcoplasm	Agranular sarcoplasm	Granular sarcoplasm
Fibrils	Many fibrils	Fewer fibrils
Mitochondria	Large mitochondria but few in number	Numerous small mitochondria
Z discs	Narrow Z discs	Wider Z discs (about 2×)
SR and T system	Sarcoplasmic reticulum abundant and well developed; T tubules (and triads) at A-I junctions	SR sparse and rudimentary; T system, when present, is found at the Z lines
Innervation	Single innervation by large somatic motor neurons with fast conduction rates	Innervation by small, slow-conducting somatic motor neurons; multiple innervation in some species. Some autonomic neutrons
End-plates	Discrete (*en plaque*) end-plates with many sarcolemmal folds	Diffuse (*en grappe*) end-plates with few or no junctional folds
Blood supply	Few capillaries except those shared with adjacent slow fibers	Dense capillary supply, located at the interstitial angles between fibers
Physiological differences		
Contraction cycle	Rapid contraction-relaxation cycle	Slower cycle (2−3×); graded contraction in some muscles
Tetanus	Rapid onset of tetanic fusion but only at high frequencies and short-lasting	Slower onset of fusion but at lower frequencies and of longer duration
Potentials	Higher resting potential Larger end-plate and action potentials	Lower resting potential Smaller end-plate and action potentials, latter often graded
Active state	Rapid initial decay of active state	Slower initial decay
Endurance	Rapid fatigue	Greater endurance
Tension capacity	Higher tension which develops rapidly	Lower tension and slower development
Elasticity	Lower coefficient of elasticity	Greater elasticity
Biochemical differences		
Metabolism	Metabolism primarily glycolytic (as indicated by high ATPase activity)	Oxidative metabolism (as indicated by high succinic dehydrogenase activity)
Myoglobin content	Low	High
Glycogen	Large glycogen storage	Variable glycogen storage
Na^+ and K^+	Less Na^+ and more K^+; rapid loss of K^+ during stimulation	More Na^+ and less K^+; rate of K^+ depletion diminishes with continued stimulation
Amino acids	Differences in concentration of various amino acids	

present in the rat gastrocnemius, with type A predominant. In the soleus, A fibers were absent but the muscle contained both types B and C.

Serial sections of human vastus lateralis muscle are depicted in Figure 6.23. Three fiber types are identified.

The significance of slow and fast characteristics of muscle fibers is discussed further in Chapter 9.

Key Observations

The following summarizes key observations relating to muscle contractility:

1. The tension exerted by active muscle is a function of its length and is maximal at about the greatest length that the muscle can assume in the living animal. Tension decreases nearly linearly above and below this length. The shape of the tension length curve for isometric contraction is similar for all muscles tested.
2. A muscle which is being lengthened while it is contracting can maintain greater tension than it can develop at any given equivalent static length. Therefore, tensions greater than the isometric maximal can be recorded in eccentric contraction.
3. Velocity of shortening of the contractile material decreases with increasing load in a hyperbolic manner. When velocity is expressed as a percentage of maximal velocity at zero force and force is expressed as a percentage of maximal force at zero velocity, the force-velocity curve is essentially the same for all muscles.
4. Quick release produces the same effects in all muscles: tension falls immediately and then redevelops to a value characteristic of the shorter length.
5. The rate of the development of ten-

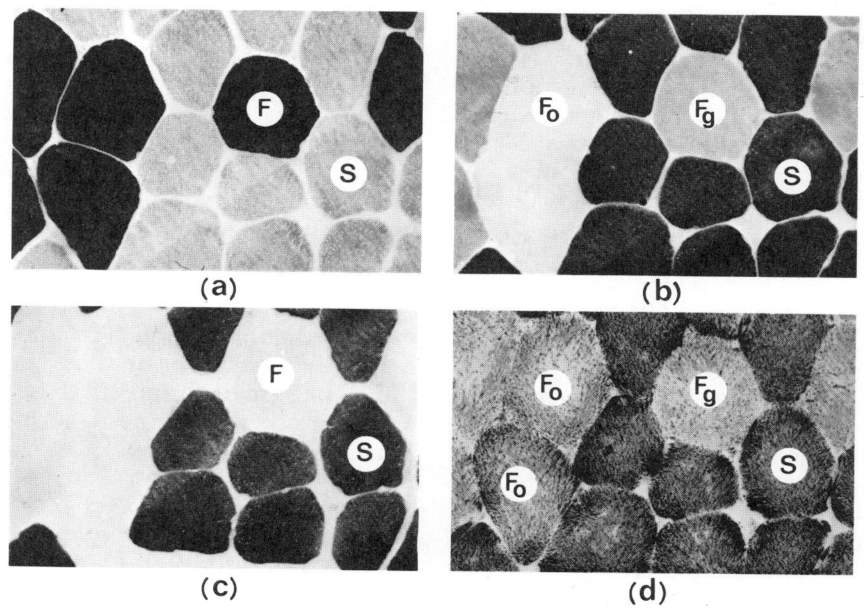

Figure 6.23. Serial sections (using light microscopy) of human vastus lateralis muscle. Magnification approximately × 240. (a) myosin ATPase pH 10.0; demonstrates two fiber types. (b) myosin ATPase pH 4.6; demonstrates three fiber types. (c) myosin ATPase pH 4.3; demonstrates reversal of pH 10.0 reaction. (d) NADH diaphorase (oxidative state); demonstrates three fiber types. F, fast twitch; F_g, fast twitch (glycolytic); F_o, fast twitch (oxidative); S, slow twitch. (Courtesy of G. Elder, McMaster University.)

sion depends on the intrinsic shortening velocity of the contractile material, compliance of the elastic components, and the rate of decay of the active state.

The existence of such extensive similarities among muscles provides a measure of confidence in assuming that information on the properties of muscle derived from animal studies may also apply to humans. However, reasonable caution and good judgment must be exercised when direct confirmation is lacking.

Hypertrophy and Atrophy

Needle biopsies of human skeletal muscle are now used as common practice to identify the structural and histological differences which can be attributed to the independent variables of concern. Investigations using this technique involve a wide variety of studies, such as those concerned with energy release and fatigue, recruitment patterns among the three fiber types, motor development studies, effects of immobilization, and responses to strength training. In an early investigation, nine healthy volunteers were studied "under control conditions and following five months of heavy resistance training and five weeks of immobilization. . . ." Needle biopsies were taken from triceps brachii and analyzed for concentrations of ATP, ADP, creatine and CP. "It was concluded that heavy resistance training results in increases in muscle energy reserves which may be reversed by a period of immobilization-induced disuse" (MacDougall et al. (1977a)).

In a subsequent paper (MacDougall et al. (1977b)) studied seven healthy male subjects using the same protocol. Needle biopsies were taken from triceps brachii so that cross-sectional fiber areas could be analyzed, and overt measures of strength were recorded by a Cybex dynamometer. Training resulted in a 98% increase in elbow extension strength and significant increases in both fast and slow twitch fiber areas (Fig. 6.24). Immobilization resulted in a 41% decrease in elbow extension strength and significant decreases in fiber area for both fast and slow twitch fibers (Fig. 6.25). A range of staining intensities can be seen in Figures 6.24 and 6.25, revealing the three fiber types: slow twitch; fast twitch oxidative; and fast twitch glycolytic.

The histological differences displayed for hypertrophied and atrophied human skeletal muscle are of major importance to the practitioner whether the application is to train the elite athlete or to rehabilitate the physically handicapped. We are only beginning to realize the implications of ultramorphological characteristics.

Muscle Classifications

Various attempts have been made to classify skeletal muscles in appropriate ways. These include the following categories: (1) red and white; (2) tonic and phasic; (3) expanders and contractors; and (4) spurt and shunt muscles.

Red and White Muscles

As indicated earlier in this chapter, physiologists and histologists have identified both red and white skeletal muscle. The red muscle contracts slowly and the white muscle contracts rapidly.

Tonic and Phasic Muscles

Most anatomists and kinesiologists generally divide muscles on the basis of function as antigravity or postural muscles and as the more rapidly contracting phasic muscles used in motor skills. More recently the inclination is to consider the antigravity-postural muscles as tonic on the basis of the continuous low level of contractile activity that is required to maintain a given posture. The muscles of the tonic group contain proportionately more red, slow-contracting muscle fibers, while the more rapidly contracting phasic muscles contain a larger proportion of white fibers.

Stockmeyer (1970) points out further

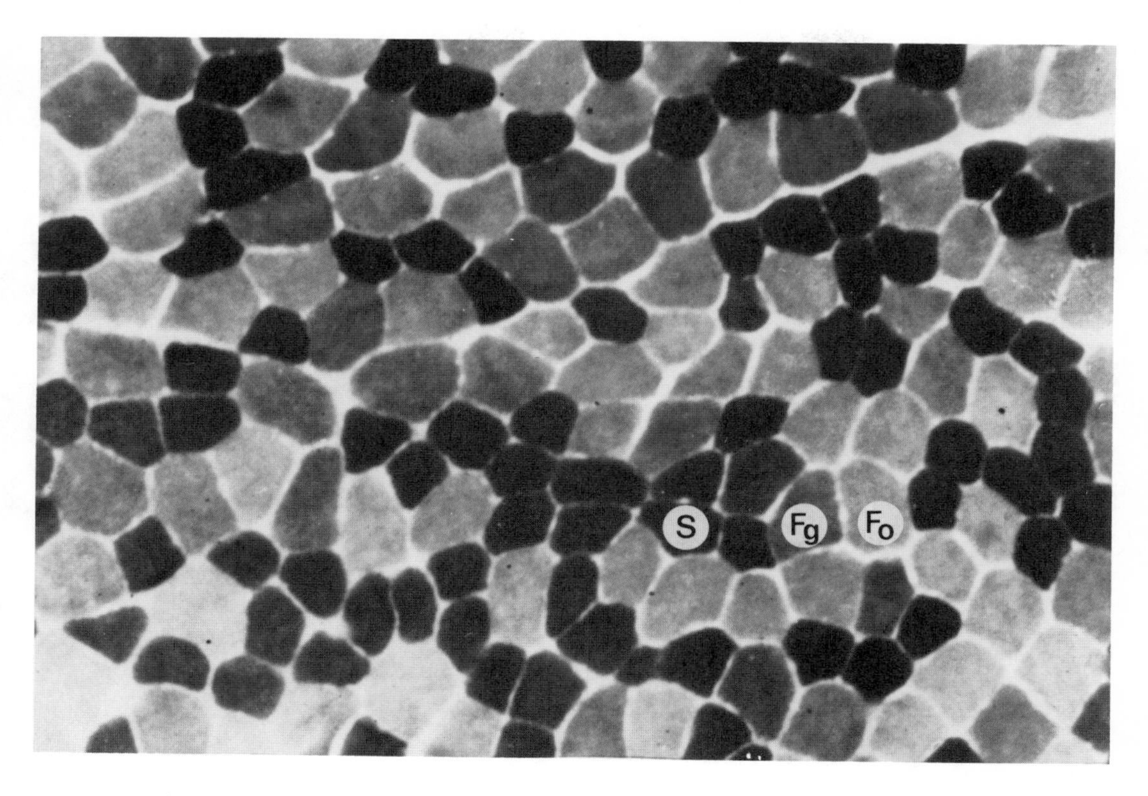

Figure 6.24. Hypertrophy of muscle: long head of triceps brachii. Magnification approximately ×200. Induced by heavy resistance weight training. Myosin ATPase, pH 4.6. Each of three fiber types is noted as follows: S, slow twitch; F_o, fast twitch oxidative; F_g, fast twitch glycolytic. (Courtesy of G. Elder, McMaster University.)

characteristics of the tonic muscles: they are mostly penniform with shorter muscle fibers; they lie deeper and more medially; they generally cross only one joint; and they belong to the extensor group functioning as abductors as well as lateral rotators. Phasic muscles are located more superficially and more laterad; they have longer fibers; they may cross more than one joint; and they generally belong to the flexor group whose functions also include adduction and medial rotation. Also, tonic muscles are classified as stabilizers and phasic muscles as mobilizers.

Contractors and Expanders

Grant and Smith (1953) point out that skeletal muscles as a whole may be divided into contractors and expanders. Those muscles which pull the body into an approximation of the fetal position, e.g., the flexors, adductors, and medial rotators, are classed as the "contractors." On the other hand, those muscles which expand or open up the body, e.g., the extensors, abductors, and lateral rotators of the limbs, are classed as the "expanders."

Spurt and Shunt

MacConaill and Basmajian (1977) offer a different concept by dividing the muscles on the basis of the relative magnitudes of their stabilizing and rotatory components. Those muscles with attachments farther from the joint axis have larger stabilizing components and are called shunt

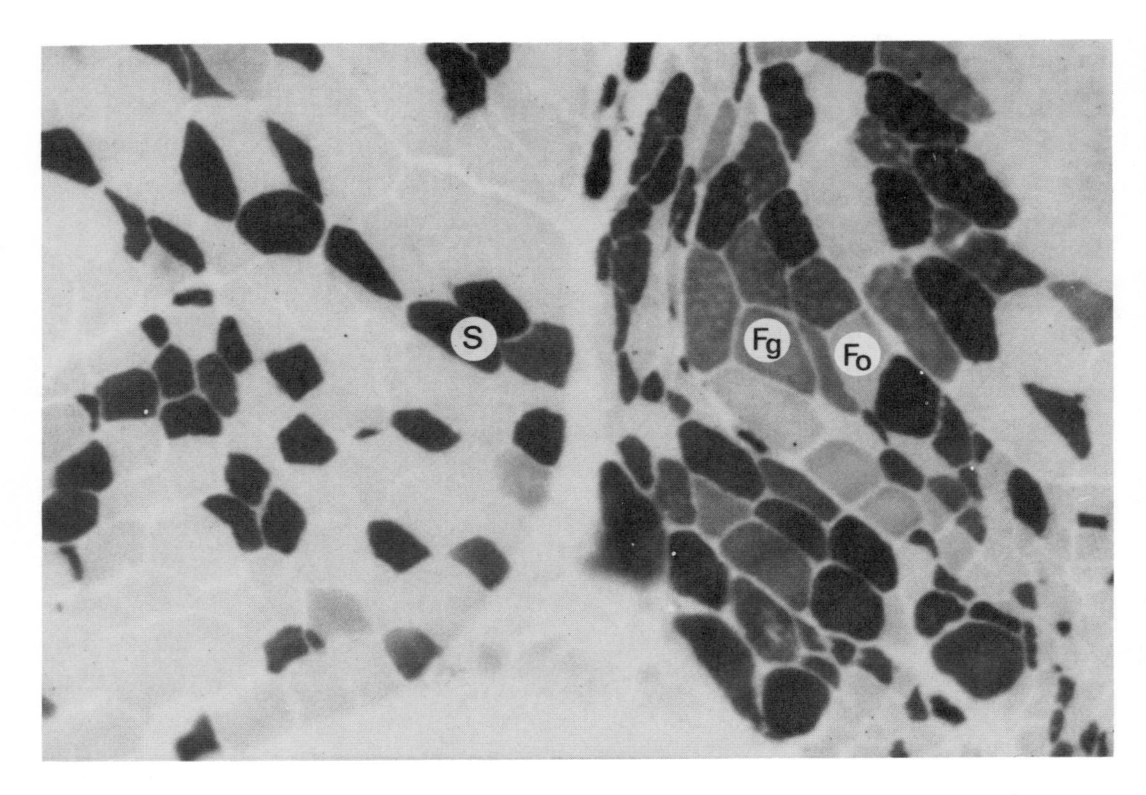

Figure 6.25. Atrophy of muscle: long head of triceps brachii. Magnification approximately ×200. Induced by 5 weeks in a cast. Myosin ATPase, pH 4.6. Normally, this reaction demonstrates only three intensities. With atrophy, the range of intensities of reaction increases. The illustration shows many fibers of all three fiber types severely atrophied. However, some fibers are relatively spared, the reason for which is unknown. Each of three fiber types is noted as follows: S, slow twitch; F_o, fast twitch oxidative; F_g, fast twitch glycolytic. (Courtesy of G. Elder, McMaster University.)

muscles, while those which attach closer to the joint axis have larger rotatory components and are called spurt muscles.

This terminology is derived from that employed by 19th-century British engineers, who used the term shunt to describe a force which prevented a body moving in a curved path from taking a tangential or rectilinear direction; the body was shunted back onto the curved track. Thus, the term shunt as used by MacConaill and Basmajian refers to the centripetal force described by the physicist or engineer. The 19th-century engineers used the term spurt to indicate the force which "provided the necessary spurt of energy that impelled the body into motion, or if necessary, kept it in motion." This is simply applying a different term to the force defined in Newton's Second Law, which states that force is equivalent to the mass of a body multiplied by its acceleration.

Muscles do not function solely as either shunt or spurt muscles. To illustrate this concept, elbow flexor muscles are used as examples. In the case of elbow flexion from the anatomical position, biceps brachii and brachialis function as spurt muscles while brachioradialis acts as a shunt muscle. The action is best described as one in which the muscles apply their force at their distal attachments. By way of contrast, if a person were hanging from a bar with elbows extended and performed a chin-up, the force from these same muscles is now applied at their proximal attachments; hence, the brachioradialis functions as a spurt muscle while both biceps brachii and brachialis act as shunt muscles.

Muscle Function

Motor skill and all forms of movement result from interaction of muscular force, gravity, and any other external forces which impinge on skeletal levers. Muscles rarely act singly; rather, groups of muscles interact in many ways so that the desired movement is accomplished. This interaction may take many different forms so that a muscle may serve in a number of different capacities, depending on the movement. Thus, at different times a muscle may function as a prime mover, an antagonist, or a fixator, or synergically as a helper, a neutralizer, or a stabilizer. A muscle that is the prime mover for one movement will become an antagonist when the movement is reversed, and on other occasions it may perform any one of the other functions listed above, depending upon the circumstances.

Prime Movers and Movers

Whenever a muscle causes movement by shortening, it is functioning as a mover or agonist. If it is believed that the muscle in question makes the major contribution to the movement, that muscle is regarded as the prime mover. Other muscles crossing the same joint on the same aspect but which are smaller or which are shown electromyographically to make a lesser contribution to the movement under consideration are identified as secondary or assistant movers or agonists.

Antagonists

Muscles whose action is opposite to and so may oppose that of a prime mover are called antagonists. This does not mean that an antagonist, as the name implies, always exerts tension against the prime mover; electromyography has demonstrated conclusively an absence of electrical activity in opposing muscles.[g] The term antagonist, however, is a carryover from the days when it was erroneously believed that whenever a muscle contracted as a prime mover, e.g., as a flexor, the opposing extensors always regulated and controlled the movement by exerting a lesser tension against the prime mover. Electromyographic records of simple, familiar movements or of highly skilled movements performed against a light resistance (such as only the weight of the body segments being moved) clearly demonstrate that reciprocal and complete inhibition of opposing muscles is the rule, i.e., the antagonistic muscles relax (Fig. 6.26). Cocontraction in normal subjects has been demonstrated electromyographically, especially when an individual is deliberately performing a "tension" movement, or when tremendous effort is being exerted. Under these latter circumstances the antagonistic activity is, in all probability, synergic in nature as it serves to stabilize the joints against the power exerted by the prime movers.

In various pathological conditions such as hemiplegia and cerebral palsy, cocontractions are observed. Often therapeutic strategies must be devised to minimize or even eliminate such cocontractions.

Synergy

Synergic action has been defined as cooperative action of two or more muscles in the production of a desired movement. A synergist, then, may be regarded as a muscle which cooperates with the prime mover so as to enhance the movement. Synergic interaction may take many forms and variations as discussed below.

Conjoint Synergists. Two muscles act-

[g] The reader is referred to Chapter 11 for an explanation of details concerned with electromyography.

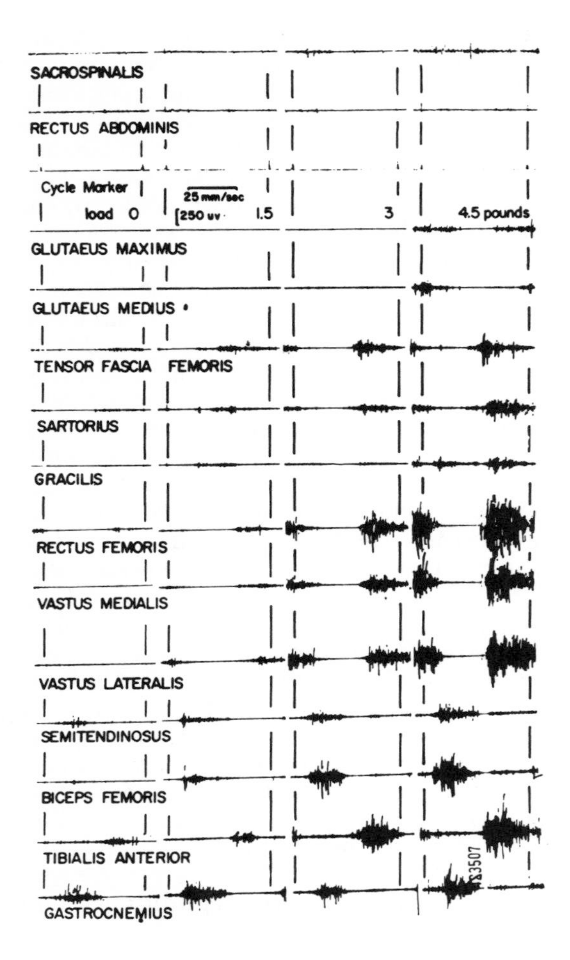

Figure 6.26. Electromyograms showing reciprocal inhibition of muscles crossing the hip, knee, and ankle joints while riding a stationary bicycle. Cycle marker (third line) indicates foot at 3 o'clock. (From Houtz, S. J., and Fischer, F. J., 1959. An analysis of muscle action and joint excursion during exercise on a stationary bicycle. *J. Bone Joint Surg. 41-A:* 123.)

ing together to produce a movement which neither could produce alone may be classed as conjoint synergists. Dorsiflexion of the foot at the ankle is an example. The movement is produced by the combined action of the tibialis anterior and the extensor digitorum longus. The tibialis anterior alone would produce a combination of dorsiflexion and inversion, while shortening of the extensor digitorum longus alone would produce toe extension, dorsiflexion, and eversion. Acting together, the muscles produce a movement of pure dorsiflexion. Another example occurs in lateral deviation of the hand at the wrist; e.g., ulnar deviation results from the simultaneous action of the flexor carpi ulnaris and the extensor carpi ulnaris.

Neutralizing or Counteracting Synergists. Some muscles cross bi- or multiaxial joints in such a fashion that they are capable of causing more than one action at that joint. The muscles on the medial aspect of the thigh are all prime movers in adduction of the thigh at the hip joint but, if unopposed, the pectineus and the adductors longus and brevis will at the same time flex the thigh and rotate it medially. Thus, whenever only adduction is desired, other muscles crossing the hip joint must become active to neutralize or counteract the undesired components. Further problems arise whenever muscles cross two or more joints. A muscle producing a desired movement at one joint frequently produces an undesirable effect at another joint (or joints). Therefore, still other muscles must contract synergically to neutralize the undesired actions. The classic example of this situation is that of the finger-flexion wrist-extension interaction. Contraction of the finger flexors to grasp an object also tends to flex the wrist. The unwanted wrist flexion is counteracted by the synergic contraction of the wrist extensors.

Stabilizing Synergists. The sine qua non of an effective coordinate movement involves greater stabilization of the more proximal joints so that the distal segments move effectively. The greater the amount of force to be exerted by the open end of a kinematic chain (whether it is the peripheral end of an upper or of a lower extremity), the greater the amount of stabilizing force that is needed at the proximal links. Thus, when throwing or striking forcefully or moving a heavy weight, the muscles of the shoulder girdle and shoulder must contract more forcefully to support the scapula and shoulder joint than they do when the hand is making a gesture or lifting food to the mouth. In such instances the opposing muscles interact, contracting simultaneously in such a manner that, while the scapula moves on the thorax during the effort, it still provides a firm base for the movement of the humerus. At the same time the rotator cuff muscles (supraspinatus, subscapularis, infraspinatus, and teres minor) all contribute their opposing tensions to the common end of supporting the humeral head against the glenoid fossa.

Other examples of stabilizing synergy may be mentioned. Executing a Valsalva maneuver (holding the breath while forcefully contracting the abdominal musculature, which exerts pressure on the viscera) while running or jumping provides a firm base for the lower extremities to push against so that the body is propelled efficiently. The abdominal muscles are exerting a stabilizing synergy on the trunk. When one foot is supporting the body weight, all of the muscles crossing the ankle joint become active in synergic contractions supporting and stabilizing the ankle joint.

Fixation

When a joint is voluntarily fixed rather than stabilized, there is, in addition to immobilization, a rigidity or stiffness resulting from the strong isometric contraction of all muscles crossing that joint. These muscles will forcefully resist all external efforts to move that joint. As fixation can be very tiring, it is seldom used—and rarely useful.

Stabilization Versus Fixation

From the above discussion one should recognize the difference between stabilization and fixation of joints. As stated, fixation denotes a rigidity or stiffness in opposition to all movement, whereas stabilization implies only a firmness. Fixation may occur inadvertently when, for example, a student learning a skill misinterprets directions given by an instructor. When a golf instructor stresses the need for a straight left arm (for a righthanded player) during the swing, many beginners will respond with maximal contraction of both the biceps and triceps, with disastrous results. Under these circumstances the long heads of both muscles are pulling on their scapular attachments above and below the glenoid fossa, which severely inhibits the free movement of the humerus so necessary to a successful swing. On the other hand, tension in the triceps just sufficient to keep the elbow extended will stabilize that joint and provide a constant lever length or radius for the free and easy swing that will stay "in the groove" and result in forceful contact with the ball. Similar instances may arise during the teaching of any skill new to the student.

Economy of movement involves the use of minimal stabilizing synergy and no fixation of joints. Teaching cues and coaching points for motor skills should be geared to this end; e.g., "Keep your arm straight, but not stiff," etc.

Therapists employ similar kinds of strategies, perhaps more intensely, as they endeavor to instill effective therapeutic ideas in patients.

MUSCLE ATTACHMENTS AND THEIR EFFECTS ON FUNCTION

The torque about any joint is determined by the muscular forces generated, coupled with several anatomical factors which contribute to the length of the moment arm or arms of the musculoskeletal elements under consideration. The contributing factors which should be recognized are: (1) attachment of the muscle to bony prominences; (2) passage of the muscle tendon over bony prominences; and (3) both attachments of the muscle distant to the joint.

Attachment to Bony Prominences

The stress exerted on bones created by the frequent tension exerted by contracting muscles stimulates increased growth of the bone in the area of the attachment. Thus, the very fact that a muscle is anchored to bone causes the protuberances at anchor points. The attachments of the adductors and the two vasti muscles create the linea aspera on the back of the femur; the pull of the glutei increases the size of the greater trochanter; the size of the deltoid tuberosity is directly related to the size of the deltoid muscle attached to it, that of a muscular man being much larger than that of a woman who is comparatively inactive. These prominences and even the bones themselves may change in size, both increasing and decreasing during the lifetime of an individual. Prives (1960) in dealing with the influences of labor and sport on skeletal structure in humans reported that X-rays taken of the same individuals over a period of 10 years showed "Variations of bone shape and structure characteristic for the given occupation or sport. . . . Change of professional trade provokes changes in the structure of bones corresponding to the new type of loading."

In this study laborers and athletes were deliberately changed to a more sedentary way of life while sedentary workers were inducted into labor gangs and their participation in athletics was encouraged. The results which followed indicated that:

1. Physical loading causes the redistribution of certain radioactive isotopes such that they accumulate in the most loaded or stressed parts of the skeleton.
2. Different forms of osseus hypertrophy occur.
3. Physical work favors the growth of the bones in length.
4. Physical work delays the aging of the osseus system, which is important for prolongation of life.

The increase in the size of the bone occurs, of course, in response to the strain put on the entire musculoskeletal system and in particular to the increased pull exerted by the increased use of muscular force. The osseus hypertrophy of a bony prominence as well as of the bone itself moves the muscular attachment that much farther from the joint axis and so increases the moment arm or arms of the muscle in question. This of course increases the effectiveness of the muscle force as well.

Passage of Tendon or Muscle over Bony Prominences

Passage of a muscle or tendon over a prominence alters the direction of the action line of the muscle, moving it farther from the joint axis, and therefore increasing the moment arm of the muscle. For example, the middle deltoid must pass over the greater tuberosity of the humerus (Fig. 6.27), and tendons of the gastrocnemius and the hamstrings pass over the femoral condyles (Fig. 6.28). The action line of the quadriceps femoris is deflected away from the knee joint by the patella (Fig. 6.29). Thus, in the instance of a patellectomy, for the same muscular force generated, the reduced moment arm results in the production of less torque about the knee joint.

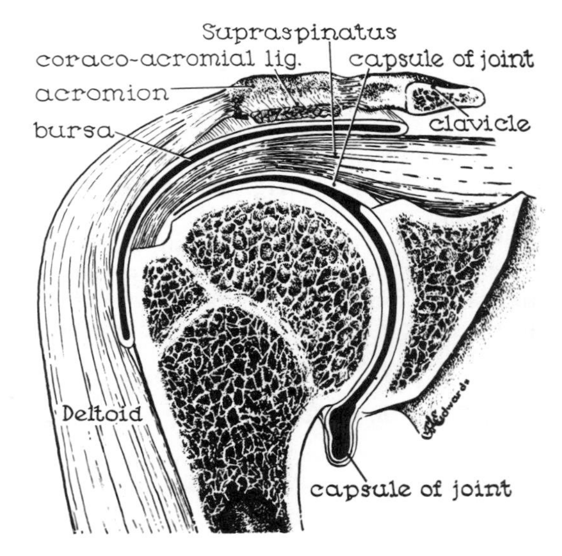

Figure 6.27. Relation of middle deltoid to greater tuberosity of the humerus. (From MacConaill, M. A., and Basmajian, J. V., 1977. *Muscles and Movements*. Baltimore: The Williams & Wilkins Company.)

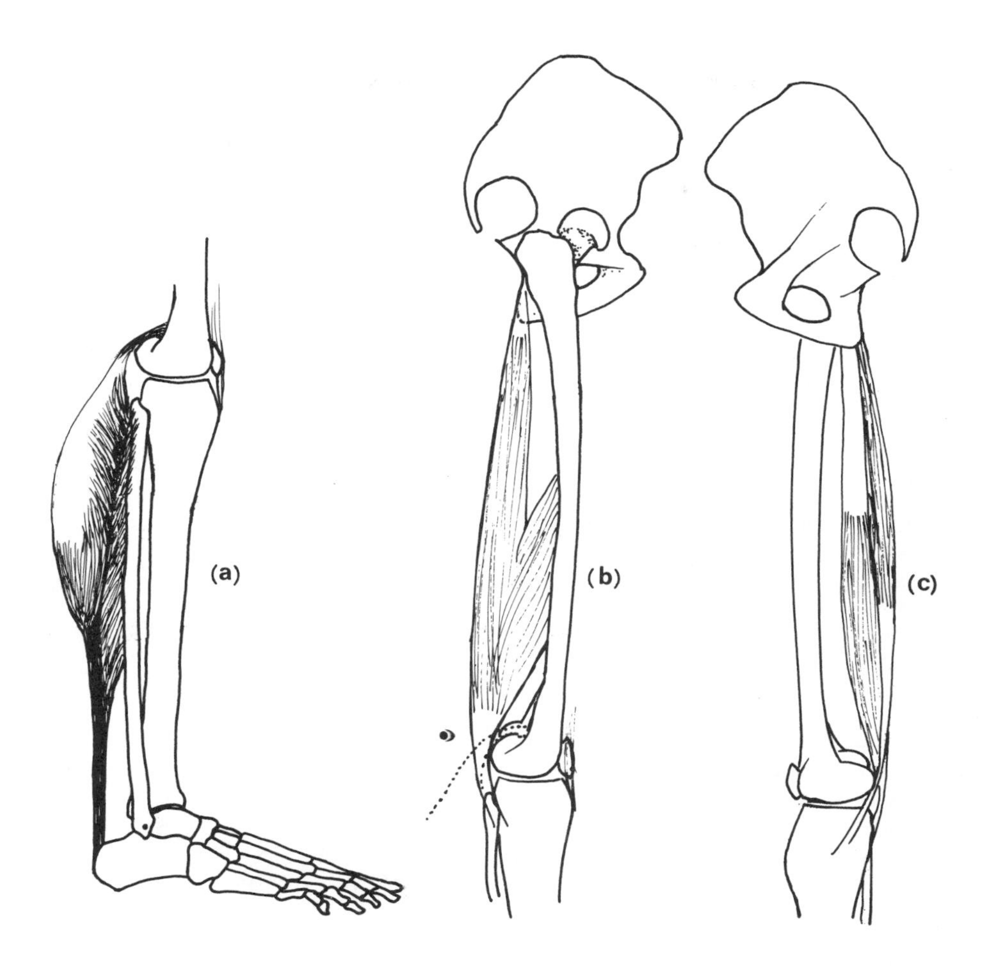

Figure 6.28. Relation of gastrocnemius and hamstrings to femoral condyles. (a) gastrocnemius, lateral view; (b) biceps femoris, lateral view; (c) semitendinosis and semimembranosis, medial view.

Both Attachments of the Muscle Situated at a Distance from the Joint

A muscle with both of its attachments at a large distance from a joint axis and without fascia or reticulum to hold the belly or tendon close to the bones would have a large moment arm. However, such an arrangement would be most ungainly and, as restrictions are always in effect so that no "bowstringing" of muscle or tendon can occur, the result is that there are no grossly exaggerated moment arms. Nevertheless, even with these restrictions, when the adult body is in the anatomical position the action lines of several muscles fall from 3 to 6 cm from the axis or axes of the joint at which they act, as a result of the fact that the closest attachment is 5 to 8 cm from the joint center. Probably the longest muscular moment arms in the body (abdominal muscles excluded) are those of the hamstring muscles at the hip. In the anatomical position their action line passes from 5 to 8 cm from the flexion-extension axis through the head of the femur at the hip. The distal

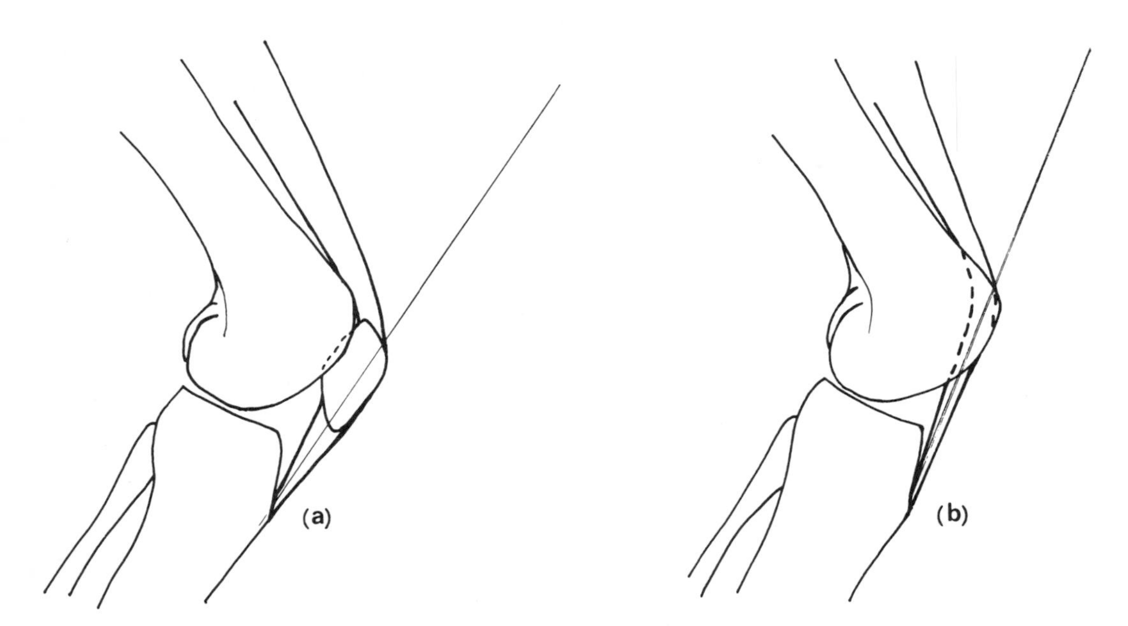

Figure 6.29. Change in angle of pull of quadriceps femoris caused by the patella. (a) patella present; (b) patella absent.

attachments of the pectoralis major on the humerus are far enough down the bicipital groove so that the flexion moment arm of the clavicular head can also be as large as 5 cm, as may be the adduction moment arm of the sternal head. The action line of the triceps surae (the two heads of the gastrocnemius and the soleus) running upward from the calcaneus can vary from 3 to 6 or more cm. These variations in moment arm lengths result from individual differences in skeletal size and bony development. Thus, the larger-boned individual has an immediate advantage over one of slighter build insofar as moment arms and therefore produced torque for the same bulk (cross-section) of muscle involved are concerned.

Muscles with long moment arms, even when the limbs are in the anatomical position, are for the most part large muscles which have the capability of exerting considerable force (e.g., the glutei, hamstrings, gastrocnemius-soleus, pectoralis major). The fact that their attachments are at some distance from the joint axis means that they must shorten more to move a segment through a given angle than a muscle which attaches closer to the joint (Fig. 6.30). Two facts are obvious from the above. (1) If the two muscles depicted in Figure 6.30 shorten at the same rate, the muscle in (a) will move the segment faster. (2) The moment arm of (b) is longer; therefore, if it exerts the same amount of force per unit shortening as does (a), the moment of force generated by (b) will be greater.

Muscle Forces Acting on Skeletal Levers

When a muscle generates tension it pulls equally against all of its attachments just as a rubber band stretched between two fingers pulls against each. In movement analysis, it may at times prove convenient to consider the muscular force as being applied only at the attachment of concern.

If a muscle or muscle group is shortening (isotonic contraction) or exerting tension while being lengthened by an outside

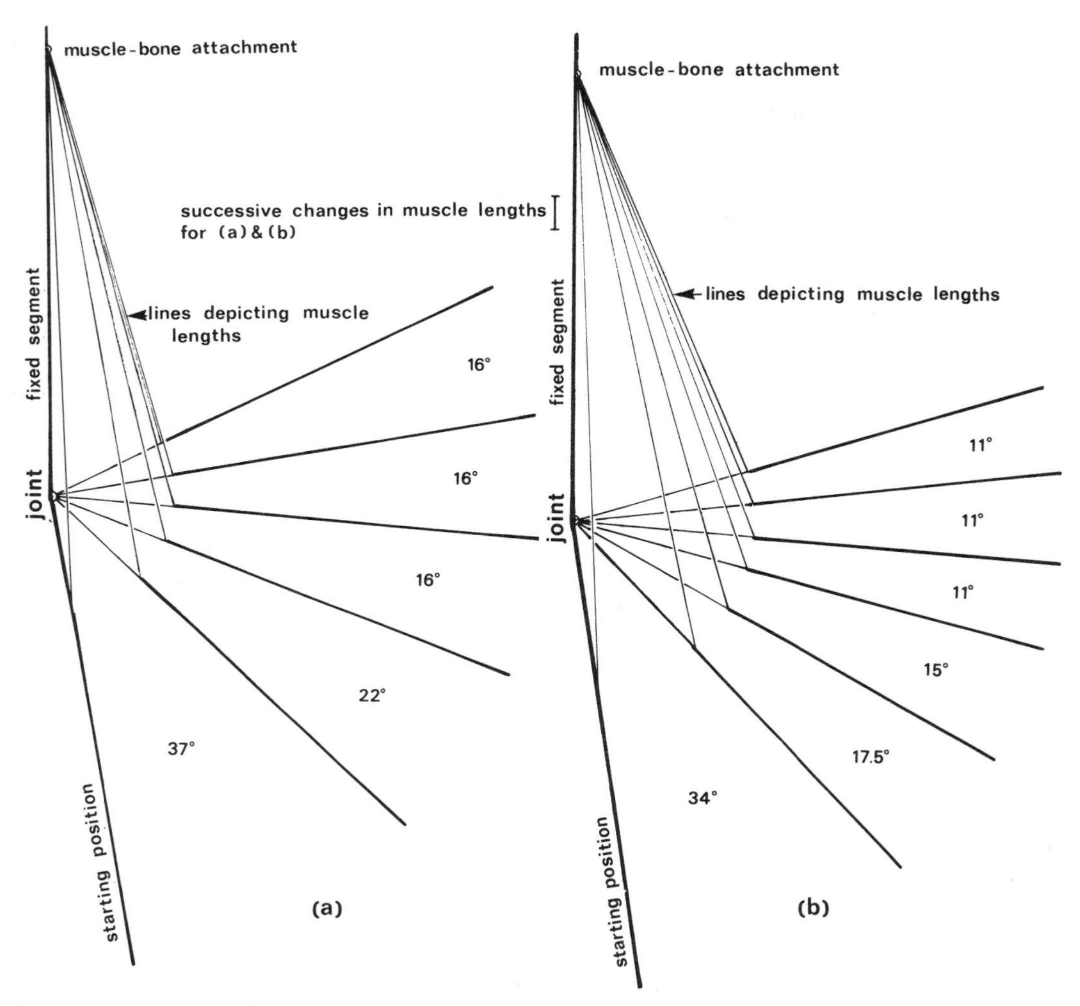

Figure 6.30. Degrees of joint motion caused by equal muscle shortening. (a) muscle attached close to joint axis; (b) muscle attached farther from the same axis. Successive changes in the muscle lengths (shortening) are indicated commencing from the common starting positions shown.

force (eccentric contraction), then one of the body parts to which that muscle is anchored is being moved on an adjacent part. Also, one of the two segments to which the muscle is attached is closer to or may even form (as is the case with a hand or foot) the open end of a kinematic chain. In the isotonic contraction the muscle contraction is the moving force. For the eccentric contraction the muscle performs a regulatory service, resisting the moving force and so controlling the rate at which the body part is being moved by the outside force. In either case the muscle force is considered as being applied to the moving segment nearest the *open end of the kinematic chain*, where its effect is clearly visible.

In Figure 6.31 the biceps brachii and brachialis are flexing the elbow. This action moves the hand toward the shoulder while the humerus remains motionless and close to the rib cage. There is no movement either of the arm or of the shoulder girdle so the tension exerted by the flexors is effective only at the distal attachment.

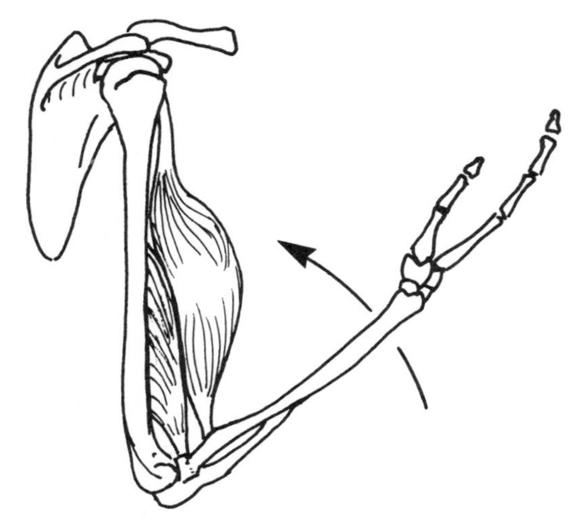

Figure 6.31. Biceps and brachialis are flexing the elbow.

Under these and similar circumstances we say the effort exerted by these muscles is applied at the distal attachments of the muscles, i.e., via the attachment to the bone of the moving segment.

Imagine a gymnast climbing a rope. To start with, consider his right hand to be supporting his entire weight as he reaches upward to grasp the rope with his left hand. After this reach when the left hand grasps the rope, the left elbow flexes and the left arm extends at the shoulder as he pulls himself higher. In this instance the hand is grasping the rope, so it and the forearm remain stationary while the body rises as a result of the combination of elbow flexion and shoulder extension. Because the body is moving on the motionless forearm, the lever for the elbow flexion is the kinematic chain made up of the body segments proximal to the elbow joint, i.e., the entire body excluding the left forearm and hand. Thus, in order to be effective and to move the body, the force exerted by the elbow flexors (biceps, brachialis, and brachioradialis) must be applied at the proximal attachments of these muscles.

Simultaneously, as mentioned above, there is extension of the left shoulder (which is also accompanied by some adduction). In other words, the angle between the elevated left arm and the trunk decreases as the body is pulled upward by the shortening of the shoulder extensors, including the latissimus dorsi and the sternal head of the pectoralis major. Anatomically these two muscles combine to form a sling which supports the body from the bicipital groove of the humerus; thus, the concentric contraction of these two muscles moves the body up *to the forearm*. Again, the effective muscular force is applied at the proximal attachments of these muscles (latissimus dorsi and pectoralis major) on the torso.

The above discussion presents illustrations involving concentric or shortening contraction, but the same principles apply when considering eccentric or lengthening contraction.

Now imagine that the gymnast is descending the rope and his elbow flexors and shoulder extensors are slowly yielding to the pull of gravity and being lengthened by this outside force. During this descent when the right hand grasps the rope, the right elbow extends slowly while the left shoulder flexes at the same rate. Here the muscles of these joints, the elbow flexors and shoulder extensors, are acting as a resistance to the outside force, and this muscular resistance must be applied to the moving end of the kinematic chain which they control, i.e., at the proximal attachments of the muscles, just as in the case of the concentric contraction.

Ranges of Muscle Extensibility and Contractility

The maximal degree of angular displacement of a body segment possible at a given joint affects all muscles crossing that joint. It determines both the greatest amount of stretch which such a muscle may be required to undergo and the maximal amount of shortening which it can be called upon to perform in situ. The full range of extensibility and contractility of a muscle has been variously termed its "functional excursion" (Brunnstrom (1964)) and its "amplitude" (Scott (1963)).

Any muscle crossing a single joint is normally capable of shortening sufficiently to move the body segment to which it is attached through its maximal angular displacement and, conversely, it is sufficiently extensible to permit a full range of motion in the opposite direction. This is not true of muscles which cross two or more joints.

Mechanics of Multijoint Muscles

A muscle crossing two or more joints has certain characteristics, capabilities, and limitations when compared with those muscles which cross only one joint. When a muscle crosses more than one joint it creates force moments at each of the joints crossed whenever it generates tension. The moments of force it exerts at any given instant depend on two factors: the instantaneous length of the moment arm at each joint and the corresponding amount of force that the muscle is exerting. The joint with the longest moment arm, and hence with the greatest moment of force, is normally the one at which the multijoint muscle will produce or regulate the most action. The hamstring muscles are a case in point: their moment arm at the hip is at least 50% longer than the one at the knee, and electromyography has repeatedly demonstrated that activity such as slow hip flexion as in toe touching is controlled by eccentric contraction of these muscles and that, when the action is reversed, they return the body to the upright posture without assistance of the gluteus maximus (Joseph (1960); Basmajian (1974)). Table 6.2 demonstrates lengths of some moment arms for the lower extremities.

Table 6.2
Muscle Moment Arms

Muscle	Hip	Knee	Ankle
	cm	*cm*	*cm*
Hamstrings	6.7	3.4	—
Rectus femoris	3.9	4.4	—
Gastrocnemius	—	2.5	5.0

If the hamstrings contract with a force of 10 kg, the moment of force at the hip will be 67 kg-cm, and at the knee it will be 34 kg-cm. At the same time, if the rectus femoris contracts with an equal amount of force, its moment at the hip will be only 39 kg-cm and at the knee 44 kg-cm. Such conditions demonstrate how these opposing muscles interact to produce extension at the hip while the knee extension is maintained; each joint responds to the greatest force moment acting on it. If greater force is needed at either location, the single joint muscles (gluteus maximus at the hip and the vasti muscles at the knee) are available. It seems that the hamstrings are more effective as knee flexors than is the gastrocnemius. The latter does not act at the knee joint unless the ankle is kept dorsiflexed. Again the moment arm of the gastrocnemius at the ankle is larger than the one at the knee, and when these moment arms are compared, the reason for the greater amount of activity at the ankle joint becomes obvious.

Muscle Insufficiency

If a muscle which crosses two or more joints produces simultaneous movement at all of the joints that it crosses, it soon reaches a length at which it can no longer generate a useful amount of force. Under these conditions the muscle is said to be **actively insufficient.** An example of such insufficiency occurs when one tries to achieve full hip extension with maximal knee flexion. The two-joint hamstrings are incapable of shortening sufficiently to produce a complete range of motion at both joints simultaneously. The individual feels considerable discomfort at the back of the thigh as the effort is made. If the individual then reaches behind and grasps the ankle and pulls the leg into full flexion at the knee, and the thigh into full extension at the hip, the discomfort disappears from the hamstrings but is transferred to the anterior thigh. The rectus femoris is not long enough to span both joints comfortably under these conditions and displays **passive insufficiency.**

Passive insufficiency of a muscle is indicated whenever a full range of motion at any joint or joints that the muscle crosses is limited by that muscle's length rather than by the arrangements of ligaments or structure of the joint itself. Thus, passive insufficiency of the hamstrings is indicated when one is unable to touch the fingertips to the floor while maintaining full extension of the knees, a situation popularly known as tight hamstrings.

The sports medicine diagnostician utilizes the concepts of active and passive insufficiency when identifying the specific muscle which has suffered a strain as a result of trauma. To illustrate these principles, the following scenario is used.

A male squash player feels a sudden pull in his right calf when he attempts to return a low shot close to the front wall. Examination at the sports medicine clinic reveals marked discomfort on palpation slightly distal and medial to the muscle belly of the calf. Question: is the soleus or gastrocnemius musculo-tendinous unit involved in the injury?

By employing the concepts associated with passive insuffiency the diagnostician initially passively forces the ankle into dorsiflexion, first with the knee flexed to allow complete stretching of the soleus muscle, and then with the knee extended, thus stretching the gastrocnemius. Discomfort associated with either the first or the second test indicates which muscle is involved.

Then, the diagnostician applies concepts related to active insufficiency. The examiner first resists plantar flexion with the knee flexed, thus principally stressing the soleus muscle; little force is available from the gastrocnemius due to its shortened position, thus active insufficiency. Then, this test would be followed by resisted plantar flexion with the knee extended. Discomfort with the latter test, in the absence of discomfort with the former test, would implicate the gastrocnemius muscle as the one that has been strained.

CONCLUSION

This chapter has covered a variety of topics concerned with properties and structure of skeletal muscle, the nature of contraction and relaxation, magnitude of contractile tension, and muscle attachments and their effect on function. It has attempted to provide the answers for questions concerned with how human body levers are usually powered. The following chapter will attempt to explain the neural basis of movement which accounts for the stimulation and control of muscle.

CHAPTER 7

Neural Basis of Behavior

Skeletal muscles are under the control of the nervous system which determines which muscles shall contract, when, how fast, to what extent, and with what changes in force and velocity from moment to moment.

THE NEURON: MORPHOLOGY AND PHYSIOLOGY

Morphology of the Neuron

The nervous system is composed of two types of cells: **neurons** and **neuroglia**. We are concerned only with the first of these. The neuron is the functional unit of the nervous system. Each neuron possesses "in miniature the integrative capacity of the entire nervous system" (Noback and Demarest (1972)).

Neurons are usually greatly elongated cells with axon diameters ranging from 0.5 μ in small unmyelinated fibers to 22 μ in the largest myelinated fibers. Some axons exceed 1 m in length. Diameters of cell bodies range from 10 to 50 μ. Neurons are specialized to receive, conduct, and transmit excitation.

A generalized neural cell or neuron consists of four morphologically and physiologically distinct portions: a receiving pole, a terminal transmitting pole, an intervening conducting segment, and a cell body or **soma**. Each is specialized for its particular role in the cell's function. Neurons possess two types of protoplasmic processes extending outward from the nucleated soma: **dendrites** and **axons**. The processes vary in length and in the amount and extent of their branching. Dendrites are usually multiple, short, and highly branched. The space occupied by their three-dimensional spread is often extensive. They constitute the receiving pole of the cell. Axons are usually single, long, and although one or more collateral branches may occur, relatively unbranched except at their ends. The axon is responsible for both conduction of excitation and its transmission to other cells.

An axon generates **action potentials** (nerve impulses) and conducts them from the receiving portion of the cell to the transmitting region. It is a delicate cylinder of neural cytoplasm with a limiting membrane, the **axolemma**. It varies in length and in diameter in different types of neurons. Axons are enclosed in a cellular sheath of lipid material, the **myelin sheath**, which serves to electrically insulate individual axons from one another and from adjacent neural components. The myelin sheath is formed by concentric wrappings of membranous processes from sheath cells, called **oligodendrocytes** within the central nervous system and **Schwann cells** in the peripheral nervous system. Small axons which are invested by only a single layer of sheath cell process are called "unmyelinated" fibers. Larger axons are enclosed in increasingly more numerous sheathing layers formed by more and more windings of sheath cell processes. As the folds become tightly packed together, most of the cytoplasm is squeezed out so that the sheath is ultimately composed of spiral layers of lipid-rich cellular membrane. The myelin sheath of the larger axons is segmented rather than continuous, and each segment is enclosed by a single sheath cell. The length of segments and the thickness of the myelin are quite consistent for neurons of a given caliber, larger axons having longer segments (1 to 2 mm long) and thicker sheaths. The diameter of myelineated fibers usually ranges from 1 to 20 μ. The segments are separated by short (less than 0.1 mm) unmyelinated gaps, the nodes of Ranvier. Collateral branches, when present, arise at nodal gaps and exit from the parent axon at approximately right angles (Fig. 7.1).

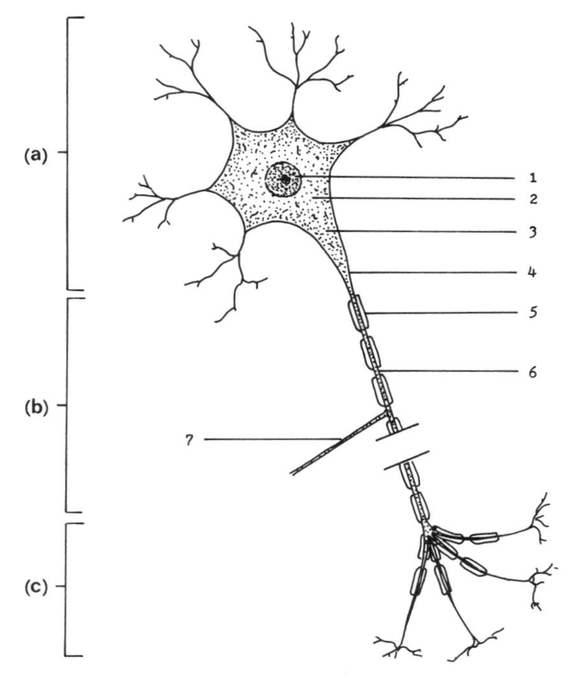

Figure 7.1. Multipolar neuron. (a) dendrites of the receiving pole. (b) axon or conducting segment. (c) telodendria and terminal arborizations of the transmitting pole. 1, nucleus; 2, cell body or soma (1 and 2 comprise the metabolic center of the cell); 3, axon hillock; 4, initial segment of axon; 5, myelin sheath segment; 6, node of Ranvier (naked membrane exposed); 7, collateral branch of the axon.

The area of axon outgrowth from the nucleated portion of the cell is known as the **axon hillock**. Nerve impulses are generated in the initial segment of the axon which, even in myelineated fibers, is unmyelineated. Axon ends usually divide distally into a spray of terminals, the **telodendria**, which lose the myelin sheath and end in naked tips.

The cell body or soma of the neuron is the metabolic center of the cell where, under control of its single nucleus, proteins and other metabolically important substances (enzymes, transmitter substances, neurohormones, etc.) are elaborated. The fact that materials are moved from the cell body into and along the neuronal processes has been well established. There are probably three channels for axoplasmic transport: the endoplasmic reticulum; the microtubules; and the neurofilaments (McComas (1977)). When severed from the nucleated portion of the cell, a nerve process will soon degenerate because it is no longer supplied with essential materials, and a new process will grow out from the cut stump which is still attached to the cell body.

The location of the soma is the distinguishing feature among nerve cells. In vertebrates the nucleated portion of most neurons, including motor neurons and interneurons, is a part of the receiving region of the cell. The somata of sensory neurons from the skin, however, have been displaced centripetally along the course of the axon where they are better protected from injury than if situated peripherally with the receiving structures (Fig. 7.2). In these cells the dendrites communicate directly with the axon, and the cell body does not participate in the reception of excitation. In other words, the part of the fiber, often myelinated, which conducts *toward* the cell body and the portion leading *from* the cell body are both parts of the axon. Only the peripheral terminals are dendrites, while the central terminals are telodendria.

Mitochondria are present in axons, especially at the nodal areas, and are numerous in the cell body and in both the receiving and transmitting portions of the neuron, being abundant in the latter. Ribosomes are mostly restricted to the cell body. Minute and unique neurofilaments are distributed throughout the cytoplasm.

Normally, excitation is conducted only from the receiving to the transmitting pole of the cell. This polarity results from nerve cells stimulated at the receiving end. An axon which is stimulated at a point along its length is capable of conduction in both directions. (Conduction of impulses in a direction opposite to the normal is referred to as **antidromic** excitation; for conduction in the normal direction, the descriptor is **orthodromic**.)

Neurons may be classified as either receptor neurons or synaptic neurons on the basis of the type of input which they receive. Receptor neurons are those which receive and transduce environmental energy such as light, sound, heat, or chemical or electrical energy. They are specialized to be excited by specific types of stimuli, and their dendritic portions are

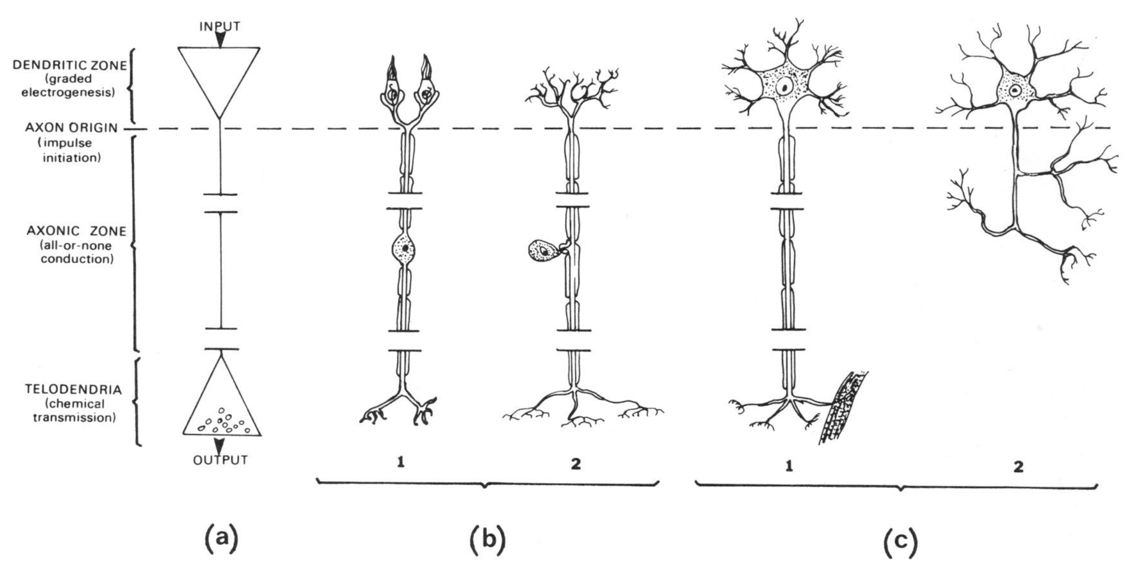

Figure 7.2. Types of neurons. (a) diagrammatic representation of the three functional portions of the neuron, showing the dendritic or receiving pole where excitation causes graded electrogenesis; the axonic or conducting segment which conducts the all-or-none impulses originating in the initial segment; and the synaptic or transmitting pole where excitation is transmitted by chemical means from the tips of the telodendria. (b) and (c): representative types of neurons. (b) **receptor neurons:** 1, special sensory neuron; 2, cutaneous neuron. Note that the nucleated portion is situated along the course of the axon. These cell bodies are located in the dorsal root ganglia or in a homologue of a dorsal root ganglion. (c) **synaptic neurons:** 1, motor neuron; 2, interneuron. These are the most common types in the nervous system. Note that the nucleated portion is located in the receptor pole of the neuron. (After Dowling, J. E., and Boycott, B. B., 1965. *Cold Spring Harbor Symp. Quant. Biol. 30:* 393.)

appropriately modified in structure (Fig. 7.2(b)). Synaptic neurons (Fig. 7.2(c)) receive information from other neurons by means of synaptic transmission. Their dendritic geometry may be extensive and complex, providing a wide field for reception of a great number and variety of inputs, all of which are already encoded in the manner characteristic of nervous system communication.

Physiology of the Neuron: Membrane Theory of Excitation and Conduction

Irritability is a characteristic property of all protoplasm, but types of cells differ widely in the extent to which they display the property. It is most highly developed in nerve and skeletal muscle cells. Excitation is induced in a cell by appropriate stimulation and is associated with chemical and electrical changes which spread over the cell membrane. In many types of cells the change is graded and spreads decrementally from the point of stimulation. In nerve axons and muscle cells, however, if the stimulus is adequate, the change is conducted without decrement as an all-or-none **action potential**. Adequate spread of excitation evokes the response which is characteristic of the cell.

In a muscle cell the response is contraction; in a gland cell, secretion. The essential function of a nerve cell is to transmit excitation to other cells, and it responds by releasing a chemical **transmitter substance** at its synaptic terminal. Neurons may be artificially excited by a number of different kinds of stimuli. The normal stimulus for synaptic neurons is the action upon their membranes of chemical transmitters released by other neurons. Stimulation of receptor neurons is normally provided by chemical, thermal, mechanical, and electromagnetic energies. In a few instances, rare among the vertebrates, a neuron is stimulated by direct electrotonic stimulation from another neuron.

Resting Membrane Potential

Action potential generation and conduction in the neuron axolemma and the muscle sarcolemma are essentially identical. The present discussion describes these events as they occur in the neuron.

The axon membrane in the unexcited or resting state is polarized as a result of a differential distribution of ions on the two sides of the membrane. The **cations** potassium (K^+ and sodium (Na^+), the **anion** chloride (Cl^-, and certain organic anions are the ions most importantly concerned. Ions are distributed so that high concentrations of K^+ and protein ions are found intervally and high concentrations of Na^+ and Cl^- ions are located external to the membrane. The differences in ion concentrations reflect the selective permeability of the membrane. The resting membrane is at least 10 times more permeable to K^+ and Cl^- than it is to Na^+ ions. Hence, while K^+ and Cl^- pass easily through the membrane, Na^+ passes only with difficulty. Despite this fact, a small amount of Na^+ leaks in but it is promptly ejected from the cell by an active transport mechanism called the **sodium pump**.

Similarly, K^+ pumps are mainly responsible for compensating for the K^+ "leaks" so that the resting state of cell is maintained. Na^+ and K^+ pumps have been shown to be powered by the free energy from adenosine triphosphate (ATP) breakdown.

As a consequence of the semipermeability of the membrane and of the sodium pump, sodium accumulates in the intercellular fluid outside the cell in a concentration which is about 10 times greater than that inside the cell. Potassium is about 30 times more concentrated inside the cell than outside, and the chloride concentration is about 14 times greater outside than inside. This unequal distribution of an ion on two sides of a permeable membrane causes chemical and electrical forces to act on that ion. First, the chemical force results from a chemical gradient whereby ions diffuse through the membrane away from the area of their own greater concentration into the area of their own lesser concentration. Consequently, in the resting nerve cell the chemical gradients tend to drive sodium into and potassium out of the cell. The second force is electrostatic attraction whereby an electrically charged area attracts ions of opposite charge and oppositely charged ions attract each other. The driving voltage for a particular ion is the difference between the value of the membrane potential and the equilibrium potential[a] for that ion. In the resting condition the ionic currents balance each other exactly and the membrane potential remains constant.

In the resting cell Na^+,kept out by the action of the sodium pump, attracts and holds a considerable amount of Cl^-. Although Cl^- being in higher concentration in the intercellular fluid, tends to diffuse into the cell, its inward diffusion force is balanced by the electrical attraction of the positively charged Na^+, and the chloride ions remain in equilibrium. Inside the cell, large, negatively charged protein molecules, which were formed within the cell and are too large to pass through the membrane, exert an electrical attraction on cations. Therefore, since Na^+ cannot remain within the cell, K^+ is drawn into the cell and to a significant extent held there, accounting for the high concentration of potassium on the inside. The inward attraction exerted on $K^+$by the organic anions is opposed by the chemical gradient tending to drive it out. Under resting conditions, these two factors balance each other but there is a deficiency of positive ions inside the cell (Fig. 7.3(a)). Thus, the net charge on the inside of the membrane is negative, while the charge on the outside, where there is an excess of positive ions, is positive (Fig. 7.3(b)).

The resting potential is often called a potassium potential because it is due to the excess K^+ in the intercellular fluid. However, it is the active removal of Na^+ at a rate equal to the net rates of entry of K^+ and Cl^- that is the means by which the cell maintains its normal concentration differences. Other ions distribute themselves in a Donnan equilibrium[b] as indicated by their concentration ratios.

The electrical potential difference between the inside and the outside of the membrane of an unexcited cell is the **membrane or resting potential**[c] and ranges from 40 to 90 mV in nerve and muscle cells.

Excitation: Generation of Action Potential

The forces acting on sodium and potassium across the resting membrane are importantly involved in cell excitation. Any change in conditions producing an alteration in membrane permeability will result in movement of these ions in response to their driving forces, and the resting equilibrium will be upset. The time course of change in permeability of the cell membrane to a particular ion is expressed as **conductance** for that ion. Any change in conductance for one ion will appreciably affect that of other ions and hence alter the membrane potential. Conversely, any change in membrane potential will influence conductances. Stimulation of a nerve axon increases membrane permeability, and hence conductance, to Na^+ at the point of stimulation. Sodium, driven by both its chemical

[a] The equilibrium potential is the electrical potential difference which must exist across the membrane to maintain the ionic concentration gradient. Its magnitude is just sufficient to equalize the influx and efflux for a particular ion and depends on the ratio of internal and external concentrations of the ion. Its value may be calculated from the Nernst equation, and is usually treated as a Cl^- or a K^+ potential, although the latter is the usual convention adopted by most researchers. The equilibrium potential for Cl^- is identical with the experimentally measured resting membrane potential of nerve and muscle cells.

[b] The conditions which exist when an irregular distribution of ions between two solutions, separated by a membrane, develops an electrical potential between the two sides of the membrane; e.g.,

$$\frac{[K]_i}{[K]_o} = \frac{[K]_o}{[Cl-]_i}$$

[c] All biological "potentials" are actually potential differences.

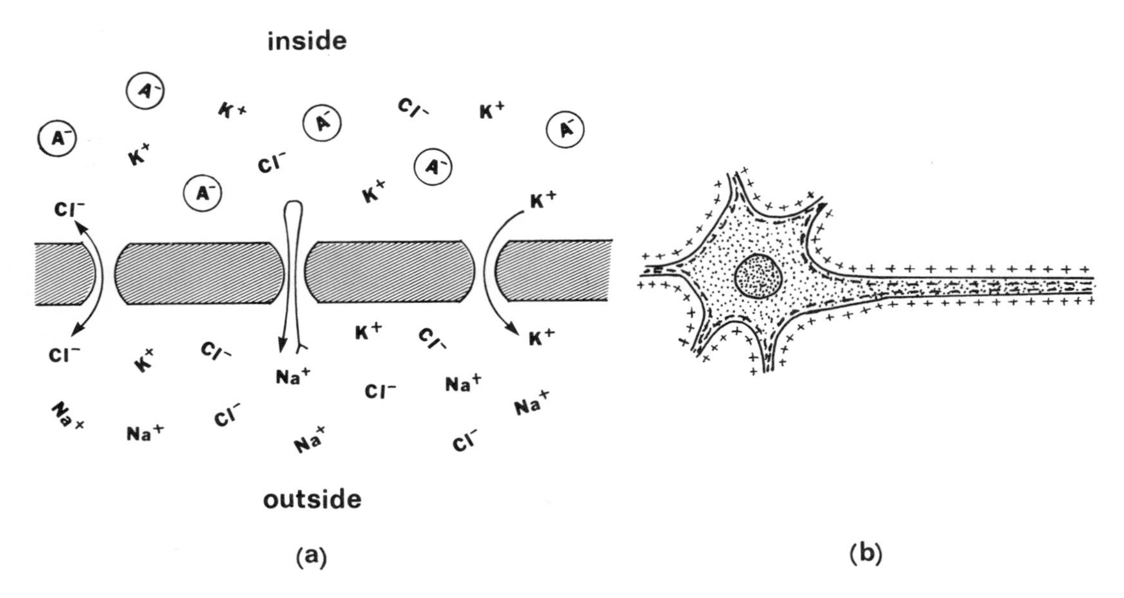

Figure 7.3. Resting membrane. (a) because of permeability properties of the membrane, K^+ and Cl^- pass readily through the membrane. Na^+ passes with difficulty and is promptly ejected by the sodium pump mechanism. As a result, K^+ accumulates inside the cell and Na^+ outside. Organic anions, each indicated by A^-, are too large to leave the cell. See text. (b) as a result of the differential distribution of ions, the interior of the membrane is negatively charged while the exterior is positively charged.

and electrical potential gradients, passes into the cell. Because sodium is a positively charged ion, its entry decreases the negativity inside and at the same time and to a similar extent decreases the positivity outside. In other words the value of the resting potential is lowered toward zero; the membrane is being **depolarized**. A change of resting potential in the depolarizing direction constitutes a local excitatory state (l.e.s.) (Fig. 7.4(a) and (b)).

The net influx of sodium is slow at first, but the depolarization caused by its entry produces a self-generative increase in permeability so that the rate of Na^+ influx and the rate of depolarization increase exponentially. This is an example of **positive feedback**. If the resting potential drops to a sufficient extent, it will reach a critical level which is characteristic and constant for each cell. In mammalian nerve and muscle cells critical depolarization levels range between 10 and 20 mV below the resting potential. At the critical level something happens suddenly which seems to throw the sodium gates wide open. Sodium rushes in to such an extent that the membrane potential in the stimulated area passes beyond zero and the polarization is reversed: the membrane becomes negative outside and positive inside at the point of stimulation. This almost instantaneous change in potential, which appears on the oscilloscope as a spike (Fig. 7.4(c)), is known as the **action potential**.[d] The more intense the stimulus, the sooner depolarization reaches the critical level and the earlier the spike appears.

The rising phase of the spike (depolarization) is accounted for by the influx of Na^+. Sodium conductance rises rapidly, in 0.1 to 0.2 msec, and then falls rapidly to a low level. Potassium conductance does not change appreciably at first but then

[d] The action potential is a sodium potential and leads to the designation of the hypothesis as the **sodium theory** of the nature of the action potential.

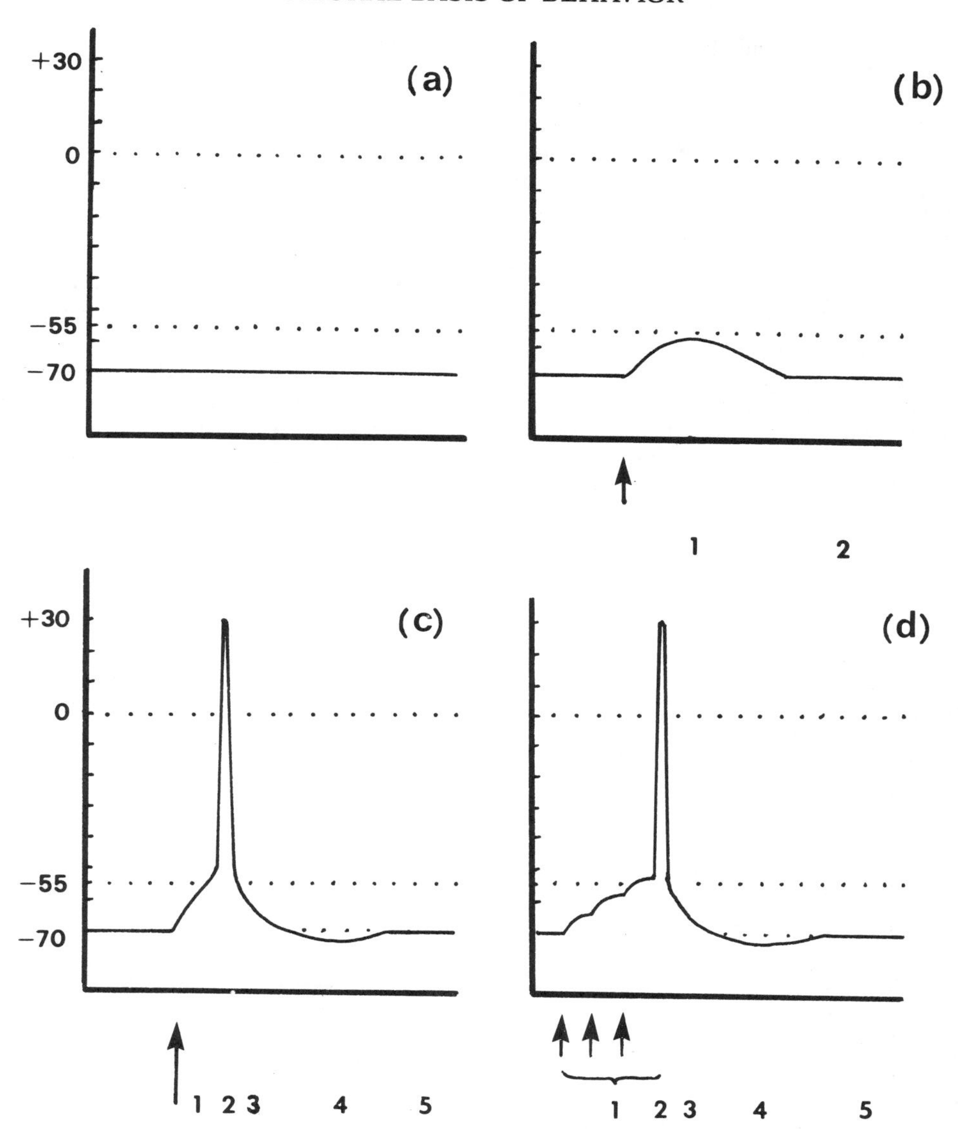

Figure 7.4. Membrane potential changes with time. Ordinate: the membrane potential in millivolts. Abscissa: time. Arrows indicate application of stimulus. (a) **resting state.** Membrane potential is −70 mV (inside). Depolarization of at least 15 mV is required to reach the critical level of −55 mV. (b) **l.e.s.** Application of an inadequate stimulus (arrow) partially depolarizes the membrane, producing an l.e.s. Since the critical level is not attained (1), no spike is generated. The l.e.s. dwindles away and the resting state is restored (2). (c) **action potential**. An adequate stimulus (arrow) induces an l.e.s. (1) which quickly reaches the critical level and the membrane potential suddenly reverses to +30 mV (inside), producing an action potential of 100 mV (2). Restoration processes act promptly and start to return the membrane to its resting potential. The process slows down, producing the short negative afterpotential (3), during which the membrane is still slightly depolarized and hence abnormally irritable. This is followed by a longer lasting positive afterpotential (4), which is an overshoot resulting in slight hyperpolarization and subnormal irritability. The resting potential (5) is gradually regained. (d) **summation of inadequate stimuli.** The l.e.s. of three inadequate stimuli summate (1) to reach the critical level (2), and an action potential spike is generated, followed as in (c) by negative (3) and positive (4) afterpotentials, and ultimate recovery (5).

becomes marked as sodium conductance is inactivated. An efflux of K^+ is responsible for the falling phase of the spike (repolarization). The separation in time of these two conductance changes accounts for the change in the membrane potential which constitutes the action potential (Fig. 7.5).

The magnitude of the action potential is the algebraic difference between the resting and the active (reversed) potential values as recorded from the inside of the membrane. In Figure 7.4(c) the membrane potential, which is −70 mV (inside) in the resting state, is reversed to +30 mV upon adequate excitation. The action potential is therefore:

$$30\ mV - (-70\ mV) = 100\ mV \quad (7.1)$$

No action potential occurs unless depolarization reaches the critical level. If the stimulus is inadequate and does not result in the entry of enough sodium to depolarize the membrane sufficiently, the local excitatory state soon dies away (Fig. 7.4(b)). Regenerative reactions, mainly K^+ efflux, restore the resting polarization. The sodium pump gradually ejects the sodium, and eventually the resting ion distribution is regained. If, however, subsequent subthreshold stimuli are applied in such rapid succession that each succeeding stimulus evokes its l.e.s. before that of the preceding stimulus has dwin-

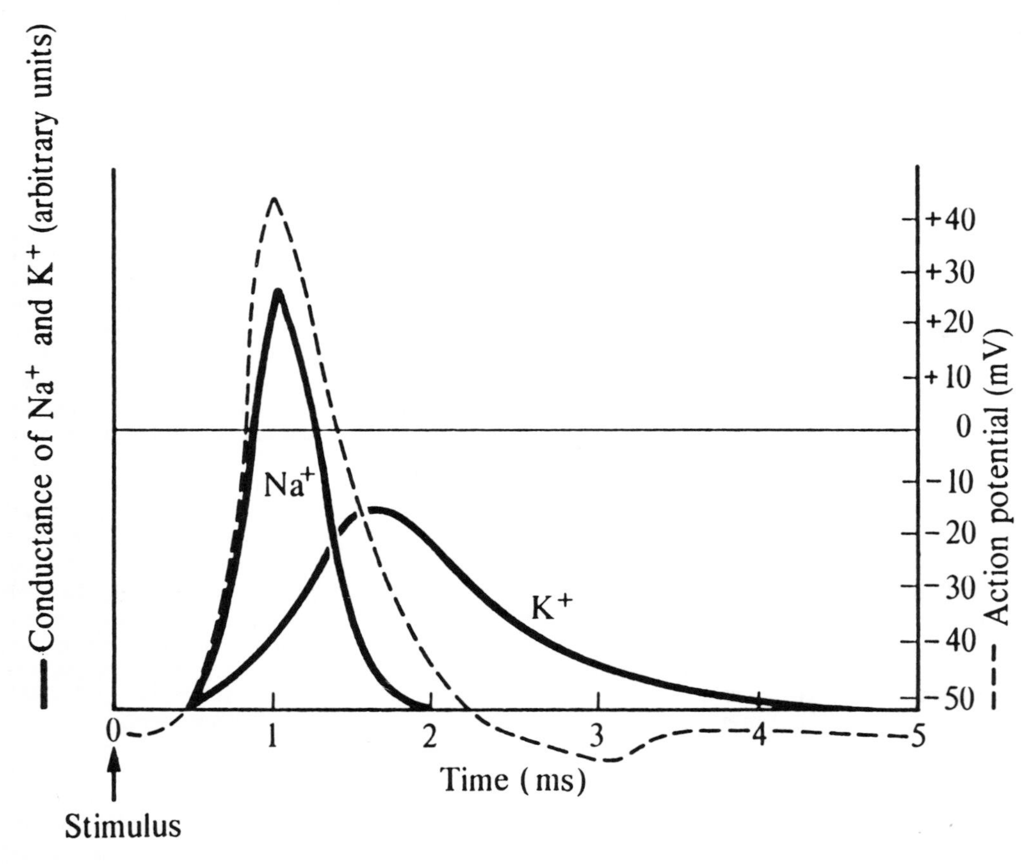

Figure 7.5. Diagram to illustrate how the sodium and potassium conductance of a nerve fiber changes as an action potential passes the recording electrodes (solid lines). The form of the action potential is given by the dotted lines (From Tribe, M. A., and Eraut, M. R., 1977. *Nerves and Muscle* (Basic Biology Course, Book 10). Cambridge, England: Cambridge University Press, p 69.)

dled away, the local excitatory states will summate. When the critical level is reached, an action potential will be generated. This phenomenon is designated **summation of inadequate stimuli** (Fig. 7.4(d)).

Recovery of Resting State

The action potential occupies a definite distance of the nerve fiber (5 to 6 cm in the largest mammalian neurons) and lasts for a definite duration (about 0.4 msec), and then the membrane potential is rapidly returned to the resting state (Fig. 7.6). Because the action of the sodium pump is relatively slow, recovery of the resting polarization requires a faster mechanism. The outflow of K^+ as its conductance increases, brings about the almost immediate recovery of positivity outside and negativity inside. Potassium is driven outward by its chemical gradient and also by the reversal of the electrical force as the cell interior becomes positively charged. Ion distributions will be eventually restored by the slower action of the sodium pump. It has been suggested that the same carrier which actively transports Na^+ out of the cell brings K^+ back.

Conduction of Action Potential

The action potential which is generated at the site of stimulation must be conducted along the fiber to the axon terminals. Conduction is accomplished by self-propagation of the disturbance away from

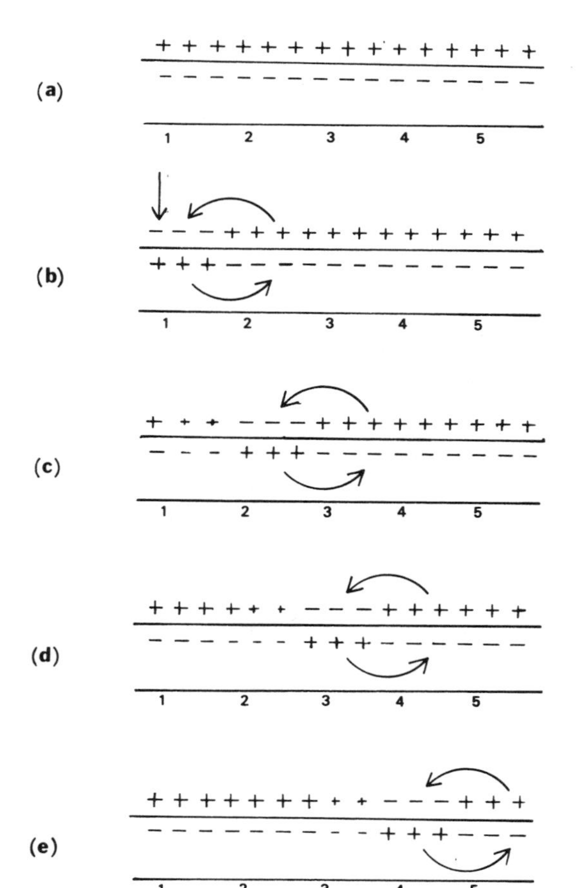

Figure 7.6. Electrical changes associated with excitation, conduction, and recovery in an unmyelinated cell membrane. (a) **resting membrane.** Resting membrane is positively charged on the outside and negatively charged on the inside. (Charges are shown on only one surface of the profile. Numbers indicate sequential areas of the membrane.) (b) **excitation**. An adequate stimulus (arrow) results in reversal of polarity (negativity) at the point of stimulation (area 1). An action current then flows from positively charged (outside) inactive area 2 to negatively charged (outside) active area 1, into and along the inside of the membrane, emerging through the inactive area. (c), (d), and (e): **conduction.** (c) emergence of action current stimulates the membrane of inactive area 2, which then becomes depolarized. The action current now flows from presently inactive area 3 to newly active area 2. (d) and (e), the action potential is self-propagated along the fiber to areas 3 and 4. **Recovery**. In (c), (d), and (e), the area behind the action potential gradually recovers its original polarization (positively charged on the outside).

the site of origin. Adjacent inactive areas on the outside of the membrane are still positively charged at the resting level. Since the extracellular fluid is electrolytic, a small current, the **action current**, flows from the positively charged inactive region to the negatively charged active area where it passes in through the membrane, through the cell fluids, and out again through the inactive region. Although small, the current is sufficiently strong to constitute a stimulus capable of increasing membrane permeability as it emerges. The same sequence of excitatory events is repeated here: sodium moves in, polarity is reversed at this point, and the action potential has progressed along the fiber. This is repeated again and again until the action potential reaches the end of the axon terminal (Fig. 7.6). A neuron can conduct for hours without any cessation of activity to provide for restoration of the original ionic distributions.

In unmyelinated fibers such as the autonomic postganglionic neurons and the afferent fibers for dull pain, conduction is accomplished by progression of the action potential down the fiber as described above. In myelinated fibers, however, the action current skips from node to node without depolarizing the internodal portions of the axon (Fig. 7.7). This is **saltatory conduction**, and the velocity of conduction is considerably greater than in unmyelinated fibers. Because the same exchange of ions occurs fewer times for a given length of fiber, saltatory conduction involves fewer ions and hence requires less energy for recovery. This is an advantage which permits myelinated fibers to continue transmitting for some time even in the absence of oxygen.

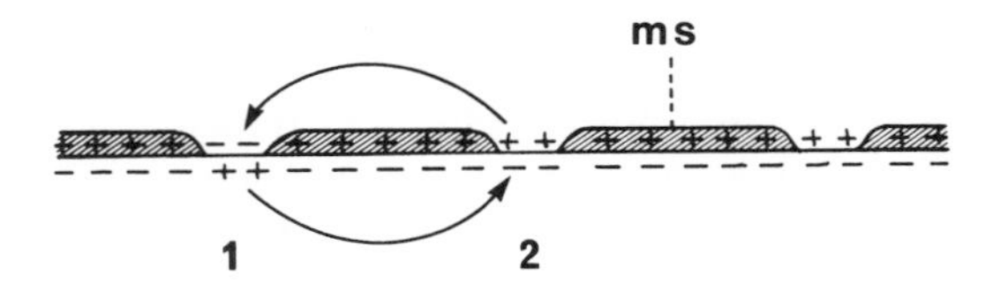

Figure 7.7. Saltatory conduction in a myelinated fiber. 1, active node of Ranvier; 2, inactive node of Ranvier; ms, myelin sheath segment. Arrows indicate flow of action current. Direction of impulse conduction is from left to right. In myelinated nerve fibers the action current skips from node to node without depolarization of internodal portions of the axon. Only a section of the surface membrane is shown in profile, with the outside of the fiber above and the inside below.

Some Characteristics of Nerve Conduction

Refractory Periods. As the action potential travels along the fiber surface, it consists of a wave of negativity followed by an area of gradually recovering positivity (Fig. 7.6). While an area is in its reversed (active) state, it is *absolutely* refractory and cannot be restimulated. During recovery, the membrane is *relatively* refractory, a state which lasts many times longer than the absolute refractory period. Intense or sustained stimuli may restimulate the original site during repolarization.

The refractory periods are due to the conductance changes. Inactivation of sodium conductance decreases excitability because a greater depolarization would be required to produce further increase in Na^+ conductance to a point where its net influx would exceed the efflux of K^+. Elevation of K^+ conductance gradually restores resting polarization and hence excitability. Thus, the inactivation of Na^+ conductance and the elevation of K^+ conductance account for both the absolute refractory period and the relative refractory period.

Afterpotentials. During the relative refractory period both the amplitude and velocity of the spike are altered, reflecting changed conditions in the fiber. In some neurons the latter portion of the downward course of the spike is considerably less rapid than its rise, showing a marked concavity before reaching its initial level. This is the **negative afterpotential** because it indicates a delay in return to the resting potential and a prolongation of some slight depolarization (hence negative). During this period of 12 to 80 msec, the membrane is hyperexcitable or supernormal and hence more easily restimulated. The recovery may continue into a hyperpolarized state, the **positive afterpotential**, which persists for a much longer time, up to 1 full second, during which the membrane is subnormal in excitability (Fig. 7.4(c) and (d)). The positive afterpotential is probably due to a delay in the restoration of K^+ conductance to its normal level.

Frequency of Impulses. In general, natural stimuli are of sufficient duration to reactivate the membrane after the absolute refractory period. For this reason neurons normally carry trains of impulses. A single electric shock may produce a single action potential but only because its duration does not outlast the refractory period of the fiber. The stronger the stimulus, the earlier it will re-excite, and the shorter will be the time span between impulses, hence, the greater the frequency.

Because each action potential is followed by an absolute refractory period, action potentials cannot summate[e] but remain separate and discrete. Neurons do not conduct impulses at rates as high as the absolute refractory periods would suggest. Cognizance must also be taken of the characteristics of the relative refractory period. A fiber with a spike duration of 0.4 ms might be expected to conduct impulses at a frequency of 2500/s but its upper limit will be closer to 1000/s. Conduction frequencies rarely approximate their possible maxima. Motor neurons usually conduct at frequencies of 20 to 40, rarely as high as 50, impulses/s, although, at the start of a maximum contraction, rates greater than 100 Hz have been recorded (McComas (1977)). Upper-limit frequencies for sensory neurons normally lie between 100 and 200 impulses/s although auditory neurons may conduct between 800 and 1000 impulses/s. Information is conveyed by the presence or absence of an action potential, as well as by the frequency of action potentials.

Velocity of Conduction. Velocity of conduction depends not only on myelination but, more importantly, on the diameter of the fiber. It can be fairly accurately predicted from the following equation.

$$\textit{Velocity in m/s} = 6 \times \textit{diameter in } \mu \quad (7.2)$$

Hence the largest motor and sensory nerve fibers, with diameters near 20 μ, have conduction velocities up to 120 m/s.[f] In small unmyelinated fibers, velocities range from 0.7 to 2 m/s. Large fibers not only conduct more rapidly than small fibers but characteristically have lower stimulus thresholds and larger spikes with shorter durations.

Classification of Nerve Fibers. As a result of the classic experiments of Erlanger and Gasser in 1937, nerve fibers are classified into three major groups, A, B, and C, on the basis of conduction velocities. Group C contains the unmyelinated postganglionic fibers and group B the small myelinated preganglionic fibers of the autonomic nervous system. Group A includes the large, rapidly conducting myelinated somatic fibers. Group A has been further divided into four subgroups: alpha (α), beta (β), gamma (γ), and delta (δ) on the basis of velocity and diameter. The fastest fibers are those with the largest diameters. Sensory nerve fibers have been separately classified by Lloyd (1943) according to diameter into groups I, II, III, and IV, with corresponding velocities. These do not correspond exactly in size and velocity to the subgroupings of the Erlanger-Gasser class, but among afferents from the skin and muscles, group I approximates $A\alpha$ and group II approximates $A\beta$ and $A\gamma$. In order to avoid confusion, use of the alphabetical designations is restricted to efferent fibers and the numerical designations to afferent fibers. Table 7.1 provides a comparison of the two classifications as related to both motor and sensory fibers. As indicated by the columns in the table for termination of motor fibers and origin of sensory fibers, it is obvious that specific structures tend to be innervated by fibers of quite specific size and conduction characteristics.

---

[e] The l.e.s. displays no refractoriness and hence summation is possible at subthreshold levels.

[f] In mammalian skeletal muscle fibers, conduction velocity is about 5 m/s.

Table 7.1
Comparison of Classification Systems for Motor and Sensory Nerve Fibers

Motor[a]					Sensory[b]			
Group		Diameter	Velocity	Termination[c]	Group	Diameter	Velocity	Origin (receptors)[d]
		μ	m/s			μ	m/s	
A	α	12-20	60-100	Muscle fibers	I	12-22	70-120	Spindle, GTO, joint
	β	6-12	30-70	Axons to spindle IFs	II	6-12	30-70	Spindle, skin, joint
	γ	2-8	15-30	Spindle IFs				
	δ	2-6	12-30	(Blood vessels?)	III	2-6	12-30	Pain
B		1-3	3-15	ANS preganglionic				
C		0.5-1	0.5-2	ANS postganglionic	IV	0.5-1	0.5-2	Pain

[a] Classified according to Erlanger and Gasser (1937).
[b] Classified according to Lloyd (1943).
[c] IF, intrafusal fibers; ANS, autonomic nervous system.
[d] GTO, Golgi tendon organ.

MORPHOLOGY AND PHYSIOLOGY OF THE SYNAPSE

While the functional unit of the nervous system is the single neuron, it is apparent that more than one neuron is traversed between receptor and effector. Usually, a chain of at least two, and more likely three or more, neurons connect receptor with effector.

Each neuron in the chain remains a separate and discrete entity. Its axon terminal ends in close proximity to the receiving structures of other neurons, but there is no protoplasmic continuity between neurons. Dendritic branches and axon telodendria interweave to form a complex network known as the **neuropil**. The region of functional contact between neurons is the **synapse**, across which excitation must be transmitted. The synapse is probably the most important aspect of neural organization; in fact, its importance cannot be overemphasized. It is responsible for the physiological continuity of conduction through the neural chains. It is also the site in the nervous system where the modification of communication occurs, without which the integrated response would be impossible.

In the neuron itself, nerve impulses are transmitted in an all-or-none fashion in both magnitude and velocity, and these properties vary only with changes in the condition of the fiber. At the synapse, however, transmission is *not* all-or-none and may be amplified, reduced, or even completely blocked. As a result, the signal transmitted by a subsequent neuron in the chain may be quite different from the original input. Furthermore, blocking of some synapses and concurrent facilitation of transmission at others serve to determine the distribution of communication by directing it into specific channels.

Morphology of the Synapse

A synapse (Fig. 7.8) consists of the specialized axon terminal of the transmitting or **presynaptic neuron**, separated by a fluid-filled space (the **synaptic cleft**), from the receiving membrane of the **postsynaptic neuron**. The axon ending and the postsynaptic membrane are closely contiguous. The two apposing membranes are strongly adherent and there is evidence that they may be held together by a special **synaptic cement**.

Each telodendron terminates in a specialized unmyelinated ending which may take one of several forms. Frequently endings are bulbous swellings known as **knobs** or **boutons**, but sometimes they are diffuse arborizations which form nests,

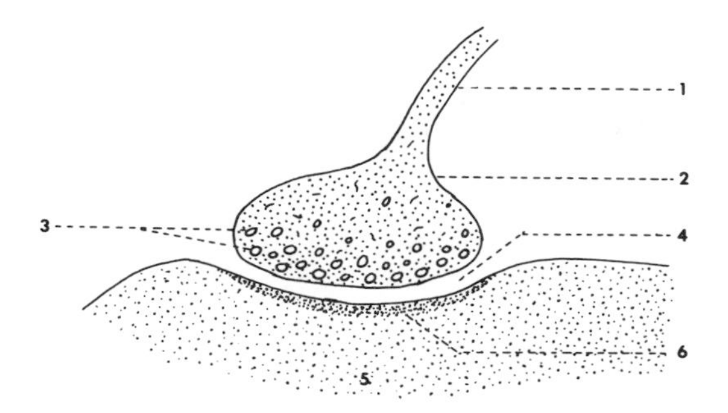

Figure 7.8. Diagrammatic representation of the components of the synapse. 1, presynaptic telodendrion; 2, bouton; 3, vesicles; 4, synaptic cleft; 5, postsynaptic neuron; 6, receptor site or subsynaptic membrane.

brushes, or baskets, and sometimes they are simply naked terminals which climb along a dendrite for some distance or cross it at right angles.

Presynaptic terminals contain mitochondria, neurofilaments, and numerous minute vesicles, 200 to 1000 Å in diameter, which are often clustered against the presynaptic membrane. Vesicles occur in a variety of shapes and sizes.

The synaptic cleft is continuous with the intercellular space. In width it ranges from 100 to 200 Å. It is usually occupied by vague dense material which forms a thin dark plate between the apposed membranes and is sometimes thicker on the postsynaptic side.

The contact area on the postsynaptic neuron may be called the **subsynaptic membrane** or the **receptive site**. While morphological specialization has not been clearly revealed, some electron microscope studies have shown what appear to be delicate hooklike fibrillar extensions which make contact with presynaptic fibrils in the synaptic cleft (Fernandez-Moran (1967)). Biochemical and physiological studies suggest the presence of ion-specific channels or pores and specific reactive chemical groups.

Synaptic contacts occur on dendrites and, in neurons whose nucleated portion is located within the receiving pole of the cell, on the soma. Synapses formed by contact of axon terminals with postsynaptic dendrites are **axodendritic** synapses; those formed by contact with the cell body are **axosomatic** synapses. Those between dendrites, **dendrodendritic** synapses, are such that messages are passed to each dendrite via separate synapses. There are also synapses in which an axon terminal makes contact with another axon terminal or even with the initial segment of another neuron; these are **axoaxonic** synapses. And lastly, it is possible to find a synaptic glomerulus, in which the axon of one neuron synapses with the dendrites of two others (Fig. 7.9).

Each presynaptic neuron makes synaptic contact with many different postsynaptic neurons, often sending several telodendria to each (Fig. 7.10). Each postsynaptic neuron receives multiple axon terminals from many different presynaptic neurons: a single motor neuron in the spinal cord may have more than 1000 synapses occupying 40% of its receiving surfaces. The scope of a neuron's influence may be further extended by collateral branches of its axon which may have destinations quite different from that of the parent axon. In some cases a collateral may even turn back into the dendritic field of its own neuron as a **recurrent collateral**.

Physiology of the Synapse

Synaptic Transmission

In an active neuron, nerve impulses travel out into all of its many tiny terminal branches and into as many synapses. Abundant evidence indicates that synaptic transmission is accomplished in most instances by a chemical process. (Electrical transmission, which is known to occur in many invertebrates such as the crayfish, squid, and annelid worm, has recently been identified in some vertebrates but is as yet unknown in mammals.) The nerve impulse itself does not cross the interneuronal gap but rather, upon its arrival at axon ends, it causes the

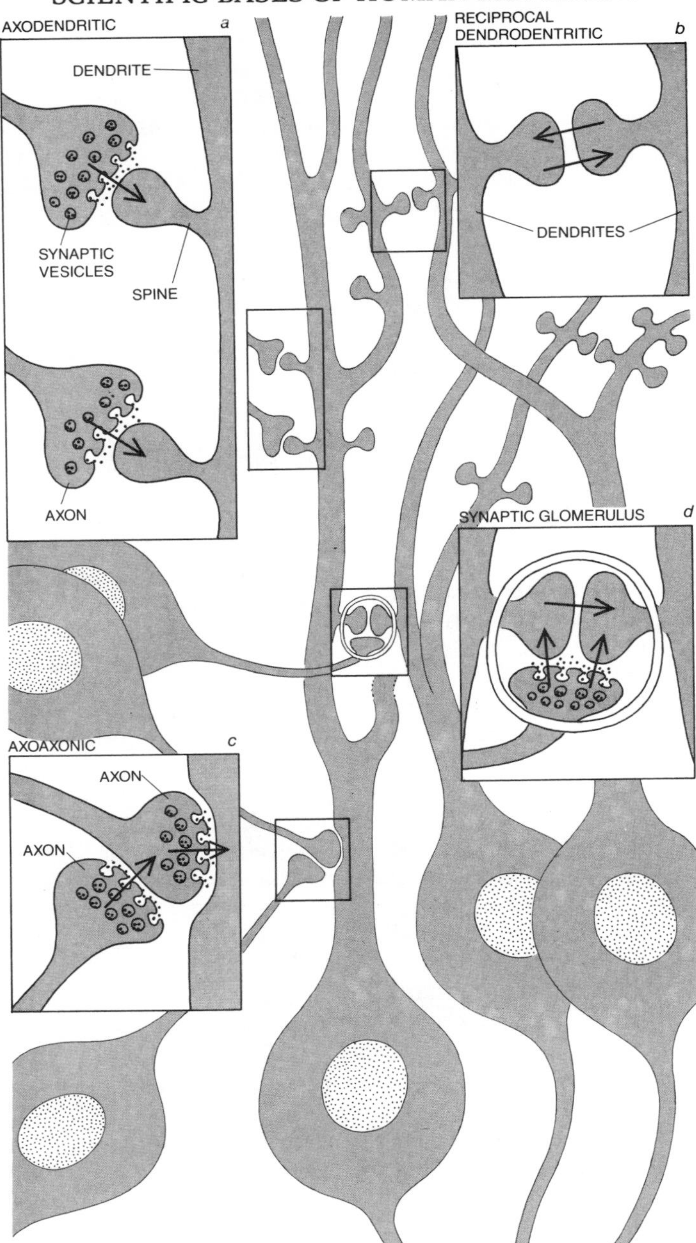

Figure 7.9. Communication between neurons takes place across gaps called synapses. In the classical axodendritic synapse (a) synaptic vesicles in the axon of one neuron release neurotransmsitter toward receptors on the dendrite of a target neuron. It is also possible for a dendrite to pass a message to another dendrite; such messages are passed by way of dendrodendritic synapses. In a reciprocal dendrodendritic synapse (b) each dendrite passes messages to the other by way of a separate synapse. In some other synapses, called axoaxonic synapse (c) the axon of one neuron passes a message through the axon of another neuron to the dendrite of a third neuron. In a synaptic glomerulus (d) the axon of one neuron passes messages to dendrites of two others; the dendrites may pass messages to each other as well. (From Snyder, S. H., 1985. The molecular basis of communication betweens cells. *Scientific American 253*, No. 4: 138.)

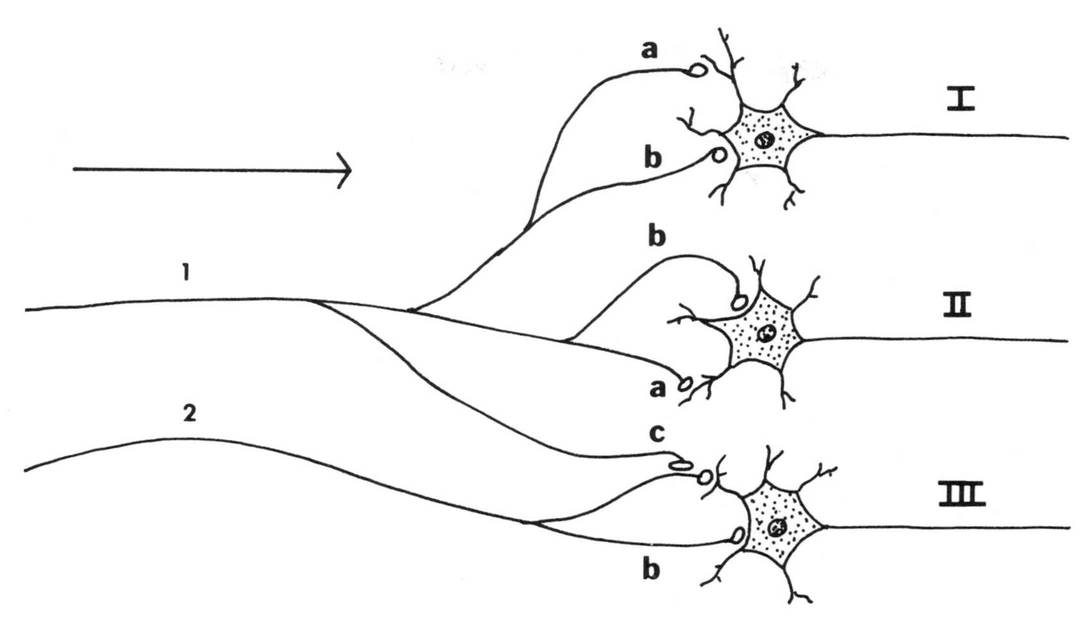

Figure 7.10. Schematic drawing of simple synaptic connections. Afferent neurons 1 and 2 make synaptic connections with interneurons I, II, and III. A collateral from the axon of neuron 1 forms synapse with a terminal of neuron 2 just before its synapse on III. Horizontal arrow indicates direction of conduction. The reader is invited to identify (a, b, or c) the synapses which are axodendritic, axosomatic, or axoaxonic.

secretion of a chemical **transmitter substance**. The minute vesicles revealed by electron microscope photographs of presynaptic terminals contain storage units of the transmitter, which may have been manufactured in the vesicles or, more likely, in the nerve cell body. Weiss (1963) and others have shown by time lapse photography and cinematography that nerve fibers are living, squirming, moving streams through which a peristaltic flow of chemical supplies is driven from the cell body at a rate of from one to a few millimeters per day. These materials may include the neurosecretions, materials to nourish and replenish the neural processes, and, in motor nerves, substances which significantly influence the fast-slow response characteristics of muscle fibers.

The depolarization produced in the membrane of the presynaptic terminals by the arrival of an impulse is assumed to trigger an excitation-secretion coupling mechanism which causes the rupture of the synaptic vesicles. A quantal amount of transmitter substance is ejected into the synaptic cleft by the bursting of each vesicle. Impulses probably do not initiate transmitter release but simply accelerate a secretory process which goes on continually at a low rate. The amount released is proportional to the magnitude of the impulse, and it has been calculated that, for each 30 mV of action potential, transmitter release is increased 100-fold. An impulse probably causes all of the vesicles in immediate juxtaposition to the membrane to rupture and also mobilizes other vesicles for subsequent release by causing them to move into the strategic area.

The transmitter substance, which diffuses across the intervening space in a few microseconds, reacts with the specific chemical groups at the receptor site of the postsynaptic membrane. Bunge et al. (1967), and more recently Nathanson and Greengard (1977), investigating the intracellular effects of neurotransmitters, used electron micrographs to show that the transmitter substance released from synaptic vesicles is geographically associated with that part of the subsynaptic membrane possessing neurotransmitter receptors (Fig. 7.11). These sites are associated with fine channels or pores which are somehow opened by the chemical reaction to permit ions to flow through the membrane at many times their normal rates. As a result, a small change occurs in the resting potential of the postsynaptic membrane at the subsynaptic site. The potential difference between this and adjacent unstimulated areas of the membrane is the **postsynaptic potential** (PSP).

Gardner (1967) derived a schematic representation of the transmitter-receptor interaction (Fig. 7.12) as it was then understood from Eccles (1964, 1965). Electron microscopy confirms the reality of Gardner's schematic, as shown by Nathanson and Greengard (1977) and Lester (1977) citing Heuser's work on the neuromuscular junction.

The action of the transmitter upon the subsynaptic membrane does not directly induce an action potential. The membranes of dendrites and soma (with some exceptions among dendrites of certain brain cells) are electrically inexcitable and incapable of generating action potential spikes. Therefore, the PSP is a local, graded, nonpropagated change in the resting potential which spreads *electrotonically* from the point of origin. The potential change gradually diminishes (decrements) as it spreads. The initial segment of the axon, however, is electrically excitable and has the lowest threshold of any part of the cell membrane. If the PSP is an excitatory change (depolarization) and if its magnitude reaches the critical level of the axon membrane, an action potential will be generated in the initial

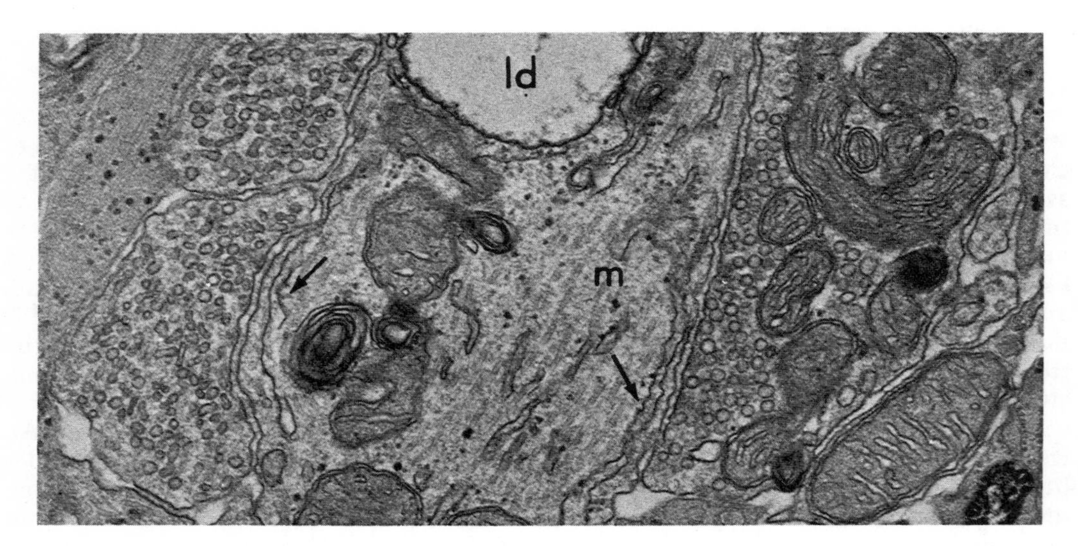

Figure 7.11. Electron micrograph illustrating synaptic junction. This electron micrograph from rat spinal cord shows a dendrite contacted by three axonal boutons. The dendrite contains microtubules (*m*), mitochondria, a large lipid droplet (*ld*), and cisterns of endoplasmic reticulum underlying its surface membrane (*arrows*). Note that the axonal terminal on the *right* contains predominantly round synaptic vesicles, whereas the two terminals on the *left* contain vesicles that are generally somewhat smaller and somewhat flattened. These differences in synaptic vesicle morphology are revealed only after primary fixation in aldehyde. × 46,000. (From Bunge, M. B., Bunge, R. P., and Peterson, E. R., 1967. The onset of synapse formation in spinal cord cultures as studied by electron microscopy. *Brain Res.* 6: 728.)

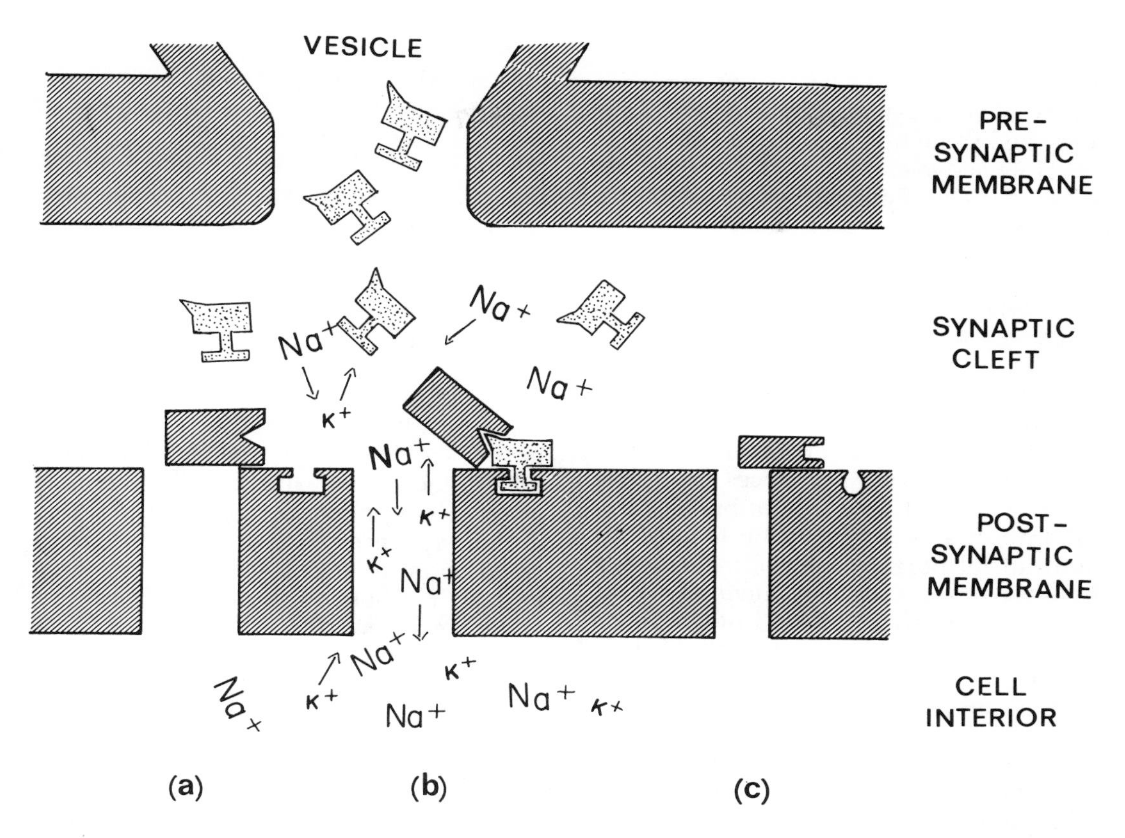

Figure 7.12. Hypothetical explanation of synaptic transmission. A synaptic vesicle is releasing excitatory transmitter substance (stippled) which diffuses across the synaptic cleft. (a), a specific group of atoms on the postsynaptic membrane (rectangle) is so oriented that it occludes the pore which passes through the postsynaptic membrane. (b), the excitatory transmitter has already interacted with chemical groups at the receptor site, producing a change in molecular configuration which has "opened" the pore. This enables sodium to enter and, later, potassium to leave the cell. (c), a narrower channel is shown which requires a different transmitter substance, presumably an inhibitory one. (From Gardner, E. B., 1967. The neurophysiological basis of motor learning—A review. *J. Am. Phys. Ther. Assoc. 47:* 1115.)

segment and conducted nondecrementally over the axon. Except for the interposition of the electrically inexcitable receptor portion of the cell, the sequence of events in synaptic excitation appears to be similar to that described for direct stimulation of the axon. The chemical transmitter substances do not remain long in the synaptic cleft but are soon destroyed, each by a specific enzyme. Almost immediate destruction of the transmitter is essential to neural regulation of activity because its persistence and accumulation would result in exaggerated and uncontrolled responses.

There are two types of transmitter substances, those which are excitatory and those which are inhibitory. A transmitter is excitatory if it exerts a depolarizing effect upon the postsynaptic membrane, thus bringing its resting potential toward the firing level. It is inhibitory if it decreases the possibility of firing either by hyperpolarizing the resting membrane or by stabilizing it, possibly by combining with the chemical groups of the receptor site in a way which prevents activation. A postsynaptic neuron has many synapses on its surface, some of which are excitatory and some inhibitory, and both types are often active at the same time. The constant interplay of excitatory and inhibitory activity results in a fluctuating membrane potential in the initial segment which, at any moment, is the algebraic sum of these depolarizing and hyperpolarizing influences.

Unsuccessful attempts have been made to correlate morphological differences among synapses with excitatory and inhibitory action. Some evidence indicates that excitatory synapses may have wide clefts and broad, continuous postsynaptic plates and may be located on more distal portions of dendrites, while inhibitory synapses may have narrower clefts and thinner, discontinuous plates and may be located upon dendritic trunks and soma surfaces. The situation, however, is not a simple one. Many intermediate and exceptional forms are found. In fact, some of the larger terminals show both synaptic types on the same postsynaptic dendrite. Attempts have also been made to correlate differences in the design of presynaptic endings (knobs, baskets, brushes, trails, etc.) with excitation and inhibition. At the present time no hard and fast conclusions can be drawn linking fine structure and synaptic function. However, the spatial distribution of active terminals in relation to each other and to the axon hillock may be important. Because of the decremental nature of conduction in the receptive membranes, synapses far out on dendrites should be expected to exert less influence than those closer to the cell body, and synapses on the soma near the axon hillock should be the most effective. The possibility exists, however, that large dendrites may have electrically excitable sections which could act as booster stations for their otherwise decremental conduction. Another interesting thought is that strategically placed inhibitory terminals could markedly alter the effectiveness of excitatory endings.

The chemical identity of the transmitters which act at neural junctions *outside* the central nervous system is well known. At the neuromuscular junction release of acetylcholine (ACh) by motor neuron terminals excites the end plate membrane of the muscle fiber. Acetylcholine is released at all autonomic ganglia by the preganglionic neurons and is the transmitter at all parasympathetic and some sympathetic neuroeffector junctions. For the majority of sympathetic junctions, the transmitter is norepinephrine (nor-E).

The transmitters which operate at synapses within the central nervous system have been identified in only one instance: ACh is known to be liberated by terminals of recurrent collaterals of motor neurons at their synapses with certain cells (**Renshaw cells**) in the spinal cord. It seems certain that both ACh and nor-E may be widely involved in central nervous system transmission, and it seems equally likely that other substances are also involved, especially in the brain. Candidates include γ-aminobutyric acid, histamine, 5-hydroxytryptamine (serotonin), and dihydroxyphenylalanine (dopamine), all of which are present in significant amounts; L-glutamic acid as an excitor and γ-aminobutyric acid as an inhibitor have been identified in the invertebrates.

An example of a chemical transmitter found in the brain which has a clearly defined motor function is dopamine. Victims of Parkinson's disease have long attested to the fact that their symptoms can be reversed or even eliminated when they are treated by a drug called L-dopa (*levo*-dihydroxyphenylalanine). This drug, an amino acid precursor of dopamine, when taken up by the bloodstream and converted to dopamine, supplies the amount of chemical transmitter needed by receptors in the basal ganglia. Normally, the neurotransmitter dopamine is secreted by a group of neurons which originate in the substantia nigra and project to the basal ganglia. The degeneration of this group of neurons characterizes the Parkinson patient, who displays symptoms of tremor, rigidity, and delay in the initiation of movement (Nathanson and Greengard (1977)).

Variations in the shape and size of synaptic vesicles may be related to the transmitter contained. Clear, nongranular vesicles, 200 to 400 Å in diameter, probably contain ACh, and dense, granulated vesicles, 800 to 900 Å in diameter, each hold a dense spherical droplet which may be nor-E. Other vesicles differing from these may contain other transmitters.

The action of ACh is fairly well understood. Its reaction with the postsynaptic membrane produces a permeability increase which results in a rapid, localized depolarization of short duration. It is then quickly destroyed by the enzyme acetylcholinesterase (ACh-ase), which hydrolyzes it to choline and acetic acid. Destruction of the transmitter is necessary to avoid persistent and convulsive responses. Several chemical substances (e.g., eserine and neostigmine) inhibit ACh-ase, preventing destruction of ACh, and much has been learned about this neural transmitter through the use of these agents. They have also proven useful in the management of myasthenia gravis, a disease characterized by weakness and extreme muscular fatigue resulting from subnormal release of ACh by motor nerve terminals.

The classic concept of synaptic function is that each neuron releases the same kind of transmitter at all of its terminals (Dale-Feldberg law) and that the transmitter has either an excitatory or inhibitory effect on all of the postsynaptic neurons upon which it acts. The unitary nature of neuron secretion is universally accepted. There is, however, considerable evidence that the sign (+ or −) of the action of a transmitter may be determined by properties of the postsynaptic cell. In the autonomic nervous system ACh is excitatory for some effectors (for example, smooth muscle of gut and bladder) and inhibitory for others (cardiac muscle). Norepinephrine exerts both effects but oppositely in the various tissues. Furthermore, instances are known in which the effect may be reversed by hormonal influences acting on the innervated tissue. For example, the smooth muscle of the pregnant uterus is excited while that of the non-gravid organ is inhibited by ACh. Also, in some simple vertebrate nervous systems clear-cut instances have been found in which the same presynaptic neuron excites some postsynaptic neurons and inhibits others, presumably by the same transmitter (Kandel and Wechtel (1968a)).

Synapses control the normal impulse traffic through the nervous system, determining the amount and pattern of information input and the consequent behavior of each neuron and group of neurons. Synaptic integrative action is based upon the interplay of antagonistic influences: facilitation and inhibition. Is it any wonder that the use of drugs is carefully monitored by our medical colleagues! We need only recognize the potent effect that results from a slight imbalance of one chemical and its predictable effect on synaptic transmission, hence, on possible behavior of the individual.

Synaptic Facilitation

Excitation in a presynaptic neuron does not necessarily result in transmission at every synapse which its terminals encounter. A certain amount of resistance is inherent in each junction and reflects the critical level of depolarization which is required to fire the postsynaptic neuron. Synaptic resistance varies from synapse to synapse and at each synapse is subject to temporary or persistant modification. If the transmitter is excitatory, the PSP which results is known to be a reduction in membrane potential, i.e., a partial depolarization. Such a decrease in the electric charge across the postsynaptic membrane is the **excitatory postsynaptic potential** (EPSP) and represents a reduction of synaptic resistance toward the firing level (Fig. 7.13(a)). This is known as **facilitation.** The action of the excitatory transmitter upon the postsynaptic membrane is thought to result in a general increase in membrane permeability, an opening of all ionic pores. The most notable ion move-

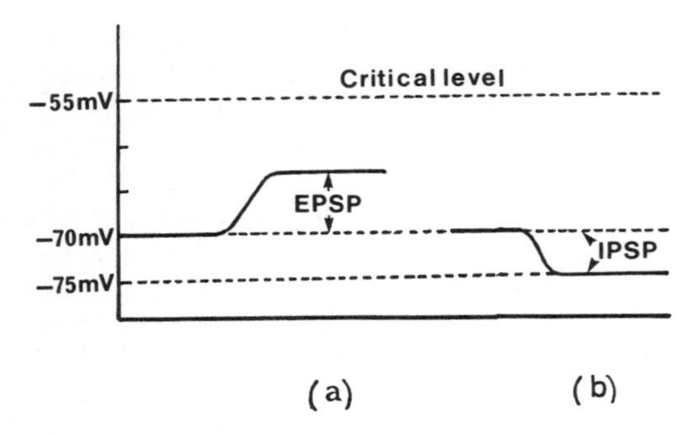

Figure 7.13. Facilitation and inhibition in synapses. The resting potential of the postsynaptic membrane is −70 mV and its critical level is −55 mV. (a) Excitatory transmitter has evoked an EPSP of about 7 mV (membrane potential is now −63 mV). Further excitation equivalent to 8 mV will be required to fire this facilitated postsynaptic neuron. (b) Inhibitory transmitter has induced an IPSP of about 5 mV (membrane potential is now −75 mV). Excitatory transmitter equivalent to 20 mV will be required to fire this inhibited postsynaptic neuron.

ment, however, is that of Na^+ because of its greater electrochemical driving force.

When a number of excitatory volleys arrive simultaneously or in close succession at several synapses of a cell, each contributes its small amount to the postsynaptic depolarization. If summation of EPSPs reaches the critical level of the neuron, a spike potential is generated in the initial segment and conducted along the fiber. The rise of the action potential wipes out the EPSP, probably by antidromic invasion of the soma. However, if the total excitatory effect is in excess of the threshold level or if the presynaptic bombardment is sustained, the initial segment will be repeatedly restimulated and the postsynaptic impulse frequency will rise accordingly. The frequency of impulses in the postsynaptic axon will therefore depend upon the amount of facilitatory transmitter substance released. The greater the amount, the earlier in the relative refractory period will another spike be generated.

Summation of excitatory effects occurring at a number of synapses on the same postsynaptic neuron and derived from terminals of presynaptic neurons from a variety of sources is known as **spatial summation.** The partial depolarization of the postsynaptic membrane by the concurrent subliminal inputs makes the neuron more ready to respond. As a result it may be fired by a subsequent input which alone would have been inadequate. Facilitation may also be accomplished by a high frequency of impulses arriving over a single presynaptic terminal. Such **temporal summation** is probably less effective, except perhaps at the synapses of receptor neurons. In both spatial and temporal summation, each quantum of transmitter contributes toward the possibility of ultimate firing. If a response is already ongoing, facilitatory inputs will cause amplification of the response by increasing the frequency of the postsynaptic impulse train.

Sources of synaptic facilitation include the following: multiple sensory inputs which provide summation as a result of differences in modalities and/or in topographical distribution within a single modality; proprioceptive feedback of information concerning body position or movement; and supraspinal influences from brainstem, cerebellum, and cortex.

Synaptic Inhibition

Postsynaptic Inhibition. Although involving different transmitters, both excitatory and inhibitory synapses are presumed to have the same general manner of function. In both, quantal packets of transmitter are released and react at receptor sites on the subsynaptic membrane, producing a momentary increase in permeability. Eccles (1964, 1965) conjectured that the action of the inhibitory transmitter differs from that of excitatory transmitter in that it produces a selective rather than a general permeability increase by opening pores for penetration restricted to small ions. The flow of current through the membrane of an inhibitory synapse is probably due to either the outflow of K^+ or the inflow of Cl^- or both, with a concomitant increase in internal negativity (Fig. 7.13(b)). The resulting hyperpolarization is the **inhibitory postsynaptic potential** (IPSP), and it opposes the EPSP. Consequently, a greater summation of excitatory transmitter is required to lower the resting polarization to firing level. This type of inhibition is known as postsynaptic inhibition. It is the basis of reciprocal inhibition of antagonistic muscles, an essential factor in coordinated motor activity.

Excitatory input from afferent neurons is transformed into inhibition at appropriate points in the neural network by the interposition of inhibitory interneurons (Fig. 7.14). These are special short-axon neurons which release an inhibitory transmitter at their synapses, thus making it harder to fire the postsynaptic neuron. Therefore, all inhibitory pathways must contain at least three neurons and all pathways involving only an afferent and an efferent neuron (monosynaptic chains) must be excitatory. Conductive delays substantiate the inclusion of at least two synapses in even the most direct inhibitory pathways in mammals.

There are two forms of postsynaptic inhibition which merit special mention: recurrent or "surround" inhibition and disinhibition.

Recurrent or Surround Inhibition. A particular type of postsynaptic inhibition, in which active cells in sensory or motor projection systems inhibit adjacent neurons, has received considerable attention among neurophysiologists. The pathway for this inhibition involves recurrent collaterals which leave motor axons before they emerge from the gray matter of the cord. They pass back into the cord and excite special short inhibitory interneurons called **Renshaw cells**. A Renshaw cell responds to a single stimulus with a high frequency burst of impulses and the release of inhibitory transmitter, with a consequent reduction of excitability in the inciting and adjacent neurons upon which its terminals impinge (Fig. 7.15). More strongly stimulated cells exert a stronger inhibition on their neighbors than that which they receive and hence the excitatory difference between them is exaggerated.

The exact function of this recurrent or surround inhibition is not yet clear. In motor neurons it presumably plays a role in localizing activity within a muscle and so may be of value in distributing motor unit activity for fine movements (Wilson (1966a)). A similar mechanism in sensory pathways may serve to sharpen contrast (Brooks (1959)).

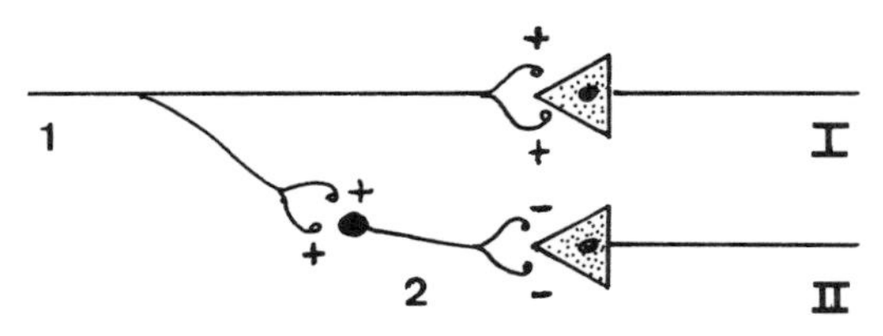

Figure 7.14. A simple inhibitory circuit. A neuron (1) synapses with an efferent neuron (I) and on a short inhibitory interneuron (2). Both of these synapses are excitatory and both neurons are activated. However, since the interneuron (2) secretes inhibitory transmitter, the efferent neuron (II) will be inhibited and will fire only if other excitatory synapses (not shown) induce sufficient EPSP to reach the critical level.

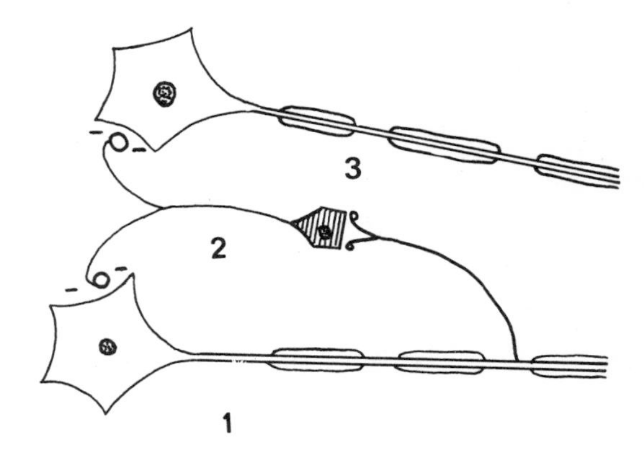

Figure 7.15. Recurrent or surround inhibition. A recurrent collateral from motor neuron (1) re-enters the ventral gray matter and synapses with a short inhibitory neuron, the Renshaw cell (2) (cell body cross-hatched). The Renshaw cell sends terminals to the inciting neuron and to surrounding motor neurons, where its inhibitory transmitter diminishes their irritability.

Disinhibition. Not only do Renshaw cells inhibit adjacent motor neurons but they may also inhibit an already existing inhibition and thereby *facilitate* neurons of the motor pool. Motor neurons are subject to a tonic inhibitory influence by some as yet unidentified interneurons, probably reticulospinal fibers. Through inhibitory synapses on these cells, the Renshaw cells depress their inhibitory action and thus release the motor neurons from the inhibition. This then is a facilitation by **disinhibition** (Fig. 7.16). The fact that this is not a usual type of facilitation is supported by both electrophysiological and pharmacological evidence: membrane potential changes are *hyper*polarizations, and the effect is blocked by strychnine and tetanus toxin, drugs which block postsynaptic inhibitory synapses but do not affect excitatory junctions.

Normally, both facilitation and inhibition are occurring simultaneously but to different extents at the multitude of synapses of a postsynaptic neuron. The postsynaptic cell will fire whenever the algebraic sum of the two antagonistic influences is sufficient to depolarize it to its critical level; the greater the sum the greater will be the frequency of impulses generated.

Presynaptic Inhibition. As the name implies, the effect of presynaptic inhibition is exerted upon the presynaptic neuron rather than upon the membrane of the postsynaptic cell. The pathway for presynaptic inhibition appears to involve neural circuits in which the inhibiting neuron synapses upon the *axon* of the presynaptic neuron close to its own termination (Fig. 7.17). The electron microscope has revealed the existence of small boutons making synaptic contact with telodendria near their large end knobs. These axoaxonic synapses are believed to be the morphological basis of presynaptic inhibition. Pharmacological evidence indicates that the transmitter substance is quite different from that of postsynaptic inhibition. First, the presynaptic inhibitory effect is not blocked by strychnine or tetanus toxin, both of which are powerful antagonists of postsynaptic inhibition, and second, it is sensitive to picrotoxin, a convulsant drug which has no action upon postsynaptic inhibition.

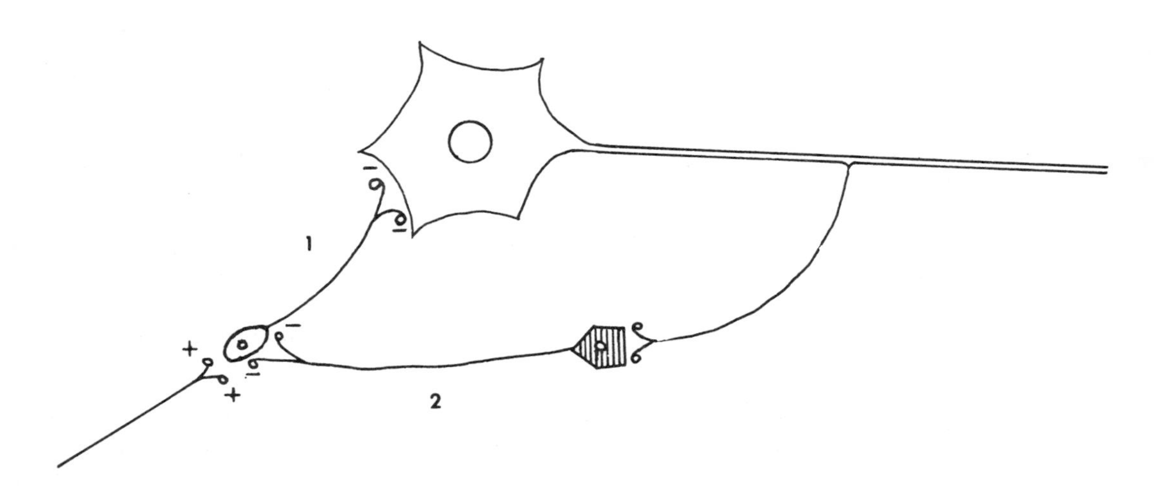

Figure 7.16. Facilitation by disinhibition. The motor neuron is under inhibition from an unknown source by way of an inhibitory interneuron (1). A collateral from the motor axon activates a Renshaw cell (2) which inhibits the inhibitory neuron. The more strongly the motor neuron is stimulated, the greater is the reduction in the incident inhibitory influence. Disinhibition would thus contribute to enhancement of the muscle's response.

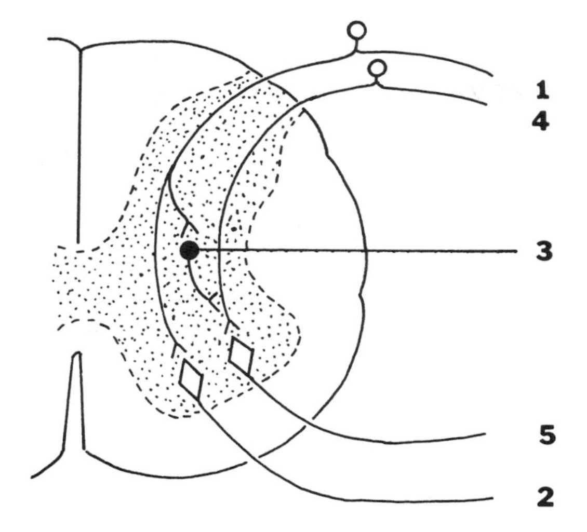

Figure 7.17. Presynaptic inhibition. A hypothetical circuit mediating presynaptic inhibition. An afferent neuron from a muscle spindle (1) is shown making an excitatory connection with a motor neuron (2) to its own extensor muscle. A collateral branch of the afferent neuron activates a short interneuron (3), whose terminals synapse with the axon of an afferent neuron (4) which is making an excitatory connection with an efferent neuron (5) to the antagonistic flexor muscle. As a result, excitation over the extensor afferent (1) will diminish the excitatory influence upon the antagonistic flexor motor neuron by presynaptic inhibition.

The distinctive characteristic of presynaptic inhibition is that EPSPs of the postsynaptic neuron are depressed without any measurable hyperpolarization of its membrane. There is good evidence that the depression is due to a partial depolarization of the presynaptic axon which reduces the magnitude of the action potential invading its terminals. For example, if a depolarization of 10 mV has been induced at the axoaxonic synapses, the spike of the presynaptic neuron will be reduced by 10 mV from its usual level. As mentioned earlier, transmitter release is proportional to the magnitude of the action potential. Consequently, when these smaller potentials reach the end knobs, less excitatory transmitter is released and the EPSP is proportionately lessened. The reduction in transmitter probably reflects a decrease in the number of ejecting vesicles, because there is no evidence that the size of individual quanta is affected. A lesser amount of transmitter results in a lower impulse frequency in the postsynaptic neuron and therefore a decreased response. When the giant axons of the squid were presynaptically depolarized by electric current, a 5% reduction in the magnitude of the presynaptic spike caused a 50% reduction in the postsynaptic response. The neurons which produce presynaptic inhibition often fire repetitively and the presynaptic spike depression may last as long as 100 msec. It is also possible that antidromic impulses traveling centrifugally in the dorsal roots may collide with the orthodromic incoming impulses and reduce their magnitude in that way.

Presynaptic inhibition provides a mechanism whereby the central nervous system can control its input by completely suppressing weak or extraneous sensory inflow and can adjust the effectiveness of signals from one part of the body in relation to conditions prevailing in another part. Most important, it can modulate or eliminate undesirable input from one specific source without altering the sensitivity of the postsynaptic neuron to input from other sources. This is in sharp contrast to postsynaptic inhibition, in which the excitability of the postsynaptic neuron is depressed.

In the central nervous system of vertebrates, presynaptic inhibition is widespread at all spinal cord levels, occurs commonly in the brain, and has been found in interactions between cord and brain. There is increasing evidence that all afferents entering the cord from the skin and other peripheral receptors may exert presynaptic influence upon adjacent neurons and upon themselves. Pyramidal tract cells are thought to reduce stretch reflex activity by imposing presynaptic inhibition upon spindle afferents (Fig. 7.17).

The existence of presynaptic facilitation through an increase of transmitter release by the presynaptic neuron is suspected though as yet unproven (Ganong (1965)).

Both recurrent (Renshaw) postsynaptic inhibition and presynaptic inhibition are feed-*back* inhibitions: an active neuron sends collaterals back to produce inhibition at an earlier point in the transmission pathway. In the cerebellum a feed-*forward* inhibition has been demonstrated. Basket cells and Purkinje cells are both excited by the same input, but the basket cells send terminals forward to inhibit the Purkinje cells. Presumably the mechanism limits the duration of excitation produced by any given afferent volley.

The total subsynaptic area, dendritic plus somatic, of a postsynaptic neuron is relatively enormous as compared with a single synapse and the number of presynaptic terminals impinging on a postsynaptic cell may be very large. Since both facilitatory and inhibitory synapses are represented, both effects may be exerted upon the cell simultaneously. The magnitude of the depolarizing current through the initial segment of the postsynaptic neuron will be determined by the number of active synapses and the algebraic sum of the two antagonistic influences. As long as the excitatory influence exceeds the inhibitory influence by at least the critical amount, the neuron will fire.

By selective facilitation of some synapses and inhibition of others, excitation may be directed into proper outflow channels. Muscles which should contract do so, and those which would interfere with the movement are caused to relax by cessation of outflow to them.

Other Properties of Synapses

The properties characteristic of synaptic transmission are compatible with the accepted chemical theory. They differ in several respects, however, from the electrochemical conduction of action potentials in the nerve fiber.

Synaptic Delay. Transmission across the width (100 to 200 Å) of the synaptic cleft requires 0.1 to 0.3 msec in humans and up to 1 msec in other animals. As compared to conduction velocities of over 100 m/s in large nerve fibers, synaptic transmission is nearly 2 million times slower. Most of the delay is consumed by transmitter release, as both diffusion and chemical interaction with the postsynaptic membrane are accomplished in a few microseconds. It is obvious that, in the conduction of information through the nervous system, the more synapses in a neural chain the longer will be the time required for impulses to travel from receptor to effector.

Polarity of Conduction. Excitation proceeds only from presynaptic axon terminals to postsynaptic dendrite or soma, never in the reverse direction across the synapse. The unidirectionality of the synapse is inherent in the functional differentiation between the receiving and transmitting regions of the cell. Impulses induced artificially in a nerve axon do travel antidromically, but they die out without arousing the electrically inexcitable membranes of dendrite and soma. Furthermore, if dendrites were stimulated, they are not structured to release transmitter. Even in axoaxonic synapses excitation does not pass backward into the presynaptic terminal.

Susceptibility to Fatigue and Drugs. While the nerve fiber requires little oxygen for conduction and is practically indefatigable, the synapse is very susceptible to hypoxia and readily fatigued. It is also more susceptible to drugs and anesthetics. During deep anesthesia synaptic transmission is completely blocked, but nerve axons conduct as usual.

Summation Requirement. In almost every instance summation of input is required to fire a postsynaptic cell. In Figure 7.18, *1*, *2*, and *3* represent neurons which form excitatory synapses on motor neurons *I*, *II*, and *III*. *1* supplies only *I*, *2* supplies both *I* and *II*, and *3* supplies both *II* and *III*. If either *1* or *2* is stimulated by a single shock, an EPSP will be generated in *I* but it will not fire because the critical level has not been reached. If, however, both *1* and *2* are stimulated simultaneously

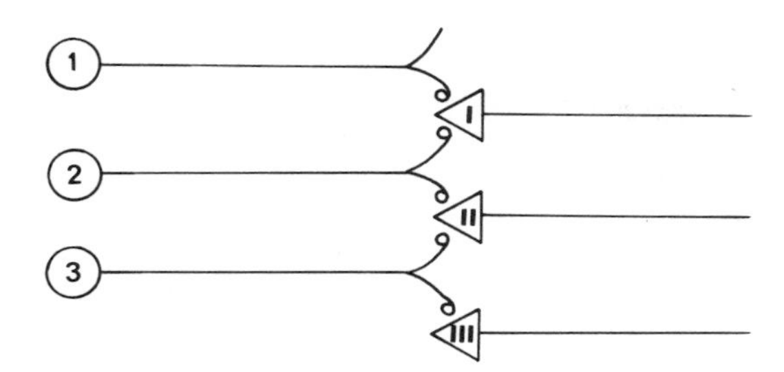

Figure 7.18. Summation requirement in synapses. Motor neuron I may be fired by impulses from (1) and (2) when firing together but will be unresponsive to either alone. Similarly, neuron II may be fired by (2) and (3).

or in close succession, the facilitation induced by one is reinforced by the other, and *I* may be fired by **spatial summation**. *2* will also facilitate *II*, which may then be fired by *3*. Repetitive stimulation of any one of the interneurons may also prove adequate to fire the postsynaptic neurons by **temporal summation**. Inhibitory synapses show similar summation effects.

Neurons with a sufficient number of active knobs are said to be in the **discharge zone**, while those which are facilitated but not fired are said to be in the **subliminal fringe**. The discharge zone increases with increased stimulus intensity, its spread being determined by the convergence of neuron activity upon the same postsynaptic cells.

Synaptic Threshold. The threshold of each postsynaptic neuron varies from moment to moment as a result of the many influences playing upon it. However, persistent increases or decreases of threshold seem to occur as a result of use and disuse of synapses, and they may be significant in motor learning. It is possible that the phrase "lowering the resistance at the synapse," while often used and misused, may adequately describe the changes which take place at a synapse after repeated use. This phenomenon would help to explain why motor patterns, once learned, are easily "called forth" by the cortex, and also highly resistant to change and modification.

Post-tetanic Potentiation at the Synapse. Motor neurons which have been subjected to prolonged repetitive stimulation display a persistent threshold decrease which may last for hours. The effect is specific as to input, relating only to the presynaptic pathway responsible for the tetanization. Although the exact nature of the mechanism involved is unknown, there is evidence that it is presynaptic in origin and probably related to either increased transmitter release or effectiveness. Hubbard and Willis (1963) suggest that the effect is due to an increase in the ability of nerve impulses to release transmitter as a result of an increased amplitude of the presynaptic potentials associated with a hyperpolarizing effect of the repetitive stimulation. Post-tetanic potentiation represents a facilitatory effect of use and perhaps is a primitive form of learning.

Afterdischarge. Response of the postsynaptic cell may persist after stimulation has ceased. This is thought to be due to reverberating circuits which continue the presynaptic input independent of the original stimulus. Firing will ultimately be terminated by fatigue or by inhibition of the neurons responsible for the input.

Nonlinearity of Response. Postsynaptic response does not usually correspond closely to presynaptic input either in intensity or rhythm. Nonlinearity of intensity is inherent in neural organization. The convergence of presynaptic fibers on postsynaptic neurons gives rise to a phenomenon known as **occlusion**. The explanation is embodied in Figure 7.18. Assume that the rate of stimulation applied to an interneuron is sufficient to evoke maximal response of the postsynaptic motor neuron, which will then elicit a maximal tension of 50 g in its motor unit. If *2* alone is stimulated, motor units *I* and *II* will both respond, yielding 100 g of tension in the muscle. *3* alone will excite motor units *II* and *III* and also produce 100 g of tension. If *2* and *3* are then stimulated at the same time, the result will be not 200 but only 150 g as the three motor units are activated. Because of the overlapping on postsynaptic neurons, the tension evoked by simultaneous stimulation of several presynaptic neurons is *less* than the sum of the tensions produced when each is stimulated alone. This is **occlusion**.

Another and more important source of nonlinearity is the modulation effected by antagonistic inputs. This is the basis of the integrative action of the nervous system.

Although the rhythm of stimulation is not usually reflected in the postsynaptic response, in certain instances a presynaptic neuron may **drive** the postsynaptic neuron, the latter responding one for one with the input over a limited range.

Neuromuscular Junction as a Specialized Synapse

The neuromuscular junction is essentially a specialized neuroeffector synapse, sharing many of the properties of neuroneural synapses but displaying other properties which are unique to it.

Generally a skeletal muscle fiber in humans receives only one motor nerve terminal, rather than a convergence of numerous inputs. Multiple innervation is not uncommon in some vertebrates, however, and is in fact characteristic of slow muscle fibers in the frog.

The presynaptic nerve terminal, although discrete, is "buried" in the specialized end-plate sarcoplasm of the muscle fiber. Nerve and muscle membranes are separated by a primary synaptic cleft, and beneath the neural ending the sarcolemma is extensively folded into innumerable secondary clefts, thus tremendously increasing the membrane area exposed to the transmitter. Mitochondria are abundant in both the terminal and the end-plate regions. Synaptic vesicles appear on the presynaptic side of the interface and the transmitter is known to be ACh (Fig. 7.19).

Transmitter release induces local depolarizations in the end-plate membrane known as end-plate potentials (EPPs). It was the occurrence at neuromuscular junctions in resting muscle of miniature end-plate potentials whose amplitudes were integral multiples of a minimal value that led to the concepts that transmitter was released in quantal amounts by individual synaptic vesicles and that random release occurred even in the absence of presynaptic action potentials. The amount of ACh released randomly into the junctional cleft varies directly with the concentration of Ca^{++} and inversely with the concentration of Mg^{++} at the end-plate region. Although ACh acts on the muscle fiber membrane, it is ineffective if introduced *under* the membrane into the cytoplasm.

When an action potential reaches the prejunctional terminal, the number of vesicles which rupture apparently release enough ACh to depolarize the end-plate sarcolemma to its critical level. In other words, a single impulse is able to produce a full-sized end-plate potential and to excite the muscle fiber membrane. Summation is therefore not a usual requirement for neuromuscular transmission, nor is there any synaptic inhibition at this junction. Motor inhibition must be accomplished centrally at the synapses of the motor neuron, and the muscle is then "inhibited" by the cessation of impulses traveling to it over its efferent nerve fibers.

The junction between receptor cell and afferent neuron may also be considered as a special type of synapse, at least in some instances. This is discussed briefly in Chapter 9.

SUMMARY

Briefly, the events in the functioning of neurons take place as follows.

1. Stimulation of dendrites results in excitation in the form of local graded depolarization which spreads decrementally over the dendrites and cell body.
2. If depolarization at the initial segment of the axon reaches the critical level, action potentials are generated and conducted nondecrementally along the axon.
3. Arrival of action potentials at axon terminals causes secretion of chemi-

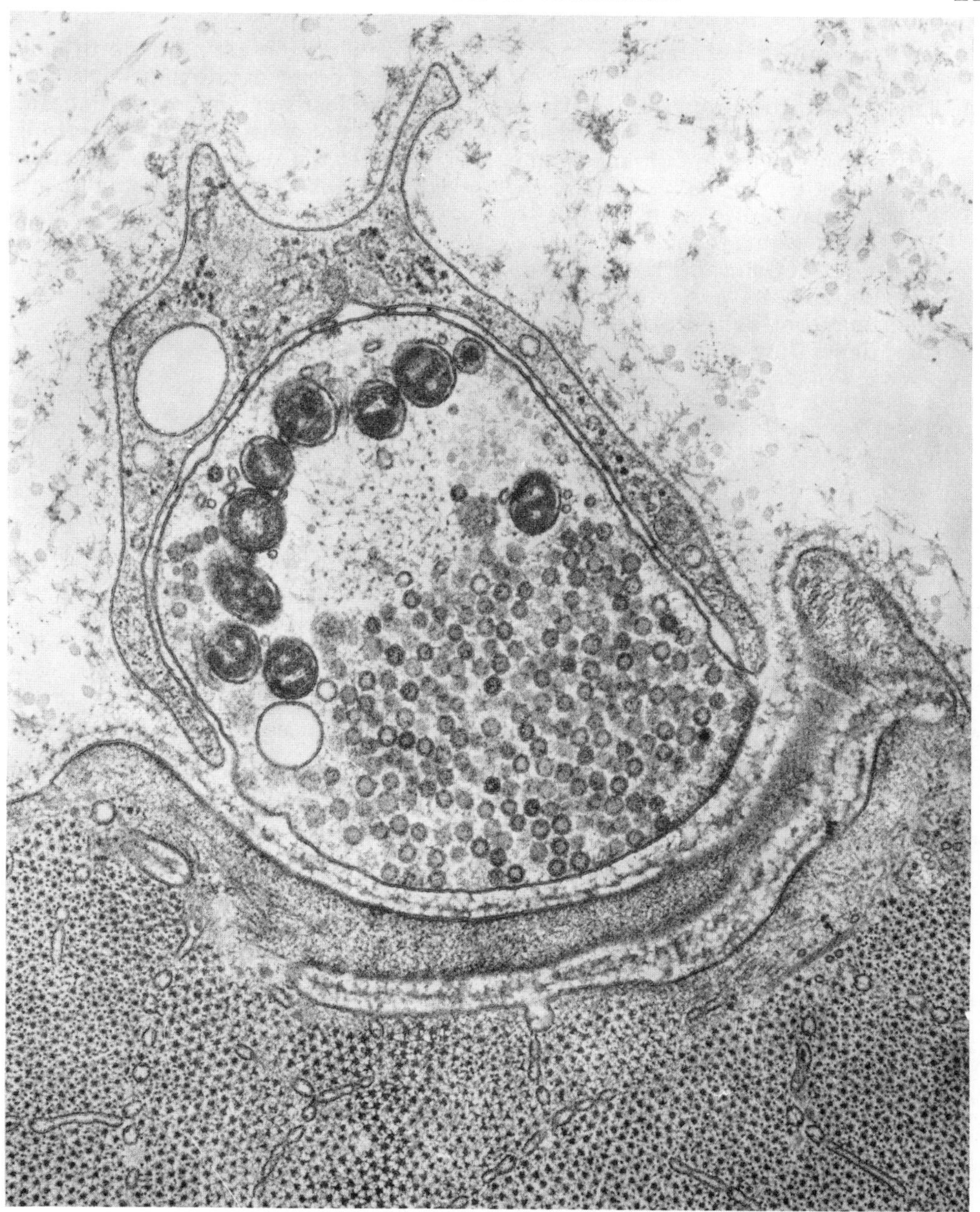

Figure 7.19. Site of acetylcholine release from a motor-nerve terminal is magnified 71,400 diameters in this electron micrograph made by John E. Heuser of the University of California Medical Center in San Francisco. The terminal is stocked with saclike synaptic vesicles containing molecules of acetylcholine; the larger dark structures are mitochondria, which generate the energy required for the activities of the nerve ending. On the arrival of an impulse the synaptic vesicles fuse with the membrane, releasing acetylcholine into the fluid-filled cleft between the terminal and the muscle cell. The molecules of acetylcholine then bind to receptors embedded in the muscle-cell membrane. Below the cleft a deep invagination in the muscle membrane, or junctional fold, is shown in partial section. (From Lester, H. A., 1977. The response to acetylcholine. *Sci. Am. 236:* 106; reproduced courtesy of Dr. John Heuser.)

cal transmitter substance into the synapses.

4. The transmitter substance diffuses across the synaptic gaps and chemically acts upon the membranes of the postsynaptic neurons.
5. Each postsynaptic neuron, if adequately activated, repeats the generation, conduction, and transmission of excitation. Its postsynaptic influence may be facilitatory or inhibitory, depending upon the type of its transmitter and the nature of the next postsynaptic neuron.
6. The last neuron in each chain, the efferent neuron, releases its transmitter at the neuroeffector junctions of the muscle fibers of its motor unit.
7. The muscle responds with tensions proportional to the total number of fibers activated and the frequency of impulse bombardment.

CHAPTER 8

Basic Neuroanatomy: The Central Nervous System

INTRODUCTION

The nervous system has often been described as being analogous to the electronic computer. The common features cited are the abilities to collect information, store information, compute new information, and process all information, converting it into some usable form for potential application. Like the computer, a programming unit is needed to direct the sequence of events, subroutines are called forth to perform specific operations, and automatic mechanisms are evident to respond to the input and execute the programs dictated by the subroutines and the programmer.

To appreciate the complexity of the nervous system, consider the not-too-common (fortunately) occurrence of the housewife touching the hot stove by mistake (Fig. 8.1.). Flexor muscles "automatically" respond to withdraw the finger from the noxious stimulus; hair follicles are facilitated to produce the "hair-raising" response, and sweat glands may suddenly become activated; a few milliseconds later, the housewife is *aware* of the pain and, depending on the "severity" of the noxious stimulus, the vocal muscles may respond to produce the "ouch" response (or other perhaps unprintable word). This series of simple events occurring within a few milliseconds is mediated by the nervous system (Moore (1969)). The immediate finger withdrawal response, followed by hair and sweat gland responses, is subserved by a series of nerve pathways at the spinal cord level; the awareness of the pain and subsequent "ouch" response occur when impulses travel, via nerves, up the spinal cord, through the brain stem and finally arrive at the cerebral cortex (the "head brain" or the programmer).

Any one nerve is simply a collection of sensory (receptor) and motor (effector) neurons. Spinal nerves are composed of the neurons which "sense" the pain, the neurons which effect the withdrawal of the finger, and the neurons which effect the hair and sweat responses. Cranial nerves, such as those which activate the vocal muscles, are importantly involved, as are other nerves, ascending and descending pathways, and brain centers. It is important to recognize that the act of withdrawal from the painful stimulus occurs first, mediated by neuronal activity momentarily entering and exiting from

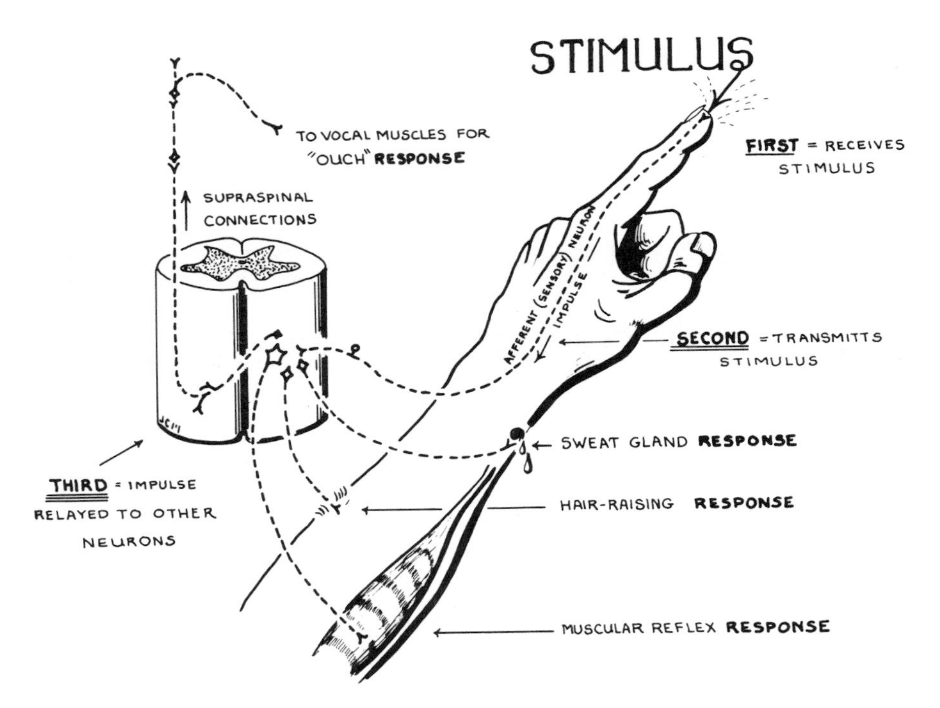

Figure 8.1. A simple stimulus-response pathway. (From Moore, J. C., 1969. *Neuroanatomy Simplified*. Dubuque, Iowa: Kendall/Hunt Publishing Company, p. 13.)

the spinal cord; the recognition of the pain occurs several milliseconds later after the act of withdrawal has taken place, indicating that longer pathways and "interconnections" with brain centers are involved. It is fortunate for the housewife that the withdrawal occurs immediately and is not dependent on her perception of the pain; a few milliseconds of additional contact with the noxious stimulus could result in a more serious burn due to the longer duration of the stimulus.

It is apparent that, even in a very simple stimulus-response series of events, all levels of the nervous system have the potential of becoming involved. This and subsequent chapters are concerned with the basic components of the somatic nervous system, highlighting those which are importantly involved in motor behavior. (The autonomic nervous system is not discussed.) The nervous system is a highly organized, extensive network designed to integrate internal reactions of the individual and correlate these with the external environment so that appropriate responses are made. After all, it is the internal mechanisms which ultimately make it possible to take advantage of the "external" mechanisms which are discussed in the first five chapters of this book. The nervous system is arbitrarily separated into two large divisions, the central nervous system and the peripheral nervous system. The central nervous system, the subject of this chapter, is composed of the brain and spinal cord. The peripheral nervous system, composed of the nerves, ganglia, and end organs which connect the central nervous system with all other parts of the body, is discussed in subsequent chapters.

To aid the reader in understanding the complexity of the central nervous system, a general overview of the major structures is presented. The descriptions are by no means complete but are intended to help the reader understand the functions attributed to each major brain and spinal cord structure and become familiar with common terminology used in most textbooks and literature.

CEPHALON MORPHOLOGY AND PHYSIOLOGY

The brain or **cephalon** is divided into five major sections, termed **telencephalon** (or "head brain"), **diencephalon** (or "primitive-brain"), **mesencephalon** (or "middle brain"), **metencephalon**, and **myelencephalon** (or "hind brain"). Man has one of the most sophisticated telencephalons, whereas lower forms of life have extensive diencephalons and limited telencephalons (Fig. 8.2). An outline of the brain, its major sections, and the principal anatomic units associated with each follows.

Telencephalon:
- Cerebral cortex

Diencephalon:
- Basal ganglia
- Thalamus
- Reticular formation (part of)
- Internal capsule

Mesencephalon:
- Midbrain (tectum and tegmentum)
- Reticular formation (part of)
- Red nucleus and substantia nigra

Metencephalon:
- Pons
- Reticular formation (part of)
- Cerebellum
- Vestibular apparatus

Myelencephalon:
- Medulla oblongata
- Reticular formation (part of)

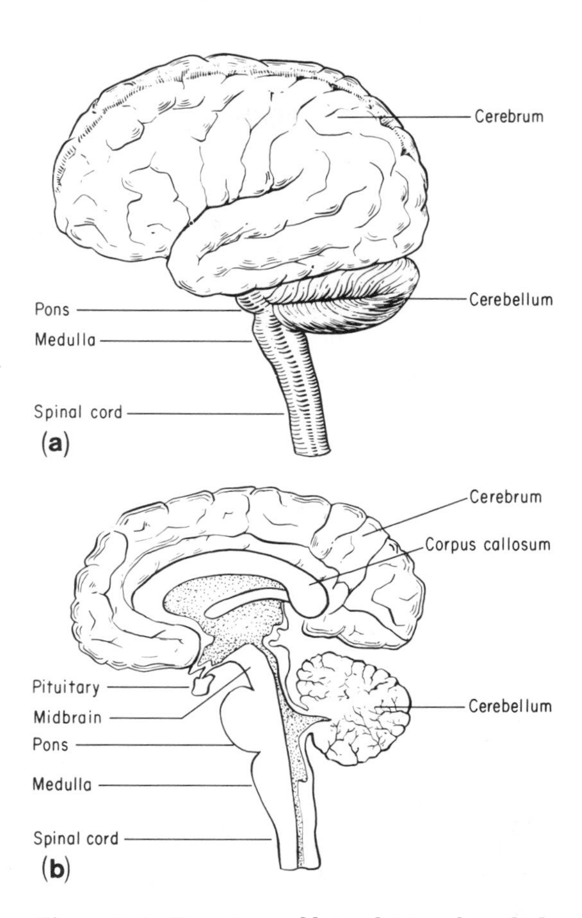

Figure 8.2. Drawings of lateral (a) and medial (b) views of the human brain. (From Thompson, R. F., 1967. *Foundations of Physiological Psychology*. New York: Harper and Row, p. 85.)

Telencephalon

The telencephalon is composed of the **cerebral cortex**, which is primarily a vast information storage area. "Approximately three quarters of all the neuronal cell bodies of the entire nervous system are located in the cerebral cortex" (Guyton (1972)). All information is collected, stored, computed, and processed in the cerebral cortex so that such information can be called forth at will to control motor behavior.

Certain areas of the cortex perform specific functions although it has been shown that the cortex has not been as discretely divided as some might believe (Guyton (1972)). Figure 8.3 shows some of the functional areas of the human cerebral cortex. The map, originally constructed by Penfield and Rasmussen in the 1950s, suggests that some areas (those labeled) have specific functions while the other areas

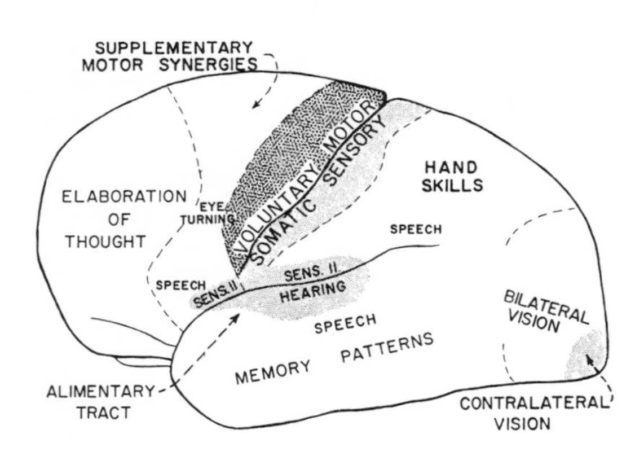

Figure 8.3. Functional areas of the human cerebral cortex. (From Penfield, W., and Rasmussen, T., 1968. *The Cerebral Cortex of Man*. New York: The Macmillan Company.)

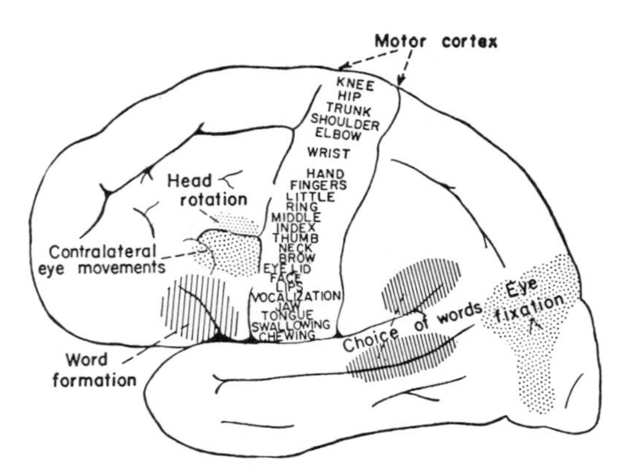

Figure 8.4. Representation of the different muscles of the body in the motor cortex and location of other cortical areas responsible for certain types of motor movements. (From Guyton, A. C., 1972. *Structure and Function of the Nervous System*. Philadelphia: W. B. Saunders Company, p. 190.)

perform in a more general way. The **primary sensory** areas receive signals from the lower centers of the brain and spinal cord and transmit the results of the "analyses" back to the lower centers and to other regions of the cortex. The **motor areas** of the cortex send impulses to specific muscle groups either directly or indirectly via other brain structures. Representation of the different muscle groups for all body parts within the motor cortex is shown in Figure 8.4. Generally, the degree of representation of body parts correlates with the complexity of movements required (Fig. 8.5). The **premotor** area is believed to be concerned primarily with the acquisition of specialized motor skills; in addition, control of complex eye and mouth movements is located here. Some of the **somatic sensory** and motor areas overlap so that sensations and muscular contractions in the same area of the body are monitored by the same area of the cerebral cortex.

In 1909, Brodmann classified the functions of the cortex according to structurally distinctive areas (Fig. 8.6); his classification is used by most neuroanatomists and is found in most modern-day texts. Brodmann's areas 1, 2, and 3 (of the **postcentral gyrus**) represent the somatic sensory region while areas 4 and 6 (of the **precentral gyrus**) represent the motor and premotor areas, respectively.

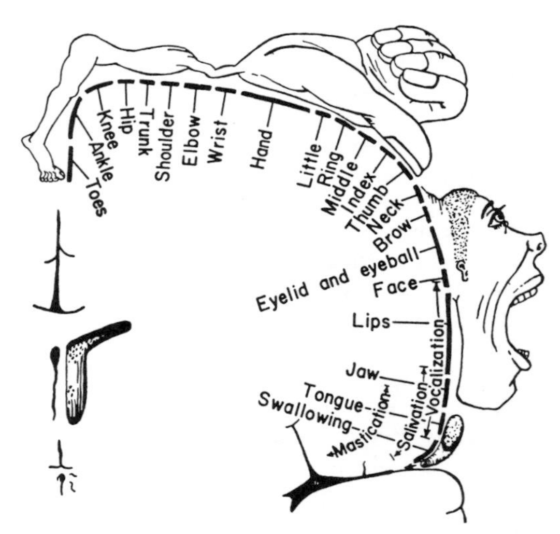

Figure 8.5. Degree of representation of the different muscles of the body in the motor cortex. (From Penfield, W., and Rasmussen, T., 1968. *The Cerebral Cortex of Man*. New York: The Macmillan Company.)

The reader should consult a textbook on the structure and function of the nervous system for details concerning all areas of the cerebral cortex. For the purposes of

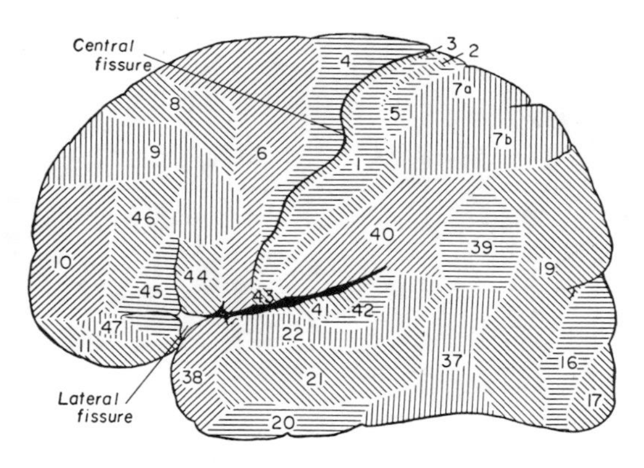

Figure 8.6. Structurally distinctive areas of the human cerebral cortex. (Modified from Brodmann.) (From Everett, N. B., 1971. *Functional Neuroanatomy*. Philadelphia: Lea and Febiger.)

this text, it is sufficient to say that movement can be initiated by the motor cortex, that patterns of human movement can be learned and then stored in the motor cortex (sometimes called an **engram**) to be called forth at will, and that motor cortex signals can be modified by impulses reaching the sensory cortex, or other brain structures, as the signals descend to the **spinal cord** (effector) level.

Diencephalon

That part of the brain called the diencephalon is composed of structures specific to modern man's needs, but is old in terms of the evolution of the species. Some of the structures of the diencephalon dominate the human body during early stages of ontogeny and will persist in directing adult body functions. They will also direct motor behavior under special conditions of stress and reduced cerebral cortex activity. The principal structures mentioned here are the **basal ganglia**, **thalamus**, part of the **reticular formation**, and the **internal capsule**.

Anatomically, the basal ganglia are composed of the **caudate nucleus**, **putamen**, **globus pallidus**, **amygdaloid nucleus**, and **claustrum**. The latter two are not directly concerned with motor function and will not be discussed. Numerous pathways exist between the motor cortex and the caudate nucleus and putamen (collectively labeled the **striate body**); the latter sends numerous fibers to the globus pallidus, which communicates back to the motor area of the cortex via the thalamus. These circular pathways operate as a kind of feedback loop or servo-control mechanism (Guyton (1972)). The functions of the basal ganglia are generally associated with the **extrapyramidal system**, one of the systems by which the cortex communicates with the final common pathways to muscle. If the cortex were destroyed, discrete movements of the body, especially the hands, would be impossible, but gross movements of a subconscious nature would still be possible; i.e., walking and controlling equilibrium (Guyton (1972)).

It is important to recognize that the basal ganglia function as a total system. Therefore, assigning specific functions to each portion will not completely describe the physiology of the system. As a whole, the basal ganglia are generally recognized as centers capable of inhibiting muscle tone throughout the body (Noback and Demarest (1972)). The striate body seems to initiate and regulate gross movements performed unconsciously. The globus pallidus is usually ascribed the ability to provide background muscle tone for movements that are initiated either by the striate body or by the cortex. Thus, the basal ganglia, together with other centers, modulate motor activities through circuits that feed back to the cortex.

The thalamus serves many important functions other than that related to motor control. Along with the **subthalamus**, **substantia nigra**, and **red nucleus**, the thalamus operates in close cooperation with the basal ganglia to exert influences on motor activity. Also, impulses from the **cerebellum** are transmitted back to the motor cortex via a specialized nucleus of the thalamus, the **ventral lateral nucleus**. The **subthalamic nucleii** are importantly involved in the total circuitry which provides the background discharge necessary for the success of fine coordinated movements.

The reticular formation is a vast network of neurons and nucleii which anatomically is located throughout most of the brain stem. A portion of it is located in the diencephalon and it extends caudad as far as the myelencephalon. Those functions concerned with motor control are the primary focus of this textbook although the reticular formation plays major roles in arousal and consciousness and various states of sleep and relaxation. It exerts powerful influences on phasic and tonic motor activities. That portion of the reticular formation located in the diencephalon is generally ascribed the ability to exert facilitatory influences on spinal motor discharge. Thus, flexor and extensor reflexes, decerebrate rigidity, and responses evoked from the motor cortex are facilitated (Magoun (1958)).

An appreciation of some of the facilitatory and inhibitory influences initiated in the diencephalon structures can be gained by studying Figure 8.7, an outline view of the brain. Areas yet to be discussed such as other portions of the reticular formation and some of the cerebellar and **vestibular nucleii,** also are shown.

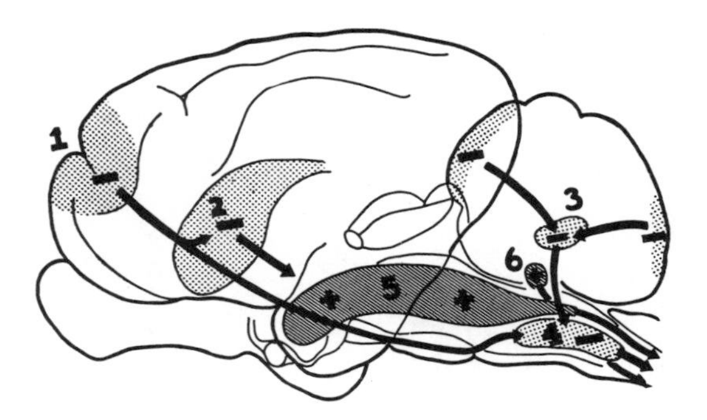

Figure 8.7. Outline view of the brain of the cat showing facilitatory and inhibitory areas. Inhibitory: 1, portion of cerebral cortex; 2, basal ganglia; 3, cerebellum; 4, reticular formation (medulla). Facilitatory: 5, reticular formation (diencephalon and midbrain); 6, vestibular nucleii. (From Magoun, H. W., 1958. *The Waking Brain*. Springfield, Ill.: Charles C Thomas, p. 16.)

The internal capsule is actually a massive bundle of nerve fibers which links the cerebral cortex with other portions of the central nervous system. If the reader can imagine the diencephalon as a structure shaped something like a fist and inserted upward "inside" the cerebral cortex, it is easier to understand that the major routes to and from the cortex travel through the diencephalon. Ascending fibers projecting from subcortical nucleii to the cerebral cortex and descending fibers projecting from the cerebral cortex to lower centers of the brain and spinal cord are massed together and include both sensory and motor pathways. Generally, the internal capsule is divided into **anterior** and **posterior limbs**, each of which has distinct pathways associated with them. For example, the **corticospinal tract** (cerebral cortex to spinal cord) passes downward through the rostral portion of the posterior limb of the internal capsule and brain stem, crossing to the opposite side at the **medulla.** The caudal half of the posterior limb of the capsule contains various projections from the many pathways of the thalamus to the cortex (Noback and Demarest (1972)). The anterior portion of the capsule includes mostly the many fibers which connect the several portions of the brain stem with each other and with portions of the cortex. Overall, the internal capsule, elaborated at the diencephalon level, can be thought of as the great "elevator" or "escalator" system of the human body because it represents the *only* means by which nervous system impulses may descend from or ascend to the cerebral cortex.

Mesencephalon

The term mesencephalon is often used synonymously with the term **midbrain**. The mesencephalon is the most anterior extension of the brain stem which still appears to have the basic structural characteristics of the spinal cord. The dorsal portion of the midbrain (**the tectum**) is easily identified by the two pairs of relay nucleii which subserve the visual and auditory systems. The ventral portion of the midbrain (**tegmentum**) contains nucleii for the third and fourth **cranial nerves**, all of the ascending and descending tracts mentioned earlier in the section concerned with internal capsule, and a portion of the reticular formation. Two important centers are also located here: the substantia nigra and the red nucleus (Thompson (1967)). The function of the substantia nigra is not well known but apparently it is the major center for excitation of the gamma loop associated with the **neuromuscular spindle** (see Chapter 10 on the spindle). Its importance is recognized because it apparently activates the **gamma efferent system** even before the **alpha motor neurons** to muscles are activated, and provides the background muscular tone so that discrete and highly coordinated movements can be performed. The red nucleus is primarily concerned with gross body movement especially as the body deviates from the standing upright posture.

Metencephalon

The subdivision of the brain called the metencephalon includes the **pons** and part of the reticular formation. The cerebellum and vestibular apparatus are usually associated with the metencephalon because of location although they are not considered to be a part of this subdivision.

The pons, like the midbrain, contains ascending and descending tracts and a large mass of transverse fibers on its ventral aspect. The literal meaning of the word pons is "bridge," which implies its function: it interconnects the two sides of the cerebellum and brain stem as well as the fibers connecting the cortex with the spinal cord. Also, several cranial nucleii are located in the pons, notably the main motor nucleus of the fifth nerve and the nucleus of the seventh nerve.

That portion of the reticular formation located at metencephalon level is concerned with facilitation and inhibition of lower spinal cord neurons.

The cerebellum is often dubbed "the great motor coordination center." It overlies the pons and presents a convoluted appearance with numerous fissures. Its specialty as a center for sensory-motor coordination is noted by the many afferent and efferent fibers associated with it. Sensory input is received from the vestibular system, spinal fibers, auditory and visual systems, reticular formation, and various regions of the cerebral cortex. In turn, it sends efferent fibers to the thalamus, reticular formation, and other parts of the brain stem. The cerebellum is often subdivided anatomically, and this type of organization appears to have some functional significance. The oldest part (phylogenetically) is the **flocculus,** which projects to the vestibular nucleii. The next oldest is the **medial region (vermis),** which projects to the **medial (fastigial) nucleus** and the vestibular nucleii. The intermediate portion projects to one of the three cerebellar nucleii known as **interpositus**, and the lateral portion projects to the **lateral (dentate) nucleus**. The lateral portion, principally the dentate nucleus, is the part markedly developed in humans and other mammals (Thompson (1967)).

An extremely important aspect of the cerebellum is its extensive connections with many parts of the brain stem and cerebral cortex. The anterior lobe of the cerebellum projects to the primary motor area of the cerebral cortex via the dentate and interpositus nucleii and the ventral lateral nucleus of the thalamus. The vermis and intermediate regions are functionally associated with the visual and auditory areas of the cerebral cortex. In addition to projections mentioned above, the cerebellar cortex also projects, via the appropriate **cerebellar nucleii**, to the reticular formation, the pons, the midbrain, the red nucleus, the basal ganglia, and the motor cortex via the ventral lateral nucleus of the thalamus. As mentioned in the section concerned with the thalamus, the cerebellum is involved in a large number of circles or loops so that it receives sensory information, projects to many parts of the brain including the cerebral cortex, and is importantly involved in the descending tracts to the spinal cord (Thompson (1967)).

Myelencephalon

The **medulla oblongata** is the main structure composing the myelencephalon. It contains ascending and descending fiber tracts and a number of cranial nerve nucleii. It is easily distinguished from other parts of the brain stem at the point where descending tracts cross and continue to the spinal cord on the contralateral side of the body (called **decussation of the pyramids**). The medulla houses the most caudad portion of the reticular formation, which, at this level, exerts inhibitory influences on spinal cord neurons.

CORONAL AND CROSS-SECTIONAL ANATOMY OF THE CENTRAL NERVOUS SYSTEM

It has already been mentioned that the means of communication between the various parts of the brain and the spinal cord involves a complex series of nerve networks which travel in specific pathways or tracts. To facilitate the understanding of the "routes" fibers take in ascending and descending within the central nervous system, selected coronal and cross-sectional views of the central nervous system are shown. A coronal section through the cerebrum is shown in Figure 8.8; note the location of the internal capsule and the caudate nucleus. A cross-section of the midbrain is shown in Figure 8.9 and reveals the location of the **cerebral peduncles**, the pons, and the origination of some of the cranial nerves. A cross-section at the level of the pons (Fig. 8.10) demonstrates the location of the reticular formation, one of the cranial nerves, the **cerebellar peduncles**, and some of the **pontine nucleii.** Figure 8.11 displays the medulla in cross-section so that some of the nucleii, cranial nerves, medial pathways, and the pyramid decussation can be seen.

At the spinal cord level, a cross-section at the cervical level (Fig. 8.12) reveals the characteristic gray formation usually described in the shape of an "H" or butterfly and the outer white portion. The gray portion represents the several neuron cell bodies in this part of the spinal cord while the outer part is white because it is totally made up of axons of nerve fibers, both ascending and descending. The "white" ascending and descending fiber tracts are called **fasciculi** and are organized into three columns: dorsal, ventral, and lateral. The "gray" matter is divided into dorsal and ventral horns: the cell bodies in the dorsal horns receive sensory information

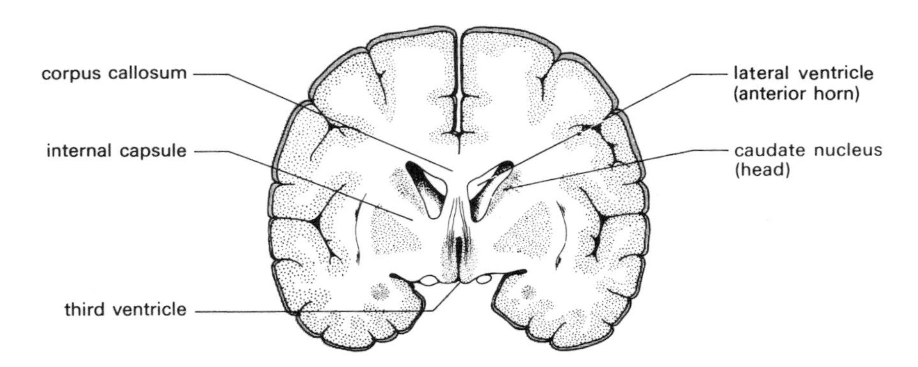

Figure 8.8. Coronal section through the cerebrum. (From Sage, G. H., 1977. *Introduction to Motor Behavior: A Neuropsychological Approach*. Reading, Mass.: Addison-Wesley Publishing Company, p. 24.)

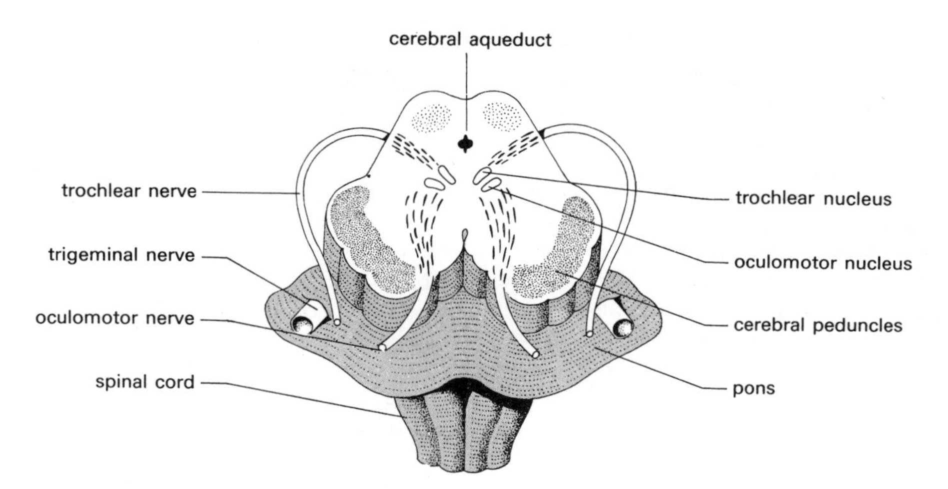

Figure 8.9. Cross-section of the mesencephalon. (From Sage, G. H., 1977. *Introduction to Motor Behavior: A Neuropsychological Approach*. Reading, Mass.: Addison-Wesley Publishing Company, p. 26.)

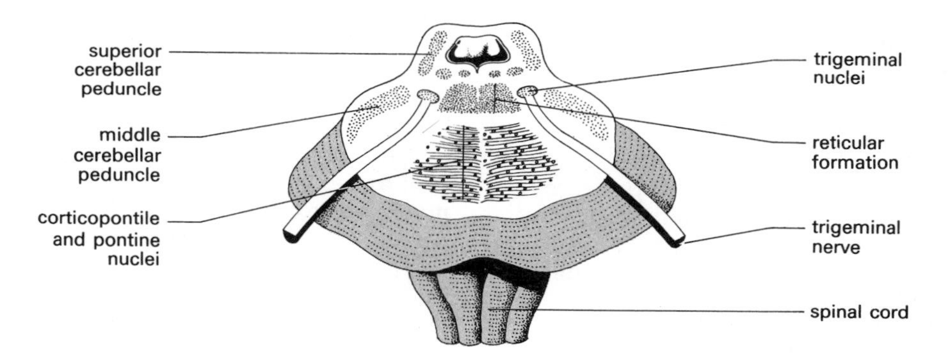

Figure 8.10. The pons in cross-section. (From Sage, G. H., 1977. *Introduction to Motor Behavior: A Neuropsychological Approach*. Reading, Mass.: Addison-Wesley Publishing Company, p. 27.)

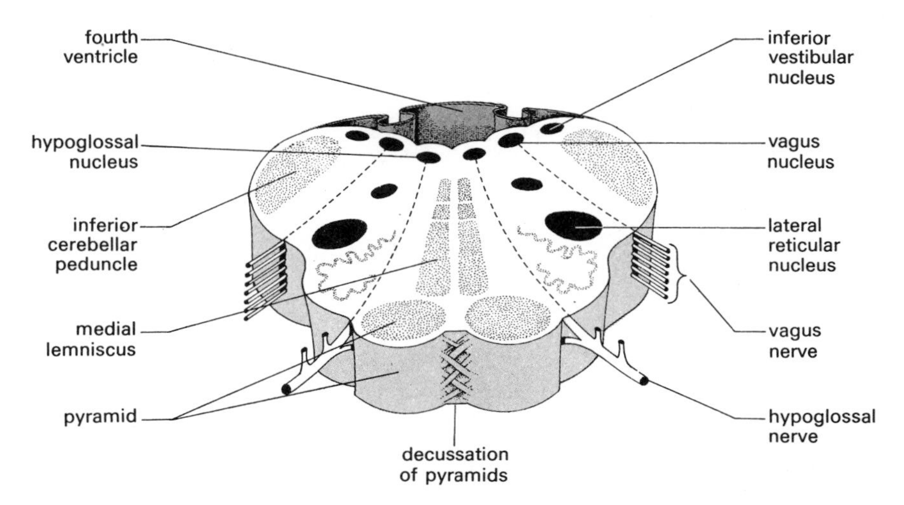

Figure 8.11. Medulla oblongata in cross-section. (From Sage, G. H., 1977. *Introduction to Motor Behavior: A Neuropsychological Approach*. Reading, Mass.: Addison-Wesley Publishing Company, p. 28.)

from the periphery via afferent neurons and transmit to the brain and other levels of the spinal cord; the ventral horns are composed of cell bodies which are the "**final common pathway**" to muscle (or to other effectors), and thus are referred to as efferent neurons. The terms "final common pathway" (an old Sherrington term), alpha motor neuron, and **lower motor neuron** are often used synonymously to refer to the final nerve cell which innervates striated muscle.

PATHWAYS

All parts of the brain and nervous system are linked together in an elaborate network of pathways. Those pathways entering and exiting from the cerebral cortex are of interest here.

Descending Pathways

The means of communication from the cerebral cortex to the spinal cord, terminating on lower motor neurons, is functionally divided into two systems: the **corticospinal system** (also called the **pyramidal system**) and the extrapyramidal system. Other descending efferent projections are used to communicate with other parts of the brain and are named for the parts with which the fibers interconnect. For example, the **corticorubro**, **corticopontine**, and **corticostriate** projections are those pathways which carry fibers from the cerebral cortex to the red nucleus, pons, and striate body, respectively.

Corticospinal System (Pyramidal System)

Roughly 60% of the fibers arise from the motor cortex (sensory fibers comprising the remainder) and descend through the internal capsule, crossing over at the medulla level and continuing down into the **lateral white columns** of the spinal cord. These pathways supply the motor nerves of the contralateral side of the body. The corticospinal system is often regarded as a very fast system because one synapse exists in the circuit between the motor cortex and the effector, at the juncture with the alpha motor neuron (Fig. 8.13). The sensory fibers project to the dorsal horn of the spinal cord for the purpose of modifying information which is entering from the periphery.

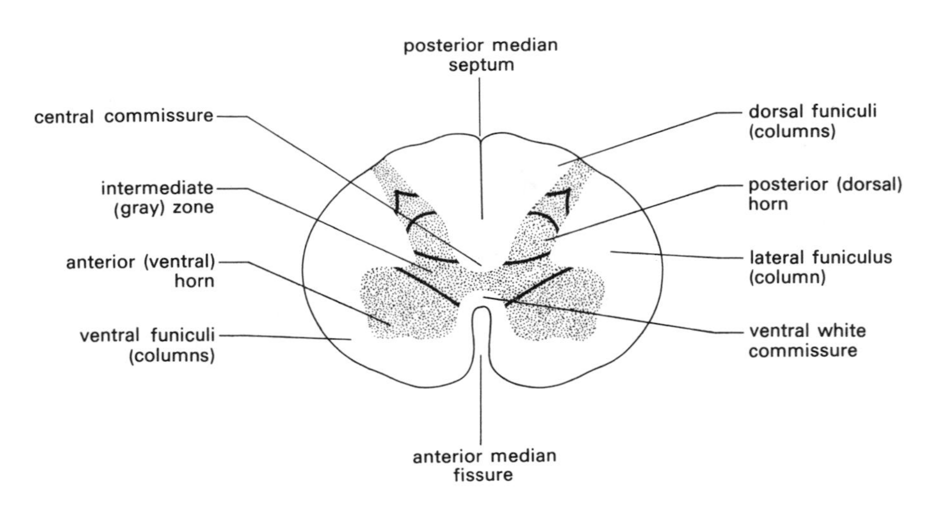

Figure 8.12. Section through the spinal cord at the cervical level. (From Sage, G. H., 1977. *Introduction to Motor Behavior: A Neuropsychological Approach*. Reading, Mass.: Addison-Wesley Publishing Company, p. 31.)

Extrapyramidal System

The second means of communication from the motor cortex (as well as other cortical areas) to the final common pathway to muscle is via a series of "side trips" which involve other parts of the brain so that they can modify the signal which finally arrives at the spinal cord level. None of these pass through the pyramids of the medulla; they can travel ipsilaterally or contralaterally. These pathways are also named according to the parts of the central nervous system with which they communicate. Thus, the names **rubrospinal**, **vestibulospinal**, **reticulospinal**, and **tectospinal** immediately convey to the reader the fact that the fibers originate in the red nucleus, vestibular nucleii, reticular formation, and the roof (tectum) of the

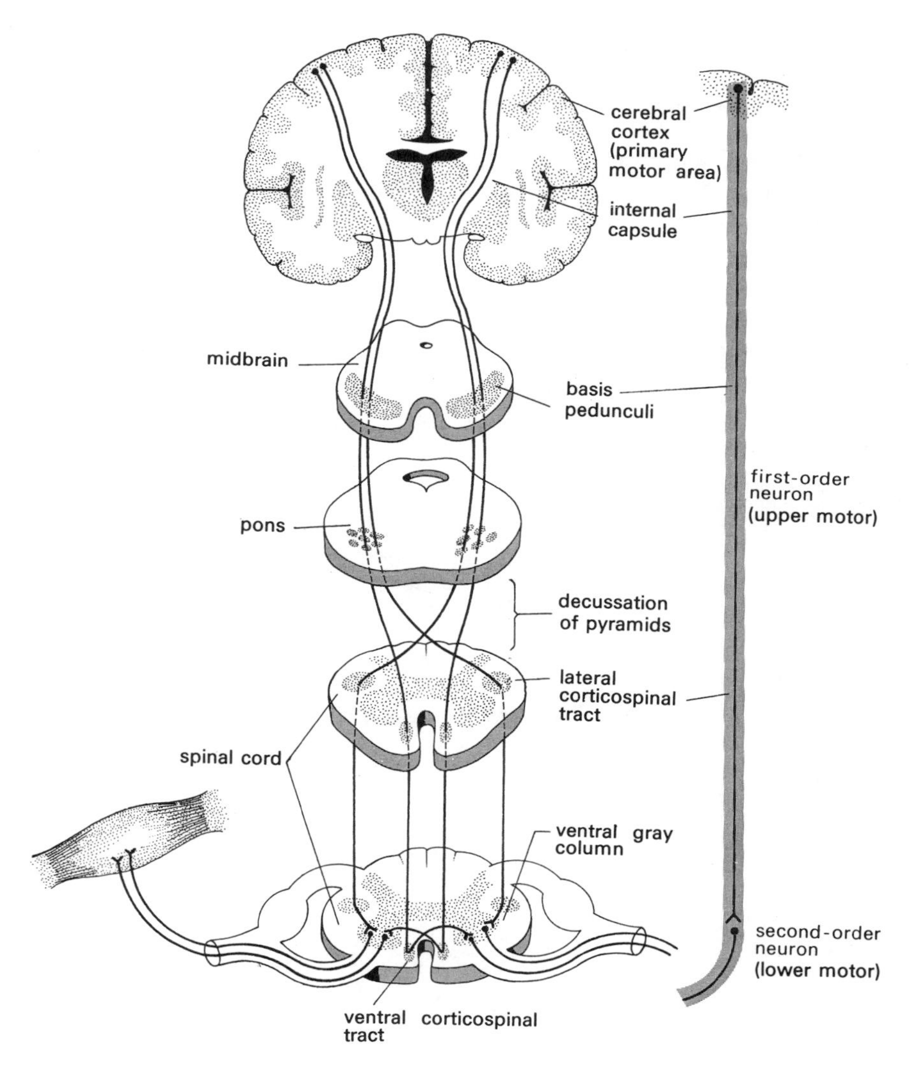

Figure 8.13. The pyramidal motor pathways. (From Sage, G. H., 1977. *Introduction to Motor Behavior: A Neuropsychological Approach.* Reading, Mass.: Addison-Wesley Publishing Company, p. 126.)

midbrain, respectively, and all terminate at the spinal cord level (Fig. 8.14).

The rubrospinal pathway, which originates in the red nucleus of the midbrain, descends in the lateral white column and innervates distal muscles. The vestibulospinal pathway, which originates in the lateral vestibular nucleus, descends in the **ventral white column** of the spinal cord. The reticulospinal pathway, which originates in the reticular formation at the pons level, descends in both the ventral and lateral white columns of the spinal cord. Both the vestibulospinal and reticulospinal pathways terminate on proximally located musculature. The tectospinal pathway, which originates in the tectum of the midbrain, descends in the ventral white column and innervates "neck" muscles.

Vestibulospinal and Reticulospinal Tracts

The functions associated with two of these extrapyramidal systems are further amplified by associating them with the specific tracts they follow as they descend to the spinal cord. Vestibulospinal pathways are divided into **medial** and **lateral tracts**. Vestibular nucleii located in the upper medulla and lower portion of the pons travel in the **medial vestibulospinal tract**, project to the spinal cord, and end in the cervical and thoracic level of the cord. The **labyrinthine response** (explained later in Chapter 10), which corrects head position with respect to gravity, is mediated by the vestibular nerve, which feeds into this tract and modifies neck and upper-extremity muscle activity. **Lateral**

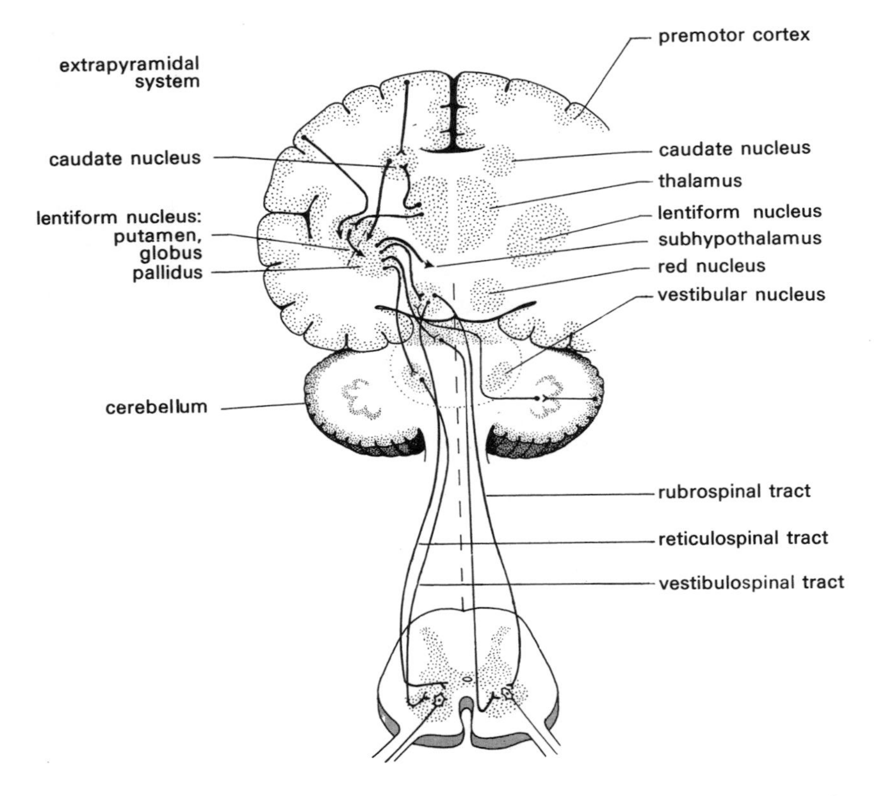

Figure 8.14. Some extrapyramidal motor pathways. (From Sage, G. H., 1977. *Introduction to Motor Behavior: A Neuropsychological Approach*. Reading, Mass.: Addison-Wesley Publishing Company, p. 128.)

vestibular nucleii are somatotopically organized and influence postural mechanisms. They project down to the spinal cord via the **lateral vestibulospinal tract** and are generally known to facilitate extensors and inhibit flexors in order to maintain the upright position.

The **reticulospinal tracts** are not somatotopically organized. They are divided into **medial** and **lateral reticulospinal tracts**: the medial is often called a **reticulo-facilitatory tract** because it facilitates extensor reflexes and inhibits flexor responses; the lateral, the **reticulo-inhibitory tract**, is involved with inhibiting stretch reflexes in extensor muscles and facilitating flexor responses. The medial reticulospinal tract assists the lateral vestibulospinal tract with problems of balance, although the vestibular pathways are more substantially involved. Both of the reticulospinal tracts act on the alpha and **gamma motor neurons** for the purpose of modifying and coordinating reflexes at the spinal cord level.

Cerebellum Feedback Loop

It was mentioned earlier that the cerebellum is regarded as the "great motor coordination center." To explain its many interconnections in a chapter of this scope is impossible. There are approximately three times as many afferent fibers as efferent emerging from the cerebellum, which is why it is regarded as a **somatic afferent organ** or, as Sherrington called it, the "head ganglion." The feedback loop to and from the cortex involves the following structures: cerebral cortex, pons, cerebellum, dentate nucleus, red nucleus, thalamus, and cerebral cortex; or, if you prefer, the loop would be called the **cortico-ponto-cerebello-dentato-rubro-thalamo-cortico** servo-mechanism. The largest cerebellar projection emerges from the dentate nucleus, proceeds to the red nucleus, and continues to the thalamus and ultimately to the cortex. Other efferents travel from **deep cerebellar nucleii**, via the **superior peduncles** to the red nucleus and thalamus or to the vestibular nucleii and reticular formation. The afferent pathways are usually divided into three parts: the inferior, middle, and superior. The **dorsal spinocerebellar tract**, a very complex path, is an example of an afferent projection belonging to the inferior division. An afferent from the pons (originating in the cortex first) is an example of a projection belonging to the middle division. The superior division is primarily an "output" center, consisting mostly of efferent pathways. In summary, the many cerebellar projections emphasize the fact that the cerebellum is important in the coordination of muscle activity but does *not* initiate activity; it acts as a monitor for other brain centers and modulates muscle activity.

Ascending Pathways

The means of communication from the spinal cord to the many brain centers, i.e., sensory information, can be divided into two functional units: those which contribute to conscious awareness and those which may not reach the cerebral cortex but play an important role in coordination. Most of the ascending pathways travel via the thalamus, especially those concerned with conscious perception; the "unconscious" pathways terminate in the cerebellum.

Conscious Awareness

Those pathways concerned with **conscious perception** usually cross to the contralateral side at some point in their "journey" to the sensory cortex. In most cases, three neurons are involved between the spinal cord and the cortex. Those ascending in the dorsal columns (called the **medial lemniscus**) are associated with specific information such as the highly specific sensations which arise from touch, two-point discrimination, and **kinesthesia**. The first-order neuron terminates in the **dorsal column nucleii** in the medulla. Two fasciculi of the medulla, the **cuneatus** and **gracilis**, are the termination points for the upper and lower extremities, respectively. The second-order neuron crosses to the contralateral side, terminating at the **ventrobasal complex**of the thalamus. The third neuron projects to the primary sensory area of the cerebral cortex (Fig. 8.15). A second ascending system subserving conscious perception travels in the **anterolateral system** and also involves three neurons. It carries nonspecific information from small fiber systems such as those for pain, temperature, and crude touch. The first-order neuron synapses in the dorsal horn of the cord; at the same level of the spinal cord, the second neuron crosses to the contralateral side and ascends in the **anterolateral tract** (also called the **ventral spinothalamic tract**) to the ventrobasal complex of the thalamus. The third neuron usually goes on to the cortex (Fig. 8.15).

Proprioceptive Pathways

The morphology and physiology of proprioceptors is discussed in detail in

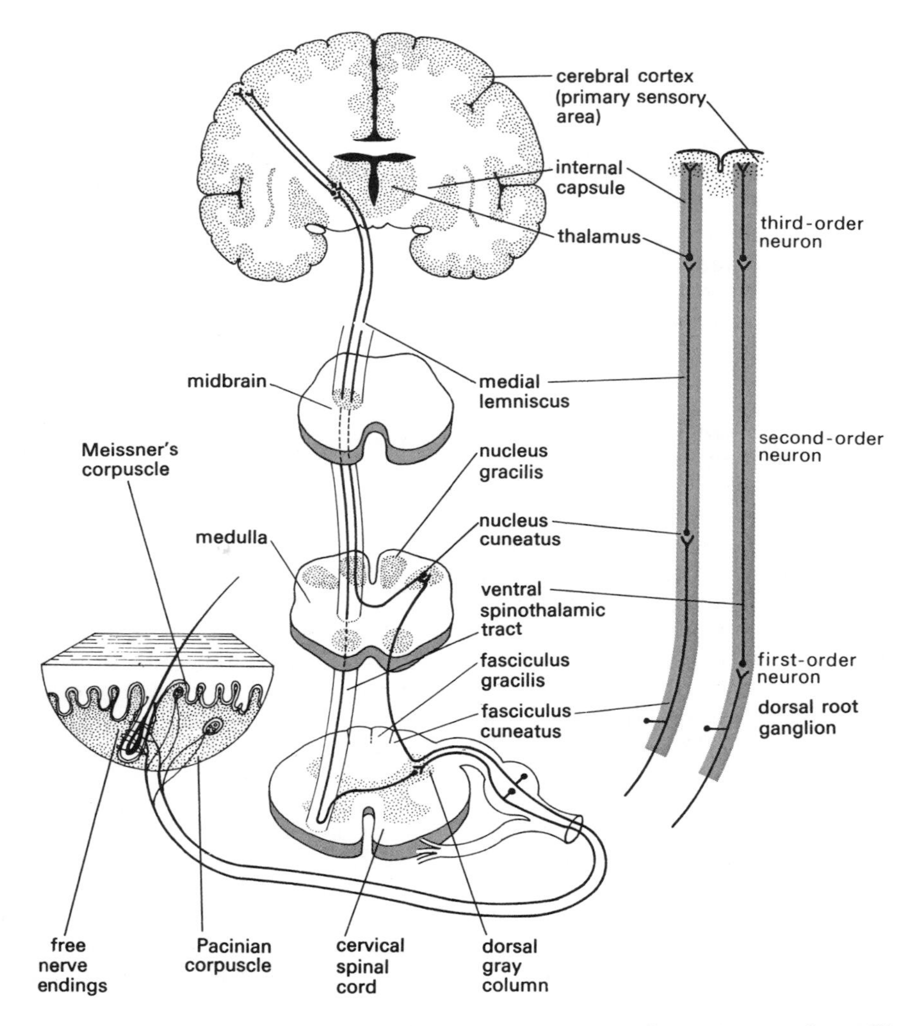

Figure 8.15. Sensory pathways for touch and pressure: the fasciculi cuneatus and gracilis and the ventral spinothalamic tracts. (From Sage, G. H., 1977. *Introduction to Motor Behavior: A Neuropsychological Approach*. Reading, Mass.: Addison-Wesley Publishing Company, p. 85.)

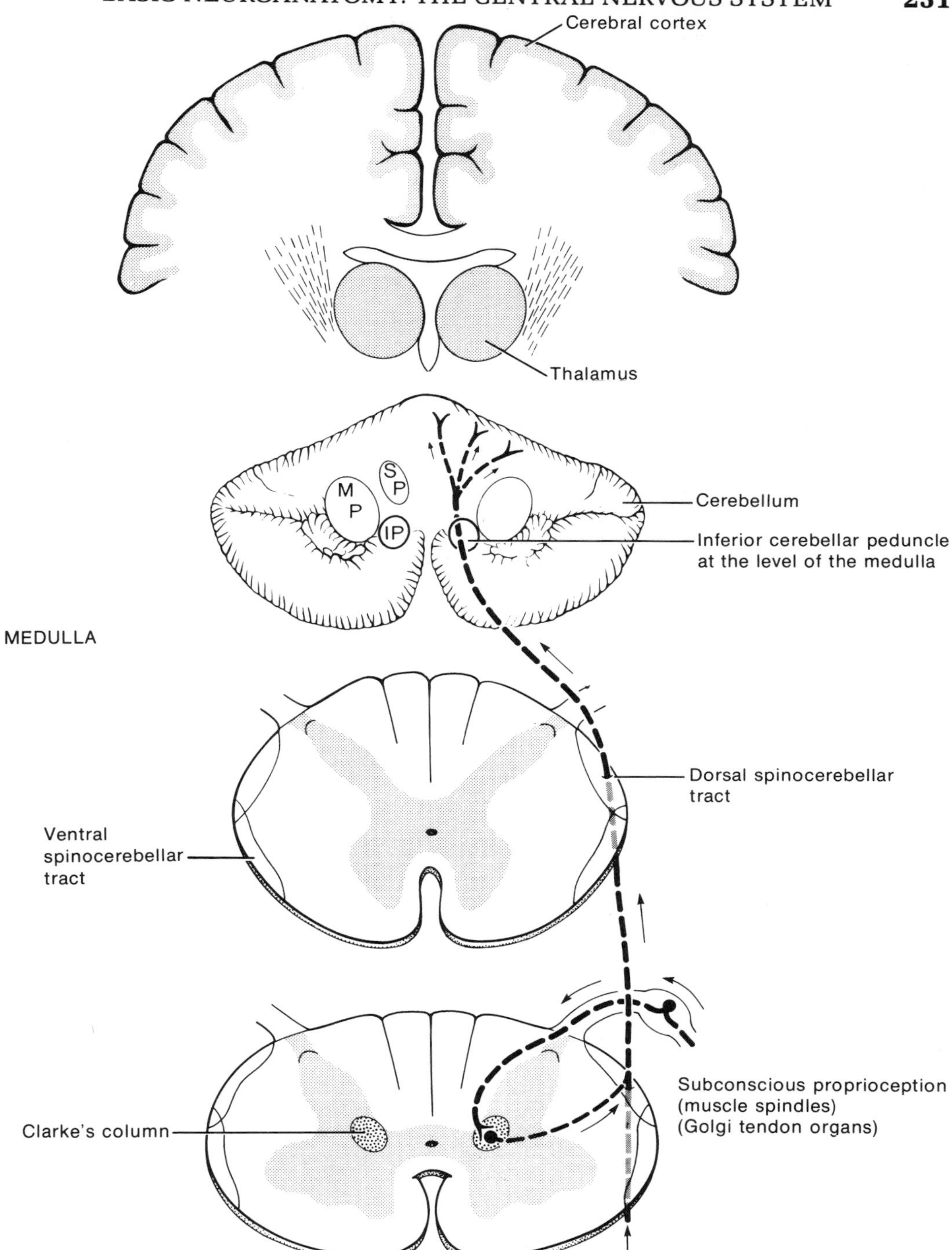

Figure 8.16. The dorsal spinocerebellar system. Note that the primary afferent neurons synapse in Clarke's column, after which the second-order neurons ascend in the dorsal spinocerebellar tracts and enter the cerebellum by way of the inferior cerebellar peduncles. *SP*, *MP*, and *IP*, superior, middle, and inferior cerebellar penduncles, respectively. (From Werner, J., 1980. *Neuroscience: A Clinical Perspective.* Philadelphia: W. B. Saunders Company, p. 47.)

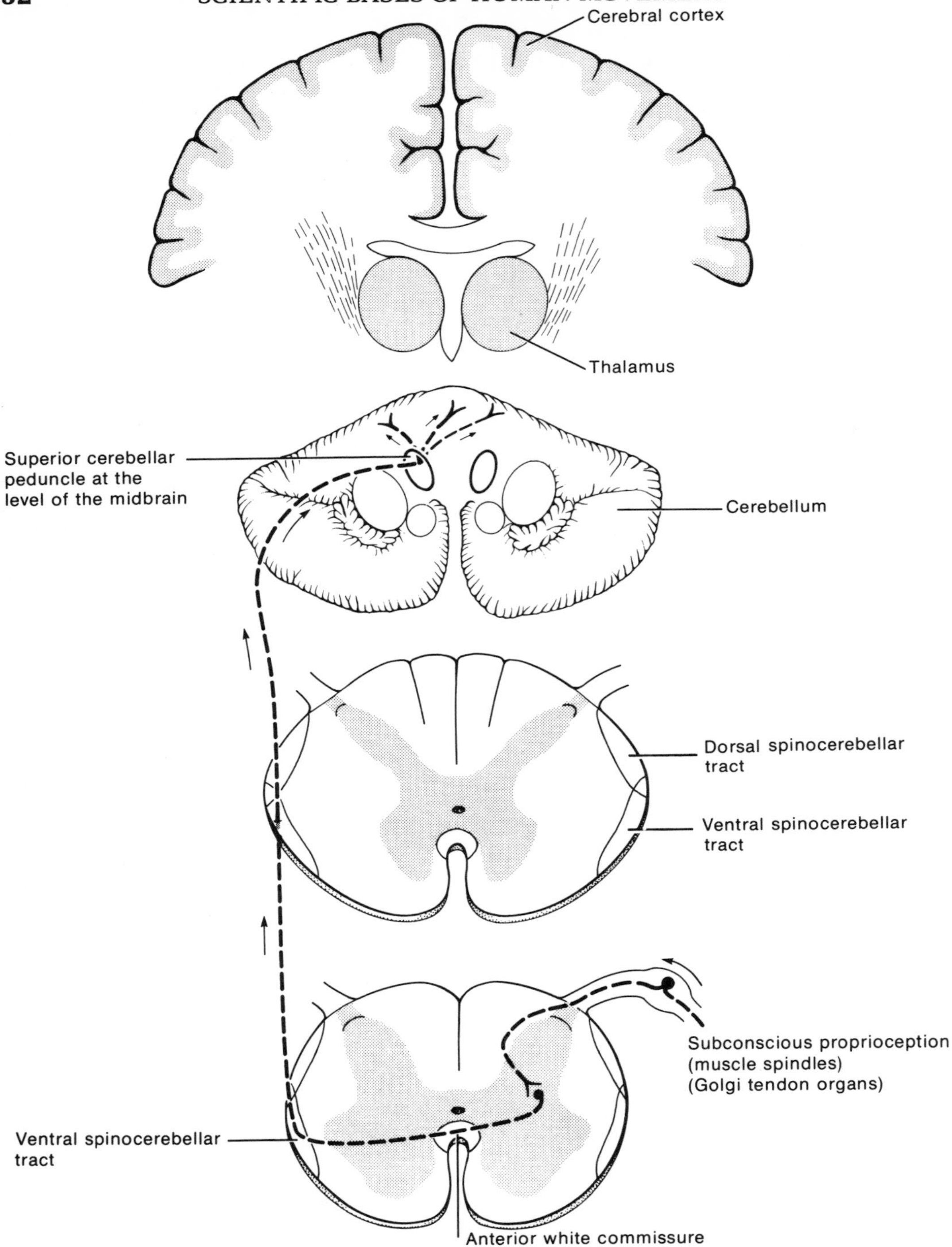

Figure 8.17. The ventral spinocerebellar system. Note that this system enters the cerebellum by way of the superior cerebellar peduncles. After entering the cerebellum, most fibers terminate ipsilateral to the side of their entry into the spinal cord; a few fibers remain on the contralateral side. (From Werner, J., 1980. *Neuroscience: A Clinical Perspective.* Philadelphia: W. B. Saunders Company, p. 48.)

Chapter 10. The intention of this section is to describe the pathways used between the receptors and the cerebellum. The **dorsal spinocerebellar tract** (Fig. 8.16) carries information from the neuromuscular spindle and the **Golgi tendon organ** via two neurons which travel ipsilaterally. The first-order neuron terminates in the cord for lower extremities and in the medulla for the upper extremities. The second-order neuron enters the cerebellum via the **inferior cerebellar peduncles** and terminates there. The **ventral spinocerebellar tract** (Fig. 8.17), which also carries information from the neuromuscular spindle and the Golgi tendon organ, has the first-order neuron ascend ipsilaterally to the medulla. The second-order neuron crosses to the contralateral side and enters the cerebellum via the **superior cerebellar peduncles**. Some fibers may recross to the original ipsilateral side.

Other Pathways

Ascending pathways subserving other functions include the **spinotectal**, which is associated with head and eye movement; the **spino-olivary**, which is presumed to be associated with movement; and the **spinovestibular**, which is concerned with postural reflex mechanisms.

REFLEXES

By definition, a reflex is an automatic act such that a specific stimulus will produce a predictable response. Psychologists and physiologists have interpreted the term "reflex" in various ways, the former tending to confine the reflex to the spinal cord level. Some physiologists have adopted a broader definition of the term and have classified reflexes according to the levels of the central nervous system involved: third-level reflexes are confined to the spinal cord, such as the simple stretch reflex; second-level reflexes include those automatic responses involving levels of the brain stem; first-level reflexes involve the cerebellum and cortex. The motor learning implications of reflex acts are germane to this discussion and have been reserved for that portion of Chapter 10 where reflexes are discussed in greater detail, as they are manifested by proprioceptors.

SUMMARY COMMENT

The salient features, organization and function of the central nervous system have been highlighted. Chapters 9 and 10 attend to some elements of the peripheral nervous system, focusing on integration of the sensory-motor systems and proprioceptive neuromuscular constructs.

CHAPTER 9

Basic Organization of the Neuromuscular System

The mammalian system is organized so that changes both in the external environment and its own internal environment can be monitored and acted upon. Receptors are necessary to detect the changes; effectors, if action is to be taken, need to be stimulated into activity. Thus, a centralized control system is necessary to coordinate all of these monitoring and decision-making activities (Tribe and Eraut (1977)).

The individual components of the neuromuscular system—the receptors, neurons, and muscle fibers—are organized into functional units which make possible the integration of activity of body parts into purposeful response patterns. Three major divisions may be recognized: the sensory or input system, the motor or output system, and the integrative system. The first includes receptors with their afferent neurons, and the second includes effectors (muscles) with their efferent neurons. The integrative system is composed of a complex arrangement of interneurons whose synapses determine the strength and direction of the signal which is transmitted from the input to the output system.

THE SENSORY SYSTEM: MORPHOLOGY AND PHYSIOLOGY OF RECEPTORS

The survival of the organism, its normal functioning, and its behavior depend upon its ability to detect and respond appropriately to changes in the external and internal environments. Receptors are the first elements in the communication pathway leading ultimately to the effectors. They function as transducers to convert stimulus energy into neural excitation.

Receptors are strategically situated over the body surface and within and among the muscles, bones, and viscera, with sensitivities attuned to specific and pertinent types of stimulus energy. Axons of afferent neurons enter the nervous system at all levels: spinal cord, brain stem, cerebellum, and cerebral cortex. Many branch divergently and, as they synapse upon interneurons with a variety of destinations, they establish the basis for extensive distribution of the input signal. Even the simplest motor behavior, such as the automatic response to a noxious stimulus, derives from conduction of action potentials over chains of neurons.

General Morphology of Receptors

Types of Receptor Structures

Receptors are specialized cells or organs highly sensitive to specific stimuli. There are three general types of receptors:

1. Relatively unspecialized **afferent neurons** which branch freely at their receiving poles. Examples are the free nerve endings in the skin (Fig. 9.1(a)).
2. Special neural cells known as **receptor cells**. Their receiving poles are highly specialized so as to present a low threshold for some particular type of environmental energy. Some resemble ordinary neurons except for the specialization of their dendritic terminals: their axons serve as afferent fibers (Fig. 9.1(b)). Others do not resemble typical neural cells, having little or no axonal segment. Their function is to set up the physicochemical conditions which will trigger an afferent neuron in close association with them (Fig. 9.1(c)). Examples of receptor cells are the taste buds, olfactory cells, the rods and cones of the retina, and the hair cells of the cochlea and labyrinth.

   In numerous sensory systems one afferent neuron serves many receptor cells, which may be spatially distributed over several square centimeters (cf. Chapter 8 on morphology of the synapse).
3. Complex **sense organs** consisting of specialized nonneural auxiliary or accessory structures surrounding or otherwise closely associated with receptor cells and afferent neurons.

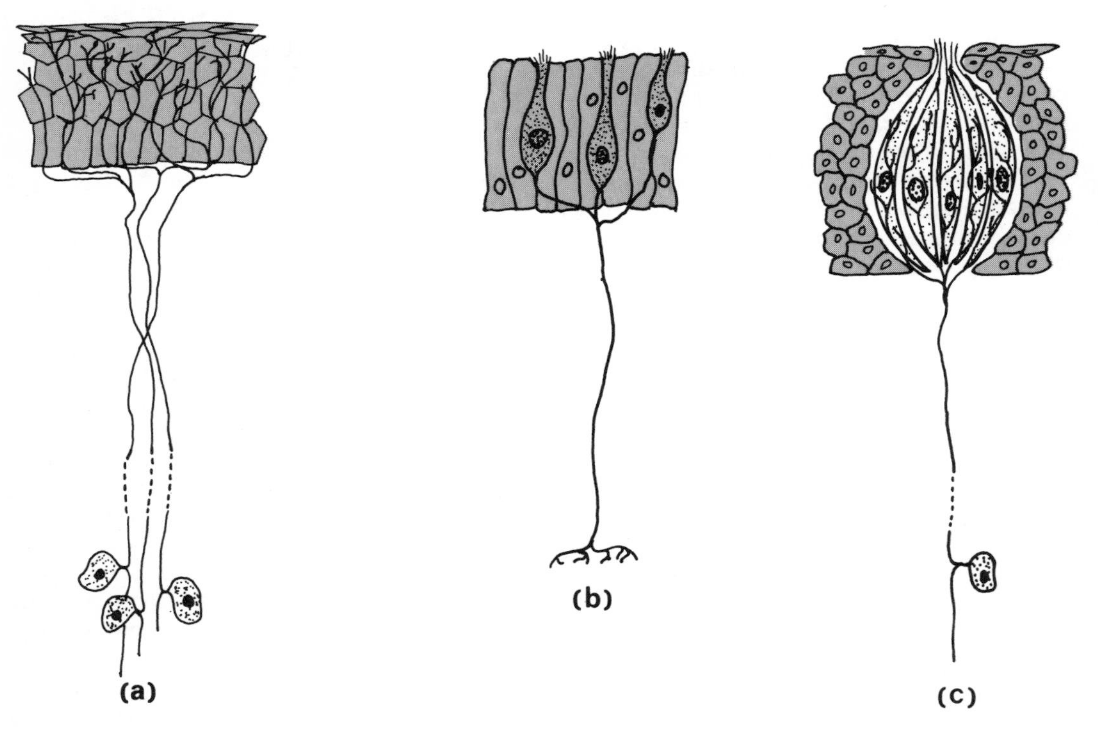

Figure 9.1. Types of receptors. (a) relatively unspecialized afferent neurons acting as receptors: free nerve endings. (b) neural cell with highly specialized receiving pole and axon serving as afferent fiber: olfactory cells. (c) specialized receptor cells having no axonal segment. These cells act to trigger dendrites of the afferent neuron closely associated with them: a taste bud.

Examples are the eye, ear, and neuromuscular spindle. The auxiliary structures assist the receptor cells in dealing selectively with their particular kind of stimulus energy. They analyze the incident stimulation and distribute it properly on the sensory surface or among the groups of receptor cells. The accessory structures of the eye are the cornea, iris, and lens, which focus light on the retina. In the ear the tympanum, ossicles, lymph-filled canals, and the basilar membrane conduct and distribute sound vibrations to the auditory receptor cells. The structures associated with the neuromuscular spindle are discussed in Chapter 10.

Receptors may be classified in a number of ways: (1) according to the specific type of stimulus to which they respond: chemoreceptors, photoreceptors, pressoreceptors (baroreceptors), nociceptors, etc.; (2) according to the type of sensation aroused: visual, auditory, cold, etc.; (3) according to anatomical criteria such as morphology, location, cortical termination, etc.; (4) according to the type of response evoked: posture, movement, visceral function. Because the response of the organism is often determined by a combination of such criteria, a generally useful scheme classifies receptors into four groups as follows.

1. **Exteroceptors:** receptors located peripherally and responding to stimuli from the external environment. The organism's responses are sensation and/or movement.
2. **Interoceptors:** receptors located within the body in association with the viscera and responding almost exclusively to stimulation from the internal environment. The organism's response is a change in visceral function.
3. **Proprioceptors:** receptors located in muscles, tendons, joints, and the labyrinth of the inner ear and responding to stimuli arising from some aspect of posture and/or movement. The organism's response is change in activity of appropriate muscles. Some exteroceptors, especially those of the skin, also evoke proprioceptive responses.
4. **Nociceptors:** receptors distributed throughout the body and responding to injury. The organism's response is usually withdrawal movement (flexion).

Physiology of Receptors

The Sensory Unit

A single sensory axon and its receiving pole with all of the receptor cells associated with its peripheral terminals constitute a **sensory unit**. For some units the receptive field may be very large. Generally the area supplied by one unit overlaps or interdigitates with areas supplied by other units. Consequently, stimulation of one small site may involve several sensory units (see Fig. 9.4).

Excitation of Receptors

In a sensory system, special receptor cells may or may not be present. The afferent neuron is the essential element, and the generation of nerve impulses by depolarization of its initial segment is the significant feature common to all. In the absence of receptor cells, the stimulus may act directly upon the receptive pole of the neuron by mechanical deformation of the nerve terminal itself. In systems where receptor cells are present, the stimulus acts on the receptor, which then fires the afferent neuron. Most receptor cells function like dendrites: they are electrically inexcitable and when stimulated produce a local graded potential, the **receptor potential**, which spreads electrotonically. The receptor potential may directly depolarize the initial segment of the afferent neuron or it may induce the secretion of a chemical mediator at a receptor-neural junction. The chemical mediator then acts upon the dendrites of the afferent neuron in a manner similar to that of synaptic transmitter. It induces in them a graded electrotonic potential, known as the **generator potential**, which is similar in every way to the postsynaptic potential at the synapse. If the generator potential is of sufficient magnitude when it reaches the initial segment, the neuron will fire and the frequency of impulses will be related to the intensity of the stimulus (Fig. 9.2). The fact that the generator potential and the impulse arise separately or are subserved by different mechanisms is proven by a variety of experiments. For example, various drugs abolish nerve impulses but leave the generator potential unaffected. In instances in which the receptor potential itself stimulates the initial segment or when the stimulus acts directly upon the afferent neuron, the receptor potential and the generator potential are one and the same.

In more complex sensory systems, which include both specialized receptor cells and auxiliary structures, each sense organ comprises two intimately related mechanisms: one responsible for its spe-

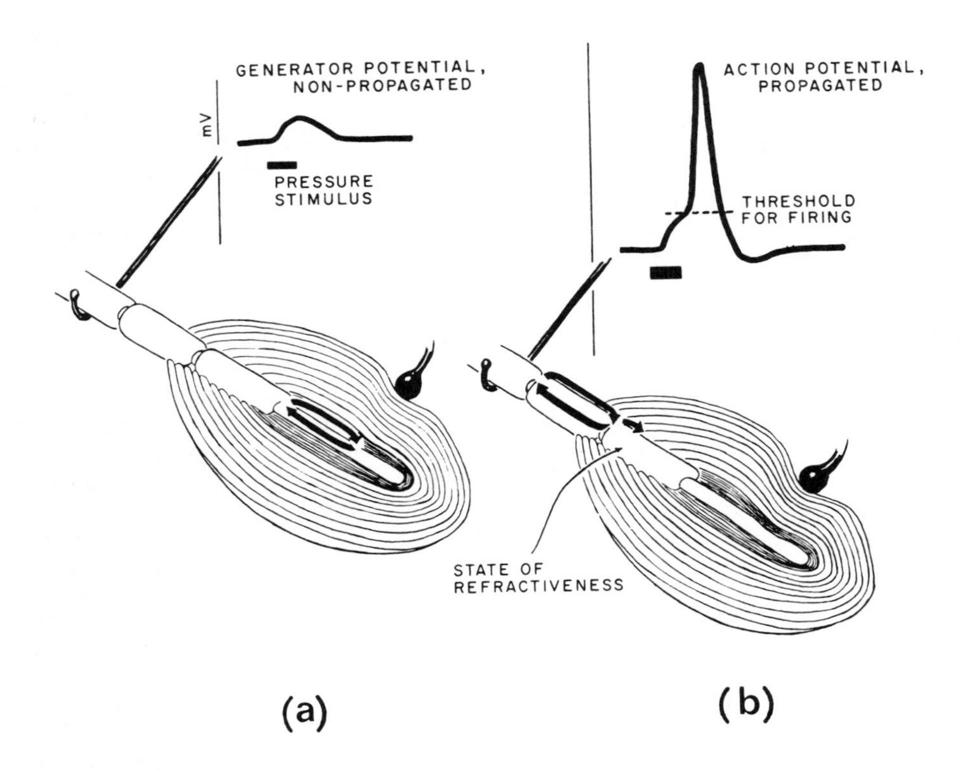

Figure 9.2. Generator potential versus action potential. Distinction between the generator potential and the action potential is illustrated by the Pacinian corpuscle. (a) the generator potential is a local, graded, nonpropagated shift in potential across the membrane of the sensory axon terminal within the corpuscle, induced by light pressure. The currents set up by the generator potential spread electrotonically and reach the first node of Ranvier, shifting the membrane potential toward depolarization. (b) as a result of stronger pressure, the membrane potential at the node reaches the critical level and an all-or-none spike develops. The spike will travel in a saltatory manner along the myelinated axon. (From Eldred, E., 1967, Peripheral receptors: Their excitation and relation to reflex patterns. *Am. J. Phys. Med.* *46*: 71.)

cific sensitivity and the other responsible for transduction of the stimulus into a form of energy capable of discharging the neuronal terminals.

In sensory systems which include receptor cells distinct from the afferent neuron, the following sequence of events has been postulated in the transmission of information to the central nervous system.

1. The external stimulus acts upon the receptor cell, sometimes directly, sometimes through interposed auxiliary structures, and sets up a graded receptor potential in the receptor cell.
2. The receptor potential spreads electrotonically over the cell and at a receptor-neural junction causes secretion of a chemical mediator.[a]
3. The chemical mediator acts upon appropriate regions of the dendritic receiving pole of the afferent neuron, evoking a graded generator potential.
4. The generator potential spreads electrotonically over the dendrites and cell body to the initial segment. If it is sufficient to depolarize the membrane to its critical level, all-or-none nerve impulses are generated and conducted along the axon.
5. At the axon's terminals, chemical transmitter is secreted at synapses of the first central interneuron.
6. If sufficient summation of input occurs, excitation continues on through the central nervous system.

Regulation of Sensory Input

Threshold. It is well established that a single sensory neuron has a threshold requirement which must be satisfied by the stimulus in order to induce the all-or-none neuron response, that each neuron has a maximal frequency limit, and that each sensory system has a limited number of neurons which may be recruited by increased stimulus intensity. Therefore, there are limits inherent in the system below which no response is evoked and above which no increased response is achieved by increased stimulus intensity.

Adaptation. Nerve fibers readily adapt to the flow of directly applied constant current. After delivering one or a few impulses, the fiber becomes silent. Adaptation is also characteristic of receptor cells. The generator potential is at first proportional to the stimulus intensity but declines gradually during steady stimulation. The rate of decline appears to depend upon mechanical coupling between the stimulus and the sensor element of the receptor and to be related more to mechanical properties of adventitious tissues than to the electrical properties of the sensor. Impulse frequency in the afferent neuron declines with the same time course as the generator potential and is probably related to depression or inactivation of the mechanisms responsible for the generator potential.

Adaptation and negative afterpotential have been shown to be related as to time course in some receptor neurons. Repetitive activity in the neuron may result in an accumulation of Na^+ or K^+ or both because of decreased activity of the sodium pump. As a result prolonged negative afterpotential is produced with concomitant reduction of irritability and decrease in the impulse frequency evoked by the continuing stimulus. In other words, adaptation may be related to electrogenic pumping (Goldberg and Lavin (1968)).

Some receptors adapt slowly and continue to maintain impulse frequencies with little decrease for as long as they are subjected to appropriate stimulation. For example, the muscle spindle afferents in posture (antigravity) muscles continue to fire as long as the muscles are under stretch. Such slowly adapting receptors mediate **tonic responses**. Rapidly adapting receptors are characterized by a sharp decline in frequency with sustained stimu-

[a] It should be remembered that not all receptors produce a receptor potential and not all receptor potentials cause the release of a chemical mediator. In fact, some authors believe that chemical transmission is the exception rather than the usual occurrence at receptor-neural junctions.

lation. They are concerned with **phasic responses**. For example, when the skin is in constant contact with clothes, receptors of hair follicles, which are movement detectors, quickly adapt to the slight pressure on the skin.

Some types of sensory systems include receptors which exhibit a range of adaptability as a function of time or some other parameter of the stimulus. For example, some sensory receptors in and around joints are excited by acceleration or deceleration of joint movement but adapt quickly when the movement ceases. Others adapt very little, maintaining a discharge rate related to the stationary angle of the joint.

Central Regulation of Sensory Input. Responses of complex sense organs are determined not only by the fundamental properties of individual receptor cells but also by the influences which they exert on each other and the control exerted over them by the central nervous system.

Regulation by Presynaptic and Recurrent Inhibition. Strongly stimulated afferent neurons inhibit themselves and adjacent neurons by presynaptic inhibition. The most strongly stimulated cell exerts the greatest inhibitory effect. Therefore, since its neighbors exert less inhibition upon it, its impulse frequency is exaggerated in comparison with that of its neighbors. For example, the perception of contrasting temperatures in the skin of a limb partially immersed in hot (or very cold) water is greatest at the water-air boundary. The water feels much hotter here than where it contacts the skin below the surface. This may be explained by the supposition that presynaptic inhibition stemming from sensory neurons in the immersed skin is being exerted upon afferent neurons originating in both the strongly stimulated receptors beneath the water and the less stimulated receptors above the water. Because the more strongly stimulated cells exert the greater inhibitory effect, inhibition of the cells above the water line will be relatively greater than their excitation and their impulse frequencies will be much more reduced than will those of the cells in the skin areas contacted by the water. The difference in inputs from the two adjacent groups of receptors will thus be enhanced by the inhibition. Furthermore the neurons serving completely submerged receptors will be subject to inhibition from all sides, while those of water line receptors will receive inhibitory influences only from the submerged side. Combination of these two effects will lead to enhancement of perception localized at the water line. In other words, contrast has been sharpened.

It has been suggested that presynaptic inhibition functions by decreasing the magnitude of the generator potential. This appears to be true in both light and sound receptors.

Regulation by Efferent Neurons. The sensitivity of some receptors is controlled by the central nervous system through efferent neurons. The best studied example is the muscle spindle, discussed in detail in Chapter 10, whose sensitivity can be "set" by the central nervous system so as to maintain a particular rate of discharge under changing conditions. Visual, auditory, and olfactory receptors are also subject to central regulation by efferent fibers.

Receptor Specificity

Anatomical Basis of Specificity

Receptor cells are distinguished by their greater irritability to some particular type of stimulus, a property described as specificity. Specificity is based on one or more of the following factors, which either restrict responsiveness to certain forms of environmental energy or contribute to a high level of sensitivity to one particular kind.

Structural Design. Most receptor cells and accessory organs are so adapted that their sensitivity is restricted to one type of stimulus. The eye is designed as an organ to handle light effectively, and the rods and cones are specialized to respond to the intensity and wavelength of the incident light energy. The cochlea is designed to receive and conduct sound vibrations, and different areas of its basilar membrane respond to different wave lengths.

Location of Receptors. The site of a receptor often determines the stimuli to which it will respond. Auditory hair cells lie deep within the skull where they are inaccessible to stimuli other than sound waves. Skin receptors, located at the body surface, are exposed to many forms of stimulation. Stretch receptors in the chest wall are strategically placed to respond specifically to inflation of the thorax at inspiration. Pressure receptors in the carotid arteries are stimulated only by pressure in those blood vessels supplying the brain.

Topographical Arrangement of Pathways. Afferent pathways in the nervous system isolate transmission lines for one kind of receptor or one body region from those of others. Discrete cranial nerves and pathways serve the major senses. For example, information regarding position and movement of the legs travels in nerve bundles which are distinct from those serving the arms.

Central Terminations. The neural path for each sensory system has a definite destination in the central nervous system. The impulses aroused by pain reach neurons to flexor muscles. Impulses aroused by muscle stretch terminate on efferent neurons to motor units of the stretched muscle. Impulses from the retina travel to the visual areas in the occipital lobes of the cortex, while impulses from the cochlea reach the auditory areas of the temporal lobes.

In a number of cases specificity cannot be accounted for by any of the above. For example, there are several histological types of cutaneous receptors but specificity is poorly defined. Some temperature-sensitive afferents arise from receptors which appear to be specific for either coldness or warmth, but for others the response seems to be related not to receptor type but to axon diameter. The skin of the ear has few morphologically distinct receptor endings yet it has all of the cutaneous sensations. Itching seems to be a distinct sensation but no special receptors have been found. Sexual sensations do not arise from stimulation of histologically specific endings. Different responses arise from receptors which appear similar. Touch, pressure, and pain sensations can vary so much in quality that little is known regarding the types of endings and axons concerned. In some instances sensations seem to be dependent on *patterns* of stimulation and may involve two or more types of receptors.

It is possible that further study may reveal physiological specificity in receptors which at present appear identical. Vibration-specific receptors, once thought to be nonexistent, have recently been identified.

Sensitivity Range

Each sensory system transmits to the central nervous system information about stimuli which fall between minimal and maximal stimulus amplitudes of its particular modality. Some receptors respond over the full range of their sensitivity, with corresponding frequency variations. Others respond only to a limited portion of the range. For example, some cold receptors display optimal response at a particular temperature, and a whole population of receptors is required to cover the full excursion of a joint.

Some receptors appear to change their specificity at different regions within the stimulus range. There are some temperature receptors which respond to cold at low temperatures, become unresponsive as temperature rises, and then, beyond a certain level, paradoxically become heat receptors.

There are receptors which fire only when the stimulus is applied ("on" receptors); some only fire as the stimulus is removed ("off" receptors). Others fire in both instances ("on-off" receptors). Examples are the Pacinian corpuscles specific for pressure, the touch receptors of the hair follicles, and certain visual receptors.

Specificity of Sensory Perception

Some but not all sensory stimulation results in perception. Although perception depends upon psychological factors related to past experience, there are certain parameters of sensation which have morphological or physicological bases.

Modality of Sensation. Different kinds of stimuli are distinguished as distinct modalities of sensation: light, sound, taste, smell, etc. Because all impulses in all neurons are qualitatively the same, in order to be correctly recognized as a particular sensation the impulses aroused by each particular type of stimulus must reach its special area of the cerebral cortex. Conversely, impulses arriving at a specific area of the cortex will be interpreted as having been aroused by that particular type of stimulus (or referred to a particular receptor location, as in the "phantom limb" phenomenon). For example, impulses arriving at the visual cortex, regardless of source, will be interpreted as light; those arriving at the auditory cortex, as sound. If the optic and auditory nerves could be cut and successfully cross-sutured, we would "hear" the lightning and "see" the thunder.

Quality of Sensation. Within a single modality we can identify different qualities: light consists of colors; smell of odors; taste of flavors; sound of tones; cutaneous sensations of touch, pressure, warmth, cold, pain. We cannot describe qualities of sensation except by comparison with some other quality; they are basically psychological. In some cases, however, the differentiation of qualities has an anatomical or physiological basis, such as the further specialization of receptors. Examples are the three types of cones in the retina, the different types of taste buds, and the maculae and ampullae of the labyrinth. Other bases may involve the overlapping of receptor fields, or certain combinations of afferent neurons from spatially separated receptors. Our knowledge, although extensive in some modalities, is still inadequate to explain fully all of the qualities of sensation.

Intensity of Sensation. Intensity is a parameter of sensation whose physiological basis is well known. Generally the areas supplied by sensory units overlap and interdigitate (Fig. 9.3). A weak stimulus activates only the receptors with the lowest thresholds but, as the intensity is increased, less irritable receptors are also activated, involving more sensory units. Some of the newly recruited receptors belong to the already active units and as a result their impulse frequencies rise. With further increase the influence of the stimulus tends to spread over a larger area, activating sense organs beyond the actual sphere of contact and adding to the afferent inflow. In short, the stronger the stimulus, the greater will be the number of active sensory neurons and the greater the impulse frequency in each. The more intense the bombardment of cortical centers, the stronger the sensation.

THE MOTOR SYSTEM: MOTOR UNITS

The impulses aroused in sensory units traverse the neural pathways of the central nervous system. Upon reaching cortical centers, some impulses produce conscious sensation or evoke discriminative thought processes which may ultimately result in movement. Following shorter routes, others directly induce muscular response without cortical intervention. Movement is the most common response to stimulation.

Like the sensory units, which are composed of receptors and afferent neurons, the units of the motor system include efferent neurons and muscle fibers. The cell bodies of the motor (efferent) neurons lie in the ventral horn of the gray matter in the spinal cord or in the motor nuclei of the cranial nerves. All of the cell bodies of the efferent neurons serving a particular muscle are gathered together topographically and constitute the **motor neuron pool** for that muscle. Their axons leave the cord through ventral roots and travel peripherally with the appropriate spinal

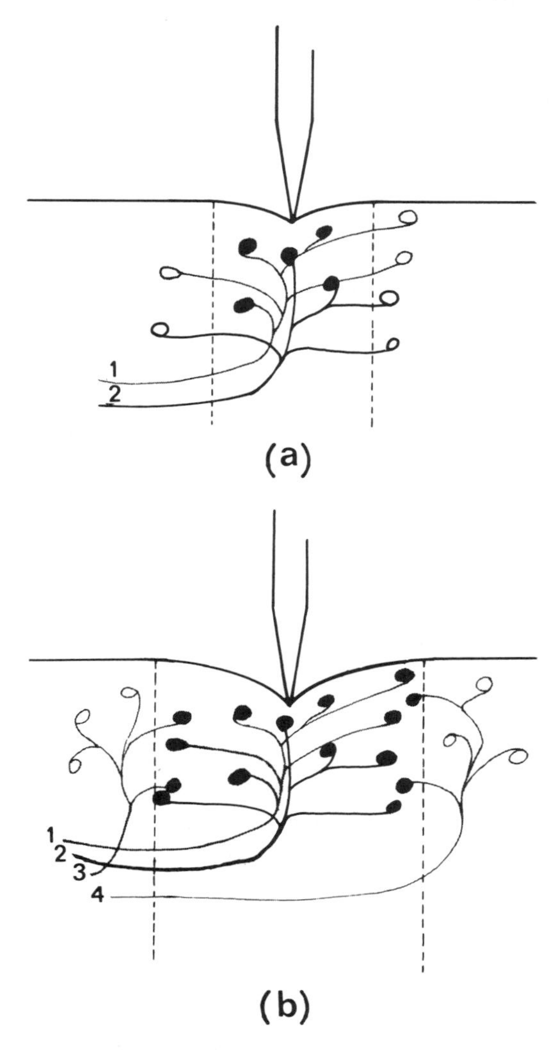

Figure 9.3. Recruitment of sensory units. Pressure exerted upon the skin surface by a probe activates receptors in the affected area. •, active receptors; ○, inactive receptors. (a) light pressure activates five receptors served by the sensory units 1 and 2. Three of unit 1's six receptors and two of unit 2's five receptors are active. The sphere of stimulus influence is bounded by broken vertical lines. (b) stronger pressure widens the area of influence and activates more receptors in units 1 and 2 (now six in 1 and five in 2) and recruits two more units: 3 (two receptors) and 4 (two receptors). The greater number of active receptors in 1 and 2 will result in greater generator potentials and hence higher impulse frequencies in their sensory nerve fibers.

nerve to the specific muscle nerve. Because a muscle nerve contains fewer motor neurons than there are muscle fibers in the muscle, it is obvious that each motor neuron must supply a number of muscle fibers. **A motor unit**, therefore, consists of one motor neuron, its axon branches, and all of the muscle fibers which they innervate (Fig. 9.4).

In each muscle, motor units vary in size within a characteristic range. The mean size may comprise as few as three to six muscle fibers, as in the extrinsic eye muscles, or more than 1700 fibers, as in the gastrocnemius muscle.[b] Mean motor unit size is related to the function of the muscle. Muscles concerned with strength or endurance, such as the trunk and leg muscles, are composed of large motor units, while those involved in fine or manipulative movements, such as the muscles of hand, face, and eyes, have small units.

The various fibers of a single motor unit are not aggregated into one bundle but are scattered among a number of muscle fascicles within a localized area of the muscle belly, interdigitating with fibers of other units. Impulses traversing the motor neuron reach all of the neuromuscular junctions where the axon of the motor neuron forms synaptic end-plates with the muscle fibers. All muscle fibers of the unit whose thresholds are reached will respond together. However, since muscle fiber thresholds are not identical, the neural discharge may fail to activate some fibers. The muscle's tension will be determined by the total number of active muscle fibers, which in turn will vary with impulse frequencies in the motor neurons. In other words, the motor unit is *not* all-or-none in its response. This can be demonstrated by applying two stimuli in such rapid succession that the second falls during the absolute refractory period of any

---

[b] Motor unit size may be calculated by dividing the total number of muscle fibers, estimated from the muscle's volume and architecture, by the number of afferent neurons in the muscle nerve. Examples: platysma, 25; first lumbrical, 95; tibialis anterior, 609; medial gastrocnemius, 1775. An average figure for all muscles of the body is about 180.

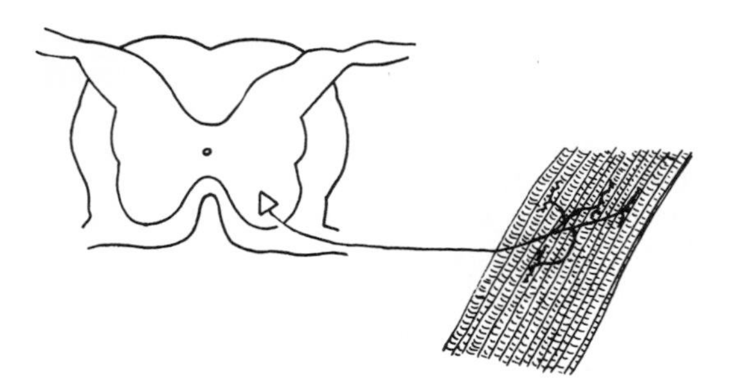

Figure 9.4. The motor unit. A single motor neuron is shown sending its axon via the ventral root to a muscle where its telodendria innervate four muscle fibers distributed within a restricted area of the muscle.

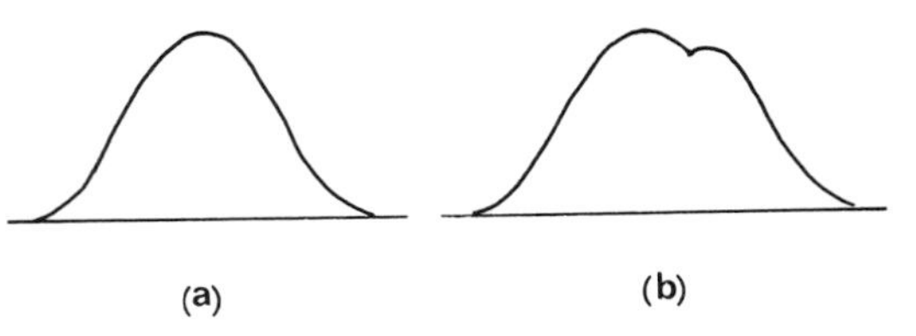

Figure 9.5. The motor unit is not all-or-none. (a) a myogram of a twitch response to a single maximal pulse. (b) two stimuli were delivered in such rapid succession that the second fell during the absolute refractory periods of the muscle fibers responding to the first pulse. The second peak is due to recruitment of fibers for which the first stimulus was inadequate. They are now responding to summation of the two stimuli. (After Ralston, H.J., 1957. Recent advances in neuromuscular physiology. *Am. J. Phys. Med. 26:* 94.)

fibers responding to the first stimulus. The twitch myogram will show a second peak, indicating that some fibers which were unresponsive to the first stimulus were recruited by summation of the two stimuli, which alone were inadequate to trigger additional fiber responses (Fig. 9.5). A further indication is found in the fact that the motor unit potentials[c] recorded from the muscle surface by the electromyograph increase in amplitude with increase of excitation or stretch tension. Recruitment of lazy fibers may be implicated in the greater tension of tetanic contraction as compared with a single twitch. It is also possible that the stretch response may be due to some favorable effect of stretch upon the irritability of muscle fibers, perhaps related to an increased release of Ca^+.

Although the motor unit does not respond in an all-or-none way, it does represent the indivisible functional unit of neuromuscular function. The greater the stimulation, the more motor neurons respond and the greater their firing rates. As more motor units become active and their frequencies rise, muscle tension increases as each motor neuron arouses its own group of muscle fibers to tetanic contraction. The contractile tension exerted by a muscle is determined by the total number of muscle fibers which are contracting, each to its all-or-none capacity. Tension increases progressively in a graded manner as more and more motor units are recruited. The increments of gradation are directly related to the mean size of the motor units composing the muscle. Muscles with relatively small motor units are capable of small variations in tension and hence of delicate movements, while the greater increments of muscles with large motor units are suited to strong, gross movements.

Response Characteristics of Motor Units

Threshold and Frequency

Some motor units begin to respond at low stimulus intensities and show a rapid increase in frequency which then levels off (Fig. 9.6, curve a). Others may have equally low thresholds but a slower change in frequency with increased stimulus intensity (curve b). High threshold

[c] A motor unit potential is the deflection on the electromyogram produced by the summation of the muscle fiber action potentials of one motor unit. Each appears as a spike on the oscilloscope or graphic record, the size of the spike being proportional to the size of the motor unit. A motor unit potential is recognized by its size and regularity of occurrence, both of which are relatively constant for any given set of conditions.

units show similar differences in frequency rise patterns (curves c and d). Frequency peaks also vary among neurons with the same thresholds and the same frequency increase rates (compare curves a and b with c and d). In general, low threshold units show greater discharge rates than high threshold units, but the latter produce motor unit potentials of higher amplitude. The motor units with high thresholds, large spikes, and relatively low frequencies tend to be recruited only when high tensions are demanded of the muscle. Frequencies in motor neurons generally range between 25 and 40/s, rarely exceeding 50/s and reaching the range of 40 to 50/s only when tension is over 75% of maximal voluntary contraction.

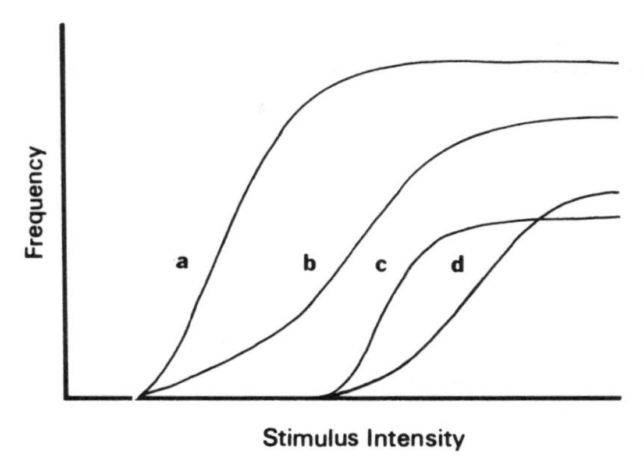

Figure 9.6. Motor unit curves, showing thresholds and frequencies. (a) a motor unit with a low threshold, rapid rise in firing frequency, and a high maximum. (b) a unit with a low threshold but a slower rise and lower maximal frequency. (c) a unit with a high threshold, rapid rise, and low maximal frequency. (d) a unit with a high threshold, slower rise, and moderate maximal frequency. Note (1) that, while a and b have the same threshold, their rise rates and maxima differ; c and d likewise; and (2) that, while a and c (and b and d) are shown with similar rise rates, their thresholds and peak frequencies differ.

Discharge Patterns

Motor unit discharge frequencies, as indicated by electromyograph potentials, are not always rhythmic. Strong stimuli may evoke bursts at peak frequency, alternating with intervals of lower frequency or even silence. Some respond to any stimulation with intermittent bursts in a regularly repeated pattern.[d]

A motor unit may fire only over a specific tension range, becoming silent as tension rises above its peak and resuming as declining tension returns to its response range. Such silent periods during strong stimulation may be due to a Renshaw-like inhibition imposed autogenically or by higher threshold neurons.

Fast and Slow Motor Units

Among motor units within the same muscle, differences of response are observed which are related to the sensitivities and response characteristics of motor neurons. The ventral (efferent) roots of motor nerves show a heterogeneous population of nerve fibers which fall into two subdivisions, alpha and gamma, of the class A category of the Erlanger-Gasser classification. (See Table 7.1 for Erlanger-Gasser classification.) More than half (about 70%) of the motor fibers have diameters and conduction velocities which place them in the A_α division, and these are known as **alpha motor neurons**. Their characteristic sizes and speeds, however, separate into a bimodal distribution with peaks toward the upper and lower ends of the alpha range. Therefore, two groups of alpha motor neurons are recognized: one containing the largest-diameter fibers with high velocities, and another group of smaller fibers with velocities which are distinctly lower but still within the alpha range. The former have been designated as **fast** motor neurons and the latter as **slow** motor neurons.

The descriptive terms **phasic** and **tonic** are also applied to these alpha motor neu-

[d] Similar neural response patterns exist among interneurons in the central nervous system and are considered to constitute one aspect of the information coding which is essential for the discriminative and integrative functions of the nervous system.

Table 9.1
Comparison of Phasic and Tonic Motor Neurons

Characteristic	Phasic Motor Neurons	Tonic Motor Neurons
Fiber size	Large	Smaller
Conduction velocity	Rapid	Slow
Threshold	Low to electricity	High
	High to physiological stimuli (i.e., stretch)	Low
Impulses		
Size of spikes	Large	Small
Spike duration	Short	Longer
Afterpotential (hyperpolarization)	Short	Prolonged
Response to muscle stretch		
Frequency	Low	High
Pattern	One or two discharges	Prolonged discharge
Rest state	Electrical silence	Tonic activity

rons. Table 9.1 compares phasic (fast) and tonic (slow) motor neurons. These two groups of alpha motor neurons constitute the neural components of motor units, and their differences are reflected in differences in the response characteristics of the motor units.

The remainder (about 30%) of motor neurons in the muscle nerve fall into the gamma division of the Erlanger-Gasser group A and are hence known as **gamma motor neurons**. They do not innervate the muscle fibers but supply the muscle spindles, performing an important function in the regulation of muscle activity which is discussed in considerable detail in Chapter 10.

In terms of twitch time course, tor unit responses fall into two classes: **fast** and **slow**. Muscles composed of a mixture of muscle fiber types contain both fast and slow motor units. Each single motor unit is homogeneous, containing either fast or slow muscle fibers but not both. No instances of heterogeneous motor units have been found. The fast units are innervated by phasic (fast) alpha motor neurons and the slow units by tonic (slow) motor neurons.

Role of Motor Neuron in Determining Properties of Its Motor Units. In many mammals, all of the limb muscles are slow at birth, although nerve conduction velocities are already either fast or slow. In the adult animal, however, the muscles also have differentiated into fast and slow. Study of the course of postnatal differentiation has shown that the distinction between the two types of muscles is brought about by a relative increase in the shortening speed of the muscles which are genetically predetermined to be fast muscles, with little change occurring in the genetically slow muscles

If a normally slow muscle is denervated soon after birth then allowed to be reinnervated by a nerve which normally supplies a fast muscle, the slow muscle is changed into a fast muscle. The opposite effect can also be accomplished and what should have been a fast muscle is converted into a slow one.

In the adult the normally slow soleus muscle can be transformed into a histochemically and physiologically fast muscle by anastomosing its nerve to the nerve that originally served a fast muscle, such as the peroneus. Conversely, the normally pale and fast gracilis can be made red and slow by cross-innervation with the normally slow crureus muscle. In both instances reinnervation by the muscle's own nerve produces no significant change, indicating that surgery is not responsible.

A large number of such cross-union and

hetero-innervation experiments have demonstrated that metabolic and contractile changes are induced by a "foreign" nerve, leaving little doubt that motor neuron influence is an important, and perhaps the prime, factor in determining the nature and properties of the muscle fibers. This conclusion is in agreement with the concept of homogeneity of the muscle fiber composition of motor units.

It has been suggested that a motor neuron influences differentiation of its muscle fibers trophically. Chemical substances that are moved by axoplasmic flow down the motor neuron and across the neuromuscular junction are postulated as mediators. Axoplasmic flow has been well established, and its presence is associated with the time of differentiation both in normal postnatal development and during reinnervation.

Several theories have been proposed to explain conversion of one type of muscle to the other. Some workers think that the change from slow to fast may be produced by an alteration of the time course of the fiber's active state. A second hypothesis implicates the mean frequency discharge of the motor neurons, for there seems to be an inverse relationship between impulse frequency and contraction speed of the muscle fibers. Relatively high frequencies are associated with slow motor units and vice versa. Experimental results have provided support for both hypotheses. Low-frequency stimulation by artificial means or absence of stimulation, as assured by appropriate surgical techniques, has been found to transform the characteristics of slow muscles in the direction of fast muscle, while continuous or high-frequency stimulations have converted fast muscles into slow ones. These relationships are consistent with the theory that the tonic activity of slow muscles is due to their constant postural stimulation by gravity. These experiments also show that skeletal muscle is capable of responding adaptatively to the type of activity demanded of it, a factor of considerable medical importance and one which may be significant in training and in the development of motor skill.

"Catch" Mechanism in Slow Motor Units. A study by Burke et al., (1970) of motor units in the cat triceps surae (the medial and lateral gastrocnemius muscles and the soleus) has revealed a phenomenon in the slow motor units resembling the "catch" mechanism of crustaceans and mollusks by which certain muscles are able to exert prolonged tensions without a large energy expenditure. It was found that in slow motor units interruption of a background stimulation frequency by introducing an additional stimulus produced a marked prolongation of normal tetanic enhancement (Fig. 9.7). Apparently the shortening of one interstimulus interval in the rhythmic frequency extended the range of output tension that could be produced by a given unit without a large change in the mean firing frequency of its motor neuron. The background frequencies which were optimal for the phenomenon were found to be in the same range observed to be characteristic of slow type motor units when activated by maintained stretch, but it was not clear whether such a pattern of firing is actually utilized by motor units participating in normal behavior.

Three Types of Motor Units. Since the identification of three types of muscle fibers (A, B, and C), evidence has been accumulating for the existence of three types of motor units. These have been designated as motor units A, B, and C or, by some authors, as types I(fast), intermediate, and II(slow). The behavior of each unit suggests that it is homogeneous for one type of muscle fiber. The type A motor units represent the classic fast or phasic units discussed above. Types B and C represent two distinguishable types of slow or tonic motor units, with type B intermediate between A and C in some of its properties and similar to C in others.

The flexor digitorum muscle of the rat is homogeneous for type A units. The rat soleus muscle, although containing only slow units, is heterogeneous for both slow types, B and C. Some muscles contain all

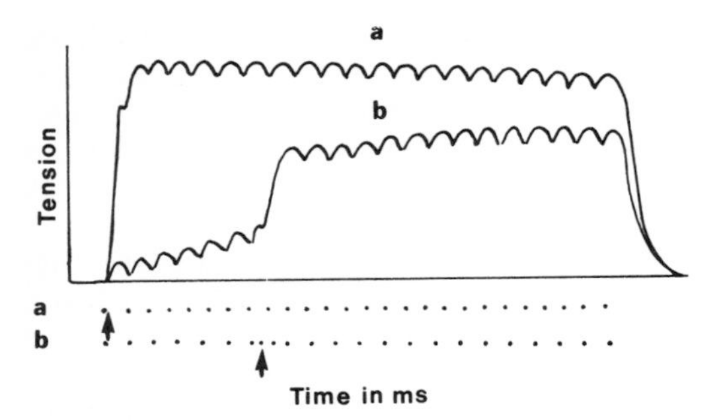

Figure 9.7. Effect of alteration in interstimulus interval. Curve (a), initial enhancement by a double stimulus (indicated by arrow on frequency trace a below curves). Curve (b), no initial enhancement; later enhanced by one short interstimulus interval (indicated by the arrow below the stimulus trace for (b)). Curves represent tension responses of a slow motor unit in the cat gastrocnemius to interruption of a basic rhythmic stimulation. Tension traces (solid lines) are labeled (a) and (b) and their pulse sequences (dotted lines) are similarly designated. Intervals of the train were about 100 msec. The arrow at the first pulse in (a) indicates a double stimulus with a very short (about 10 msec) interstimulus interval used to produce strong initial mechanical enhancement (twitch fusion). The arrow in (b) denotes the introduction of an added pulse, resulting in one interstimulus interval which is shorter than that of the basic train (but longer than the 10 msec interval of the initial summation). In (b) the initial double stimulation was omitted; hence there was no large initial enhancement. (Adapted from Burke, R.E., Rudomin, P., and Kajac, F.E., 1970. Catch property in single mammalian motor units. *Science 168:* 122.)

three types of motor units in various proportions. The cat and rat gastrocnemius muscles include types A, B, and C, with the fast type A units predominating (Close (1967); Wuerker et al. (1965)). These structural differences between the gastrocnemius and soleus muscles appear to be correlated with their functional roles. Although both plantar flex the ankle, sharing a common tendon, the gastrocnemius is designed for uneconomical rapid, powerful contraction, and the soleus for endurance. The power and speed of ankle plantar flexion, which are required for running and jumping, can be supplied by the gastrocnemius but this muscle cannot appreciably shorten owing to the shortness of its fibers and their oblique arrangement in the muscle. Also it fatigues quickly. The soleus, with longer parallel fibers and greater endurance, "covers" for the inadequacies of the gastrocnemius. It can shorten enough to accomplish full ankle plantar flexion and has sufficient endurance to oppose gravity for prolonged periods. Henneman and Olson (1965) have suggested that separate heads were developed in response to need, each designed for its own special purpose and each complementing the other; nature had discovered that two heads were better than one!

The three types of motor units, when all are present, seem to participate in contraction in a definite sequence, each being recruited at its own tension level and contributing to a particular tension range. The cat gastrocnemius muscle provides a good example. It contains all three types of motor units, with type A predominating. Weak stretch was found to arouse the low threshold tonic motor neurons and their type C motor units became active. With increased stimulation, other units were activated as their thresholds were reached. The order of recruitment was inversely related to velocity, type C units being followed by types B and A at respectively greater intensities. As more and faster units were added, the previously active units dropped out. Because new motor units are recruited before previous ones have reached their maxima, smoothness of contraction is assured.

Henneman and Olson (1965) have pointed out the probable significance of the recruitment order. First, the C-B-A order makes possible relatively small tension increases at all tension levels. At low tensions the small-fibered type C motor units provide small progressive increments. At higher tensions, although the motor units are composed of larger muscle fibers, their tension increments still contribute only a small percentage of increase in the total tension already present at the time of their recruitment. As a result, fine control is possible at all tension levels. Second, because the small-fibered type C units act first, they are subjected to the most frequent use. Their slower contractions and oxidative metabolism are thus a distinct advantage, consistent with the early and frequent demand placed upon them by their recruitment primacy.

THE INTEGRATIVE SYSTEM: NEURAL CIRCUITS

Essentially the integrative system is composed of neural circuits interconnected in all possible combinations. Circuits range from simple two-neuron pathways to devious networks of infinite complexity. The brain receives an extensive input of information compounded from all of the stimuli playing upon the body's receptors. By some means not yet understood, incoming signals are analyzed, appraised, and interpreted against a background of past experience. In some remarkable manner a decision is made as to the appropriate response, and the outgoing signal is correspondingly fabricated so as to regulate all the muscles concerned in that response. Limitless variation in impulse transmission is possible by means of the simple expedient of varying synaptic facilitation and inhibition within the circuitry.

Integrative Function of the Neuron

The neuron itself is undoubtedly capable of some degree of integrative action. Experimental evidence is accruing which suggests that the neuron is much more than a passive conductor of information. Rather it probably has far greater potentialities for the analysis and abstraction of the afferent patterns impinging upon it than was thought. Apparent integrative capacity has been noted in individual neurons of sense organs, interneurons, and motor neurons. Multiple trigger zones, rather than a single trigger area restricted to the initial axonal segment, have been identified in some types of neurons. As many as four or five different kinds of circumscribed loci in various parts of the neuron may function to exercise some evaluative action, so that the cell does not pass on the information it receives merely in a one-to-one ratio.

Dendrites, which for some time were thought capable only of nonpropagated electrotonic response to stimulation, have been shown to have definite trigger zones in some instances. Considering the extensiveness and complexity of dendritic trees, especially in the brain and cerebellum, multiple trigger zones and propagated impulses in this region of the nerve cell would greatly increase the potentiality for integrative processing of afferent input than is possible in the case of a single-trigger cell. Furthermore, the pattern of synaptic firing then becomes a parameter for integration. Still other parameters are input frequency, number of active endings, pattern of firing bursts, and coincidence or succession of firing of different pathways, all of which must be considered in a complete analysis of the neuron's integrative role.

Coordinative Centers

During embryological development, nerve cell bodies migrate to form nerve centers strategically located within the central nervous system so that specific types of sensory input may readily be translated into motor outflow which will produce appropriate responses. Centers are interconnected to permit the coordination of responses and to allow for variation as may be required by changing environmental conditions. In the reflex centers of the spinal cord, extensive and important integrative activity occurs, and the fundamental interregulation of individual muscles is accomplished automatically there. Centers of the brainstem, cerebellum, and subcortical nuclei integrate activity in the various parts of the body and adjust muscle tone for balance and equilibrium. Conscious centers of the cerebral cortex are capable of imposing voluntary modification upon any part of and at any point in a movement pattern by suitably altering the descending signal. The reader is referred to Chapter 8 for information on the organization and function of the central nervous system.

So far as is presently known, the complex circuitry of the integrative system provides the mechanism for all of the functions of the nervous system, both conscious and unconscious, even including the more abstract faculties such as thought, judgment, sensation, and emotion.

Types of Neural Circuits: Number of Neurons

Two-Neuron Circuit

While most circuits include at least three neurons (an afferent neuron, an interneuron, and an efferent neuron), a simple spinal circuit consisting only of an afferent and an efferent neuron plays a significant role in regulating muscular contraction. This is the pathway of the passive stretch reflex and includes only the afferent neuron from the muscle spindle stretch receptor and the efferent (motor) neuron to the contractile elements of the same muscle fibers, the two neurons synapsing in the spinal cord without an intervening interneuron.

Three or More Neuron Circuits

Except for the stretch reflex circuit, pathways in the spinal cord and brain contain all three types of neurons. Circuits of only three neurons, one of each type, are the exception rather than the rule, however. Generally the central portions of the circuit, the interneurons, are multiple, forming chains and networks. As a result, basically simple circuits are converted into complex pathways.

There are essentially three basic circuit types: (1) divergent circuits, by which a single receptor may influence many effectors; (2) convergent circuits, by which a number of different receptors may influence the same effector; and (3) repeating circuits, by which a single input is multiplied a number of times.

Divergent Circuits

Figure 9.8(a) shows diagrammatically a simple **divergent circuit**. The axon of the

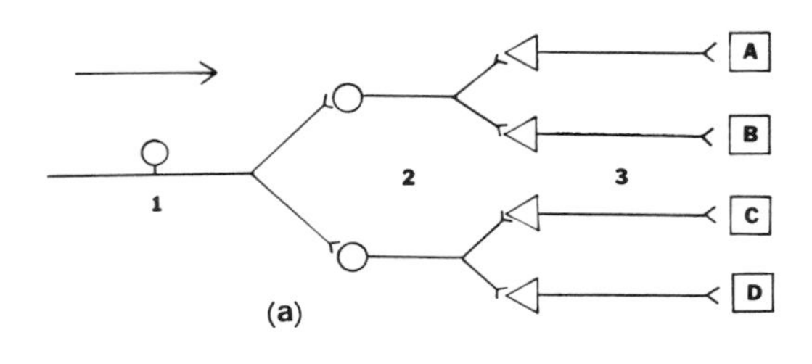

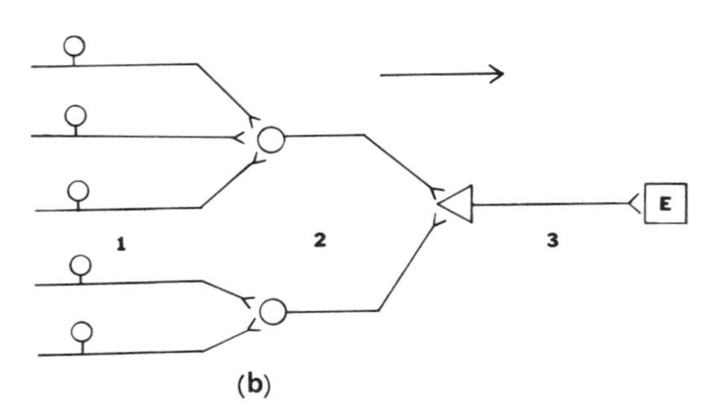

Figure 9.8. Basic circuits: divergent and convergent (diagrammatic). (a) a simple divergent circuit. The axon of an afferent neuron (1) branches to synapse with two interneurons (2) which send collaterals to synapse, each with a different efferent neuron (3). As a result, impulses originating in the single afferent neuron reach four different effectors: A, B, C, and D. (b) a simple convergent circuit. Axons of several afferent neurons (1) synapse with two interneurons (2) whose axons in turn synapse with a single efferent neuron (3) serving effector E. As a result, impulses arising in a number of receptors converge upon a single effector, amplifying the signal. These schematic diagrams are intended merely to convey the basic principles of circuitry within the nervous system. Arrows indicate the direction of transmission through each circuit.

afferent neuron (1) branches to synapse with two interneurons (2) which also send collaterals, each synapsing with a different efferent neuron (3). As a result, impulses originating in the single afferent reach four different effectors: A, B, C, and D. The circuit pictured is overly simplified and geometric in form. The central components, namely the interneurons, characteristically occur not singly but in chains of different lengths and with a variety of branching patterns so that a stimulus impinging upon a single receptor organ, e.g., the eye, can evoke responses involving many parts of the body. Further illustrations include a noxious skin stimulus which evokes a mass withdrawal response, or a sudden loud sound which may produce a total body response.

Convergent Circuits

Figure 9.8(b) presents a simple **convergent** pattern. Axons of several afferent neurons (1) synapse with two interneurons (2) whose axons in turn synapse with a single efferent neuron (3) serving the effector (E). As a result, impulses arising in a number of receptor organs converge upon a single effector, thus amplifying the signal to the muscle. Again the diagram is oversimplified but serves to present the basic principle.

Repeating Circuits

Two basic types of repeating circuits are known, reverberating and parallel circuits, in both of which a single input results in repetitive firing of efferent neurons.

Figure 9.9(a) illustrates a **reverberating circuit**. The afferent neuron (1) synapses with a chain of interneurons (2, 3, and 4), with the fourth neuron transmitting the signal to the efferent neuron (5). Impulses traveling this circuit, however, reverberate through an axon collateral from one of the chain (3) to restimulate a neuron (2) situated earlier in the chain. Repetitive firing will continue to activate the effector (F) until terminated by fatigue or by inhibition imposed through another circuit.

The **parallel circuit** is shown in Figure 9.9(b). A linear chain of interneurons (2, 3, and 4) connects the afferent (1) with the efferent neuron (5). Collaterals from these interneurons (2 and 3) also project to synapse on the efferent neuron, however, so that in the case illustrated a single input will stimulate the efferent not once but successively (three times), with synaptic delay determining the order of the arrival of the impulses. Complexity is readily introduced into the repeating circuits by collaterals supplying additional efferent pathways or interconnecting with other circuits of the same or different type. Interposition of inhibitory neurons assures suitable direction of excitatory flow.

Combination of Circuits

The possibilities of combination are obviously limitless. Consider the relatively simple combinations suggested by the diagrams in Figure 9.10 and determine the response which will be induced in each effector. With further modification by appropriate facilitation and inhibition, these combinations of simple circuits provide a basis for the observed complexity of response patterns.

SUMMARY

The neurons of the nervous system are arranged in a complex but orderly manner. Afferent and efferent neurons are combined to form the peripheral nerves which connect receptors and effectors with the central nervous system. The interneural chains, which complete the circuits from receptor to effector, compose the spinal cord and brain. Cell bodies serving specific functions are aggregated into clusters (nuclei or centers) from which axons travel in groups (tracts) to distant destinations. Collateral branches leave the longer tracts

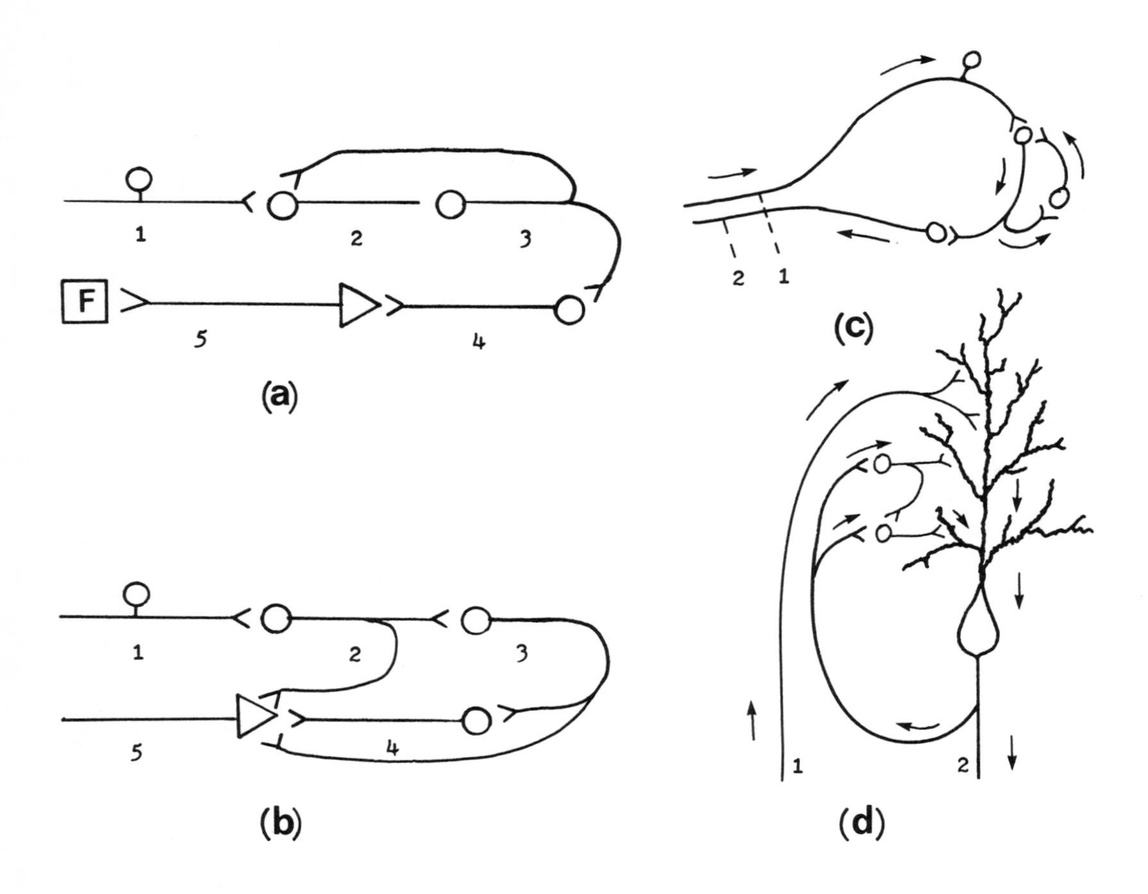

Figure 9.9. Basic repeating circuits (diagrammatic). (a) a simple reverberating or oscillating circuit. An afferent neuron (1) synapses with a chain of interneurons (2, 3, and 4) with the last (4) transmitting the signal to an efferent neuron (5). Impulses traveling through the circuit, however, reverberate through an axon collateral of one of the interneurons (3) to restimulate a neuron (2) situated earlier in the chain. Repetitive firing will continue to activate the effector F until terminated by fatigue or by inhibition imposed through another interconnected chain. (b) a simple parallel circuit. A linear chain of interneurons (2, 3, and 4) connects the afferent (1) with the efferent neuron (5). Collaterals from the interneurons also project forward to synapse on the efferent, however, so that in the case illustrated a single input will stimulate the efferent neuron not once but three times. Because of synaptic delay the arrival of impulses from collaterals 2 and 3 will precede the arrival of 4. (c) a reverberating reflex circuit. A simple reverberating circuit which may be responsible for continuation of a response after the stimulus ceases (the dog's scratch reflex, for example). Such circuits require damping by cerebellar function to obviate oscillation. 1, afferent neuron; 2, efferent neuron. (d) a reverberating cerebral circuit. An ascending fiber from the thalamus (1) excites a large pyramidal cell whose descending corticospinal fiber (2) gives off a collateral branch. The collateral stimulates two or more interneurons which have excitatory terminals synapsing upon this same pyramidal cell. Interconnection between the interneurons further increases the repetitive firing of the pyramidal cell. As a result, a single ascending volley may be multiplied many times in a second. The number and complexity of such circuits may be important in the rate of learning of which an individual is capable. ((c) and (d) from Gardner, E.B., 1967. The neurophysiological basis of motor learning—A review. *J. Am. Phys. Ther. Assoc. 47:* 1115.)

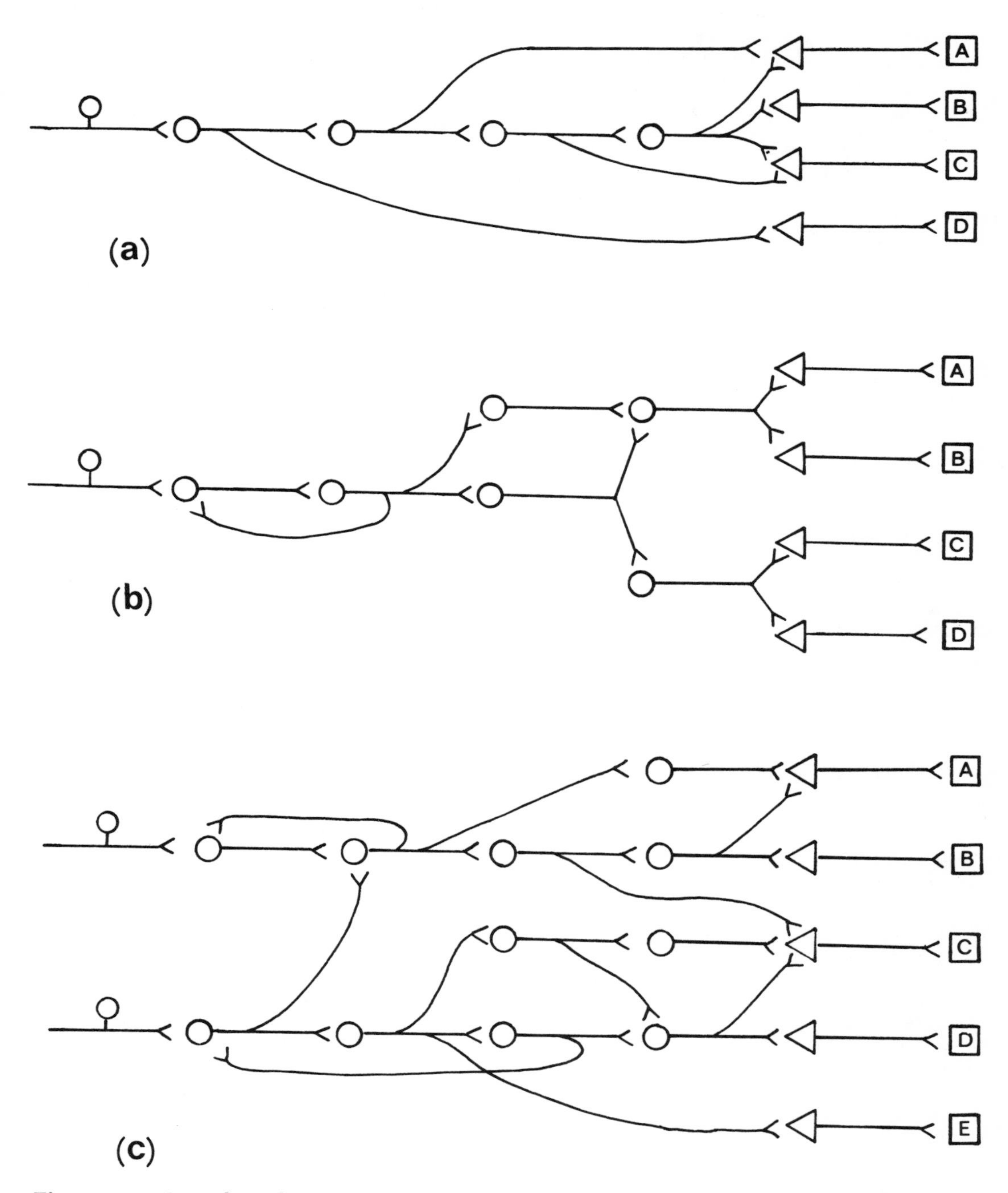

Figure 9.10. Some hypothetical combinations of basic circuits. These relatively simple diagrammatic combinations suggest the manner in which complexity may be introduced in central nervous system circuitry. The student should determine the response which will be induced in each effector and the order of sequential responses. The possibilities of combinations are obviously limitless. Further modification by interposition of appropriate facilitatory and inhibitory circuits provides an adequate basis for the observed complexity of reflex response patterns.

to provide interconnections with centers at various levels of brain and cord. Interconnections between circuits, properly modified by facilitatory and inhibitory influences, provide the framework for coordinated movement patterns.

CHAPTER 10

Proprioceptors and Allied Reflexes

INTRODUCTION

All animals, including humans, are born with genetically built-in neural circuits which are preprogrammed by modification of synaptic transmission to produce stereotyped response patterns useful to the species. These are not learned. They are present at birth or appear as the developing nervous system progresses to completion. They represent the heritage of the species. Genetic material prescribes and ontogeny builds the components for the types of movement characteristic of human behavior. These components include not only the bones, joints, muscle attachments, and nerve supply, both afferent and efferent, but also appropriate interconnections and patterns of facilitation and inhibition. A child is born with a repertory of a few hundred movements which compose the raw material of motor learning. The modification and recombination of these in all possible ways result in the acquisition of additional motor patterns, some of which are very different from inherited patterns. These are learned motor skills.

Stereotyped responses in the form of simple human reflexes are well known, such as the stretch reflex, withdrawal (flexion) reflex, extensor thrust and the positive supporting (extension) reflexes, crossed extensor reflex, righting reflexes, placing reactions and others. Some of these appear to be very simple, others must be amazingly complex. Some are fully formed at birth, while others develop during the maturation of the neuromuscular system. It should be noted that, a variety of meanings have been ascribed to the term **reflex**. Neurophysiologists tend to include automatic patterns mediated at all levels of the central nervous system, psychologists usually indicate that the "true" reflex is spinal cord mediated and all other motor patterns, mediated at higher levels, cannot truly be called reflexes. The former description most adequately conveys the authors' use of the word.

Each of these reflex patterns consists of a coordinated combination of several-to-many joint movements, and each joint movement further consists of a coordinated combination of muscle actions: contraction of prime movers, relaxation of antagonists, and supportive contractions of synergists and stabilizers. All of these must be precisely regulated in regard to their intensity, speed, duration, and sequential changes in activity from the beginning to the end of the movement. This requires a considerable amount of integrative function which is largely automatic and unconscious.

Muscle-to-muscle integration is accomplished by basic reflex reactions which are initiated by receptors strategically located to feed back information to the central nervous system. Information must be received continuously regarding body position, muscle length and tension, speed, range and angle of movement, acceleration of the body or its parts, and balance and equilibrium. This information must then be integrated by cord and lower brain centers and converted into a suitable modification of the impulse outflow to produce immediate adjustment of each muscle concerned. As the state of a muscle changes, the information input will also change, evoking remodifications in never-ending succession. Much of the information also becomes available to centers in the conscious areas of the brain, where it may be sorted, analyzed, interpreted, and converted into an outflow of signals to modify voluntary body movements appropriately as occasion demands.

Voluntary movement requires a foundation of automatic responses which assure a proper combination of mobility and stability of body parts. Since activity occurs in many muscles simultaneously or sequentially, precise regulation is essential. Fortunately, neural control of muscles, whether activity is unconscious or deliberate, is mostly involuntary: muscles are smoothly regulated by reflex mechanisms. The voluntary contribution to movement is almost entirely limited to initiation, regulation of speed, force, range, and direction, and termination of the movement. Volition does not normally include control of individual muscles, although the human capability of doing so and even of controlling single motor units has been amply demonstrated. For example, reaching for an object is voluntarily prescribed as to direction, speed, and the object sought; but the functional features of shoulder girdle fixation, elbow extension, wrist stabilization, and finger movement are regulated by subcortical mechanisms.

Studies of the control of movement and posture have been made traditionally from either a mechanics of movement perspective or a cellular physiology perspective; and only recently are these beginning to be integrated (Hasan et al. (1985)). (The authors would like to think that the integrated approach to both biomechanics and neurophysiology found in this textbook, now in its third edition, may have catalyzed some of the integrative processes.) Hasan et al. (1985) indicated that hypotheses concerning motor control may be classified into three broad categories: those with a postural emphasis, those with a kinetic emphasis, and those which combine facets of both. Whatever the appropriate category, the acquisition of motor skill depends on augmented information which may be of equal value to anyone along the "spectrum" from the highly skilled athlete to the patient learning activities of daily living. Augmented information in the form of biological feedback from proprioceptors is the thrust of this chapter.

Although little is known regarding the exact structure of vertebrate nerve nets, a large amount of information has accumulated regarding proprioceptors and their structure, mode of function, and reflex effects upon the activity of muscles. Some of this knowledge has been derived from clinical and laboratory studies on humans, but the greater portion and the most detailed information has come from the study of other mammals, especially monkeys and cats. It is not wise to transpose indiscriminately from one species to another, but where interspecies similarities and instances of parallel evidence exist, speculation regarding the operation of the same mechanisms in humans is justified, especially as the basis for the formulation of hypotheses and for the design of experiments.

According to Sherrington (1906), proprioceptors are those end organs which are stimulated by actions of the body itself. They are somatic sensory organs located so as to secure inside information and to bring about cooperation and coordination among muscles effectively. The nervous system uses these sensory receptors to modify and adjust muscle function so that peripheral automatic (subconscious) regulation will dominate in most of our so-called voluntary or volitional movements. When proprioceptors are stimulated by movement or position, impulses traverse neural chains to act upon muscles in diverse and interrelated ways. By exciting various proprioceptors, contraction of any muscle tends to organize other muscles to cooperate with it. In the parlance of the electronic engineer, these reflexes operate as negative feedback loops by means of which motor activity becomes in large measure self-regulating. In other words, aspects of the movement process such as muscle tension, absolute muscle length, velocity of change in muscle length, joint angle, joint movement, head position, and contact with surfaces act as stimuli to initiate signals in nerve fibers which are fed back into the central nervous system. In some way as yet unknown, this information is compared with the desired pattern which nature, past experience, or conditioning has established. If the afferent signal indicates any divergence from this pattern, centers in the nervous system modify efferent signals so that the activity of the proper muscles is appropriately increased or decreased to correct the difference.

Proprioceptors may be conveniently classified into three groups: the muscle proprioceptors, the proprioceptors of the joints and skin, and the labyrinthine and neck proprioceptors.

MUSCLE PROPRIOCEPTORS

The muscle proprioceptors are the **neuromuscular spindle** and the **Golgi tendon organ**, both of which are incorporated into the gross structure of the muscle itself. Powers (1976) also includes the **Vater-Pacinian corpuscle**, located in muscle.

Neuromuscular Spindles

The term neuromuscular spindle is the one usually considered to be most descriptive of the receptor, although the terms muscle spindle or spindle are often seen and are equally acceptable.

Neuromuscular spindles are highly specialized sense organs which are distributed among the bundles of contractile fibers in the muscles. They are found throughout the mass of the muscle but tend to be more concentrated in the central portion. There are more spindles in human phasic muscles than in human tonic (postural) muscles, as would be expected since the former require more precise control. The neuromuscular spindle is probably the most important and surely the most complex of the proprioceptive receptors. One usually expects, and finds, structural complexity associated with functional complexity. In the case of the muscle spindle, its structure presents an outstanding duality which is also reflected in its function.

The muscle spindle is sensitive to length and, when stretched, responds to both constant length, as in maintained position or posture, and changing length, as during movement. The firing of the sensory neurons of the spindle reflect both the rate of change in length (**phasic response**) and the ultimate length finally achieved and maintained (**tonic response**). Both aspects of muscle length are signaled by variations in the firing frequency of the afferent neurons serving the receptor.

Morphology of the Spindle

Structural details vary slightly from spindle to spindle, depending upon the particular function of the muscle in which the spindle lies. In general, each consists of a fluid-filled capsule 2 to 20 mm long and enclosing 5 to 12 small specialized muscle fibers (Fig. 10.1). These are known as **intrafusal fibers** (i.e., within the spindle capsule), to distinguish them from the contractile or **extrafusal fibers** of the muscle. The latter, when stimulated by their

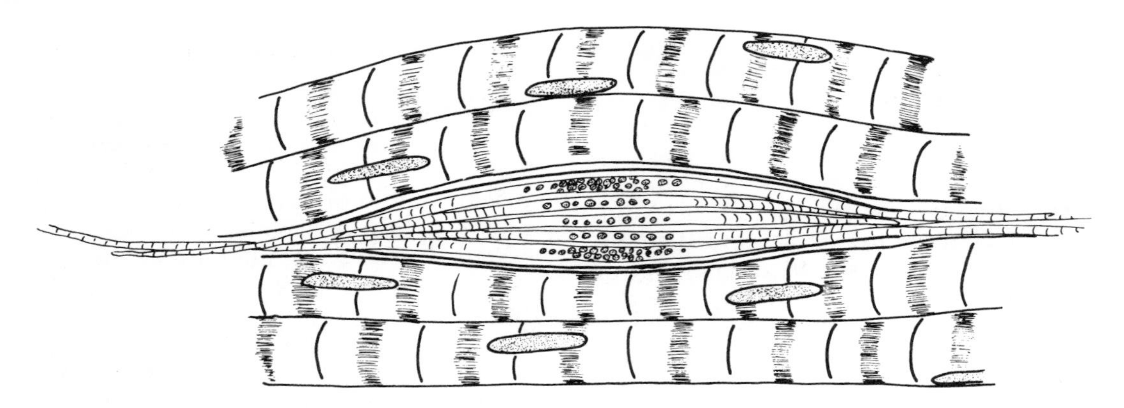

Figure 10.1. The neuromuscular spindle. A diagrammatic representation of a muscle spindle *in situ*, longitudinal section, lying in parallel with the extrafusal or contractile muscle fibers. A spindle consists of a fluid-filled capsule containing small intrafusal muscle fibers of which there are two major types. The two outer intrafusals represent nuclear bag fibers which are percapsular. The three lying centrally represent nuclear chain intrafusal fibers and are intracapsular. Innervation is omitted. (From Gardner, E. B., 1969, Proprioceptive reflexes and their participation in motor skills. *Quest XII:* Fig. 1-B, p. 5.)

large alpha motor neurons, contract to produce the muscle's tension.

Intrafusal fibers differ from contractile muscle fibers in several ways. Their diameters range from one-tenth to one-fourth the diameter of the contractile fiber, their nuclei are concentrated in the central or **equatorial region** rather than being distributed throughout the fiber, and the contractile material is restricted to the **polar ends**. There are many sizes of intrafusal fibers (Fig. 10.2). Each spindle contains one to three large intrafusal fibers ranging from 12 to 26 μ in diameter and one to eight smaller intrafusal fibers with diameters ranging from 4 to 12 μ. These two types of receptor cells differ not only in diameter but in length. The smaller fibers are usually contained entirely within the spindle capsule (**intracapsular fibers**), while the larger fibers pass well beyond the capsule (**percapsular fibers**). In the large fibers the centrally located nuclei are aggregated into a swollen baglike region in the equatorial portion and hence these are called **nuclear bag fibers**. There may also be a single-file projection of several nuclei known as the **myotube**, extending outward from the bag region on either side. The smaller intrafusal fibers contain only a single column of nuclei through their equatorial region. These are known as **nuclear chain fibers**. The large bag fibers, extending well beyond the capsule, attach to the connective tissues and endomysia of the contractile muscle fibers. In some instances, a single nuclear bag fiber may pass through several capsules, each having a nuclear bag. In such cases the structure is known as a **tandem spindle**. Although the significance of tandem spindles is not yet clear, they are probably concerned with the intricate role of the spindle in muscle regulation. The chain intrafusals, fully contained within the capsule, attach to the inner surface of the capsular connective tissue at either end. The two types of intrafusal fibers also vary in viscosity, the bag fibers being more viscous than the chain fibers. Viscosity may determine the type of contraction of the intrafusal fibers, a characteristic which has an important influence upon its function. The innervation of the two types of intrafusal fibers also differs.

Only recently, a few investigators have chosen to distinguish three types of intrafusal fibers. There are two types of

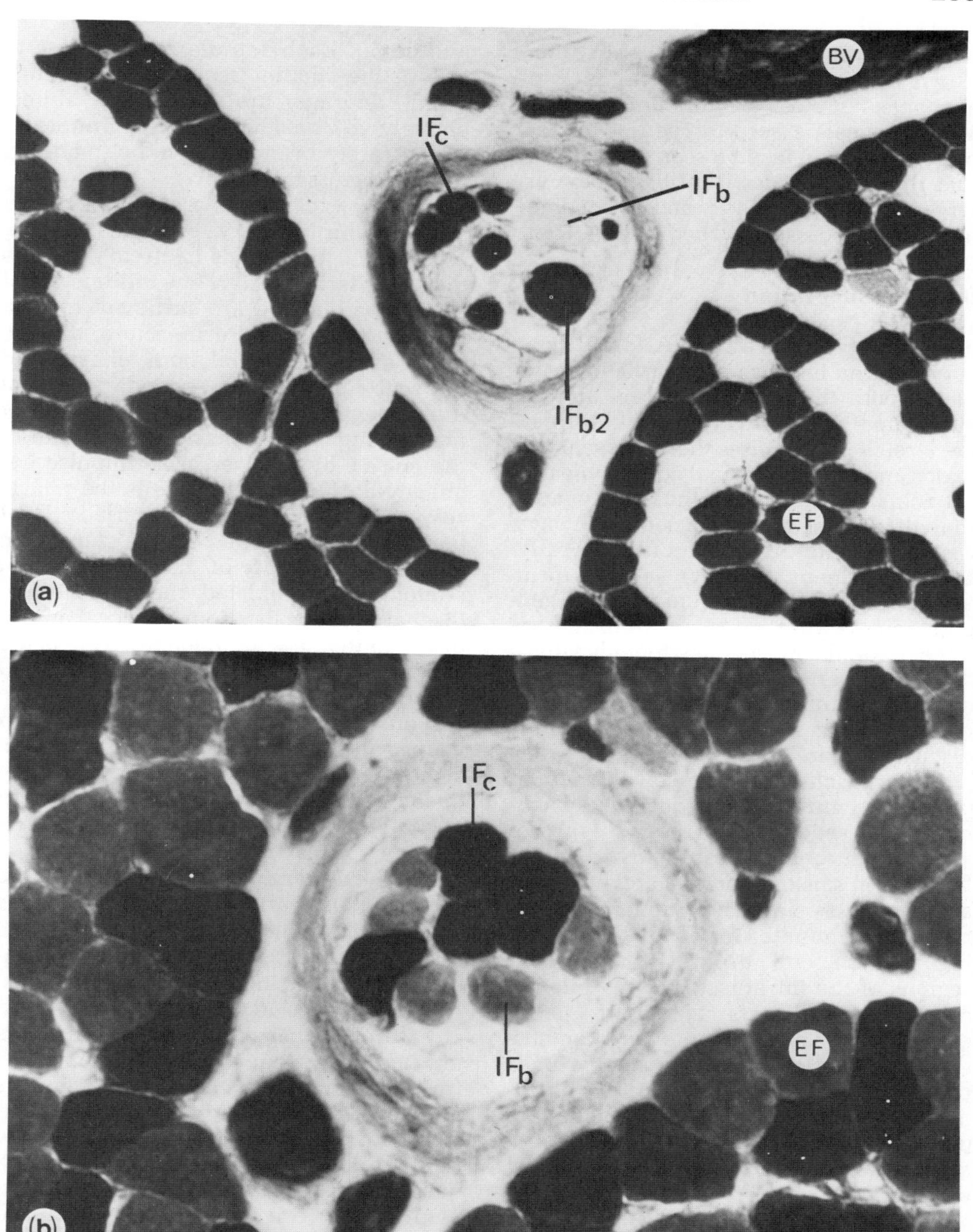

Figure 10.2. The neuromuscular spindle in cross-section. Demonstrates three intrafusal fiber types as follows: nuclear bag fiber (IF^{b}), low ATPase pH 10.3 activity; nuclear chain fiber (IF^{c}), high ATPase pH 10.3 activity; second type of nuclear bag fiber (IF^{b_2}), high ATPase pH 10.3 activity. (Courtesy of G. Elder, McMaster University.) (a), magnification approximately ×200. The equatorial region of the spindle is seen surrounded by its multilayered capsule. Extrafusal muscle fibers (EF) and a blood vessel (BV) are identified. (b), magnification approximately ×400. A second neuromuscular spindle, demonstrating that some chain fibers have larger diameters than bag fibers.

bag fibers and one type of chain fibers (Boyd et al. (1975)). Very little data are available, but close examination of Figure 10.2(a) reveals a second type of nuclear bag fiber. It can also be seen (Fig. 10.2(b)) that the popular view that bag fibers are large and chain fibers are smaller in diameter is not necessarily true (Elder (1978)).

Spindle Innervation

Afferent Innervation. Afferent (sensory) neuron endings are intimately associated with the intrafusal fibers and are stimulated mechanically when the fibers are stretched. Impulses then pass over the axons and enter the spinal cord by the dorsal roots. Within the spinal gray matter they distribute to a number of pathways, rostrally, caudally, and contralaterally. Most prominent, however, is their influence upon their own muscle group. In general, spindle afferents exert an excitatory effect upon the muscle in which they lie, a facilitatory effect upon synergistic muscles, and an inhibitory effect upon antagonistic muscles. An important exception is discussed later.

Most muscle spindles receive two types of afferent innervation, designated **primary** and **secondary**. The two types of afferent neurons are distinguished by differences in sensory endings and by differences in axon size. The primary afferent neurons terminate in **annulospiral** endings which coil around the nuclear regions of the intrafusal fibers, while the secondary afferents terminate juxtaequatorially, i.e., farther toward the striated polar regions, either in smaller coils or in **flower-spray** endings. Axons of primary afferents are large group I fibers (known as Ia), while the axons of secondary afferents fall into group II of the Lloyd classification.

The two types of afferent neurons distribute differently to the two types of intrafusals (Fig. 10.3). Each spindle has only one primary afferent. It enters the spindle and branches to supply an annulospiral ending to each of the intrafusal fibers of the spindle, both bag and chain. Each spindle receives one to five of the smaller secondary afferents. Their endings are restricted almost entirely to the chain fibers, and their axons rarely branch, the axon-to-ending ratio being essentially 1:1.

The two types of afferents differ in sensitivity, the primary afferents having much lower thresholds to stretch than do the secondaries. Only a few millimeters of stretch per second are sufficient to activate the primaries. Furthermore, the primary afferents signal both phasic and tonic stretch, while the secondaries signal tonic length only. The primary afferent neuron signals the phasic length state of the muscle by changes in its impulse frequency *during* stretch. This is the phasic response. The frequency reflects, not length as such, but *rate of change* in length, i.e., **velocity** of the stretch. When stretching is completed, the frequency of discharge drops to a constant level appropriate to the new tonic length. This is the tonic response (Fig. 10.4). When a small stretch is rapidly imposed, the phasic response frequency rises sharply and then drops markedly when stretching ceases. The difference between the maximal frequency attained during the phasic portion of the stretch and the level to which frequency settles in the tonic response is called the **dynamic index**. The tonic value is taken at 0.5 s after the final position has been reached. The contrast between the responses of primary and secondary spindle afferents during a slow stretch is illustrated in Figure 10.5.

Efferent Innervation. Intrafusal fibers are supplied with motor innervation in the form of small gamma-sized neurons known as **gamma motor neurons** or **fusiform neurons**, whose cell bodies lie in the anterior horn of the gray matter of the spinal cord. Axons of these neurons leave the ventral root, travel to the muscle in the appropriate spinal nerve, and terminate in motor endings on the contractile polar end regions of the intrafusal fibers. Each spindle receives 7 to 25 (average 10 to 15) efferent neurons. Impulses traveling over these fusimotor neurons evoke contraction of

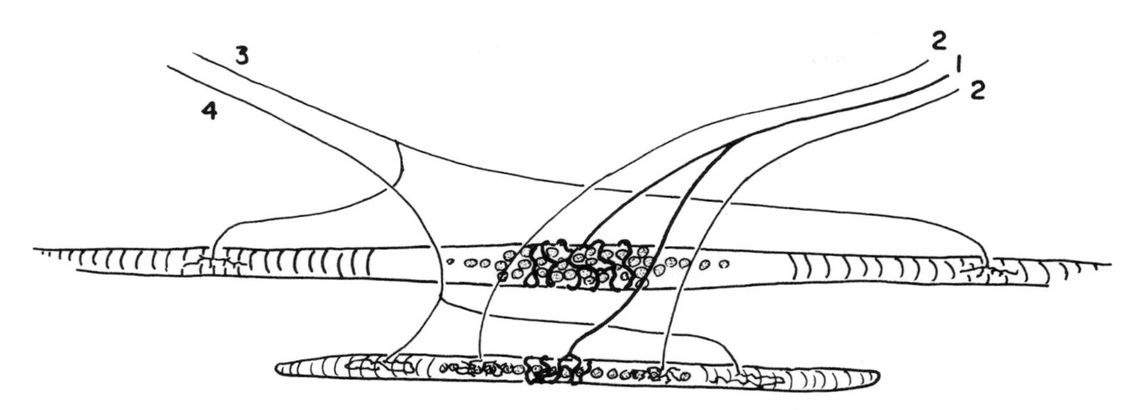

Figure 10.3. Intrafusal fibers and their innervation. A large nuclear bag intrafusal fiber is shown above with equatorial region filled with nuclei and with contractile polar ends extending beyond the limits of the picture. A nuclear chain fiber is shown below. The nuclear chain fiber is smaller in diameter and shorter, with the characteristic single row of nuclei in its central region. Afferent innervation is pictured on the two types of intrafusal fibers. The single large primary afferent neuron (1) ends in coiled terminals (annulospiral endings) on the nuclear region of each intrafusal, while the smaller secondary afferents (2) have branched endings (flower sprays) located on the outer parts of the nuclear region and appear only on the chain intrafusals. Efferent innervation is also shown. The gamma (fusimotor) neuron (3) ends in gamma plates located distally on the polar regions of the nuclear bag fiber. The nuclear chain fiber receives another type of gamma neuron (4) which terminates in gamma trail endings situated closer to the equatorial region. (From Gardner, E. B., 1969, Proprioceptive reflexes and their participation in motor skills. *Quest XII:* Fig. 1-B, p. 5.)

the polar ends of the intrafusal fibers just as impulses in the large alpha motor neurons evoke contraction in the large contractile fibers. Contraction of the intrafusals exerts no detectable influence on muscle tension. Since an intrafusal fiber is connected at both ends either to the interior wall of the capsule or to the connective tissues of the extrafusal muscle fibers, the contraction of its polar ends imposes a stretch upon the nuclear region (Fig. 10.6). Afferent endings are activated just as they are during passive stretch of the whole muscle. Stretch produced by gamma innervation may be referred to as **internal stretch**, while stretch imposed by gravity, an outside force, or shortening of an antagonistic muscle is designated as **external stretch**. Impulses generated in afferent neurons, whether by internal or external stretch, traverse the usual neural circuits to the muscles.

The structural duality of the muscle spindle is also reflected in its motor innervation: there are two types of gamma neurons (Fig. 10.3). Existence of the two types is supported by anatomical, physiological, and pharmacological evidence. *Anatomically* there are two types of endings differentially located on the intrafusal fibers. There are endings found on the polar regions of nuclear bag fibers which resemble smaller versions of motor end-plates, and these endings are known as **gamma plates**. On the nuclear chain fibers are more diffuse endings known as **gamma trails**. Gamma trails are situated more centrally on the polar regions, i.e., closer to the equatorial region, sometimes overlapping the secondary sensory endings, while the plate endings occur more distally. Furthermore, in the muscle nerve there occur two distinct types of gamma-sized axons: one type is thickly myelinated, the other thinly myelinated. The two types show no separable differences in their axon diameters, overlapping over the whole range.

Physiologically, some gamma neurons conduct rapidly and may be called **fast**

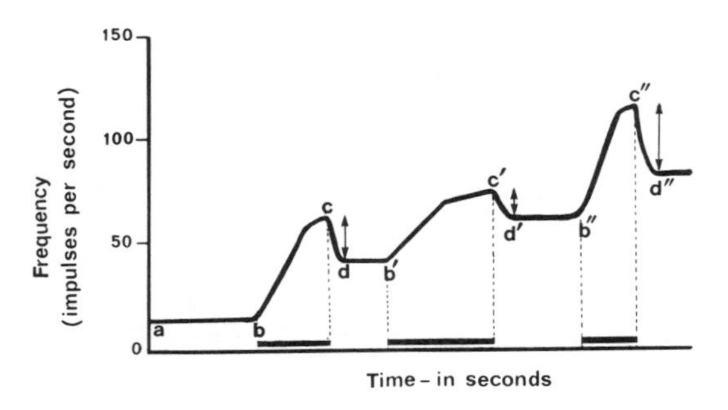

Figure 10.4. Phasic and tonic responses of the spindle primary afferent neuron to interrupted stepwise increases in muscle length. Each stretch, denoted by the solid bars on the time scale, involved the same amplitude of stretch but was imposed at a different velocity as indicated by the time. After each stretch the new length was maintained. a, the initial frequency before stretch was applied; b, the beginning of the stretch; c, frequency at the completion of the stretch; d, frequency after the new length was attained; b′, c′, d′ and b″, c″, d″, responses to subsequent stretches. Note that frequency of impulse discharge increased during each stretch (b to c, b′ to c′, b″ to c″, the phasic response) but dropped back to a new constant level consistent with each new length attained (d, d′, and d″, the tonic response). The small double-ended vertical arrows represent the dynamic index for each stretch. Hypothetical. (Based on Matthews, P. B. C., 1968. Central regulation of the activity of skeletal muscle. In *The Role of the Gamma System in Movement and Posture*, Revised edition. New York: Association for Aid of Crippled Children.)

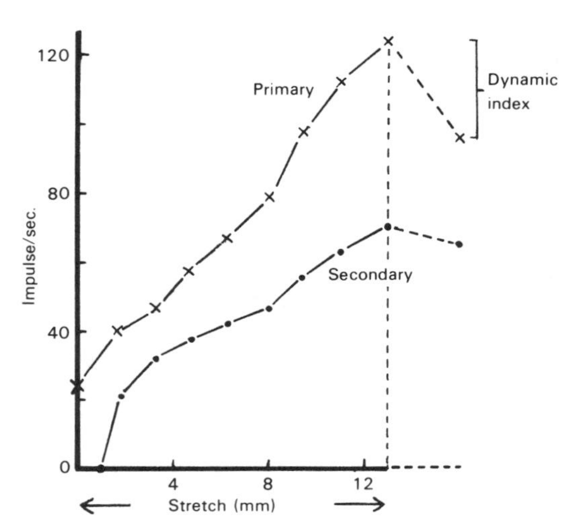

Figure 10.5. Responses of primary and secondary spindle afferents. Upper curve: the phasic frequency response of a primary spindle afferent neuron during and directly following a slow stretch of the muscle. Lower curve: response of a secondary spindle afferent under same conditions. Broken vertical line indicates point at which stretch was completed, and the muscle was held at this new length. The last point on each curve is the frequency recorded after the new length had been maintained for 0.5 second. Note the marked drop (dynamic index) in the response of the primary afferent, indicating that it is responding to velocity of stretch, while the secondary afferent has responded only to absolute length. (Adapted from Harvey, R. J. and Matthews, P. B. C., 1961. The response of de-efferented muscle spindle endings in the cat's soleus to slow extension of the muscle. *J. Physiol. (Lond.)* *157:* 370. Cf. Matthews, P. B. C., 1968. Central regulation of the activity of skeletal muscle. In *The Role of the Gamma System in Movement and Posture*, Revised edition, Fig. 17, p. 32. New York: Association for Aid of Crippled Children.)

fusimotor neurons, while others conduct more slowly (**slow** fusimotor neurons). There is little overlapping in conduction velocities between the two groups. Indirect evidence suggests that the thickly myelinated axons are the fast-conducting ones, although this needs further substantiation.

Furthermore, the gamma neurons display two different types of influence on the primary afferents. Some gamma neurons, when stimulated concurrently with external stretch of the muscle, appear to produce an increase in the dynamic index. In other words their action exaggerates the phasic response of the primary endings. Such neurons are called **dynamic fusimotor neurons**. Other gamma neurons markedly *decrease* the dynamic index. These are known as **static fusimotor neurons**. They may also increase the tonic response somewhat, but this effect is less noticeable than is the reduction of the phasic response (Fig. 10.7).

Pharmacologically, evidence for two separate gamma systems is found in

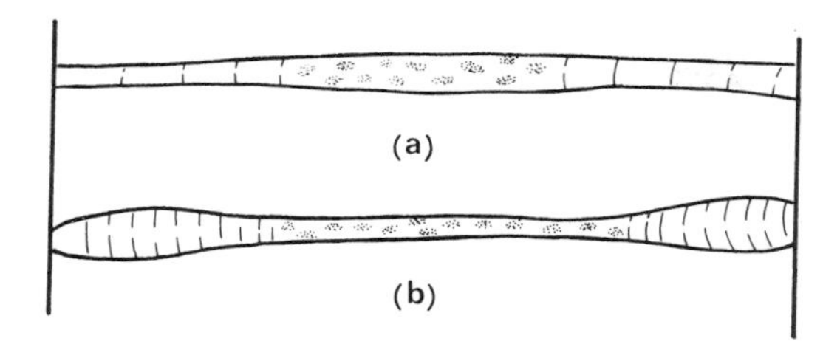

Figure 10.6. Internal stretch of an intrafusal fiber. The vertical bars on either side represent tendons to which the intrafusal fibers are attached. (a), a nuclear bag fiber in the neutral state with polar ends uncontracted; (b), the same fiber under gamma (fusimotor) stimulation. Polar ends have contracted, putting the nuclear region under stretch.

Rushworth's (1968) investigation of drug effects on phasic responses as compared with tonic responses. He found that barbiturates depressed the tonic response but left the phasic response unaffected, and that procaine completely suppressed the tonic response and reduced the phasic response. Moreover, Rushworth found that the phasic response was exaggerated in cerebellar disease, while the tonic response was absent. Other studies have shown that the anterior lobe of the cerebellum normally inhibits dynamic gamma activity, suggesting that in disease of the cerebellum this inhibition is reduced or absent. Such differential effects on the two types of response support their correlation with separate structural properties.

It is not known for sure which gamma axon is related to which type of motor ending and, coincidentally, which type of intrafusal fiber. Indirect evidence seems to relate the dynamic gammas to the plate endings, and therefore the nuclear bag fibers, and the static gammas to the trail endings and the chain fibers. For example, the primary afferent neurons have their endings on both types of intrafusal fibers and are influenced by both gammas; however, the secondary afferents, whose endings are restricted almost entirely to chain fibers, are activated only by static gammas. Examined from another viewpoint, if dynamic gammas end in the plate endings on the nuclear bags as suggested, they should influence the primary afferents but not the secondaries, which is the actual experimental finding. If static gammas end on chain intrafusals, they should affect both types of afferent endings, and, in fact, the static fusimotor neurons can **drive**[a] both primary and secondary afferents. So it seems likely that the dynamic fusimotor neurons distribute to the nuclear bag intrafusal fibers, and the static fusimotors distribute to the chain intrafusals.

Evidence is accumulating that the type of contraction of the two kinds of bag fibers differs from that of the chain intrafusal fibers differ. This is significant because the nature of intrafusal contraction may determine the afferent response, while the particular type of gamma neuron simply activates a particular type of intrafusal fiber. Smith (1966) has shown that bag fibers contract slowly in a local, graded manner. This would be consistent with the phasic response, in which frequency increases with the rate of stretch. The smaller chain fibers, however, contract in a faster, twitchlike manner, and complete tetanus can be evoked in them by a stimulating frequency of about 15 impulses/sec. This would be compatible with the tonic response to a maintained stretch, the extent of tetanus in the intrafusal fiber determining the frequency in the afferent neuron. Since the primary afferents serve both types of intrafusal fiber, they would be expected to signal both phasic and tonic stretch, while the secondaries, associated mainly with the chain fibers, would signal only tonic length; this is in fact the case. Further, if it is the dynamic type of gamma which serves the bag fibers and evokes a contraction appropriate to phasic response (as suggested), it is not surprising that stimulation of dynamic gammas during external stretch enhances the phasic response, increasing the dynamic index as it does

[a] A neuron is considered to drive another neuron when it causes the other neuron to respond one for one to its frequency over a limited number of cycles.

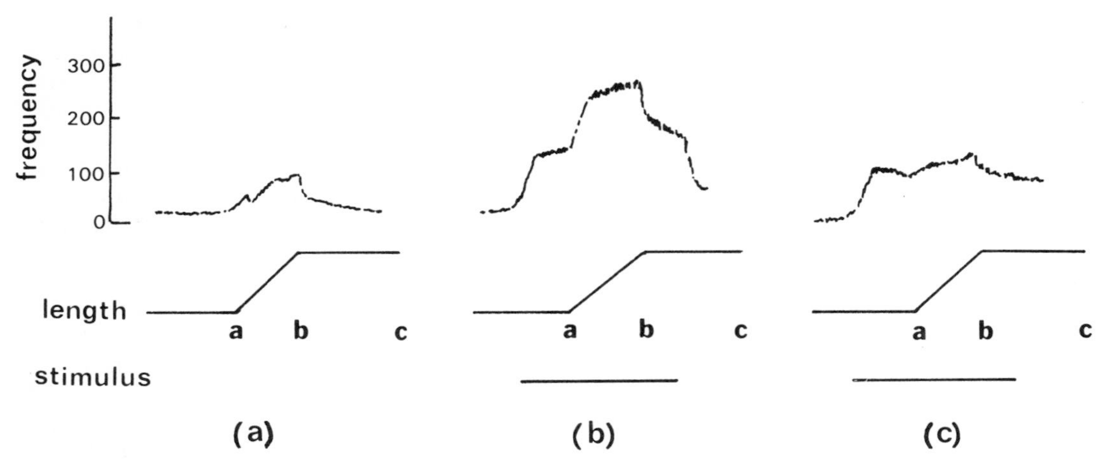

Figure 10.7. Effects of dynamic and static fusimotor stimulation on response of a spindle primary afferent to stretch. (a), no gamma stimulation; (b), stimulation of a dynamic gamma neuron; (c), stimulation of a static gamma neuron. The application and duration of stimulation is indicated by the bars (under (b) and (c)) in the lower trace. The middle trace shows the time course of the stretch: a, beginning of stretch; a to b, phasic stretch; b to c, static stretch. The top trace shows the frequency response in the spindle primary afferent neuron under the three conditions. (a), without gamma stimulation, the frequency rise of the phasic response closely paralleled the phasic stretch (a to b) and was followed by the tonic response (frequency decrease) as the new length was maintained. (b), under dynamic gamma stimulation the resting frequency in the primary afferent was markedly increased even before the stretch was begun. The phasic response during the stretch was almost twice that in (a). When stimulation ended, the tonic response frequency returned to about the same level as in (a). (c), the initial or resting frequency was increased, as in (b), by stimulation of the static fusimotor neuron. However, activity of this gamma neuron markedly reduced the phasic response as compared with (a). In fact, it was hardly greater, if at all, than the initial gamma background discharge. (Adapted from Matthews, P. B. C., 1962. The differentiation of two types of fusimotor fibre by their effects on the dynamic response of muscle spindle primary endings. *Q. J. Exp. Physiol. 47:* 324. Cf. Matthews, P. B. C., 1968. Central regulation of the activity of skeletal muscle. In *The Role of the Gamma System in Movement and Posture*, Revised edition, Fig. 24, p. 46. New York: Association for Aid of Crippled Children.)

(Fig. 10.8). Finally, if bag fibers contract at both "slow" and "fast" rates, *perhaps* some bag fibers act like chain fibers, indicating only tonic length.

After gamma denervation, chain fibers atrophy quickly, whereas bag fibers are slow to degenerate. This suggests that the chain fibers have a greater dependency upon gamma innervation, thus indicating their tonic sensitivity, while the phasically sensitive bag fibers should be more dependent upon external stretch. Greater sensitivity to stretch is consistent with the fact that the nuclear bag fibers are percapsular.

Evidence has indicated that spindles also receive branches from the alpha motor neurons to the contractile fibers. These have been designated as **beta axons**. The presence of skeletomotor as well as fusimotor axons has been demonstrated histologically by Adal and Barker (1965) and electrophysiologically by Bessou et al. (1963). Beta innervation is relatively rare in cats and humans but more common in rats, in which it seems to derive from slow alpha motor neurons. When present, beta axons terminate in typical extrafusal-type end plates on nuclear bag fibers close to the ends of the spindle poles, and they appear to have a weak dynamic effect. While the function of these beta axons is still uncertain, such a contribution to intrafusal motor innervation would obviously be useful in the integration of muscle activity.

Physiology of Spindle Innervation

Functions of Primary Afferent Neurons. More is known about the effects of the firing of the primary endings than about those of the secondary endings. Firing of the spindle primary afferent neurons reflexively excites motor units of the muscle in which the spindle lies, simultaneously facilitating synergists and inhibiting antagonists. Figure 10.9 shows diagrammatically the simple divergent circuit presumed to mediate the effects produced. It has been well established that the primary afferent enters the dorsal root of the spinal nerve. Its cell body lies in the dorsal root ganglion, and its axon passes into the gray matter of the dorsal horn, makes it way to the ventral horn, and synapses directly upon an anterior horn cell of the motor pool of its own muscle. In other words, the reflex arc from spindle to muscle consists of only two neurons: a **monosynaptic** pathway. When stimulated by stretch of its endings, the primary afferent neuron fires signals into the central nervous system which evoke a contraction just sufficient to relieve the stretch. In any muscle contraction it is essential not only that the prime mover be activated but that the activity of the muscles of its functional group (those which operate upon the same joint) be appropriately modified. Synergists must support the activity of the prime mover, and antagonists must not oppose it. Collateral branches from the axon of the primary afferent make connections through interneurons with the motor neurons of the synergists. These are excitatory synapses, and the synergists are facilitated. Other collaterals synapse through inhibitory interneurons with antagonists, producing reciprocal inhibition. There is some indication that the inhibition of antagonists may be accom-

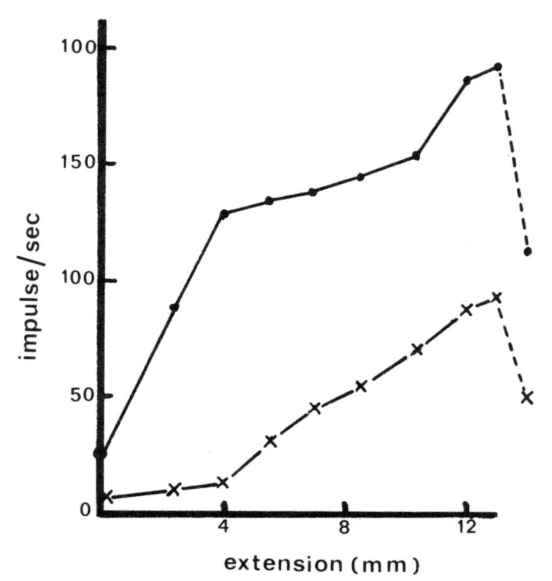

Figure 10.8. The influence of fusimotor stimulation on the response of a spindle primary afferent neuron. Both curves represent the dynamic response of the same primary afferent neuron. Upper curve (•), ventral roots were intact; therefore the spindle was under fusimotor influence. Lower curve (×), ventral roots were cut, and as a result the spindles were de-efferented. The last point on each curve (joined by a broken line) shows the frequency recorded at 0.5 second after completion of the phasic part of the stretch. Note the early increase in frequency which occurred with gamma firing. In this case the dynamic index was also increased. (After Jansen, J. K. S., and Matthews, P. B. C., 1962. The central control of the dynamic responses of muscle spindle receptors. *J. Physiol. (Lond.) 161:* 357. Cf. Matthews, P. B. C., 1968. Central regulation of the activity of skeletal muscle. In *The Role of the Gamma System in Movement and Posture*, Revised edition, Fig. 21, p. 42. New York: Association for Aid of Crippled Children.)

plished by presynaptic inhibition exerted upon their own primary afferents (see Fig. 8.17). This seems reasonable, since the contraction of any muscle will automatically put its antagonists on stretch; and if the antagonist should respond to its own stretch reflexes, it would oppose the contraction of the prime mover muscle and

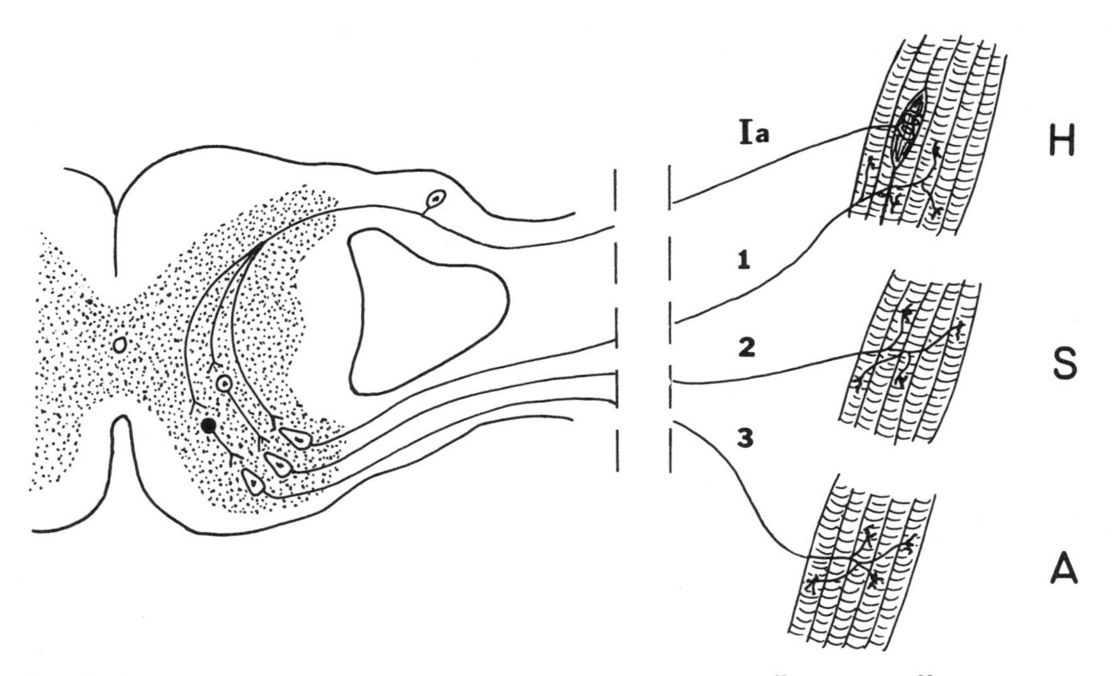

Figure 10.9. Circuits of the spindle primary afferent neuron. Ia, spindle primary afferent neuron: 1, alpha motor neuron to homonymous muscle (H); 2, alpha motor neuron to synergist (S); 3, alpha motor neuron to antagonist (A). The primary afferent neuron (Ia) of the spindle is shown entering the spinal cord (seen in cross-section) by the dorsal root. In the ventral horn its axon makes monosynaptic connection with the alpha motor neuron (1) to its own muscle (H). A collateral connects with the alpha motor neuron (2) to the synergist (S) through an excitatory interneuron (shown with a cell body as an open circle). Another collateral connects through an inhibitory interneuron (shown with a cell body as a filled circle) with the alpha motor neuron (3) to the antagonist (A). These circuits assure proper cooperation among the muscles of the functional group.

produce cocontraction. Connections with synergists and antagonists are probably disynaptic, since the time consumed in transmission from spindle afferent to alpha motor neuron is compatible with the interposition of two synapses. An inhibitory interneuron in the path to the antagonistic muscle converts the signal from excitatory to inhibitory. Other collaterals from the primary afferent axon travel upward and downward through the spinal cord, making ipsilateral intersegmental and supraspinal interconnections through appropriate interneurons. Still other projections cross the cord to influence contralateral muscles. Here, then, is an example of a simple but highly important circuit preprogrammed to produce appropriate excitation and inhibition within a functional muscle group.

As already described, the primary ending is sensitive to both phasic and tonic stretch. When a muscle is stretched, the primary neurons fire a rapid burst of impulses *during* stretching, with frequency directly related to the *velocity* of stretch: the phasic response. When the muscle reaches a length which is maintained, the frequency drops to a lower level appropriate to the new length: the tonic response.

Example of Phasic Response. The knee jerk represents an isolated phasic response evoked by primary afferents. The stimulus, a single sharp blow on the patellar tendon, constitutes phasic stimulation only, because no new length is attained and held. The blow on the tendon puts a small momentary stretch upon the quadriceps muscle group and coincidentally upon its spindles. The stretching of the nuclear regions of the intrafusal fibers, slight though it is, is sufficient to cause distortion of the primary endings and evoke a burst of impulses in the primary afferent neurons. This burst of impulses travels into the spinal cord and is conveyed across synapses to the cell bodies of the alpha motor neurons of the same muscle. Their axons return the discharge to the same muscle, indeed to the vicinity of the muscle where the excited spindles are located. As a result the muscle contracts quickly (i.e., the knee joint momentarily extends), the magnitude of the contraction being determined by the velocity of the stretch. Because the shortening of the muscle relieves the original stretch, the spindle is "unloaded" so the primary neurons stop firing. Since the alpha motor neurons are no longer excited, the muscle relaxes and the limb drops immediately to its original position. The knee jerk is a purely artificial response, useful to the physiologist as an example of an isolated stretch reflex and to the neurologist as an index of the general state of the nervous system. The jerk as such has no functional value in neuromuscular activity.

Example of Tonic Response: Postural Reaction. An example of a purely tonic response may be found in the postural reaction to stretch, i.e., the simple **stretch** or **myotatic reflex**. Slow stretching of muscles as a result of the shifting of the center of gravity causes tonic response in spindle primary afferents and a contraction of the stretched muscles which will correct the displacement. For example, the vertical projection of the center of gravity of the human body usually falls in front of the medial malleoli (ankle joint "centers") by a distance of 5 to 8 cm, when in a normal standing posture. This concentration of weight (force) on the ventral side of the ankle joints places a tonic stretch on the calf muscles, especially the soleus, which involves the myotatic reflex in these muscles. By servo-action, they contract that amount necessary to counter the forward force and "pull" the body backward over its base of support. Such postural adjustments go on continually in the living animal.

Example of Phasic Plus Tonic Responses: Adjustment to Load. Another illustration of the activity of the primary afferents, one that is more complex and more functional than the knee jerk, is found in the automatic adjustment of muscle contraction to an added load. This is a normal functional reaction related to both phasic and tonic changes in muscle length and evoked by corresponding phasic and tonic signaling of the primary afferent neurons. The muscle reactions to the two primary afferent neural responses, one to the velocity of spindle stretch and the other to the length per se, can be readily distinguished in electromyograms (EMGs) recorded during the automatic adjustment of an outstretched arm when a load is added (Fig. 10.10). Before the addition of the load, the electromyographic record shows a steady discharge in the forearm flexors consistent with the maintained muscle length and the weight of the arm. Addition of the load puts stretch on the muscles, and the EMG first shows a sudden burst of electrical activity in the muscle, reflecting the phasic response in the primary afferents of the spindle and directly related to the velocity of the stretch (abruptness of the load addition). This is followed by a lower steady firing rate consistent with the new length and load and is a result of the tonic response of the spindles. These spindle influences are also reflected in the overt movements of the forearm. As the load is dropped suddenly into the basket, the forearm is depressed by the added weight, and the flexor muscles and their spindles are stretched. The muscles, reacting to the velocity-dependent phasic response of the spindles, contract more strongly than the load requires. As a result the forearm and basket are raised momentarily. Except when the load is light, muscle response more than compensates for the added

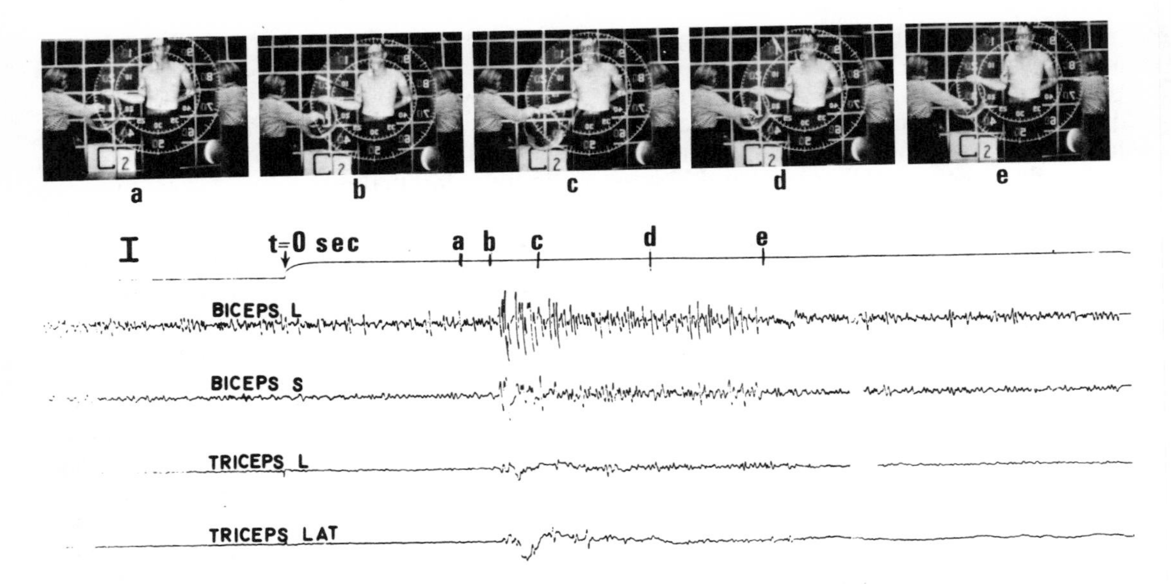

Figure 10.10. Adjustment to load. Electromyograms with syncrhonized cinematography. Subject is supporting basket in the hand, forearm supinated, with the elbow flexed approximately 90° and instructed to maintain this position as load is added and removed. Upper trace synchronizes camera and EMG; rise in baseline indicates start of clock: time = 0.000 second. EMGs recorded from biceps and triceps; L, long; S, short; and LAT, lateral heads. Photograph identification letters indicate situation at corresponding lettered points marked by vertical lines on the EMG. Before b the EMG records tonic activity in muscles to sustain weight of forearm, hand, and basket in prescribed position, reflecting gamma bias under voluntary control. I. b, load is dropped abruptly into basket, load is removed at e.

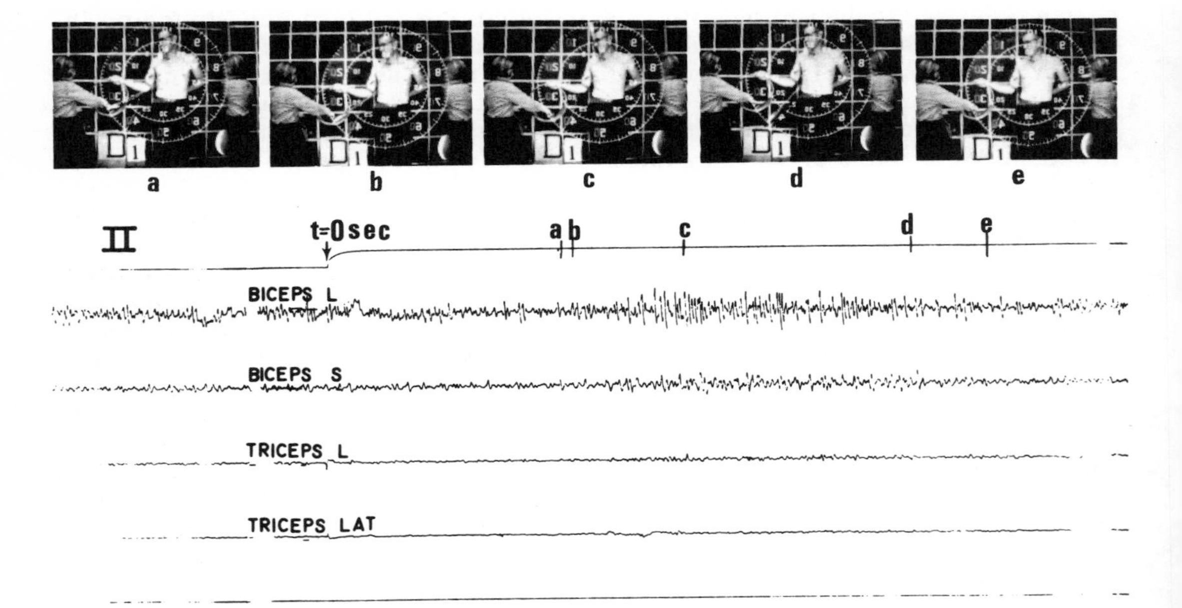

Figure 10.10. II. b—c, load is added gradually to basket; d—e, load is removed. See text.

load, thus giving evidence of the phasic response in the primary afferent neurons. The more abrupt the application of the load the greater the phasic response to the same load, hence the greater the extent of the muscular contraction and therefore the greater the overcompensation. Because the overcompensation partially unloads the spindles, the impulses following the reflex pathways to the muscle are decreased, the muscle's contraction lessens, and the limb begins to drop. This again puts stretch on the muscle, which will again respond appropriately to the velocity of the drop. A tendency to oscillate develops but is quickly damped, and the limb shortly comes into equilibrium at a level consistent with the new load. This is then maintained by the tonic response. However, if the load is placed slowly into the basket, the spindles respond tonically only, with no phasic burst but with sufficient increase in activity to support the increased load. There is little or no overcompensation.

Vibration Response. A unique method of applying stretch to muscle spindles is by vibrating the belly of the muscle. Vibration appears to excite the spindles by deforming their equatorial regions. The vibrated muscle contracts slowly and maintains the contraction throughout the vibration, a purely tonic response. The phasic response may be reduced or absent if the muscle is tested shortly thereafter for the jerk response. When applied to the muscle *belly*, low-frequency vibration seems to be a specific stimulus for the primary afferents. High frequencies will excite the secondary endings as well, except that when vibration is applied to the *tendon*, regardless of frequency, only primary endings respond. It appears that only the chain intrafusals are involved in the vibration response, but it is too early to make positive statements on this subject. Vibration also is an effective stimulus for the Vater-Pacinian corpuscle and is discussed under that section.

Functions of Secondary Afferent Neurons. Secondary afferent neurons, as already mentioned, are less sensitive than primary afferent endings and produce only the tonic response. They reflect the mean length of the muscle at any instant. Their influences upon their own and other muscles differ markedly from those of the primary afferents. The effect of secondary afferent excitation is to facilitate flexor muscles and inhibit extensor muscles regardless of the type of muscle in which the spindle lies. In other words, if the stimulated spindle lies in a flexor muscle, then its effect is to reinforce both the activation of the muscle itself and the inhibition of antagonistic extensors, supplementing the action of its primary afferents. If, however, the spindle lies in an extensor muscle, it will tend to inhibit its own and other extensor muscles and facilitate the antagonistic flexors. Nevertheless, stretch of an extensor muscle which is sufficient to activate its secondary endings will at the same time so strongly excite its primary endings that excitation of the muscle will be maintained in spite of the autoinhibition. Consequently the facilitation of the flexors results in cocontraction, i.e., simultaneous contraction of flexors and extensors. This is especially true of stretch applied to a single joint extensor muscle. Most of the single joint extensors are primarily antigravity muscles, and their spindles are richly supplied with secondary afferents. Rather than being a useless interference with muscle activity, cocontraction acts to provide stabilization of joints for both weight bearing and movement. It is particularly important at proximal joints.

In the case of a muscle which passes over two joints, acting as a flexor of one and an extensor of the other, the spindles tend to act as though in a flexor muscle, the secondary afferents facilitating their own muscle and inhibiting antagonistic single joint extensors.

Functions of Efferent Innervation. The importance of gamma neurons in spindle function cannot be overemphasized. If the spindle had afferent innervation only, all of the responses that it evoked would be of the jerk type with a tendency to produce oscillation. However, impulses over the gamma neurons can, by maintaining contraction of the polar ends, "set" the spindle at any prescribed firing level, thus adjusting its sensitivity independently of absolute muscle length or load. The gamma-innervated spindle, with nuclear regions already under internal stretch and primary afferents already firing in a manner appropriate to the initial length, will be far more sensitive to any further length change which may be imposed by external forces (Fig. 10.8). This setting of spindle sensitivity by the gamma neurons is known as **gamma bias**. At rest, gamma bias holds the spindle just below motor neuron firing level, but, if gamma outflow is raised sufficiently, the muscle will be activated.

The internal stretch of spindles under gamma bias obviates the jerkiness of muscle response which would otherwise result and makes possible smoothness and steadiness of movement or position. Figure 10.11 compares diagrammatically the responses of a spindle primary afferent neuron in a weighted muscle with and without gamma stimulation. In (a) the afferent neuron was firing (upper trace) in response to the external stretch of the load. In (b) stimulation of fusimotor neurons caused increased firing of the primary afferent without any change in the muscle tension (lower trace). In (c) stimulation applied to alpha motor neurons resulted in a twitch contraction of the muscle (lower trace), accompanied by temporary cessation of firing in the primary afferent neuron as the contraction unloaded the spindle. In (d) alpha and gamma neurons were stimulated simultaneously. A contraction occurred, but the spindle continued to fire throughout the twitch because the internal stretch of the spindle was maintained by the fusimotor discharge.

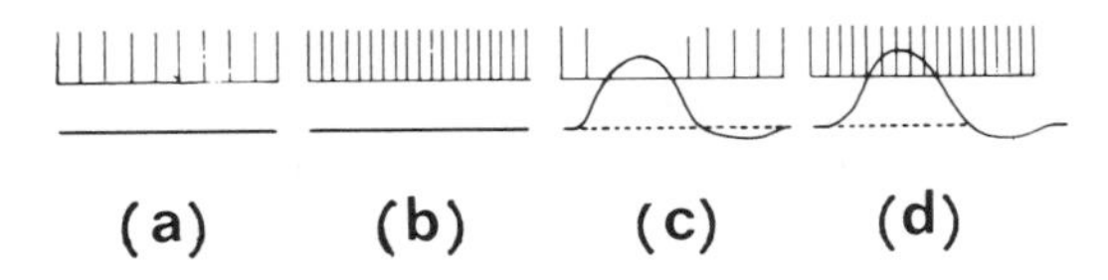

Figure 10.11. Response of the primary afferent neuron of the spindle to stretch under various conditions. The muscle was weighted with moderate load. Upper trace: frequency response of the primary afferent; lower trace: tension in the muscle. (a), response of the unstimulated muscle to the stretch imposed by the load. (b), stimulation of fusimotor (gamma) neurons induced no change in muscle tension but caused an increase in the firing frequency of the spindle afferent. (c), stimulation of alpha motor neurons caused contraction of the muscle, and the "unloaded" spindle ceased firing. (d), stimulation of alpha and gamma neurons simultaneously resulted in tension development in the muscle but without unloading the spindle. (After Hunt, C. C., and Kuffler, S. W., 1951. Further study of efferent small-nerve fibres to mammalian muscle spindles, multiple spindle innervation and activity during contraction. *J. Physiol. (Lond.) 113:* 283. Cf. Eyzaguirre, C., 1968. Some functional characteristics of muscle spindles. In *The Role of the Gamma System in Movement and Posture,* Revised edition. Fig. 11, p. 22. New York: Association for Aid of Crippled Children.)

The shortening and lengthening reactions, long recognized in muscle, are explained by gamma bias. In resting muscle, gamma innervation maintains a tonic firing at low frequency in the primary afferents. If the limb is passively put into a position which shortens the muscle, the primary discharge decreases for a short time, and this is reflected in a reduction of the muscle's contractile capacity. Soon, however, the primary afferent discharge is resumed at the original resting rate and functional capability is improved. This is known as the **shortening reaction**. The change in position has caused a temporary slack in the spindle and hence a reduction in afferent firing. Almost immediately, however, gamma outflow is increased, causing the intrafusal fibers to contract and take up the "slack." In other words, the gamma bias, and hence the spindle sensitivity, is re-established at the new, shorter length. If a limb is passively placed so as to put the muscle in a lengthened state, the firing of the primary afferents increases at first, as a result of passive stretch, producing an increased resistance to stretch, but then drops back to its resting frequency as gamma bias is reset. This is the **lengthening reaction**. Note that in each case the spindle response briefly opposes the change in length.

When a subject voluntarily holds a load in a prescribed position, the spindle gammas are already under cortical stimulation to produce appropriate gamma bias to maintain that position. When a load is added gradually, the static response alone adjusts muscle activity to the increased load. However, if that load is added abruptly, the velocity of stretch produces a phasic response which evokes overcontraction, as described earlier under "Example of Phasic plus Tonic Responses: Adjustment to Load." In addition to the spindle responses to the rate and extent of external stretch, information regarding limb position is fed back into the central nervous system by other sensory receptors (visceral, joint, skin). If the new position deviates from that voluntarily prescribed, cortical outflow alters gamma bias appropriately until the desired position is regained. Gamma activity acts as an important factor in damping out the tendency to oscillate resulting from overcompensation.

Efferent innervation confers on the spindle its capacity for minute and precise control of muscle contraction in accordance with the requirements which unconscious coordination and conscious direction prescribe. Fusimotor neurons receive impulses from corticospinal nerve fibers, from neurons of reticulospinal tracts, and from collaterals of all sensory pathways, including those of the proprioceptors. As a result they discharge tonically, holding intrafusal muscle fibers in a continuous but highly variable state of gamma bias. The greater the state of gamma bias, the greater the sensitivity of the spindle afferents to external stretch. Wooldridge (1963) states that the reticular formation of the brainstem acts on the gamma efferents "to adjust the operating parameter" of the muscles by setting the zero point of the spindles so that they will fire minimally when the muscle is at a desired length. As a result the muscle "automatically seeks out just the degree of contraction that will minimize firing frequency of its stretch receptors." Since the two types of gamma neurons exert distinguishable influences upon spindle activity, the two gamma systems would be expected to have separate supraspinal controls.

Role of Spindle Innervation in Voluntary Movement. Granit (1955, 1975) proposed that cortical signals for the programming of voluntary movement may reach muscles over two pyramidal pathways: (1) directly to the alpha motor neurons and thence to the muscle or (2) indirectly via gamma motor neurons to the spindles, from which they are relayed back to the alpha motor neurons by spindle afferents. The simultaneous activation of both the alpha and gamma motor neurons by the same signal is termed alpha-gamma coactivation. Figure 10.12 shows diagrammatically the indirect pathway, which is known as the **gamma loop**. Signals from the pyramidal tract activate the proper gamma neurons. Impulses passing out over the gamma neurons cause intrafusal fibers to contract, putting stretch upon the afferent endings. The activated afferents transmit impulses back to the spinal cord and across the synapses to the alpha motor neurons of the muscle. These convey the signal to the muscle fibers, which then respond to a degree commensurate with the amount of stimulation introduced by the corticospinal neurons. Granit points out that, by appropriate cortical influences, the spindle can be used to regulate the amount and rate of muscle shortening. In isometric contraction (i.e., position) the cortex sets the spindle and the muscle follows by servo-action. This position-controlling servomechanism will maintain muscle length regardless of changes in load or resistance. In isotonic contraction (i.e., movement) the cortex progressively increases the downflow to the gamma neurons, their outflow keeps the spindles progressively activated, and the

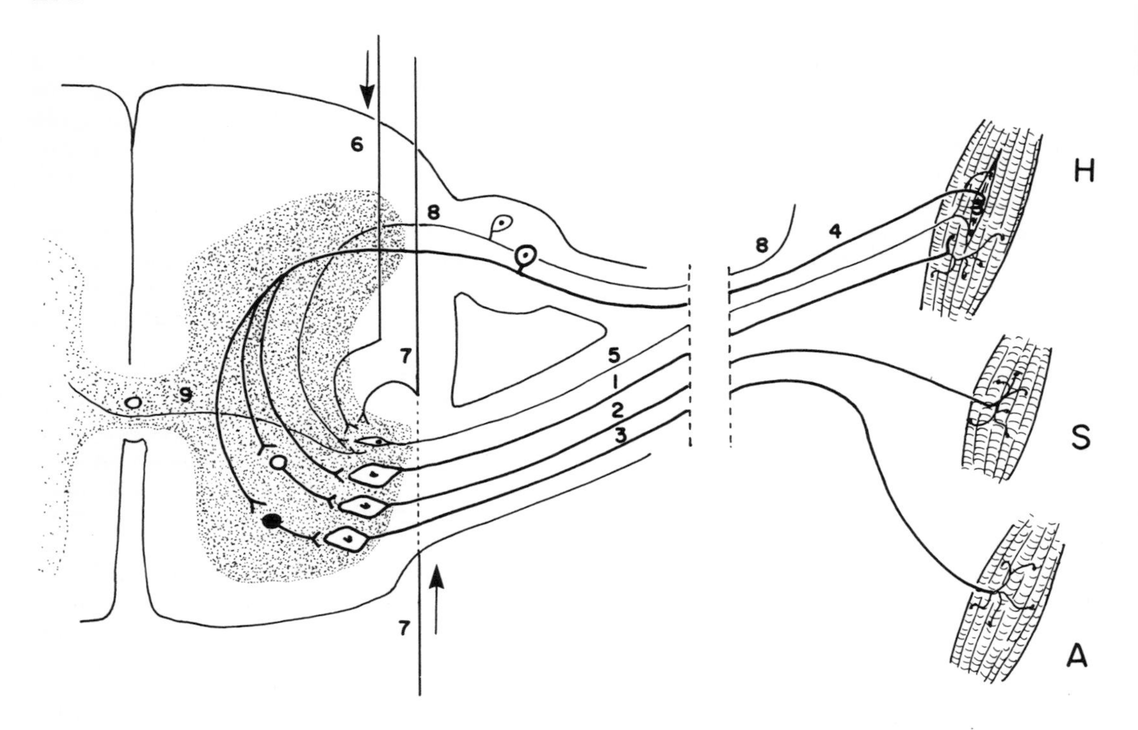

Figure 10.12. Spindle primary afferent circuits and the gamma loop. The spindle primary afferent circuits (as in Fig. 10.9) are shown with the gamma circuits added. As in Figure 10.9, 1, 2, and 3 are alpha motor neurons to the homonym, synergist, and antagonist, respectively. 4, primary afferent from the spindle; 5, gamma (fusimotor) neuron to the spindle; 6 and 7, fibers descending from supraspinal centers; 8, ipsilateral afferent neuron from a joint or skin receptor situated near another muscle; 9, interneuron from a contralateral source. (From Gardner, E. B., 1969. Proprioceptive reflexes and their participation in motor skills. *Quest XII:* Fig. 2, p. 6.)

muscles follow the spindles, thus serving the dictates of volition. Cortical impulses can also hold gamma bias just below firing level in anticipation of the stimuli which will finally trigger activity.

Partridge (1961) gives experimental support to Granit's hypothesis. He concludes that the central nervous system controls voluntary movement by sending simple length-setting signals of proper sign (+ or −) via the gamma system to the spindles of both agonists and antagonists and then allows the stretch reflex to determine the necessary adjustment of muscle force to produce the desired position or movement.

Although transmission over the gamma loop is obviously slower, it is probably the preferred route for voluntary activation and cessation of muscle activity. Rushworth (1968) has shown that the gamma delay, about 50 msec, is not inconsistent with human's fastest repetitive movement (piano trills at 6 to 8/sec), whose latency is about 120 to 150 msec. Voluntary activation via gamma neurons also has the advantage that reciprocal activity of the muscles concerned is assured from the very first contraction.

Houk (1979) hypothesized a different type of servo-control theory which is linked to the negative tension feedback from the Golgi tendon organ. According to Houk, signals from the cortex, sent directly to the alpha motor neurons, supply a bias upon which a net proprioceptive signal from both spindles and Golgi tendon organs regulates the stiffness of the homonymous muscle.

Loeb (1984) summarized the many control theories which have been advanced and indicated that they seem to be specific to particular classes of muscle function. The first scheme advanced historically was Merton's hypothesis that the gamma system was the primary initiator of muscular contaction. Loeb pointed out that Stein (1974) had identified such a system as "slow and prone to oscillation in the face of inertial loads." According to Loeb, the Houk theory has merits when the objective is to maintain a position with a variable degree of compliance, while the Granit theory has merits for the control of single muscles shortening against a load. To help explain how the human body adapts to the many diverse tasks of motor function, Loeb organized the conditions into task groups based primarily on velocity of movement and degree of extrafusal activity. Essentially, he set the tasks into four quadrants as follows: (1) active shortening of a muscle is best controlled by a pattern of alpha-gamma coactivation; (2) active lengthening of a muscle requires selective recruitment of alpha motor neurons with little or no gamma recruitment; (3) passive shortening of a muscle is best served by selective recruitment of gamma motor neurons; and (4) passive lengthening should be viewed as alpha-gamma coinactivation. Loeb also pointed out that the conditions of the four tasks fall into categories which describe the three servo-control systems of Merton, Granit, and Houk, outlined earlier, namely, a system primarily gamma-mediated, an alpha-gamma coactivation, and a system primarily alpha-controlled.

It is not clear whether impulses from muscle receptors contribute to kinesthetic perception. Muscle afferents do project to the cerebral cortex: they relay directly from the thalamus to the *motor* portion of the sensorimotor area where the effect may be facilitatory or inhibitory. An extensive and diffuse convergence occurs upon these cortical cells from muscles of different joints and from antagonists at the same joint. Since these cortical cells are not activated antidromically by stimulation of pyramidal tract fibers, they cannot be pyramidal cells. It is assumed that they are the cells which *activate* the pyramidal cells. Because the region of convergence rules out the possibility of a role in kinesthetic perception, it is suggested that this afferent projection provides the cortical motor area with information regarding the state or tone of muscles, and this information is used in adjustment of the voluntary descending signal. The spindle afferent projection would thus provide the ascending limb, and the pyramidal tract would provide the descending limb of a large servo-loop. Over this loop impulses would travel from motor cortex to muscles (probably via both alpha and gamma motor neurons) and back to the cortex from the spindle afferents. Although the information thus relayed would probably be crude owing to the diffuse nature of the convergence, nevertheless it would serve as an important feedback for the regulation of voluntary activity (Oscarsson (1966); Swett and Bourassa (1967)).

Of particular interest to students of motor learning should be the work of Buchwald and Eldred (1961) in which classic conditioning of the gamma efferent system was demonstrated in cats. The investigators found that gamma fibers developed a conditioned response, as shown by accelerated discharge, after 5 to 10 presentations of a conditioned stimulus (a normally ineffectual audible tone) with the unconditioned stimulus (an electric shock to the toe sufficient to induce flexion withdrawal). In a follow-up series of experiments (Buchwald et al. (1964)) further evidence was obtained suggesting that conditioning in the gamma system may play an essential role in the learning of new motor responses. Subsequent work of Hagbarth and Vallbo (1968) in unanesthetized humans also provide evidence for uniform and consistent patterning of spindle function.

Speculations and Questions on Spindle Innervation

Spindle afferents provide two routes for feedback to the central nervous system: movement information (phasic length changes in muscles) and position (static length). The two gamma systems can provide separate systems for activation of muscles. If the voluntary objective is movement, the central nervous system may use predominantly the dynamic gamma system, while if the objective is position, the outflow may use the static gamma system. Granit et al. (1955) demonstrated that the two systems are under separate supraspinal controls.

Matthews (1968) points out that alpha excitation alone evokes a load-dependent tension which is length independent and produces uncontrolled movement if alpha activation exceeds the load. (An example of this is found in the unnecessary speed and extent of movement which occurs when one picks up a very light object which was expected to be heavy.) On the other hand, gamma stimulation results in a cortically determined amount of shortening which is independent of load. Therefore, the central nervous system may control muscle length, both for position and during movement, through the gamma system.

The secondary afferents activated by the static gammas may provide a simple signal of misalignment between muscle length and gamma bias which is fed back to the central nervous system and which may be used to appraise the success of the voluntary muscular activity. The primary afferent would be useless for such appraisal because its phasic response is not related to length.

Basic control theory emphasizes the need for damping the tendency to oscillate in a servo-loop. This is important in the adjustment of muscle to sudden changes in position or load and in pendular movements which tend to overshoot and which if undamped would develop positive feedback. The dynamic response of the spindle probably serves to damp oscillation in the servo-loop through gamma control of spindle sensitivity to the phasic response. Both Matthews (1968) and Eldred (1967b) present good discussions of this subject. Damping not only reduces oscillation but minimizes jerkiness and smooths out contractions.

There are a number of questions which still need answers before the correlation of the two types of gamma fibers to the two types of endings on the two types of intrafusal fibers can be finally determined. We must first ascertain that the intrafusals are exclusively and separately innervated, each by a different type of gamma neuron. Second, there must be more evidence regarding differences in contractile properties, if any, in the intrafusals. There are some spindles which have both kinds of intrafusal fibers but no secondary endings, and yet they still evoke the two responses (phasic and tonic) in their primary afferents. This is not difficult to reconcile with the theory previously discussed that the two responses arise as a result of contrasting contractile characteristics in bag and chain fibers. Third, investigation is needed of the possibility that different types of motor endings may produce different types of contraction in the same intrafusal fibers. Fourth, if the primary ending responds to acceleration as well as to velocity of lengthening, this would add further complexity. At present this does not seem to be the case. Finally, the role of the collaterals from alpha motor neurons, the so-called beta innervations of the spindles, needs elucidation. It seems reasonable that alpha neurons should be concerned in the regulation of spindle activity and that the beta collaterals supply such a pathway.

Instances of Spindle Activity

Spindles serve to coordinate the activity of muscles through the entire course of a movement. As the movement progresses, feedback from spindles changes in precise relationship to the changing lengths of their muscles. Thus, contraction in any muscle is always appropriate to the conditions of the moment.

There are many instances in daily living in which proprioceptive reflexes show up inadvertently, isolated from the context of any complex skill pattern. Some familiar examples of spindle responses follow.

Jerk Responses. Almost everyone has experienced simple jerk responses (knee, ankle, elbow, wrist, etc.) evoked by external stretch applied as a sharp, light blow to a tendon. Although these responses are functional artifacts, they are nevertheless coordinated responses and as such involve reciprocal inhibition and synergic cooperation of the muscles in the functional group concerned. They are thus excellent examples of simple automatic intermuscular regulation, demonstrating that a single proprioceptive reflex not only activates the responding muscles but simultaneously assures coordinated cooperation of muscles without intervention of conscious nervous mechanisms. Muscle coordination in a jerk response appears to be accomplished entirely by feedback from the primary afferents of the stretched muscles. Since the limb is hanging relaxed before its tendon is struck, there is no gamma involvement other than the resting background discharge.

Reciprocal Inhibition. a. When a person standing with the knees slightly hyperextended is kicked lightly in the popliteal space (behind the knee), the stretch reflex thus evoked in the knee flexors excites them, at the same time inhibiting the knee extensors so that the knees buckle. Fortunately, the buckling stretches the extensor muscles which then contract and save the subject from falling. If, however, the buckling of knees is extreme and the stretch of the extensors becomes sufficient to activate the secondary afferents of the extensor spindles, extensor inhibition and flexor facilitation will result and the subject may actually fall.

b. Cramps in the leg muscles are readily relieved by strong, active contraction of antagonistic muscles, whose spindles will evoke reflex inhibition of the affected muscles.

Self-inhibition of Extensors by Their Secondary Afferents. a. Inhibition of extensors by their own secondaries may be a prime factor in causing the novice to crumple to the floor if the knees become extremely flexed on landing from the vault.

b. Therapeutic technique involving passive manipulation of limbs for the purpose of reducing muscle spasm in extensors often includes slow and sustained stretch of the affected muscles; it is possible that the threshold of the spindle secondaries is reached, which reflexly inhibits the extensors, hence the spasm, and restores joint flexibility to acceptable ranges.

Evidence of Anticipatory Settings of Gamma Bias. This may be seen when one reaches to push open a heavy door. Voluntary impulses set the gamma bias to hold spindles just below firing level. The touch of the fingers on the door will trigger the gamma neurons to deliver just the amount of cortically programmed outflow to the muscles which experience has shown to be required in such circumstances, and the muscles will respond with suitable force of contraction. Occasionally someone pulls the door from the opposite side just as one's fingers touch the surface. Since the contracting muscles do not meet the resistance anticipated, the force is converted into forward propulsion. At the least, one's balance is upset; at the worst, one goes flying through the door and collides with the person on the opposite side.

Pressure as a Spindle Stimulus. Pressure applied to the belly of a muscle has been shown to activate primary afferents. Advantage is unconsciously taken of this fact by older persons, especially when arising from a chair. Firm pressure on the ventral surface of the thighs increases the contractile force of the quadriceps muscles, markedly assisting in extension of the knees. Conversely, when sitting down, similar pressure on the quadriceps will aid in controlling the rate of knee flexion by facilitating the eccentric or lengthening contraction in these muscles.

Positive Supporting Reaction. It should be mentioned that muscle spindles are responsible, at least in part, for the positive supporting reaction. The weight of the body pressing upon the feet spreads the interphalangeal joints, stretching the interosseus muscles. Their spindles transmit signals which reach the extensor muscles of the limbs and induce a tonic contraction response, probably by means of gamma facilitation of extensor spindles. Evidence for the existence of this reflex is found in the situation that occurs when it is absent. For example, if one sits cross-legged on the floor for a long time, afferent nerves from the feet may become anesthetized as a result of circulatory deficiency. Upon standing, the lack of the feedback from the receptors in the feet results in loss of the positive supporting response and the legs flex under the body weight. The person stumbles and will fall unless visual feedback and/or the stretch reflexes evoked in leg extensors by the flexion come to the rescue.

Another instance illustrating the effect of the loss of the positive supporting reaction is supplied by an experiment carried out in the laboratory by the authors of the first edition of this text (O'Connell and Gardner (1972)). A swing seat was suspended from the ceiling with a pulley arrangement which enabled a subject seated in it to be raised and lowered. At some time during the lift the seat was suddenly tilted so that the subject slid off to land on a gymnasium mat on the floor. During control "spills" subjects had no difficulty regaining the erect stand. In a second set of lifts and spills the subject, blindfolded, was alternately raised and lowered several times and to varying distances until he lost an accurate sense of distance from the floor. When dropped, he regained equilibrium readily but with some slight delay as compared with the controls. Finally the subject's feet were immersed in ice water for a period (20 minutes) sufficient to produce local anesthesia in the feet. Immediately after immersion the subject was raised, lowered, raised, etc., and spilled as before. In this case, however, there was no extension of the lower limbs upon landing and he crumpled to the mat. It is assumed that chilling interfered with the sensory feedback essential for the positive supporting reaction and also for the extensor thrust reflex.

Golgi Tendon Organs

A second type of proprioceptor intimately incorporated into the gross muscle structure is the Golgi tendon organ. Unlike the spindle, its effect upon its own muscle is inhibitory. It consists of an encapsulation of tendon fibers (Fig. 10.13) located at the musculotendinous junction and hence lying in series with the contractile muscle fibers. Figure 10.14 provides a comparison of the locations of spindles and tendon organs in relation to the extrafusal contractile muscle fibers. While the muscle spindles are situated *in parallel* with the contractile fibers, the tendon organs lie *in series* with them. A tendon organ usually involves several to many muscle fiber tendons.

Golgi tendon organs may be excited by strong passive stretch but are much less sensitive than the muscle spindles. They are, however, highly sensitive to the stretch imposed upon them by the *contraction* of the muscle in which they lie. Whereas contraction of its own muscle tends to relieve the stretch on a spindle and hence to result in a decrease or cessation of its discharge, the Golgi tendon organ, because of its location in series with the contracting muscle fibers, is increasingly stimulated. Its discharge accelerates as contractile tension mounts (Fig. 10.15). Houk and Henneman (1967) found that some of the Golgi tendon organs in the cat soleus had absolute thresholds of less than 0.1 g and were caused to fire during contraction even in a shortened muscle where no tension was developed. In other words, these receptors were sensitive to internal forces developed in the muscle. Apparently the Golgi tendon organs continuously feed back to the spinal cord a filtered sample of the active forces operating in the muscle, information which is vital to coordinated response.

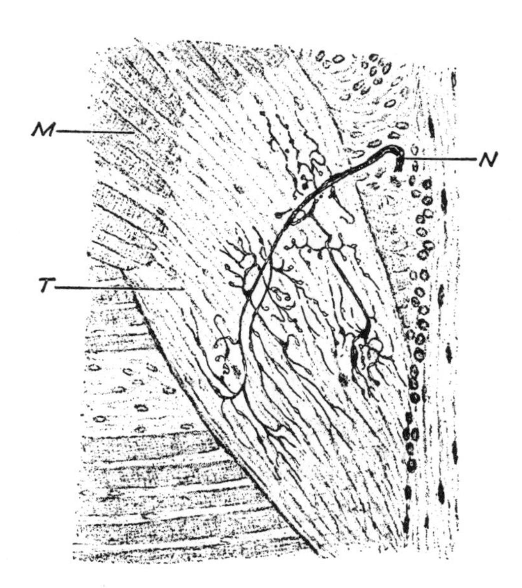

Figure 10.13. Drawing of a Golgi tendon organ. The Ib afferent neuron (N) supplies flower-spray endings lying among tendon fibers (T). Muscle fibers (M) attach to the tendon fibers. (From Truex, R. C., and Carpenter, M. B., 1969. *Human Neuroanatomy,* Ed. 6, Fig. 9—15, p. 187. Baltimore: The Williams & Wilkins Company.)

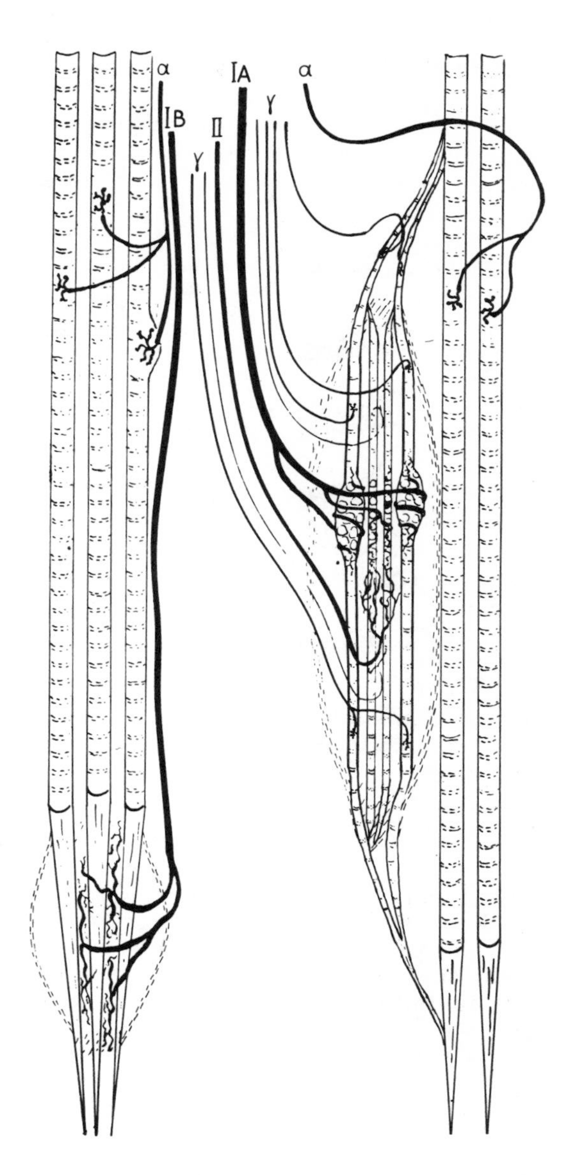

Figure 10.14. Diagrammatic drawing of muscle spindle and tendon organ. The muscle spindle is seen on the right attached to extrafusal muscle fibers and tendon. It consists of small diameter intrafusal muscle fibers which are largely enclosed in a connective tissue capsule. Longitudinally the drawing is not to scale (the length of a spindle may be 50 times its width). Transversely in the drawing the width of the extrafusal muscle fibers represents a diameter of 40 μ; the intrafusal fibers are drawn to the same scale and represent diameters of about 20 $\mu$for the two long fibers with nuclear bags at the equator of the spindle and about 10 μ for the two short fibers with nuclear chains at the equator. The group of nerve fibers shows the relative diameters of these fibers to each other. The largest fiber, IA, supplies the main primary afferent ending lying over the nuclear bags and chains. Fiber II goes to a secondary afferent ending on the nuclear chain fibers adjacent to the primary ending. Six small gamma motor fibers of varying sizes supply motor endings on the intrafusal muscle fibers. The motor endings on the extrafusal muscle fibers are supplied by larger alpha nerve fibers. The remaining IB nerve fiber goes to the encapsulated tendon organ on the left; the branches of the afferent nerve ending lie between the tendons of a group of extrafusal muscle fibers. (Drawing by Sybil Cooper. From Bell, G. H., Davidson, J. N., and Scarborough, H., 1969. *Textbook of Physiology and Biochemistry,* Fig. 40.4, p. 912. Baltimore: The Williams & Wilkins Company.)

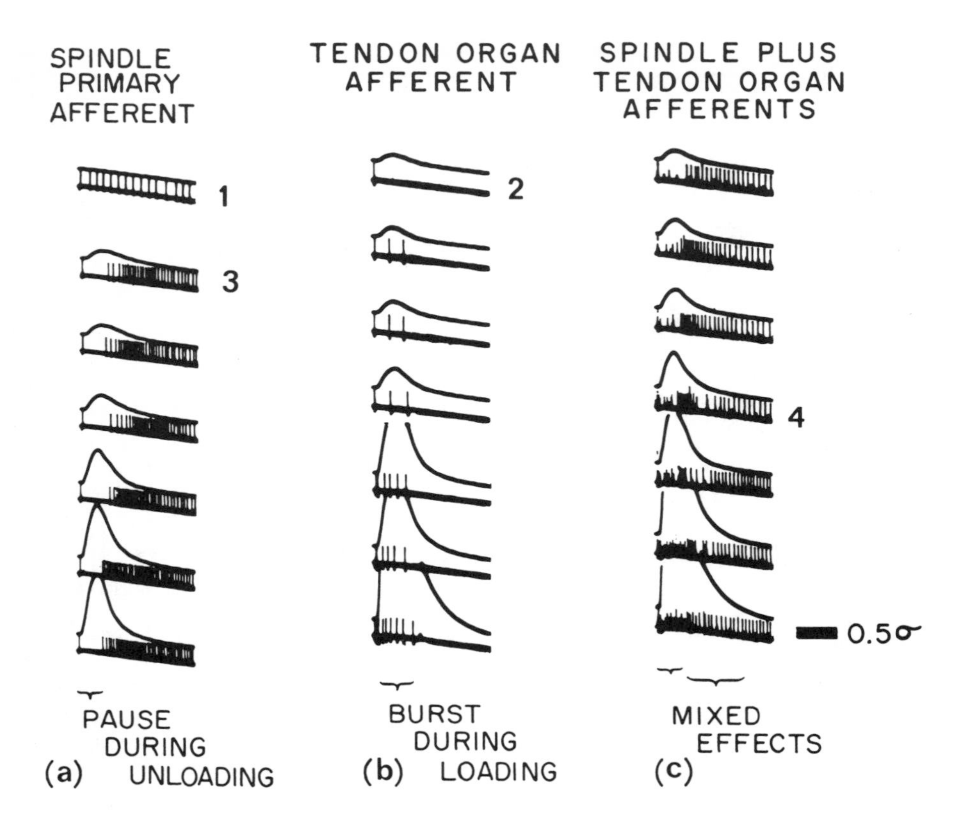

Figure 10.15. Afferent discharge responses from a spindle and a Golgi tendon organ during a graded twitch contraction. Muscle shortening is recorded by the tension trace. The application of the stimulus to the muscle is reflected in the single spike at the left of each record. (a), during contraction, the spindle primary afferent discharge ceases as the spindles are unloaded, resuming again during relaxation. 1, resting discharge; 3, note the phasic burst which is characteristic of the primary afferent during resumption of spindle firing and which may be followed by a secondary pause, 4. (b), as muscle contraction exerts a pull on the tendon organs, the tendon afferent shows a burst of activity which subsides and disappears during muscle relaxation. This is in direct contrast to the response of the spindle. The higher threshold of the tendon organ which is seen by comparing spindle and tendon organ traces, is also evidenced by the lower firing frequency of the latter. The tendon organ's high degree of adaptation is indicated by its failure to persist in firing throughout the twitch, even when tension was increased. (c), a simultaneous record from spindle and tendon afferents reflects a mixture of the responses from the "in parallel" and "in series" receptors. (From Eldred, E., 1967. Peripheral receptors: Their excitation and relation to reflex patterns. *Am. J. Phys. Med. 46:* Fig. 15, p. 84.)

The tendon receptors are supplied by large group I afferents similar in size and conduction velocity to the primary spindle afferents (Fig. 10.14). Tendon afferents may be identified in spinal nerves, however, by two response characteristics which distinguish them from the spindle afferents: (1) they are *not* unloaded by contraction of the muscle, and (2) they are unaffected by gamma stimulation. Spindle afferents have been designated as Ia and tendon afferents as Ib neurons.[b] While a single Ib neuron may supply more than one Golgi tendon organ, more usually each is restricted to a single receptor, the average being 1.2 tendon organs per Ib afferent.

Afferents of tendon organs enter the cord and make di- or polysynaptic connections with alpha, and probably gamma, efferent neurons to their own muscle and to synergists and antagonists. Their effects are opposite to those of spindle primary afferents: Golgi tendon organs inhibit their own muscles (**autogenic inhibition**) and synergists and facilitate their antagonists. It has been suggested that Golgi afferents may exert their inhibitory effects by means of presynaptic inhibition of the primary spindle afferents at their synapses on the anterior horn cells. This idea is not completely accepted.

If, as seems likely, all contractions activate tendon organs, their inhibitory effect must be offset during voluntary movement, either by voluntary exertion in excess of the opposing resistance or by some built-in reflex mechanism. Some evidence exists that during rapid voluntary movements in human the interneurons in the autogenic inhibitory circuits from the tendon organs to their own muscles are inhibited, thus canceling the inhibitory effect. Hufschmidt (1966) found that a stimulus sufficient to excite both Ia and Ib afferents produced only facilitation, indicating that the inhibitory action of the Ib neurons had been removed. Muscles have been shown to be capable of greater contractile force than their own structural makeup can withstand. Therefore it may be that the inhibitory reflex initiated by the tendon afferents is a safety measure to protect the muscle from dangerously high tensions. Loofbourrow (1960) points out that it is logical that the action of Golgi tendon organs inhibits not only the alpha motor neurons but the gamma system as well.

Four examples of Golgi tendon organ reflexes may be cited.

1. The immediate relaxation of muscles when volition ceases is probably due to the fact that the full inhibitory effect of the Golgi tendon organs is exerted as soon as their inhibition, which is associated with the voluntary movement, is no longer operative.
2. In Indian arm wrestling the loss of the contest usually occurs abruptly, when mounting tendon organ inhibition finally overcomes the voluntary effort to maintain contraction.
3. The "breaking point" in muscle strength testing is probably due to tendon organs; if this is so, it suggests that maximal strength is dependent upon the individual's ability to voluntarily oppose the inhibition of his own tendon organs.
4. In patients with upper motor neuron lesions, increased muscle tone with passive stretching is common. This increased tone is attributed to the neuromuscular spindle facilitation. As the passive stretching force is increased, the muscle tone also increases. Suddenly, there is a reduction in the muscle tone, presumably because the threshold of the Golgi tendon organ is reached.

The phenomenon is sometimes called the clasp-knife reflex.

Vater-Pacinian Corpuscles

The Pacinian corpuscles located in muscle have been identified by Gray and Matthews (1951) and Hunt (1961). Pacinian corpuscles are widely distributed throughout the body and are usually

[b] The symbols 1A and 1B are also used for designation of the afferents to spindles and tendon organs. The authors consider the Ia and Ib designation as more applicable: "I" indicates the size and conduction characteristics of the neurons (Lloyd classification of sensory neurons), and "a" and "b" serve simply to distinguish large afferents from two specific receptors, obviating any erroneous association with the A and B groups of the Erlanger-Gasser classification.

associated with joint receptors (where they will be discussed more fully). The corpuscles located in muscle respond to vibration and pressure (Powers (1976). In response to steady pressure, they respond with a rapidly adapting discharge; if they are stimulated in vibratory fashion, they apparently respond by following the stimulus frequency (as evidenced by the action potentials recorded from large myelinated afferent fibers). It would appear that the Pacinian corpuscles supply information to the central nervous system concerning the contractile state of muscle and can, therefore, contribute to the reflex control of movement (Powers (1976).

Involvement in Motor Skills

Spindle and Golgi Tendon Organ Reflexes

Spindle and tendon organ reflexes can be recognized contributing their effects to facilitation, reinforcement, or inhibition of muscle contraction in almost any movement. The advantages which accrue from spindle activity by putting a muscle on stretch before its activation are obvious. There are many skills in which preparatory movements are made in a direction opposite to the anticipated functional movement. Common examples are backswings and crouches. These put stretch upon the muscles about to be used. Contractile capacity of the muscle tissue is increased by the well-known length-tension phenomenon; however, more importantly, the external stretch increases spindle activity, reinforcing the resting background of gamma discharge. As a result, impulse outflow over afferent neurons to the motor pools of the functional or prime mover muscles is increased. The speed and extent of the preparatory movement determine, through the phasic and tonic response frequencies of spindle primary afferents, the amount of involuntary discharge to the muscle. When the voluntary command is given, impulses descending by way of supraspinal pathways are added to this background; impulse frequency rises and a stronger contraction results. (Of course, other physical and mechanical advantages are involved as well.)

In sports which involve a single alternating movement, the nature of the backswing should have certain characteristics consistent with the main objective of the forward swing, whether to achieve force or to obtain precision. When the objective is force, spindle function suggests (1) that the backswing should be fast to gain the advantage of the phasic increase in spindle discharge; (2) that there should be minimal hesitation between backswing and forward drive movements to capitalize on the maximal frequency attained as a result of the velocity of the stretch; and (3) that the swing should be as long as practical to increase the tonic response of the spindles. In anticipation of the forceful hit, an increased outflow of voluntary impulses may be expected to occur over fusimotor neurons, enhancing spindle sensitivity to both phasic and tonic stretch. Consequently, as the muscle is rapidly stretched in the backswing, spindle frequency will rise sharply. The faster the backswing, the higher the frequency of impulses transmitted to the muscles and the stronger the force of the forward driving movement. This principle finds application in the long shots of golf, assuming, of course, that a good swing has already become standardized by practice.

If the primary objective of the forward movement is something other than force, such as accuracy in golf approach shots (also tennis placements, billiards, etc.), spindle physiology would recommend that the backswing should be relatively slower and shorter, with velocity and length both carefully tailored to produce only that amount of force consistent with the distance to the green and the club being used.

In target throwing or shooting, a short hesitation at the end of a preliminary movement of optimal length should be advisable to allow the phasic frequency of spindle afferents to subside to the tonic level. Coincidentally, it would permit time for discriminative last-minute adjustments to be made for accuracy, distance, and direction. It should be mentioned, however, that readjustment can be faulty and too much cerebration may hinder rather than help.

A centipede was happy, quite,
Until a frog in fun said
"Pray which leg comes after which?"
Which raised her mind to such a pitch
She lay distracted in a ditch
Considering how to run!

In tennis and other racket games the player intuitively times his backswing with the ball's approach so that the ballistic movements of back and forward swings become a smooth continuum. He may thus take advantage of the phasic stretch effect to achieve maximal force in his drive or smash. When he wants to diminish the force of his return, as when playing at the net, he may hold his racket in a static "ready" position and allow the simple stretch reflex to drain most but not all of the force from the oncoming ball and drop it just over the net.

In instances in which force is the main factor but a preparatory position must be held to await an outside event, the advantage of the stretch response is still available. In the case of the baseball batter awaiting a pitch, he can seek the advantages of the tonic spindle response to supplement his own muscle power by taking as long a backswing as possible within the limits of mechanical efficiency. As a basketball player awaits a tie ball toss-up, he should take as deep a crouch as is consistent with the strength of his antigravity muscles and the mechanical factors and force vectors involved.

Golgi tendon organs may be used to contribute to increased strength of contraction in their antagonists during alternating movements. For example, the greater the contraction of the backswinging extensor muscles, the greater will be the activation of their tendon organs, whose effect upon the forward-swinging flexor muscles is one of facilitation. The tendon organ facilitation added to that of the stretch effects discussed above results in an increased strength of flexor contraction as the movement reverses.

For beginners, follow-through in a swing often requires specific voluntary effort to offset the inhibitory influence which the Golgi tendon organs impose upon their own shoulder flexor muscles, especially as the forward swing proceeds. Furthermore, this autogenic flexor inhibition is being simultaneously reinforced by inhibition from the spindles in the stretched extensor muscles. This is probably why many students must be taught to extend the swing voluntarily beyond the impact point, until with practice follow-through finally becomes part of the learned motor pattern.

The detrimental effects of applying force too early in a striking movement may be related to the fact that the tendon organs become activated at a point where their effects interfere with the continuation of contraction. When the ball is finally struck, inhibition has already reduced the forcefulness of the swing.

In sedentary individuals, in whom muscle weakness and imbalance are of common occurrence, spindle and tendon organ reflexes may be used for corrective purposes. By invoking maximal contraction in the lengthened or weak muscles, reciprocal inhibition may be imposed upon their shortened and too-strong antagonists. Maintaining the contraction in the weakened muscles for several minutes will cause gamma bias to be reset, increasing the sensitivity of their spindles and improving resting muscle tone (Stockmeyer (1967)). Diminished proprioceptive feedback is characteristic of sedentary living and results in a reduction of the sensitivity of spindle secondary afferents in antigravity muscles. Without the stabilizing effect of cocontraction, the joints are unprotected and the jars of walking and exercise may produce discomfort and pain. Corrective procedures consist of exercising the extensors against resistance to increase their spindle sensitivity and to improve joint stability.

Vater-Pacinian Corpuscle Reflexes

"The laying on of hands," performed by the therapist or medical doctor for a specific diagnostic or rehabilitative purpose, should be done so that every effort is made to facilitate the intended movement without introducing unwanted tensions. If facilitation of muscle is intended, then, pressure on affected muscle or vibratory techniques sufficient to reach the threshold of the Pacinian bodies in muscle, could be used. If, however, one is testing for relaxed muscles, then every attempt should be made to place hands so that reflex facilitation from the Pacinian corpuscles does not interfere with the subject's effort or intent to relax muscles.

Summary of Muscle Proprioceptors

Eldred (1965) summarizes the importance of the muscle reflexes by saying that many of the finer attributes of sensitivity, restraint, and coordination which distinguish the activity of muscles must be derived from the feedback of data on the moment-to-moment status of muscle, much of which is provided by the three afferent transducers: spindle primary, spindle secondary, and Golgi tendon afferents. She further states, "ultimate understanding of the composition of movement must take into account the interplay by triple sensory influence, dual motor-sensory modulation, and the reaction of the alpha motor neuron final common pathway."

JOINT AND CUTANEOUS RECEPTORS

Joint Receptors

Another group of reflexes which play upon both alpha and gamma neurons to modify muscle activity are initiated by receptors in the joints and in the skin. Most joint receptors resemble either Ruffini spray receptors or small Pacinian corpuscles, although several size and shape variations are identifiable. Receptors consist of nerve endings, straight or branching, that are usually enclosed in a connective tissue capsule (Fig. 10.16). They are found in or near ligaments, joint capsules and adjacent connective tissues, and muscle fascia. Some, which appear to be acceleration detectors, are highly sensitive to movement, adapting quickly when movement ceases. Paciniform corpuscles in the neighborhood of tendons and joints also respond to deformation by firm pressure.

Static joint position produces firing in slowly adapting receptors which are insensitive to movement but which discharge in proportion to joint angle. Among these are flower-spray organs, known as Ruffini endings, found widely dispersed throughout joint capsules and able to maintain firing rates for as long as 90 minutes without decline. Their sensitivity is such that a change of 2° in joint angle is sufficient to alter rate of discharge. In any given position of the joint, some sensors are under intensive stimulation, some are less stimulated, and others are unstimulated (Fig. 10.17). Impulse frequencies and discharge patterns vary accordingly. Sensitivities of joint receptors are such that there is no position in which all joint receptors are silent. The integrative significance of such specificity in sensory coding is obvious. Receptors in joints have structures similar to those found in other tissues. In order to avoid drawing erroneous analogies, Freeman and Wyke (1967) proposed a grouping of joint receptors according to their functional properties. Table 10.1 summarizes the characteristics of joint and cutaneous receptors.

Joint receptors, especially those of the interphalangeal joints of the feet, contribute to the positive supporting reflexes, along with the spindles of the interosseus muscles. Apparently the complex combination of stimuli entering the cord facilitates appropriate stretch reflexes of extensor muscles to convert the limb into a firm but compliant pillar. If the phalanges are squeezed together or flexed instead of being abducted, all joints of the limb tend to flex.

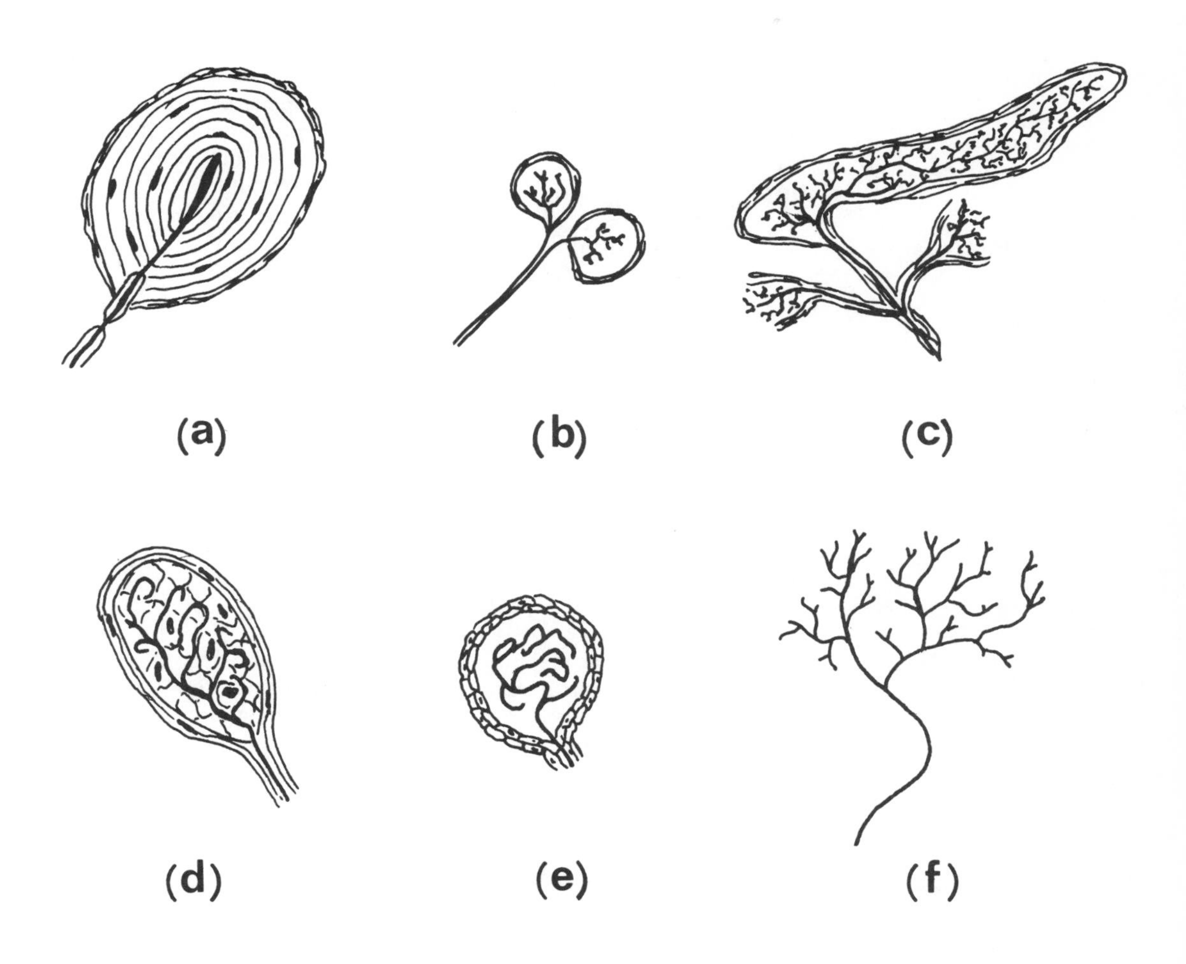

Figure 10.16. Receptors of joints and skin. Many of these receptors play a dual role as both exteroceptors and proprioceptors. (a), Pacinian corpuscle. A lamellated capsule enclosing an unbranched ending of a group I sensory neuron. These are fast-adapting receptors found near joints and tendons and in the deeper parts of the dermis; they are sensitive to quick movement, vibration, and pressure, depending in part on location. Smaller paciniform corpuscles are also present around joints. (b) Golgi-Mazzoni corpuscle. These are nerve sprays commonly situated near joint capsules, each with its own afferent neuron. They are pressure sensitive. (c), Ruffini end organs. These are slowly adapting flower-spray receptors in joint capsules and in the middle layers of the dermis. They signal joint position and may be sensitive to warmth. They are served by group II neurons. (d), Meissner's corpuscle. An encapsulated nerve ending located just under the epidermis; sensitive to touch. Touch stimuli affect skin areas of variable size and may involve many end organs. The pattern of input thus evoked provides the central nervous system with coded information for discriminative and integrative processing. (e), Krause's end bulb. Situated in the outer regions of the dermis, this may be sensitive to cooling. (f), free nerve endings. Widely distributed through the outer areas of the dermis and around joints, these probably contribute to pain sensation and to flexion reflexes.

Table 10.1
Joint and Cutaneous Receptors[a]

Type	Resemblance	J or C[b]	Description	Location	Adaptability	Type of stimulus	Afferent Neuron
I	Ruffini	J	Flower-spray within ovoid corpuscle with thin connective tissue	Fibrous joint capsules	Fast & slow	Direction & speed of movement; joint position; intra-articular pressure change	II
	Ruffini	C	" "	Middle layers of dermis	Fast & slow	Warmth?	II
II	Pacinian	J	Single terminal within a thick, laminated capsule; small	Near tendons and joints	Fast	Accelerated movement, vibration, deep pressure	II
	Pacinian	C	Larger	Deep layers of dermis, ligaments, muscle	Fast	" "	I or II
III	Golgi-type	J	Large and fibrous; spindle-shaped (looks like G.T.O.[c])	Instrinsic ligaments of joint capsules	Slow	Position of joints	II
	Golgi-mazzoni	J	Laminated capsule; looks like Pacinian but smaller with contained fibers more extensively branched	Loose, pericapsular tissues	Fast	Rapid, accelerated movement	II
	Golgi-mazzoni	C	" "	Subcutaneous tissue and on tendons	Fast	Deep pressure and vibration	II or III
	Meissner	C	Oblong, encapsulated corpuscle	Dermal papillae	Fast	Touch and two-point discrimination	II or III
	End bulb of Krause	C	Bulbous, connective tissue capsule	Outer region of dermis	?	Cooling?	III
IV	"Free" nerve endings	J	Fine, branching unmyelineated fibers	Fibrous joint capsules	Slow	Pain	II
	"Free" nerve endings	C	" "	Outer region of dermis	Slow	Pain, temperature, diffuse touch	II

[a]Constructed from material provided by Freeman and Wyke (1967) and Werner (1980).
[b]J, Joint; C, cutaneous.
[c]G.T.O., Golgi tendon organ

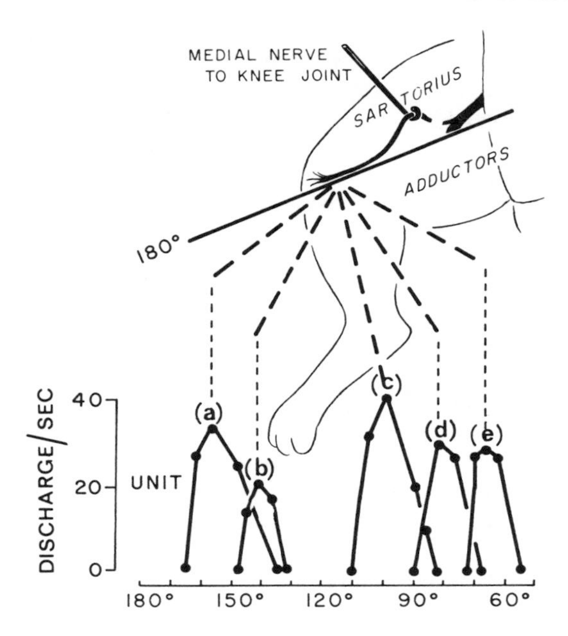

Figure 10.17. Range fractionation in the signaling of joint position. Discharge curves are shown for five joint receptor units isolated from the medial nerve to the knee joint of a cat. Each unit fired only over a restricted range of joint angles with a maximum at one fairly sharply defined point. Discharge was constant at a stable position of the joint but became elevated during movement within the response range for that unit. (From Eldred, E., 1967. Peripheral receptors: Their excitation and relation to reflex patterns. *Am. J. Phys. Med. 46:* Fig. 16, p. 85.)

When any one of the three joints of a limb is flexed, the two joints also flex. When one is extended, the others also extend. Compression of a joint, as by weight bearing, reflexively evokes contraction of extensor muscles, while traction, as in hand suspension, excites flexors. These responses are mediated by joint receptors.

Quadriceps femoris muscle inhibition in subjects with knee joint swelling has been consistently observed by many investigators (DeAndrade et al. (1965), Jayson and Dixon (1970), and Spencer et al. (1984)). This inhibition has been attributed to stimulation of the Ruffini-type joint receptor. Further, this inhibition appears to be greater in the close-packed joint position (this position which allows minimal joint volume) rather than the open-packed position (the position which allows maximal joint volume) (Stratford (1982)). This latter finding may be of importance to those persons faced with the challenge of strengthening (or at least, maintaining) the quadriceps femoris muscle in subjects with swollen knee joints. It is postulated that, in the acute stage, isometric exercises performed at approximately 30° of knee flexion (the open-packed position) may be more effective than the traditional straight leg-raising or isometric contractions performed at zero° of knee flexion (the close-packed position).

Since the afferent neurons of joint receptors interconnect in the spinal cord with interneurons of broad distribution, their influence is more diffuse than that of the muscle proprioceptors. They are especially effective in modifying activity in contralateral muscles. Their effects vary with the existing state of the joint. In other words, if the joint is in flexion, a given stimulus will produce a different pattern of response than if the joint happens to be in extension.

Interlimb joint receptors evoke responses of mutual facilitation. Sway to one side unbalances the force couples at the proximal joints, decreases the angle on the side of the sway, and increases the angle contralaterally. These changes evoke corrective action in the abductor and adductor muscles of the hip and the inverters and everters of the foot to neutralize the force couples. If, however, the sway exceeds the corrective capacity of the limb responses, equilibrium reflexes are elicited and the animal steps or hops sideward to realign center of gravity and base of support.

Signals from receptors in the joints between individual vertebrae and between the vertebral column, pelvis, and shoulder are coordinated to bring about cooperative muscle action to support the body. In combination with signals from limb joint receptors, they contribute importantly to the forces necessary to support the body weight and distribute it properly among the limbs.

The forelimbs act powerfully on the hind limbs (e.g., clenching the fists to jump) through both spinal and supraspinal circuits and both ipsilaterally and contralaterally. In contrast, the hind limbs are less effective in their action on the forelimbs; they appear to act only through supraspinal routes and to induce mainly ipsilateral responses.

Manipulation of ipsilateral joints facilitates labyrinthine responses and contributes increased stability and strength to the corresponding limb during standing and walking. Simultaneous work executed by a contralateral limb facilitates and improves performance.

Impulses arising in joint receptors are projected by multisynaptic pathways to the sensorimotor cortex, and the information thus provided appears to be the major factor in kinesthetic sensation.

Cutaneous Receptors

Most sense organs of the skin consist of a lamellated connective tissue capsule surrounding a soft cellular core in which the ending of the nerve axon is embedded. They vary somewhat in the complexity of this general design. Those responding to touch, light touch, pressure, or pain function both as somesthetic exteroceptors and as proprioceptors. In the former capacity they contribute to sensation by transmitting information to the thalamus and sensory cortex. In the latter capacity they initiate basic reflexes and contribute to the body-righting reflexes. Their signals combine with those from receptors of joints and muscles to coordinate body movements. Cutaneous reflex pathways include both spinal and supraspinal loops and, at least in some instances and to some extent, responses require an intact cerebral cortex.

Cutaneous receptors initiate many of the fundamental inborn reflexes, such as the following.

Extensor Thrust Reflex. A strong but transient extension of the limb follows touch stimulation of the sole of the foot (or palm of the hand). This reflex may play an important role in running. It is a spinal reflex and can be elicited in the hind limb of a dog with a completely transected spinal cord.

Withdrawal or Flexion Reflex. Noxious stimulation to any part of the skin evokes withdrawal of the part from the stimulating agent, usually involving flexion.

Magnet Reaction. If a timulus is moved across the sole of the foot, reflex movements are made so that the foot follows the stimulus, maintaining contact with it.

Grasp Reflex. Stimulus to the palm of the hand is followed by grasping of the stimulating object.

Placing Reactions. When a foot is placed upon an uneven receiving surface, corrective movements occur to position the foot so that it is ready to support the body weight (Fig. 10.18). These responses are considered "reflex" because of their stereotyped nature, but there is some evidence that they are learned rather than inborn.

Stimulation of the Skin. Pinching or squeezing the skin of the back produces generalized inhibition of limb extension. Stimulation of the skin of the abdomen facilitates extension of the limbs.

The location of a receptor is important in determining what muscles are activated or inhibited, and the most important factor determining direction of reflex movement is the location of skin stimulation (Sherrington's "**local sign**"). The significance of receptor locus is evident in the fact that the response tends to do something about the stimulus which aroused it. The dog's scratch reflex is an example. Hagbarth (1952) demonstrated in the cat that stimulation of the skin over an extensor muscle produced facilitation of extensors with inhibition of ipsilateral flexors, while stimulation of most other skin areas produced the opposite effect.

Impulses arising in pressure receptors as a result of contact with a surface play an

Figure 10.18. Placing reactions in a puppy and a kitten. (a), the puppy reaches for the table with his left front foot. (b), the kitten extends both front legs to place them on the table surface. (c), the puppy extends all four legs ready for weight bearing.

important part in the orientation of the body in space, evoking the body-on-head and the body-on-body components of the righting reflexes (discussed in the next section). Touch receptors also facilitate muscular adjustments for the maintenance of equilibrium, especially in subjects who are in poor physical condition. In the O'Connell-Gardner laboratory some individuals, if permitted to touch a piece of vertically hanging paper, could maintain a one-foot stance significantly longer than without touch stimulus. In fact, some blindfolded subjects stood longer when permitted the touch stimulus than they did with eyes uncovered but lacking touch. Slightly tipsy persons use touch, without taking any real support from the contact, to reinforce their depressed equilibrium reflexes. Perhaps the touch stimuli act to damp oscillation in postural correction and to avoid positive feedback.

Both skin and joint receptors exert a broad influence, acting in the interregulation of muscles at joints other than the joint acted upon by the prime mover muscles and including contralateral joints. Their afferents affect the outflow to muscles via both alpha and gamma routes, the gamma neurons having lower thresholds than the alpha neurons. Holmqvist (1961), however, found the contralateral actions by joint neurons to be stronger than those of either skin or muscle afferents. Since facilitation of both flexion and extension was common, she suggested that there must exist alternative pathways by which one afferent fiber can act upon a contralateral motor pool. Other studies support the complexity of reflex interconnections and the fact that a muscle's response is dependent upon its state at the time of stimulation, i.e., upon the sum total of the afferent influences converging upon it as a result of position and movement in any and all parts of the body.

Involvement in Motor Skills

Joint and Cutaneous Reflexes

Proprioceptive influences from joint receptors are so integral a part of movement that it is not easy to isolate specific examples of their exclusive operation. Facilitative feedback by irradiation may be used to increase the strength of contraction, as in clenching the fists to jump high or gritting the teeth while striking or lifting forcefully. Strong contraction in another part of the body has been shown to increase impulse outflow over subconscious channels to the motor synapses of prime mover muscles. Hellebrandt and Waterland (1962), who studied the phenomenon of irradiation extensively, said, "irradiation is probably a normal concomitant of pushing volitional effort into unfamiliar zones of activity and probably the excessive innervation thus aroused is just as necessary for the development of motor skill as overloading is for the development of strength."

The use of starting positions to direct impulse flow is common. Most starting positions place joints in the position opposite to that of the upcoming movement. This not only puts the muscles on stretch but also favors irradiation of impulses to the proper muscles.

Since feedback from joint receptors is important in the coordination of arms with legs, instruction may be given to foster limb reciprocity in takeoff for apparatus vaulting, diving, trampoline, etc. The limbs on one side of the body may be used to influence contralateral limbs. Irradiation from joints is useful in augmenting the flow of impulses to weak muscles in the poorly skilled.

Feedback from joints may contribute consciously as well as unconsciously to the guidance and timing of movements in patterned sequences. Gymnasts rely strongly on kinesthetic perception for assuring good form. The diver times his maneuvers by the "feel" of body position in the air. Mirrors permit the dancer to see herself in action. By visual feedback she adjusts limb and trunk movements to her satisfaction. Joint receptors feed back information on the desired angulation and movement, and this kinesthetic image is eventually impressed on her memory. As a result, she can later repeat the same positions and movements without the visual stimuli. Perfecting form in any skill depends strongly on feedback from joint receptors.

Cutaneous stimuli are often used to evoke specific parts of a movement pattern. Stimuli are supplied by contact with the environment (floor, trampoline, diving board, water, horse, box, rings, etc.); contact with a sports implement (racket, golf club, ski, ball, bat, etc.); and contact between parts of the body. Such stimuli inform the nervous system of the instant of impact (in tumbling, apparatus, vaulting, diving, dancing, trampolining, swimming, racket games, golf, baseball, football, etc.); of the changing contact with surface or implement (in skiing, skating, diving, golf, swimming, etc.); and the chance or deliberate contact between the body parts in all activities. These feedbacks are employed consciously or unconsciously (depending upon the learning stage) to time the abrupt changes in a movement sequence and to achieve the smooth transitions in force and direction which are basic components of a skilled performance.

Knowledge of certain facts regarding skin stimulation should be kept in mind when assisting a gymnast to perform. Strong stimulation of the skin over a muscle is excitatory to the muscle, but stimulation over a nonmuscular area excites extensors only. (Pressure on the belly of a muscle also activates its spindles.) Therefore the spotter should be careful to place his hands so that he does not elicit undesirable reflex response but, if possible, reinforces the desirable muscle action. When extension of the arms is essential (parallel bars, horse, box, etc.) avoid pressure upon flexor muscles and, if feasible, place the hands upon extensors or upon bony areas so as to reinforce extension. When flexion is required, the opposite hand placement is recommended.

LABYRINTHINE AND NECK RECEPTORS

The fact that the head exerts important influences upon movements of the trunk and limbs can hardly be overlooked by anyone who deals professionally with muscular activity. These orienting and reinforcing effects are mediated by two groups of proprioceptors: the labyrinthine and the neck receptors.

Labyrinthine Receptors

Labyrinthine receptors are specialized proprioceptors located in the inner ear (Fig. 10.19). They consist of complex arrangements of cells with projecting hairs whose orientation is displaced either by the pull of stony concretions or the movement of endolymph. They include the semicircular canals, which are stimulated by angular acceleration, and the utricles and saccules, which are stimulated by linear acceleration and by changes in the orientation of the head in relation to gravity. The semicircular canals serve a major function in the maintenance of equilibrium during movement, while the utricles and saccules appear to be concerned primarily with postural reflexes and in the differential distribution of muscle tone.

Utricles and Saccules

In the early life of the child, reflexes arising from the utricular and saccular receptors are present in primitive and highly predictable form known as the **tonic labyrinthine reflexes**, or TLRs. Stimulation induced by body position or head inclination produces stereotyped effects (Fig. 10.20). The supine position or corresponding orientation of the head with gravity facilitates extension in all limbs and inhibits flexion, while the prone position or corresponding head orientation produces the opposite response. Side lying or an equivalent head position induces extensor facilitation of the under limbs and flexor facilitation of the top limbs with reciprocal inhibition of antagonists. As growth and development proceed, the TLRs are supplanted by more complex reflexes, the **labyrinthine righting reflexes**, whose purpose seems to be to orient the head correctly with gravity. This is accomplished mainly by contraction of the neck muscles, although secondary influences are also exerted on trunk and limb muscles to supplement the raising of the head. These labyrinthine reflexes cooperate with other reflexes stemming from stimulation of skin, neck, and visual receptors to constitute, as a group, the **righting reflexes**, whose normal operation is essential in the child's achievement of vertical posture. The righting reflexes (Fig. 10.21) include the following components.

Labyrinthine Righting Reflexes. Stimulation of the labyrinthine receptors evokes contractions of neck muscles to orient the head in relationship to gravitational force (Fig. 10.22(a) and (b)).

Body-on-Head Reflexes. Asymmetrical stimulation of skin receptors resulting from differential contact with the supporting surface (side or back lying) results in activity of trunk and limb muscles which raises the head into an upright position. If skin stimulation is equalized by placing a board on the upper side of the animal, no righting occurs.

Neck Righting Reflexes. Impulses arising in joint receptors in the neck produce contraction of limb and body muscles which align the body with the head (Fig. 10.22(b) and (c) and Fig. 10.24, (a) and (b)).

Body-on-Body Righting Reflexes. Asymmetrical stimulation of skin receptors causes contraction of trunk muscles, which raise the body toward the upright position.

Visual Righting Reflexes. Visual feedback is used to orient the head and body correctly with the environment and is especially important when other sensory input is deficient or excluded.

Righting reflexes are later dominated by the **equilibrium reflexes.**

Semicircular Canals

The equilibrium reflexes are evoked by stimulation of the semicircular canals. Appropriate muscular responses maintain or regain body balance either by redistributing body segments to keep the center of gravity over the base of support or by

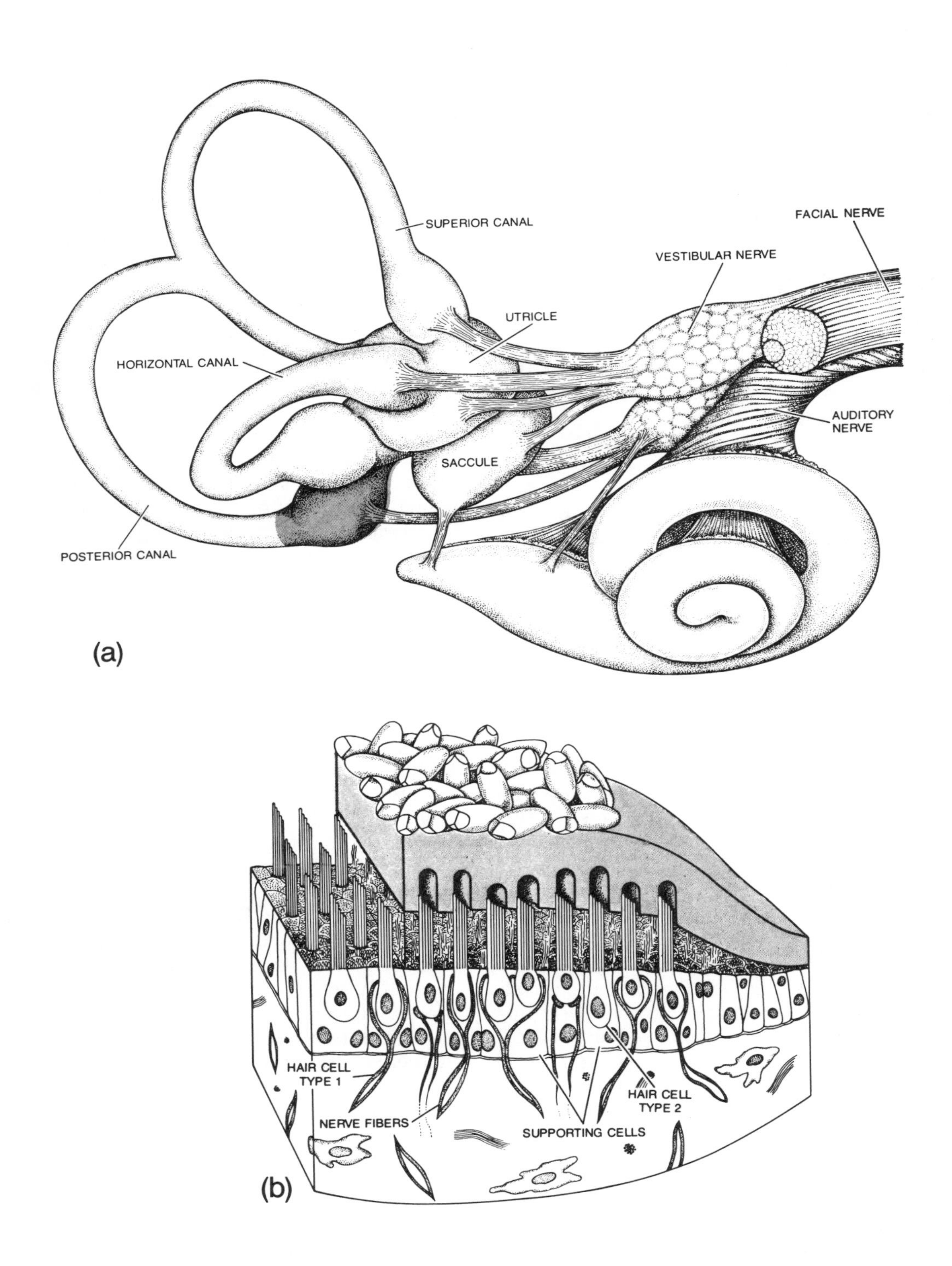
SUPERIOR CANAL
FACIAL NERVE
VESTIBULAR NERVE
UTRICLE
HORIZONTAL CANAL
AUDITORY NERVE
SACCULE
POSTERIOR CANAL
(a)
HAIR CELL TYPE 1
HAIR CELL TYPE 2
NERVE FIBERS
SUPPORTING CELLS
(b)

Figure 10.19

shifting the base of support to keep it under the center of gravity.

Movements of endolymph during angular or rotatory motions of the head stimulate the ampullae of the canals which are situated in line with the movement. Impulses traverse pathways to antigravity muscles, particularly the extensors of the limbs and neck, to evoke alterations of tone in these muscles and to produce movements which oppose the angular or rotatory acceleration of the head.

In the following account, we shall ignore the coincidental utricular responses of the neck muscles for the sake of simplicity. The reflex responses to a diagonally forward tilt are induced by the stimulation of ampullae of the ipsilateral anterior and contralateral posterior vertical canals. These responses consist of increased extensor tone on the same side, accompanied by decreased extensor tone on the opposite side. Thus in a four-legged animal tipped toward its right front leg, that leg extends strongly while the left hind leg partially flexes.

Lateral tilting of the supporting surface excites ampullae of both ipsilateral vertical canals, anterior and posterior, and produces extension of the ipsilateral limbs with concurrent reduction of extension contralaterally so that the body retains its original position in space. A straight forward or backward tilt excites the ampullae of both corresponding anterior vertical or posterior vertical canals, respectively, and produces suitable changes in extensor tonus.

Whirling about a vertical axis involves both horizontal canals and induces changes in all four limbs, which act to oppose the acceleration of the head and to keep the head stationary in space. The acceleration evokes extension of the limbs on the side of the direction of rotation; i.e., rotation to the left is accompanied by extension of the limbs on the left side. If the rotation is suddenly stopped (deceleration), inertia reverses the flow of the endolymph, producing reactions directed opposite to those during the rotary acceleration. Hence the post-rotatory movements seen in past-pointing and in staggering occur in the direction of the preceding rotation, indicating that the deceleration has evoked extension in the limbs of the side opposite to the rotatory direction (the right side in our example).

The acceleratory reflexes provide trunk and limb movement which will have a velocity-damping effect upon the rotational movements of the head. The objective seems to be to maintain a constant orientation of the head in space. However, because movement of the head upon the neck is restricted by anatomical limitations, a nystagmic jerk to return the head to its central position occurs when the limit is reached. Jerks are noticeable in the dancer's pirouette. Their absence in the

Figure 10.19 (a). vestibular apparatus consists of a series of fluid-filled sacs and ducts. In this drawing of the human vestibular apparatus, the three semicircular canals are at the left; clockwise from the top they are the superior, the horizontal and the posterior canal. They are oriented in the three dimensions of space and respond to angular accelerations of the head. In the center of the drawing are the two otolith receptors: the utricule (top) and the saccule. The fluid known as endolymph fills the apparatus; in the semicircular canals the endolymph functions as an inertial mass analogous to the otoconia crystals in the otolith receptors. Each semicircular canal has a bulge, the ampulla, one of which is shaded gray. At the lower right in the drawing is the cochlea. (b), otolith receptor has bundles of hairs that project into a gelatinous membrane. The kinocilium, the longest hair in each bundle, is attached to the side of an opening in the membrane; the shorter stereocilia extend into the opening and do not make contact. Otoconia crystals rest on top of the membrane; the membrane in turn rests on a spongelike surface, the filamentous base. Under the base is a layer of cells. Near the top of this layer are the hair cells; separating the hair cells and extending to the bottom of the layer are supporting cells. Attached to the hair cells are the threadlike nerve fibers that transmit impulses to the central nervous system. Curvature of the bottom layer of tissue corresponds to inside wall of utricle and saccule. (From Parker, D.E. 1980. The vestibluar apparatus. Sci. Am., 243; No. 5:120 and 125.)

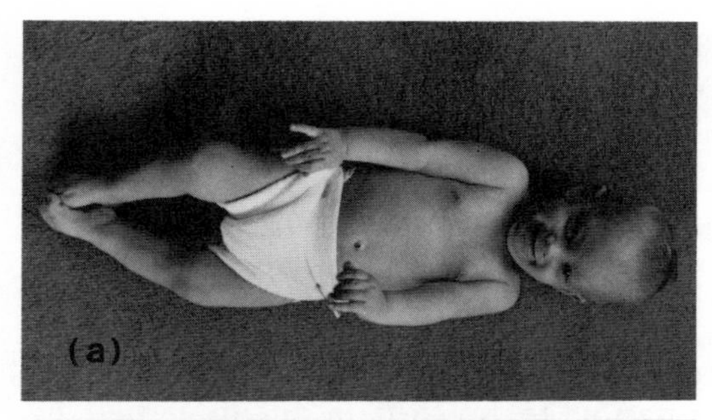

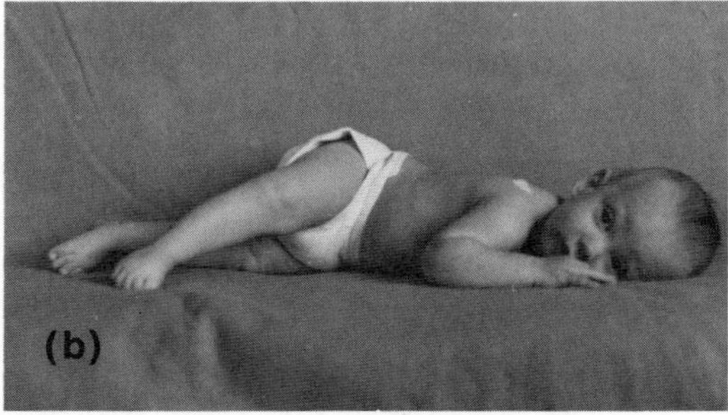

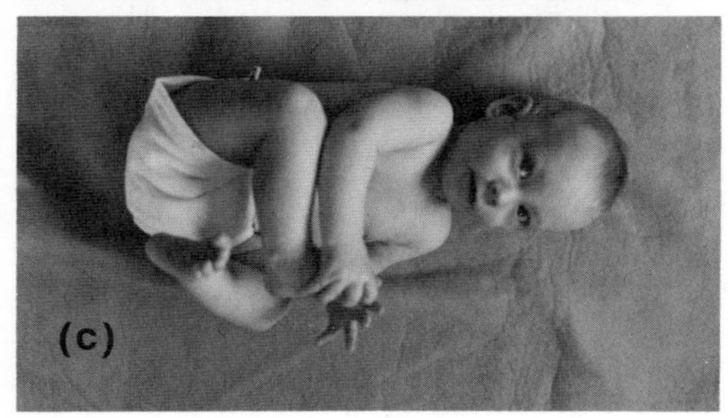

Figure 10.20. Tonic labyrinthine reflexes. The "primitive reflexes" often persist beyond the first 2 or 3 months of life, as demonstrated by this child of 7 months, and can even be elicited in the adult. The tonic neck (Fig. 10.23) and tonic labyrinthine reflexes are inseparably interrelated; they reinforce each other in their effects on the upper extremities and oppose each other in their effects on the lower extremities. However, the tonic labyrinthine reflex has been acknowledged as the more dominant reflex for lower extremity patterning, which may explain the knee flexion in (a) and the right knee extension in (b). (a), supine lying: all limbs extended. (b), side lying: under limbs (left upper and lower extremities) extended, upper limbs (right upper and lower extremities) flexed. (c), prone lying: all limbs flexed; the child was encouraged to play with his feet; a much younger child, when placed face down, would adopt a similar position with the elbows alongside the trunk and the knees drawn up under the hips.

spinning of a figure skater is a result of his training to hold the head erectly immobile voluntarily. Doing so restricts the labyrinthine responses to those of the horizontal canals, avoiding any responses which might otherwise be induced by labyrinthine reflexes arising from accelerations in forward and lateral planes. Furthermore, by obviating the changes in neck angulation which would result from stimulation of the utricles, the erect posture of the head provides a stable point of reference for the body in maintaining vertical equilibrium.

Semicircular canals are stimulated by angular acceleration about any axis and in any direction. Excitation is not limited to a particular pair of canals, but presumably all may be concerned in any angular movement, with the extent of involvement determined by the resolved components of the planes involved in the acceleration. Therefore, it is difficult to correlate precisely individual canals with specific muscles during the complex movements of locomotion and motor skills. To complicate the problem further, acceleration of the head usually simultaneously affects the utricles, thus evoking the attitudinal or postural reflexes at the same time.

Neck Reflexes

The neck reflexes arise as a result of stimulation of joint receptors in the cervical spine (especially around the atlantooccipital and atlantoaxial joints) when the head is inclined forward, backward, or sideward, or rotated to either side. These reflexes, like the TLRs, are also present at birth in a stereotyped form, the **tonic neck reflexes** or TNRs, which persist postnatally in compulsive form for a short period. The following are typical tonic neck reflex responses (Fig. 10.23): ventri-

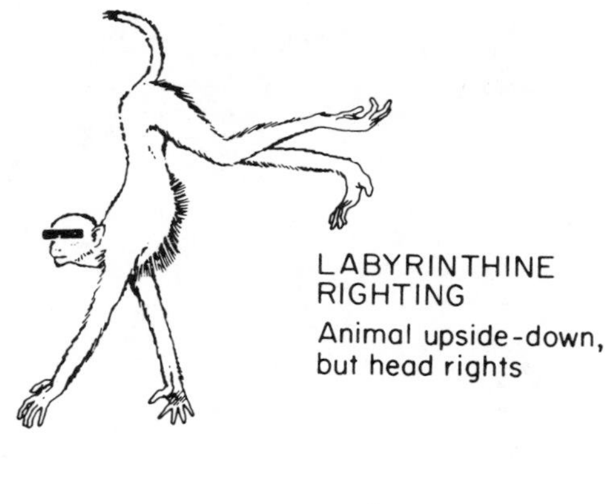

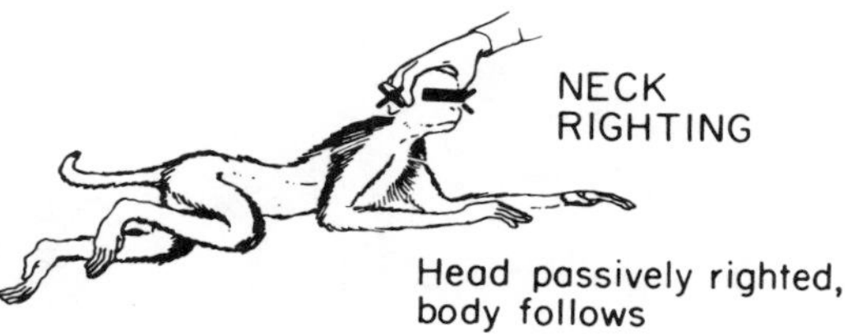

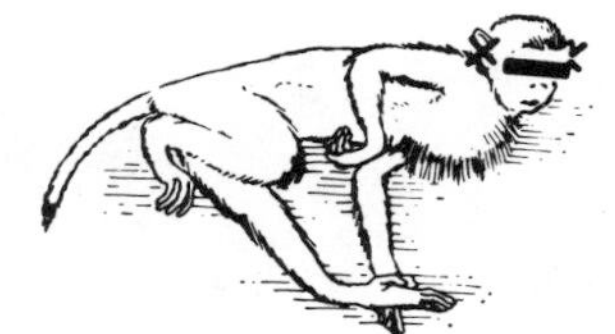

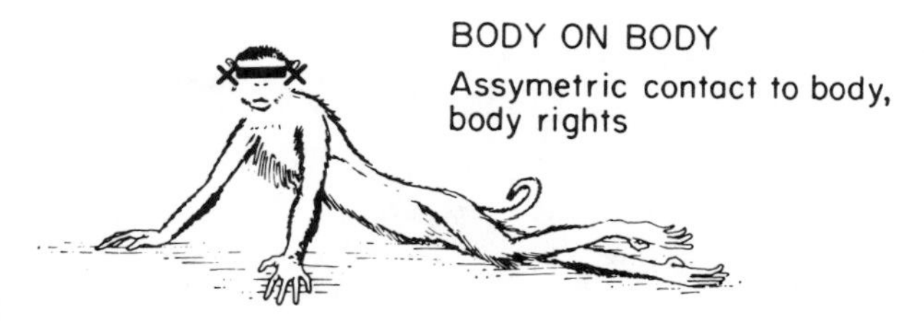

Figure 10.21. Righting reflexes. Labyrinthine righting reflexes (top): a blindfolded animal rights its head with respect to gravity; reflexes mediated by labyrinthine receptors. Neck righting reflexes (second from top): the body of a labyrinthectomized and blindfolded animal follows the head when it is passively righted; reflexes mediated by receptors in the neck joints. Body-on-head righting reflexes (middle): a labyrinthectomized and blindfolded animal rights its head from a side-lying position; reflexes mediated by asymmetrical stimulation of skin receptors of the two sides of the body. Body-on-body righting reflexes (second from bottom): as a result of asymmetrical skin stimulation, a labyrinthectomized and blindfolded animal rights its body from a side-lying position. Optical (visual) righting reflexes (bottom): a labyrinthectomized animal rights its head, using visual feedback. (From Twitchell, T. E., 1965. Attitudinal reflexes. In *The Child with Central Nervous System Deficit*, Fig. 3, p. 81. Washington, D. C.: U. S. Government Printing Office.)

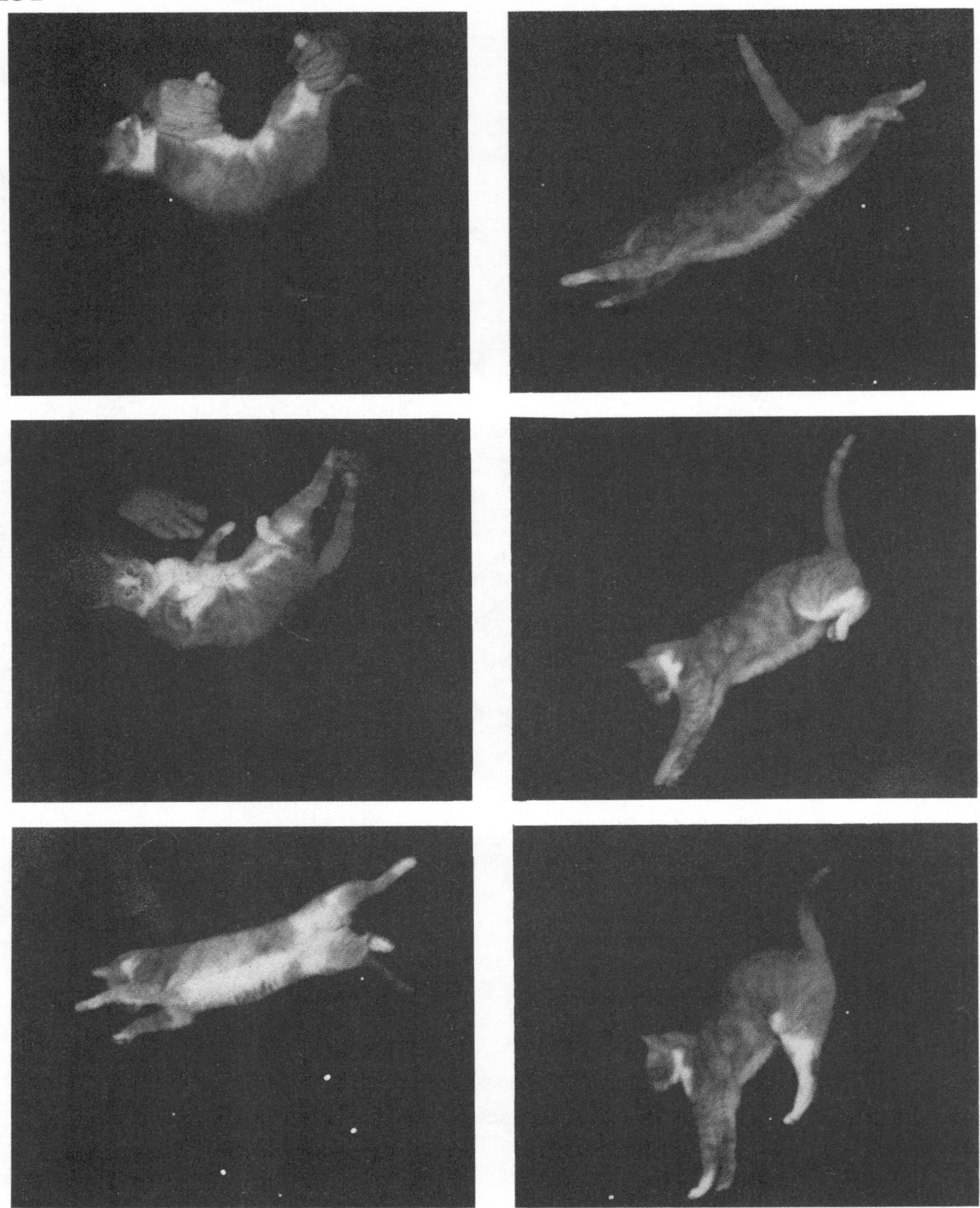

Figure 10.22. Labyrinthine righting reflexes in a cat. Cat turning over in midair is shown in this composite sequence of photographs. The cat turns over in an eighth of a second for a half-second free fall of four feet. An experimenter held the cat upside down and released it with zero angular momentum. Since air resistance is negligible, no torques are applied to the cat in the course of its descent. In the absence of torques the angular momentum of a body is conserved, and so the cat's angular momentum remains zero throughout its fall. The cat is nonetheless able to turn over in midair because a body need not be moving linearly in order to have zero angular momentum. (From Frohlich, C., 1980. The physics of somersaulting and twisting. Sci. Am., 242: No. 3:154.) The labyrinthine righting reflexes probably reorient the head into proper relationship with gravity; then the neck righting reflexes correct the relation of body to head by aligning the forelimbs first. In preparation for landing, placing reactions in the four limbs are also evident.

flexion of the neck evokes flexor facilitation of the front limbs with concomitant inhibition of antagonists (Fig. 10.24(a)), while dorsiflexion of the neck produces opposite responses (Fig. 10.24(b)); rotation of the head facilitates extension and abduction of the limbs on the face side and flexion and adduction of the contralateral limbs (a posture reminiscent of the fencer's en garde position), with reciprocal inhibition of antagonistic muscles. The tonic neck reflexes become less apparent as motor development proceeds and are no longer compulsive by the 6th to 8th week. In the older child they finally assume their mature role as **neck righting reflexes**, which, by evoking suitable activity in body and limb muscles, assure that the body follows the head.

The TNRs and TLRs are inseparably interrelated, continually modifying each other, sometimes reinforcing, sometimes diminishing the effect of the other. For example, in ventriflexion, front limb flexion is facilitated by both (reinforcement), whereas the hind limb effects oppose each other: the TLRs favor flexion but the TNRs favor extension.

The TNRs and TLRs are not completely eradicated in the adult. Their circuits are still intact; only their pattern of synaptic facilitation and inhibition has been altered as they have become dominated by later developing and more useful patterns. Twitchell (1965) says, "wholly new and distinct reactions are not added at successively higher levels of the nervous system but more primitive reactions become modified and elaborated as the stimulus for their response becomes more discriminating." Brunnstrom (1964) mentions the work of a group of Japanese scientists whose electromyographic studies in 1951 showed the presence of TNRs in normal subjects. Their presence in normal adults was also demonstrated by Hellebrandt and coworkers in 1962. Her work with Waterland (1962) indicated that the primitive reflexes operate normally during stressful activity to reinforce muscular contractions and to extend endurance.

It seems a reasonable speculation that

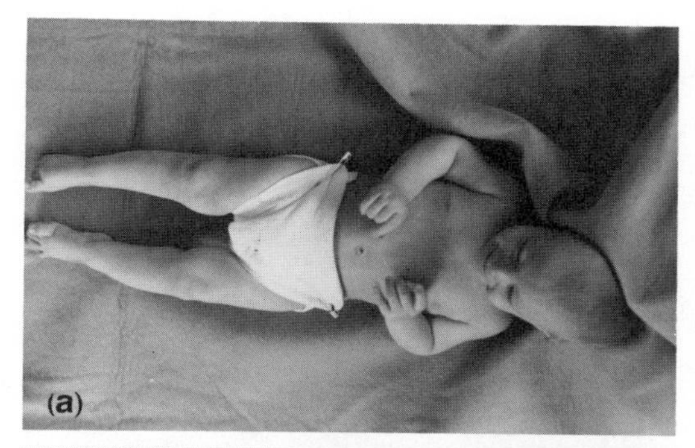

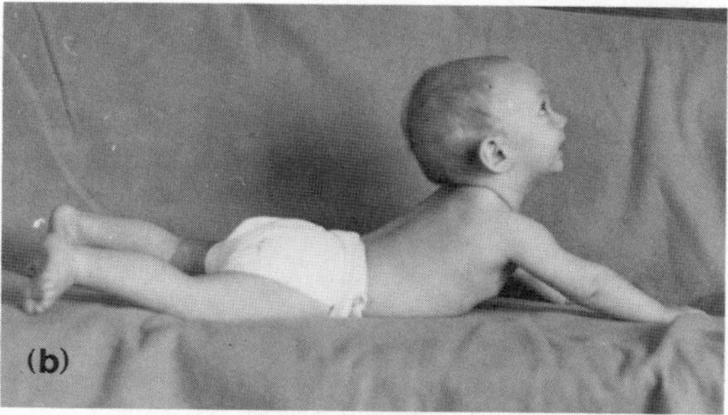

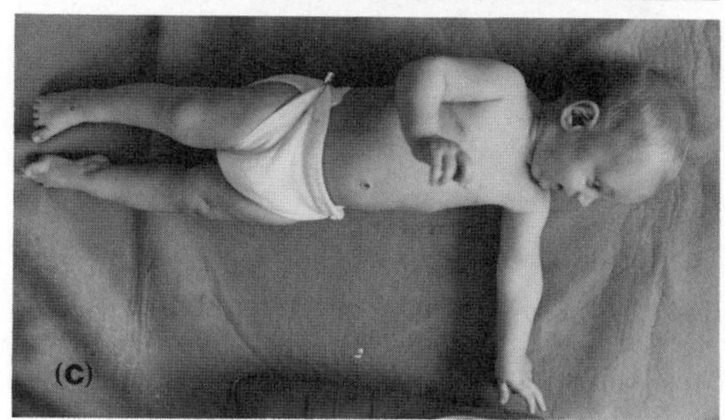

Figure 10.23. Tonic neck reflexes, demonstrated by a child of 7 months of age. The tonic neck and tonic labyrinthine reflexes (Fig. 10.20) are inseparably interrelated; they reinforce each other in their effects on the upper extremities and oppose each other in their effects on the lower extremities. However, the tonic labyrinthine reflex has been acknowledged as the more dominant reflex for lower extremity patterning, which may explain the lack of flexion in the right lower extremity of (c). (a), head ventriflexed: upper extremities flexed, lower extremities extended. (b), head dorsiflexed: upper extremities extended, lower extremities flexed. (c), head rotated to the left: left upper and lower extremities abducted and extended, right upper and lower extremities adducted and flexed.

Figure 10.24. Tonic neck reflexes in a kitten and a puppy. (a), ventriflexion induces flexion of the front limbs and extension of the hind limbs (kitten). (b), dorsiflexion induces extension of the front limbs and flexion of the hind limbs (puppy).

the simple labyrinthine and neck reflexes are concerned at all times with the integration of limb movements into a total body pattern. Their presence in the primitive form is detectable when other more complex responses are not masking, inhibiting, or precluding their automatic modification of limb muscle activity. It is difficult to separate the TLRs and TNRs since any movement of the head unavoidably activates the receptors of both the labyrinth and the neck. However, since it has been shown that neck reflexes act more effectively upon the upper limbs and the labyrinthine reflexes on the lower limbs, we may reasonably attribute influences on the arms to neck reflexes and those on the legs to labyrinthine reflexes. A few examples follow.

1. There is a stunt in which the performer stands in a doorway, pressing the backs of his hands forcefully against the door jambs by shoulder abduction for about a minute, and then steps forward out of the doorway. In subjects who are able to relax, the arms spontaneously rise to at least 90° abduction. If the head is rotated to one side, a clearly defined TNR pattern occurs in responsive subjects.
   The abduction may be explained physiologically as follows: neural transmission to the abductors persists in reverberating circuits, probably at the spinal level, and the muscle contraction, now unresisted, is converted from isometric to isotonic. (If the arms are dropped to the sides as the subject steps from the door frame, there is a brief pause before abduction begins.)
2. When a water-filled ice cube tray is placed into the freezer compartment at the top of a refrigerator, TNR may cause one to spill the water; i.e., if the head is turned away from the hand holding the ice tray to open the freezer door, inadvertent flexion tends to spill water down the arm, while if the head is turned toward the tray-holding hand, extension will spill water on the floor.
3. Anyone who has taught diving to beginners has seen the labyrinthine reflexes thwart a diver's volitional attempt to enter the water head first. In a forward dive the labyrinthine righting reflexes can raise the head as balance is lost, causing a "bellyflop," or, in response to equilibrium reflexes, the diver may jump in feet first. In a backward dive, labyrinthine righting reflexes can cause ventriflexion of the head so that the diver

lands on his back or trunk flexion so that he sits into the water.

4. The presence of head righting movements is obvious in maintaining and regaining balance. These movements occur in combination with the equilibrium movements to bring about proper alignment of the head, center of gravity, and base of support.

Involvement in Motor Skills

Neck and Labyrinthine Reflexes

The fact that the head exerts important influences upon the trunk and limbs can hardly be overlooked by anyone who deals professionally with muscular activity. With the possible exception of the spindle reflexes, neck reflexes are probably the most important single reflex mechanism used in sports skills, and they provide proprioceptive influences commonly employed by both performer and teacher.

Positions and movements of the head are used to reinforce contractions of arm muscles by invoking tonic neck and labyrinthine reflexes. A ventriflexed head favors bilateral elbow flexion; a dorsiflexed head favors extension. The symmetrical bilateral influences of the neck reflexes are obvious in weight lifting. One sees ventriflexion reinforcing the arm flexion of the lift and dorsiflexion reinforcing the shoulder flexion and elbow extension of putting up the barbell. Ventriflexion combined with contralateral rotation of the head favors unilateral pulling (flexion) movements, while dorsiflexion combined with ipsilateral rotation favors unilateral pushing (extension) movements.

Hellebrandt et al. (1962) showed that a lowered position of the head amplifies the effects of the tonic neck reflexes. In the handstand, therefore, dorsiflexion of the head probably contributes to the maintenance of arm extension as well as to balance.

Use of head and shoulder maneuvers is especially noticeable in rebound skills to accomplish turns about the long body axis. Reflex facilitation probably reinforces the mechanical advantages involved.

Neck reflexes facilitate essential trunk and limb movements in somersaults. The diver and tumbler use strong movements of the head to initiate and maintain a spin. Ventriflexion reflexively facilitates trunk and limb flexion for the forward somersault; dorsiflexion facilitates extension for the backward somersault. Resumption of the normal head-neck angle automatically assures that the body in general will be aligned with the head for a correct finish.

In balance work, fixation of the head by focusing the eyes on a point stabilizes gravitational effects upon the labyrinths so as to maintain an unchanging inflow from these receptors. The neck reflexes then make any necessary adjustments of the body to keep it aligned with the correctly postured head.

ROLE OF REFLEXES IN SKILLED MOVEMENT

A motor skill is a group of simple, natural movements combined in a new or unusual manner to achieve a predetermined objective. Skilled movement requires both mobility and stability of body parts. Each movement consists of a coordinated combination of several to many joint movements, and each joint movement further consists of a coordinated combination of muscle actions: contraction of prime movers, relaxation of antagonists, and supporting contraction of synergists and stabilizers. These are mediated by afferent information received and processed in the central nervous system and converted into appropriate signals which finally converge upon the motor neuron pools, evoking the proper activity in each muscle. All muscles must be interregulated as to intensity, speed, and duration and as to their changes in activity from the beginning to the end of the movement. This demands precise integrative function, too precise to be left to humans' conscious efforts. Hence that regulation has been delegated to automatic mechanisms, among which the proprioceptive reflexes occupy a place of importance.

The proprioceptive reflexes mediated at spinal and lower brain levels, however, must be integrated into larger, coordinated patterns at centers in the brainstem, cerebellum, and cerebral cortex. It has been demonstrated that the reticular formation exerts powerful influences, both excitatory and inhibitory, upon the spindle reflexes, the pons-midbrain areas being excitatory and the medulla areas being inhibitory to the fusimotor neurons. The gamma biasing of the spindles appears to be accomplished largely via the descending reticular system. Of special importance is the role of the cerebellum in maintaining the proper balance of muscle activity for coordinated movement. And finally, all reflex responses are subject to control by the sensorimotor areas of the cerebral cortex. While a thorough discussion of supraspinal regulation of movement is beyond the scope of this book, the reader should keep in mind the fact that the proprioceptive reflexes, important as they are, require interregulation from superior sources. The very nature of skilled movement makes this self-evident.

Motor skills are learned. They are the result of operant (instrumental) conditioning whereby inborn and proprioceptive reflexes are conditioned to be triggered by other than their natural (unconditioned) stimuli, to occur in new combinations or sequences, or not to occur at all under the usual unconditioned circumstances (to be inhibited). Conditioning involves a rechanneling of impulse flow within the central nervous system by the establishment of new patterns of synaptic resistance. Eventually the new (conditioned) response substitutes for the natural (unconditioned) response.

Consider the catching of a ball as an example. The visual stimulus of an object flying through the air toward one's head naturally evokes an *avoidance* response; one ducks, closes his eyes, and raises his hands protectively to ward off possible impact. With conditioning, however, the approach of a flying ball elicits a *catchin*greaction. One no longer ducks, that movement having been inhibited. The eyes are kept open and the hands are purposefully raised to entrap the ball. The new, acquired skill, rather than the old, unconditioned avoidance response, then occurs automatically whenever a ball approaches.

The ultimate aim of motor learning is automaticity, the performance of a skill with a minimum of conscious involvement. To that end, the proprioceptive feedback from each movement or position becomes the cue (conditioned stimulus) for the next; and the movement proceeds smoothly with a minimum of conscious direction. The cortex sits back and monitors the situation, ready to impose modification as need arises, and unconcerned with directing individual movements unless correction is required, and, relieved of integrational details, it is free to concern itself with strategy.

MOTOR LEARNING IMPLICATIONS

Each new motor skill requires a rechanneling of impulse flow between receptors and effectors to produce new spatial and temporal relationships among the component movements. A different pattern of movement sequences must be established, these sequences being precisely triggered by feedback mechanisms which have been conditioned to evoke the new responses. During the early stages of learning a difficult skill, the learner must be consciously concerned with the individual parts of the total pattern. Because conscious direction often interferes with the smooth operation of subcortical mechanisms, the attempted performance usually includes a number of movements which serve no useful purpose and which may even interfere with the achievement of the objective. As learning progresses there is a gradual reduction in the amount of cortical involvement as one after another of the pattern components becomes conditioned, until finally a single thought may be sufficient to initiate the whole train of motor events.

Once fully established, the skill becomes a feedback-regulated pattern which is somehow stored in the central nervous system to be called forth in its entirety by a simple but specific (within some permissible range of variation) input signal. Once initiated, the sequence of muscular actions proceeds almost entirely under the control of subcortical mechanisms. The major portion of movement control has been shifted by the learning process from cortical to peripheral mechanisms. Instances of the operation of proprioceptive responses can be readily identified in motor skills; their advantageous use may be inferred from specific portions of skilled movement patterns. Their involvement is implicit in many instructive cues. Some examples are discussed below.

Conscious Inhibition of Reflexes

The essential role of the proprioceptive reflexes in skilled movement is unquestionable, but in motor learning there is a negative side to be considered. Many skills involve movements in opposition to the natural inherited reflexes, especially those directed by the labyrinths. Body inversions, somersaults, cartwheels, etc., are outstanding examples. It seems obvious that a large part of the difficulty in learning such stunts relates to the need *consciously* to impose inhibition upon these responses. It is often necessary at first to emphasize a position or movement antagonistic to the natural movement and to elicit active muscular contraction against the movement. Paillard (1960) states, "learning requires before anything else the disrupting of some preexisting functional units,. . . then the selective choice of useful motor combinations and finally assembling them into a new working unit." Instructions such as "keep your head down," "hold your chin on your chest," "lock your knees," etc., are frequently used in teaching of diving.

Basmajian (1978) further confirms the concept of inhibiting the unnecessary movements and the muscles producing them. In his work with biofeedback, he uses the technique of electromyography to demonstrate for the patient and the therapist the electrical activity present in muscles not recognized as contributors to the intended movement. Patients can learn to inhibit the unnecessary muscles when they are provided with a visual or auditory display of the action potential which is unwanted. As the action potential is reduced, the muscle tension also is reduced, offering significant reduction in and relief of pain.

The need for deliberate inhibition is not limited to the labyrinthine reflexes. Examples involving other receptors are common. In the golf swing inhibitions are reflected in the need for conscious effort to keep the left elbow straight, to keep the left hand on top of the club, to pivot without swaying, and to keep the head down throughout the swing. In all of these, voluntary effort is required to inhibit "what comes naturally." As learning progresses, the need for conscious contractile antagonism diminishes as synaptic inhibition of the reflex circuits becomes an integral part of the conditioned response. Instances of involvement of conscious inhibition are abundant in the learning of most sports.

NEUROKINESIOLOGICAL ANALYSES

Although the physical education student may have had some undergraduate training in neuromuscular physiology, all too frequently the student has not been made aware of its practical applications, nor has the student consciously applied such information. Understanding the neuromuscular, physiological reasons for taking a long backswing to achieve a good hit or a big wind-up to make a good throw might be of value when attempting to modify performance on the playing field or court.

The student of kinesiology should be aware of the neural mechanisms by which motor performance is integrated and controlled. The student should be familiar with the various reflexes which may help or hinder a performance so that practical use may be made of them. Practice in trying to detect the operation of reflexes in simple responses such as postural response and in complex motor patterns may include direct observation, study of motion pictures and electromyograms, and combinations of these methods.

Analyses of Simple Responses

Adjustment to Load

Films and electromyograms taken during the adjustment to load (discussed earlier in this chapter) provide an objective basis for analysis of the spindle reflexes pertinent to that response (Fig. 10.10). The subject was asked to hold a basket in his hand, palm up, so that a 2-kg weight might easily be added and removed, and instructed to maintain the elbow at an angle of approximately 90° throughout. The muscular responses were recorded by bipolar surface electrodes placed on the long and short heads of the biceps and on the lateral and long heads of the triceps. Motion pictures, synchronized with the EMG, were taken simultaneously to record the gross movements of the limb. The load, introduced at b on the EMG record, was added abruptly in I and gradually in II. For abrupt loading (I), the starting position is shown in photo Ia. The activity in the muscle which was recorded in the EMG before b (loading) was that set by gamma bias to maintain the prescribed position under the dictation of the cortex, plus the spindle response to the weight of forearm and basket. When the load was dropped into the basket (photo Ib), there was momentarily a slight increase in elbow extension (Ic), followed by a return of the forearm to the original position (Id). Upon addition of the load (with a slight lag partially attributable to the inertia of the pens of the ink-writing polygraph), the EMG exhibited a sudden burst of activity in both heads of the biceps, showing that the sudden stretch of the muscle had evoked a phasic response in the primary afferents of its spindles. The muscle responded with an appropriate contraction. (Contraction of the muscle will momentarily unload the spindles, causing contraction to lessen and the arm to drop. The tendency to oscillate which results is, fortunately, damped out by the gamma bias. The magnitude of the oscillatory activity with this subject and this load is not sufficient to show in these gross records.) Muscle activity then settled down to a lower level consistent with the new load and reflecting the tonic spindle response. This was maintained until the load was removed at e, at which time muscle activity returned to its initial level. During removal, the small unexpected and transient increase in activity was probably due to momentary pressure on the load exerted by the assistant in grasping the weight.

In II, the same load was placed gradually into the basket, beginning at b (photo IIb) and completed by c (photo IIc). Activity in the biceps gradually increased to a new tonic level but without any phasic burst. Upon removal of the load (d to e and photos IId and IIe), the initial tonic level of activity was resumed.

Postural Reflexes: The Scale

In the gymnast's scale or the dancer's arabesque the performer stands poised with the weight balanced over one foot. Because the pose does not permit adjustment of the base of support, equilibrium must be maintained by proper counterbalancing of muscle tone at the hip, trunk (intervertebral), and shoulder joints. In order to assume and maintain this pose, a number of reflexes are functioning. These are controlled by the visual, labyrinthine, and neck righting reflexes and by interplay of joint reflexes to balance the force couples at the joints. Postural corrections are made continuously, mediated by the myotatic (stretch) reflex. The labyrinthine and visual righting reflexes maintain the head's orientation with gravity; as the pose is taken and held, extension of the neck (dorsiflexion of the head) is obvious, and this in turn aids in maintaining the arched back (spinal extension). Eye fixation helps to stabilize the head so that any change in the orientation of the body in relation to the head will be corrected by neck righting reflexes.

The performer in Figure 5.19 is balancing approximately 50 kg on the head of the right femur, 59 kg on the right knee, and 63 kg on the head of the right talus (see Chapter 5). Compression of these joints evokes a strong extensor response so that the joints of the lower extremity can support the weight. At the same time the pressure exerted by the total body weight (64 kg or 624 N) on the right foot stimulates pressure receptors in the skin of the sole and the spindles of the intrinsic muscles of that foot, thus eliciting contraction of the antigravity muscles of the extremity (the positive supporting reaction).

Analyses of More Complex Skills

The muscle response patterns of well-learned motor skills are largely integrated and controlled at the subcortical level, and consequently they involve the integrated action of many reflexes and usually the inhibition of some. Although the movements are more complex, neurokinesiological analysis of these skills is accomplished by the same procedure. There will be, however, an increase in the requirement for inhibition of natural reflexes in most instances. The following are some sample discussions taken from Figures 10.25, 10.26, 10.27, and 11.4.

Standing Long Jump

The presence of the labyrinthine head righting reflex is evident in the standing long jump (Fig. 10.25) in frames 1 through 24 as the performer prepares for takeoff; as the angle of trunk inclination increases, the head becomes more and more dorsiflexed and cervical extension is evident. During the early part of this same period, frames 1 through 17, hips and knees are flexing and ankles are dorsiflexing, evoking stretch reflexes in the antigravity muscles which facilitate their contraction in takeoff. The coincident movement of the upper extremities as they are swung from hyperextension to flexion (frames 1 to 28) contributes facilitation to the lower limbs by irradiation of neural impulses arising from the resulting stimulation of the shoulder joint receptors.

At frame 29, as the jumper's toes leave the floor, the labyrinthine head righting reflex has been inhibited and the head is ventriflexed. This inhibition of the head righting (labyrinthine) reflex is undesirable during the early stages of the flight as it could elicit a body-following-head response which can cut down the height of the trajectory and so decrease the distance of the jump. The inhibition is not total during the jump itself as the head can be seen to dorsiflex slightly in both frames 34 and 44.

Badminton Smash

Figure 10.26 depicts a few of the positions relative to time adopted by the badminton player during the force-producing and follow-through phases of the forehand smash. The first plate, (a), is taken from a "side-view" position while the concomitant frames of the "back-view" are shown in the second plate (b). Time between frames is approximately 100 msec.[c] Since one of the objectives of the smash is to hit with power, it seems logical that the player makes use of all elements and reflexes which augment the force of the muscles involved. It can be seen from frames 1 through 12 that the medial rotators and adductors of the shoulder joint and the pronators of the radioulnar joints have been put on stretch. Further, these same frames show that the arm (and racket) are whipped back quickly into this position

[c] Since the arms are almost fully flexed at the shoulders as the performer reaches for the floor, they are moving into extension as the hands accept the body weight.

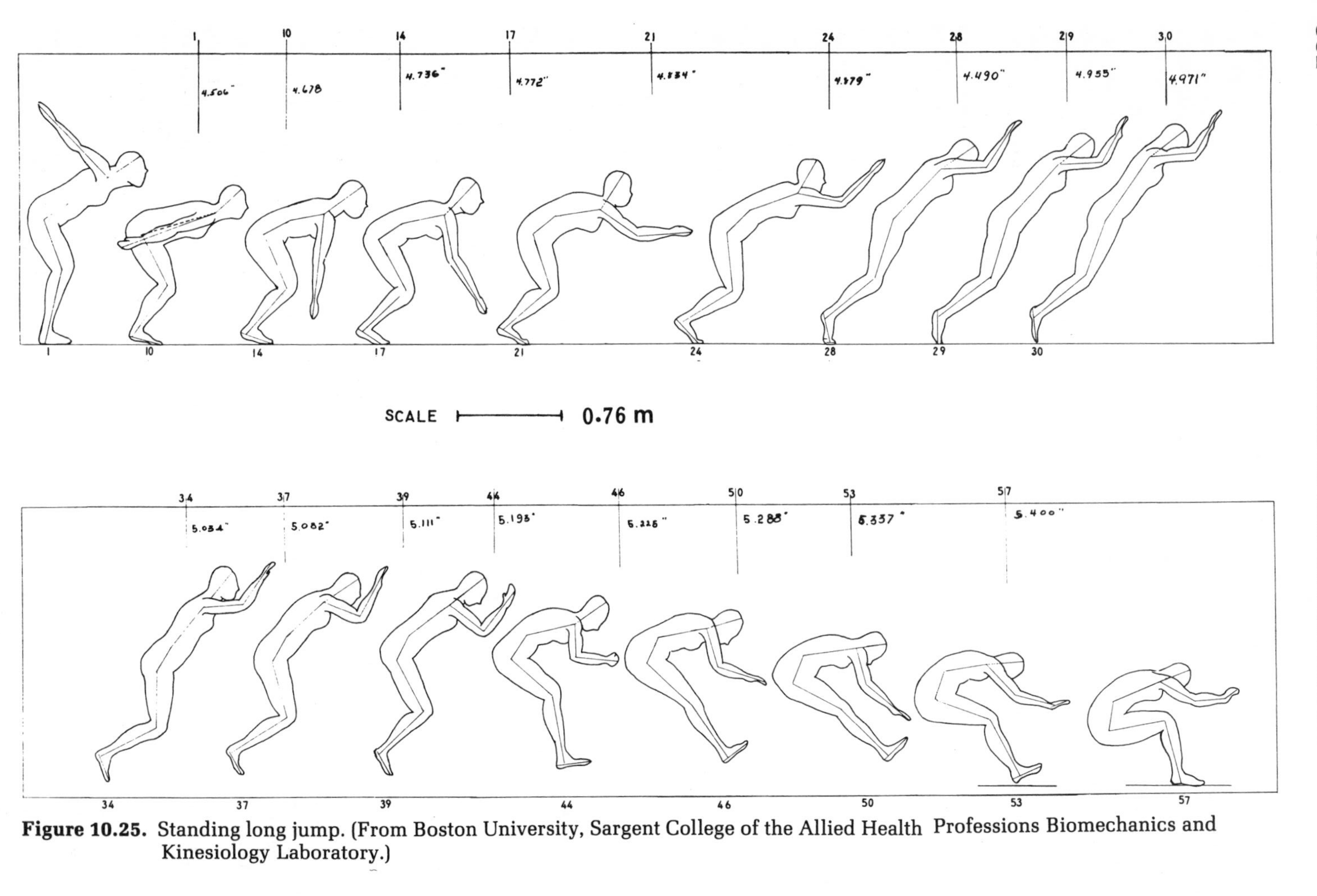

Figure 10.25. Standing long jump. (From Boston University, Sargent College of the Allied Health Professions Biomechanics and Kinesiology Laboratory.)

(i.e., in a time span of 0.03 s). By stretching the muscle groups that are about to become involved in the forceful part of the stroke and stretching them quickly, the player gains force from the length-tension relationship of muscle and the phasic and tonic components of the neuromuscular spindle.

Gowitzke and Waddell (1979) have shown that a major force component for the smash emanates from the medial rotation of the arm at the shoulder joint and pronation of the forearm at the radioulnar joints. These two actions start at approximately frame 13 (0.02 s before contact), continue through contact with the shuttlecock (frame 20), and have not stopped 0.01 s after contact (frame 24). At the same time, elbow extension occurs, changing to flexion just before contact. This combination of movements (medial rotation at the shoulder joint) suggests that the second diagonal arm extension pattern, discussed in Chapter 2, is involved. The badminton player uses elements of muscle mechanics and neuromuscular reflexes to advantage when force is the objective.

Headspring

Figure 10.27 provides still another example for analysis, and is discussed in detail later in this chapter.

Dive for Height

Figure 11.4 is a series of tracings made from a film of a dive for height. From these tracings it is possible to make postulations concerning reflex activity which assists the performer to accomplish the dive. Frames 1 and 5 show the subject in the air and landing on both feet for the takeoff. This indicates that there is a strong stimulus to the pressure receptors in the skin of the soles of the feet, which should trigger an *extensor thrust* reflex as an initial part of the takeoff. Also, in these same frames, the fact that the hips and knees are flexed and the ankles are dorsiflexed indicates that all the antigravity muscles of the lower extremities are put on stretch, thus activating the spindles in these muscles. The resulting *stretch reflex* contributes to the forcefulness of the voluntary contraction of these same muscles projecting the body forward and upward in frames 8 through 11.

During takeoff and early stages of the dive (frames 8 to 14), there is little change in the head and its cervical angulation with the trunk but shortly thereafter, by frame 18, the head is dorsiflexed, probably as a result of an effective *labyrinthine head righting reflex* resulting from stimulus to the utricle receptors, and also to some extent as a result of visual feedback. The angulation of the head (neck extension) in turn stimulates the joint receptors in the upper cervical region, which triggers a *neck righting reflex,* facilitating extension of the trunk and lower limbs to ensure a *body-following-head* response. By frame 39, as the performer approaches landing, the head is ventriflexed in preparation for a forward roll. This ventriflexion may evoke a *tonic labyrinthine reflex* which in turn facilitates the flexion of the trunk and upper extremities necessary for the performance of the roll and again stimulates the joint receptors in the cervical region which, by the *neck righting reflex,* ensure that the body will follow the head.

Contact of the hands with the floor (frame 41) initiates an extensor thrust reflex through the pressure receptors in the skin of the palms of the hands, a reflex which in this case is probably limited to an eccentric contraction of the shoulder flexors[d] and elbow extensors as these muscles aid in absorbing the force of landing.

Practice Exercises

The student's interest in and understanding of subcortical neural function in complex sequential skills may be fostered by studying films or film sequences, or tracings made from films as in Figures 10.25, 10.26, 10.27, and 11.4. After such

[d] Two high-speed cameras were operated simultaneously at a film transport speed of 400 frames/sec and a shutter speed of 1/1200 second; every fourth frame was selected to display in Figure 10.26.

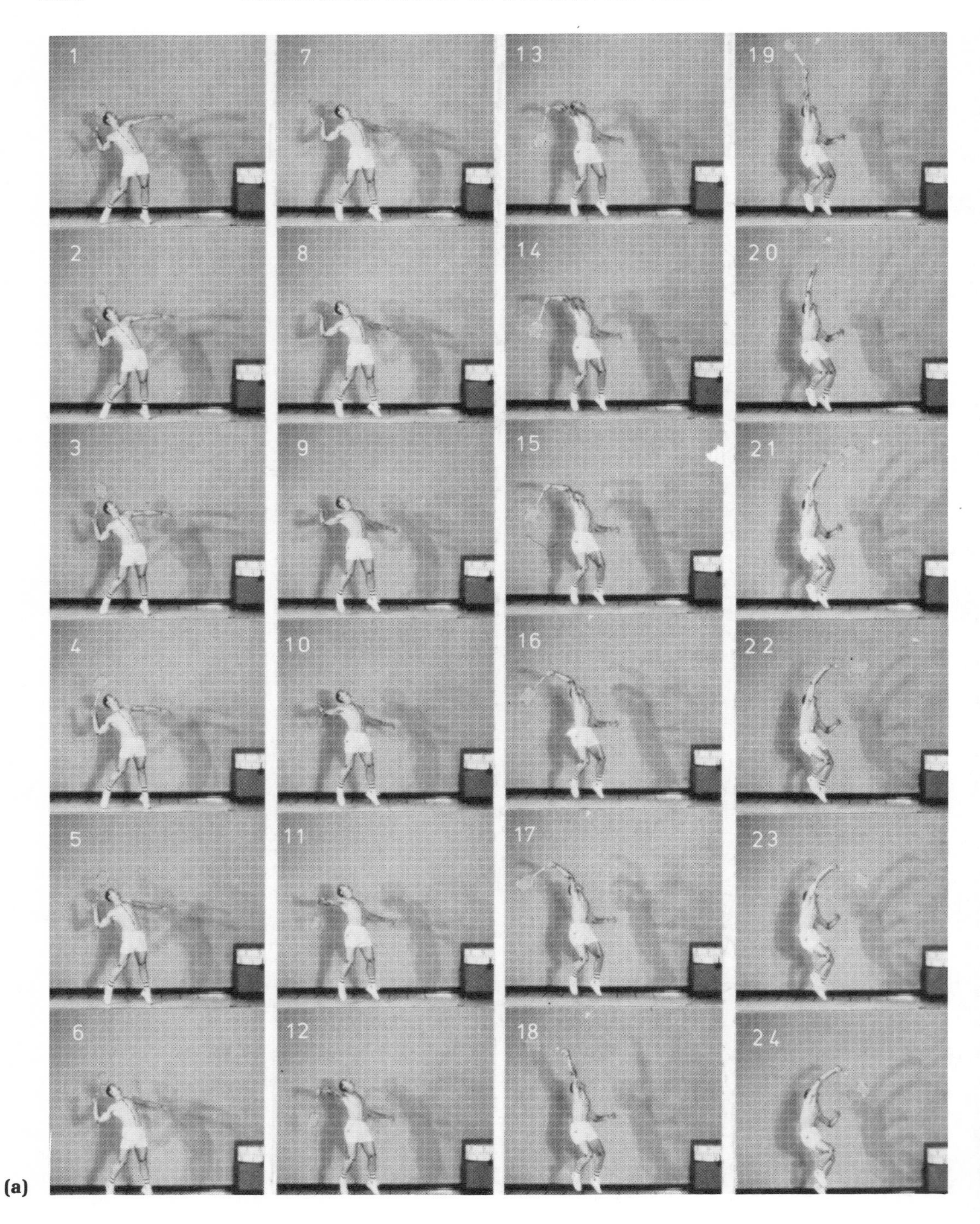

(a)

Figure 10.26(a). badminton forehand smash, "side-view": contact with the shuttlecock at frame no. 20. (From McMaster University Kinesiology/Biomechanics Laboratory, School of Physical Education and Athletics.)

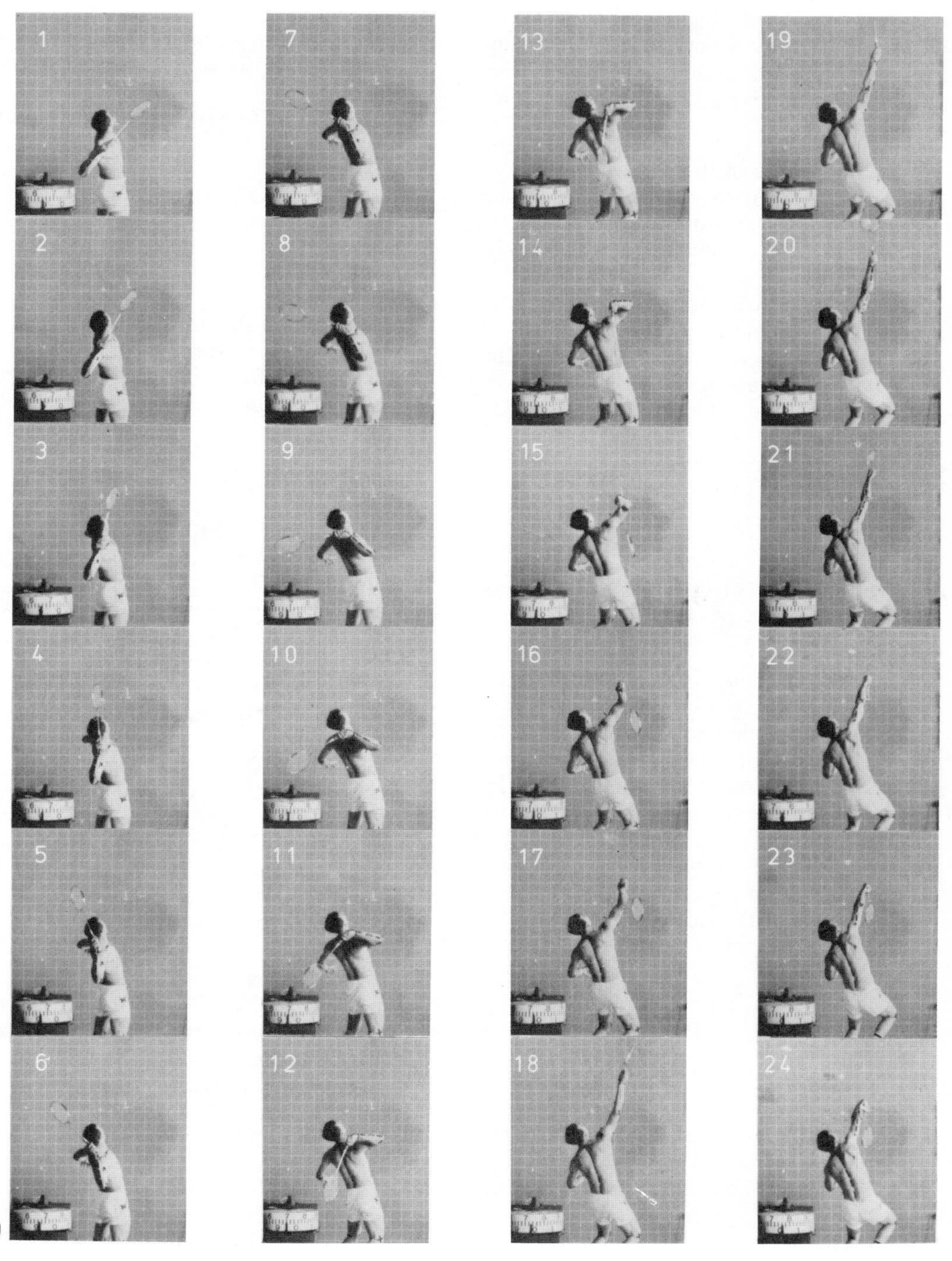

Figure 10.26(b). badminton forehand smash, concomitant frames of the "back-view." (From McMaster University Kinesiology/Biomechanics Laboratory, School of Physical Education and Athletics.)

perusal the student should be able to identify, explain, and discuss the reflexes which are or may be operating. This discussion can be somewhat simplified if the student follows an outline format such as the following.

1. a. The name of the reflex.
   b. Identification of the frame or frames (by number or time) in which evidence of the reflex is found.
2. The receptors involved and where they are located.
3. The evidence which suggests the presence or absence of the reflex.
4. The expected effect of the reflex if it occurs.
5. The actual result as seen in the subsequent frame or frames.
6. Discussion.

Types of Reflex Activity to be Identified In Motor Skills

The following list suggests a number of reflexes which the student may identify as helping or hindering the performance in his neurokinesiological analysis of skilled movement patterns. The student should follow the outline for the presentation of each reflex. The discussion (item 6 in the outline above) is especially important since it is here that the student displays his understanding of the reflex and its participation in the skill under study.

Spindle reflexes
- Stretch reflex
- Primary afferent responses: phasic and/or tonic
- Reciprocal activity in antagonists and synergists
- Secondary afferent responses: supplementary facilitation of flexors; inhibition leading to co-contraction and joint stabilization in simple extensors
- Gamma bias: setting the voluntary limits of the performance
- Positive supporting reaction; weight bearing

Golgi tendon organ reflex
- Feedback of muscle tension level
- Relaxation at the end of volitional movement (termination of active contraction)
- Limitation of use of force by autogenic inhibition

Joint reflexes
- Kinesthetic feedback
- Limb-trunk angles
- Compression
- Traction
- Positive supporting reaction
- Interlimb facilitations: forelimbs on hind; contralateral; ipsilateral
- Balancing of force couples

Cutaneous responses
- Extensor thrust
- Grasp reflex
- Placing responses
- Local sign
- Body-on-head righting
- Body-on-body righting

Labyrinthine reflexes
- Labyrinthine righting reflex
- Equilibrium reflexes
  - Shift of base of support
  - Shift of center of gravity
- Inhibition of righting reflex

Neck reflexes
- Tonic neck reflexes
- Neck righting reflexes

Visual righting reflexes

The use of such an outline for neurokinesiological analysis is illustrated below with examples drawn from tracings of the headspring (Fig. 10.27).

Headspring

1. a. Reflex: stretch reflex
   b. Frame 127
2. The receptors are the nerve endings in the muscle spindles located in the hip extensors (gluteus maximus and hamstrings).
3. Evidence: the hips are flexed more than 90°.
4. Expected effect: there should be a strong contraction of these muscles.
5. Actual results: frames 146 to 152 show a sharp extension of the hips; the reflex is effective.
6. Discussion (this should be an essay

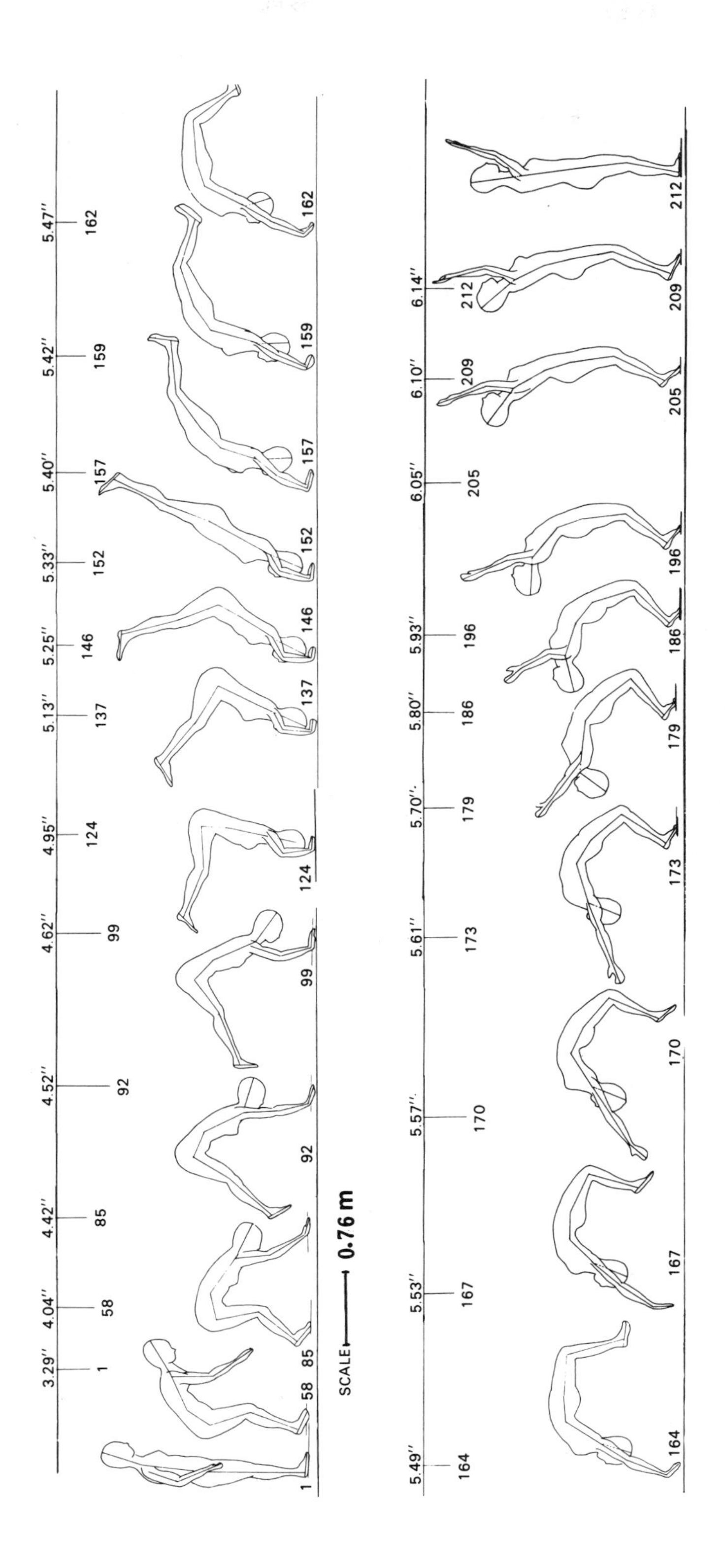

Figure 10.27. Headspring. Series of tracings made from 16 mm film.

discussion which displays the student's comprehension of the reflex). (N.B. Other stretch reflexes can be similarly identified in frame 170, rectus abdominis; frame 173, quadriceps femoris; and frame 205, the anterior tibial muscles.)

APPLYING KNOWLEDGE OF PROPRIOCEPTION

The ultimate aim of motor learning is the performance of a skill with a minimum of conscious concern for the details of the movement. Smooth and harmonious coordination of the body segments participating in the pattern must be assured by the relatively automatic action of muscles under the self-regulating control of proprioceptive feedback. The proprioceptive responses must become conditioned so that each movement supplies the stimulus for the next part of the pattern, leaving the cortex free to direct strategy rather than movement.

Most, if not all, teaching cues make direct or indirect use of proprioceptive reflexes, but this is usually more by accident than by design. However, proprioceptive reflexes are better understood than most other components of the motor learning process and they are readily accessible to manipulation. Physical therapists have successfully used techniques of proprioceptive facilitation and inhibition in the rehabilitation of patients with neuromuscular disabilities since World War II. The inclusion of neurokinesiological analysis in the study of motor patterns could contribute to the student's repertory of teaching and coaching techniques. Teaching cues and preliminary exercises based on understanding of neuromuscular mechanisms and reflex control as well as on sound principles of biomechanics and applied anatomy have been deliberately used in the training of the undergraduate student (Gowitzke (1968)). The student (prospective teacher, coach, therapist) uses this type of approach as a catalyst for evolving teaching, coaching, or therapeutic methodology. As a result, new methods evolve which have the potential for improving motor performance, shortening the learning time and the extent of trial-and-error learning, and enhancing the skilled performer's accomplishments possibly beyond presently accepted levels.

CHAPTER 11

Analysis of Movement: Measurement and Instrumentation

INTRODUCTION

In virtually all of the biomechanical examples presented in the material of preceding chapters, we have been concerned with descriptions of forces and their related reactions associated with static situations. The body or its segments have been taken to be in static equilibrium for most examples. Static analyses become the starting point for moving into dynamic analyses. A few examples are given which examine the geometric and temporal qualities of human motion, i.e., kinematics. The specification of displacements, velocities, and accelerations of a body or body parts should be accompanied by identification of the forces involved with the production of the observed phenomena, i.e., kinetics. Methods of describing motion and measuring forces are the main concern of this chapter. Instrumentation and current, state-of-the-art measurement techniques are discussed.

ON DESCRIPTIVE TECHNIQUE

The ex-football player or football fan will follow the sequence of events after the ball has been snapped and, as the action unfolds, he will know almost immediately what play is being made long before it is completed. If asked, he could diagram the play, indicating the line shifts and moves of each player in technical football-ese such that any coach could present the play to his squad and they could reproduce it. However, this onlooker cannot do the same for a single skill, such as a punt or a tackle. The kinesiologist/biomechanicst, who may be ignorant of football, can reconstruct in technical anatomical terms the exact movements occurring at each of the performer's joints and the probable forces that produce them.

Being able to describe a skilled performance, or even an unskilled one, requires a precise vocabulary as well as a complete understanding of the skill. To say that the ballet dancer performed an entrechat, a fencer made a riposte in tierce, or that a gymnast performed a dislocate on the rings would have instant meaning only to the initiate of the discipline. The initiate would immediately visualize the entire series of bodily movements that were executed by the performer. However, such generic terminology is insufficient for the kinesiologist, who should use those words or phrases most widely understood and applicable.

Adjectives are needed in order to describe anything, and the adjectives and verbs used in kinematics of movement analysis are terms used to describe joint movements and/or resulting body positions. Probably the most widely accepted terminology is that used in anatomy and medicine. The American Academy of Orthopaedic Surgeons (1965) published a booklet which they hope will lead to a standardized terminology in the area. A recent publication by Esch and Lepley (1974) is of value. There have been efforts directed at creating movement notations to facilitate movement description on the one hand, and on the other, scripting of movements to be reproduced (Eshkol and Wachman (1958); Birdwhistell (1970)).

Kinesiologists and other students of human movement are often dissatisfied with the use of standard anatomical terms inasmuch as they lack precision. This is particularly true of the engineers designing space suits, who have built their own vocabulary of descriptive terms (Fig. 11.1). This latter terminology, combined with suitable angular measures, is as precise as the angles and directions measured. Anatomists and kinesiologists may in time come to adopt this or a similar system of notation. However, in the meantime it is still possible, using the more familiar anatomical terminology as presented in Chapter 1, to give an adequate description of any sequence of movements.

MOVEMENT ANALYSIS

Movement analysis requires recordings of body movements and many technological developments have contributed to making effective studies in the field. The most popular technique has been to acquire motion picture sequences that may be analyzed frame by frame.

Any movement analysis should start with a description of the bodily movements which take place during the performance of the activity, and an identification of the probable forces involved as follows:

1. The name of the movement and the time or frame number at which the movement starts and finishes.
2. The joint(s) at which movement occurs.
3. The lever, i.e., the segment or segments making up the kinematic chain being moved as a result of the joint actions.
4. The force(s) producing the movement, muscular shortening (i.e., isotonic or concentric contraction), gravity, or some other imposed force.
5. Where this force is applied.
6. The force resisting the movement, if any: gravity or eccentric muscular contraction (muscle being lengthened while exerting tension).
7. Where this force is applied.
8. Stabilized and/or relaxed joints in the lever.
9. Forces that stabilize these joints.
10. Stabilized and/or relaxed joints outside the lever.
11. Forces that stabilize these joints.

Many students are taught to think of the skeleton as a system of levers acted on by muscles, and that each body segment is represented by its core of bone and so forms a single lever. This is a very useful concept, but it may tend to circumscribe one's thinking when analyzing even a simple motor skill. The student is soon involved with any number of moving segments; one segment moving on another, which in turn is moving on a third, etc. He is expected to determine the levers involved in each joint action and to determine the types of internal and external forces involved in moving them. Under these circumstances the concept of the body as a system of links forming one or more kinematic chains becomes very useful as it is evident that no proximal link of a chain may be moved without causing movement in one or more adjacent links as well.

Consider the actions occurring during a football punt, those of the kicking leg in particular as it starts the forward swing to contact the ball (Fig. 11.2). There are three different actions involved: (1) flexion of

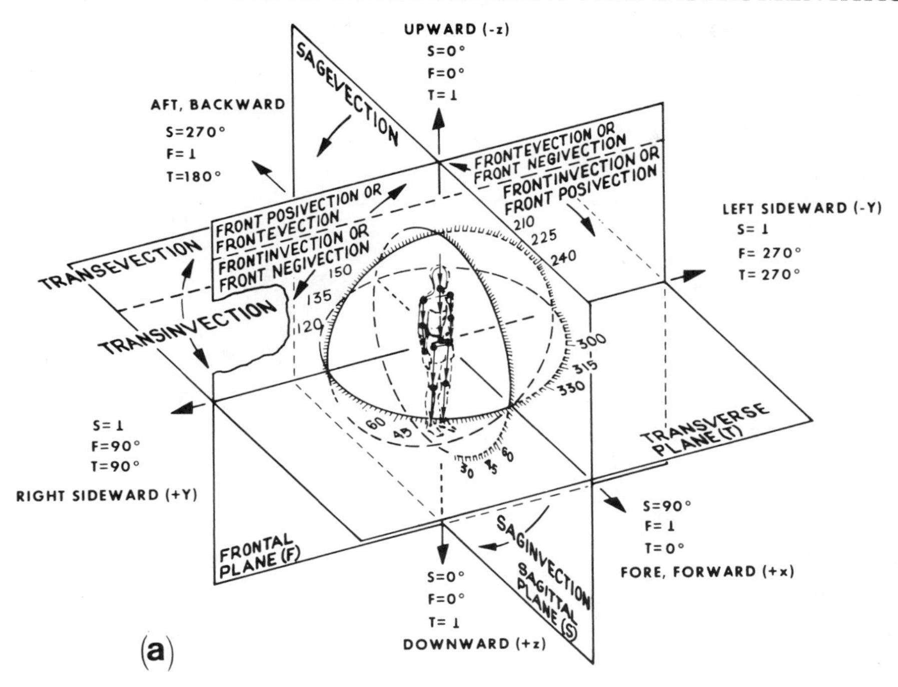

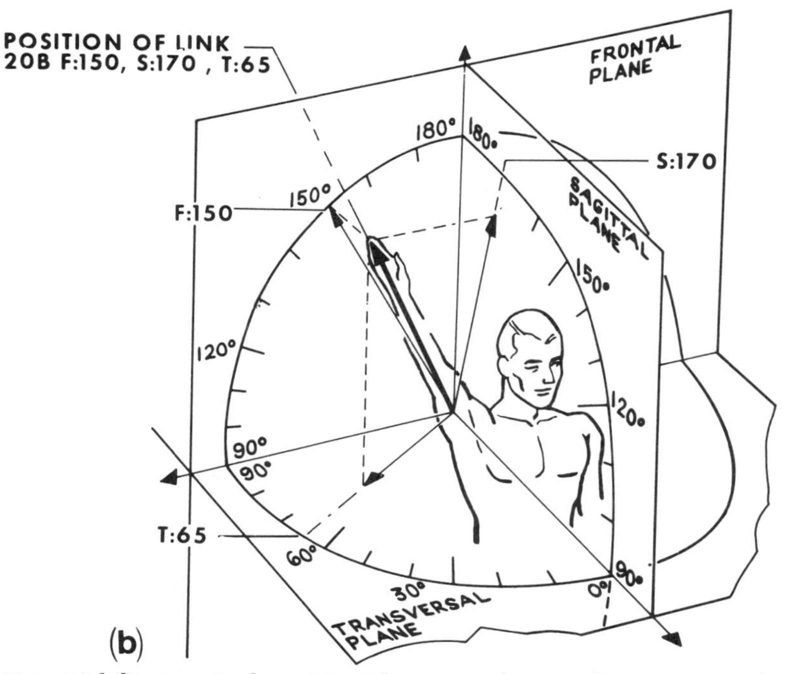

Figure 11.1. Mobility terminology. (a) triplanar angular coordinate system. (b), example of shorthand notation using terminology presented in (a). (From Roebuck, J. A., Jr., 1966. Kinesiology in engineering. Paper presented at the Kinesiology Council, Convention of the American Association for Health, Physical Education and Recreation, March 1966.)

the hip, (2) extension of the knee, and (3) plantar flexion of the ankle. In order to identify the skeletal levers involved in these actions, the student must determine the body segments that are being moved by the action given. Flexion at the hip joint moves the femur on the pelvis, but the thigh is not the only segment moved by this action. The contracting muscles flexing the hip must move the weight of all of the segments attached to the distal end of the femur as well. If the student thinks in terms of an open kinematic chain made up of three links (thigh, leg, and foot), being moved by the shortening of the hip flexors, there should be no trouble in understanding that the "lever" in this case is the kinematic chain made up of these same three links. Under these conditions he will not be tempted to say that the skeletal lever for hip flexion is only the femur or thigh. Similarly if he considers the knee extension, the chain is made up of the leg and foot links and is moved by shortening of the quadriceps femoris, so he knows that the segments forming these two links also form the "lever" in question.

Figure 11.2. Football punt.

Complete movement analyses of this type are normally made from motion picture sequences such as those shown in Figure 11.4. Multiple, synchronized views might be necessitated and three views, front or back, side, and from overhead are often sufficient (Fig. 11.9). It is often important for the body segments of concern to be in view in two planes at all times. The films may be studied by running them through a time-motion study projector, a viewer-editor, or a microfilm reader such as a Recordak. The entire movement sequence may then be studied one frame at a time.

If the film background contains a grid with lines a known distance apart, this grid may serve as a simple reference for scaling distance information and the camera frame speed provides a time reference. Appropriate optical criteria have to be met to ensure the precision of scaling distances, i.e., the camera should be located much farther from the background than is the performing subject; also, the camera should be carefully set up so that the film plane is parallel with the plane being observed. If the camera speed is 50 frames/sec each frame provides a sample of movement every 1/50 second. (The shutter speed is usually less than the camera speed; higher shutter speeds facilitate "freezing" of the action, but necessitate greater film sensitivity and/or illumination.) When timing information is of consequence, the film transport speed must be precise. Either a well-made camera may be used, or else a suitable clock or timer can be simultaneously photographed by using an additional optical pathway to the camera.

Extracting information systematically frame by frame can lead to the production of data formats which may be more amenable to detailed analyses. For example, the *X-Y* coordinates of various anatomical landmarks can be plotted; joint angle variations as a function of time can be derived. From movement-time characteristics, velocities and hence accelerations of body segments can be derived. Coupling the masses of the moving members with their velocities permits the calculation of the moments of these members. When accelerations of body segments are allied with their respective masses, the corresponding segment forces follow.

It will be evident that there is much tedium involved in extracting and processing the data acquired in the aforementioned manner. Considerable efforts have been made to introduce pertinent technology in studies of human movement so as to minimize or eliminate this tedium. Special data digitizers, sensors, computers, computer programs, and display devices have emerged to facilitate these efforts. While they are alluded to in what follows, basic principles are stressed.

Value of Motion and Static Analyses

A major aspect of movement analyses concerns the control and activity patterns of the muscles involved in a particular maneuver. The activity patterns are indicated by multichannel electromyograms (EMGs) such as those shown in Figure 11.3. When these are synchronzied with the motion analysis data, e.g., the film described above, the kinematic analyses may aid in the interpretation of the muscle control.

All too often beginning investigators are content to point out that muscles are contracting during a certain phase of a performance, but they fail to determine, or even hypothesize, the purpose of that contraction. The beginning investigator, and even the expert, may well find gaps in his analysis when he compares it with the electromyographic activity recorded from the various muscles. He may find that some of this activity may not be accounted for in the preliminary analysis. The investigator then returns to the movement data (film) to determine the situation that accounts for the unidentified muscular activity recorded on the EMG.

Once this part of the analysis has been satisfactorily completed, the investigator has the basis for the selection of one or more critical components for further study. These components include the muscular, gravitational, or other forces that may be acting, limb or segmental accelerations, and the contribution of these and of segmental velocities to the skill or activity. When the selection has been made, the appropriate portion of the movement record (film) is re-examined for the necessary data.

In essence, when using filmed data, link diagrams such as those depicted in Figures 5.20 and 5.21 are constructed. The number of frames so diagrammed depends upon the type of investigation selected. Only one such frame from each camera is necessary for the study discussed in Examples 5.11 and 5.12.

If the investigation is concerned with angular acceleration patterns of various body parts during the performance of the skill or of a particular phase of the performance, a series of such diagrams made from a regular sequence of frames is necessary. The nature of the movement decides the timing of the frames; the more rapid the movement, the greater should be the number of frames per second to ensure that all of the pertinent movement information is captured.

Figure 11.4, while not so evenly spaced, is derived from such data.

EXAMPLE 1–Derivation of Linear Translational Displacement, Velocity, and Acceleration from Processed Film Data

It is desired to extract the linear translational displacement velocity and acceleration components in the forward (x) and vertical (y) directions of the midpoint of the link connecting the hip and shoulder of the subject depicted in Figure 11.4.

This problem is best tackled using enlarged diagrams so as to minimize errors when extracting data. The relative x and y values for the midpoint of the requisite link are plotted as a function of frame number (i.e., time) as in Figure 11.5(a). The starting position of the pertinent point in frame 1 is taken as the x origin for this

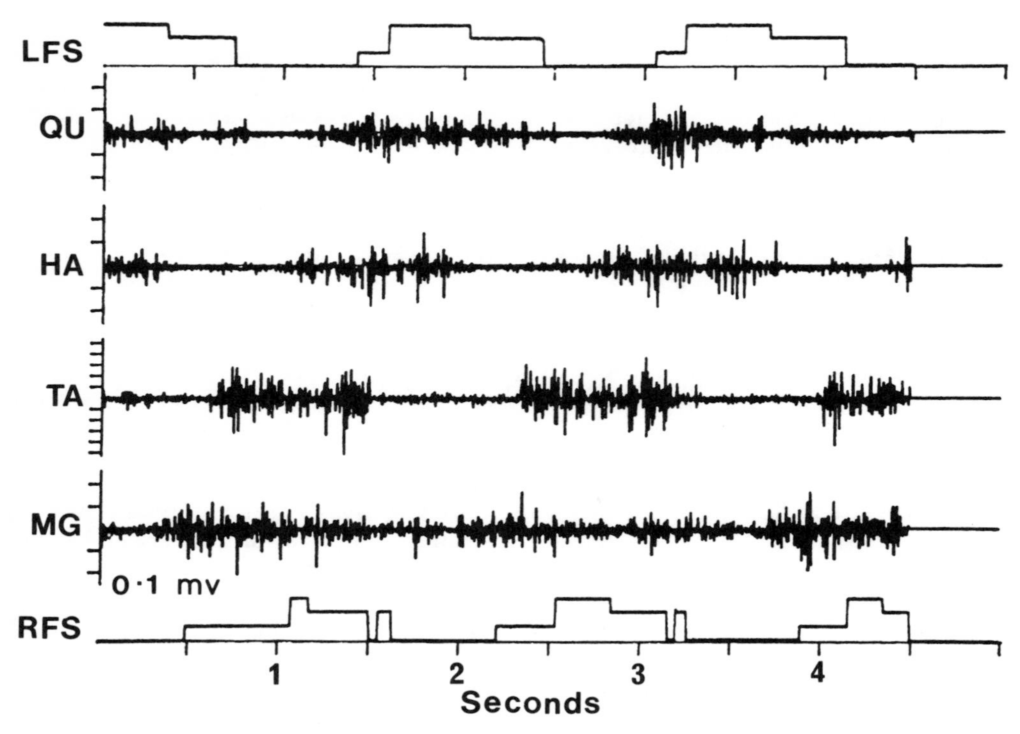

Figure 11.3. Electromyographic patterns of the left quadriceps (QU), hamstrings (HA), tibialis anterior (TA), and medial gastrocnemius (MG) muscles for a left hemiplegic subject walking at comfortable speed. The footswitch records (LFS and RFS) are shown at the top and bottom of the figure. EMG hash marks are for 0.1 mV increments.

plot, and the relative scale of 1 unit corresponds to the distance from the position in frame 1 to that in frame 58. The y origin is taken as the lowermost point reflected in frame 54. Smooth curves are drawn through the acquired data points.

At first glance, it is seen that the x coordinate varies practically linearly with time, while the y value at first decreases, then increases, remains steady, and then declines to the lowermost level. Taking the slopes of these two curves as the frame number increases, the velocity graphs of v_x and v_y, shown in Figure 11.5(b), are produced. (Velocity units are expressed as distance units/frame.)

From Figure 11.5(b) it is seen that v_x commences at its highest level, dropping to a lower level between frames 11 and 18 and remains practically constant at this level until about frame 46, after which it declines rapidly to zero. v_y has negative and positive components as time progresses. Initially v_y has a negative value since the link segment center is moving downward with time. The segment center is then momentarily stationary at about frame 5, when the y movement changes direction at increasing velocity, until about frame 12, when the maximum positive velocity is attained. Then the velocity v_y proceeds to decline to zero and this value persists for about 10 frames. The link segment center then drops, causing a negative velocity profile with a negative peak at about frame 50; finally the velocity passes through zero again.

Again, taking the slopes of the curves of v_x and v_y as the frame numbers increase gives the acceleration components a_x and a_y as depicted in Figure 11.5(c).

A burst of acceleration a_x seems to

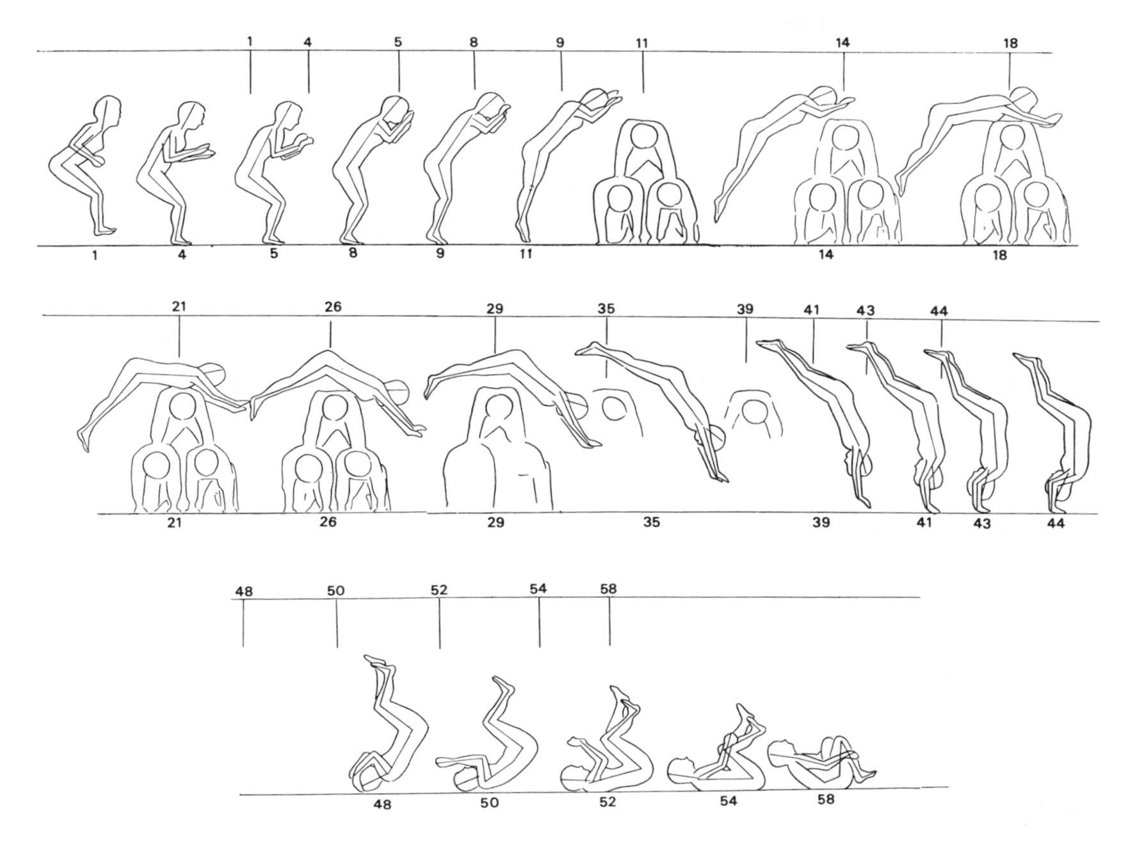

Figure 11.4. Dive for height. Series of tracings made from motion picture film. Numbered intersections of the vertical line with the upper horizontal line mark a fixed point in the background. A number marks the location of the fixed point with the figure bearing the same number. (Courtesy of University of Wisconsin Department of Physical Education for Women.)

occur between frames 11 and 18. At this time the body has reached its maximum elevation with the hips flexing so that the trunk is virtually horizontal (Fig. 11.4, frames 11-18). A burst of deceleration a_x is produced between frames 48 and 54. A component of "impulsive" force on landing taken up by the rolling back is evidently acting to brake the forward movement.

a_y reflects an acceleration component acting for the first 12 frames. Lower limb extensions coupled with the initial impulsive reaction force when the feet strike the ground serve to provide the vertical propulsive forces. The subsequent deceleration phase acting until about frame 18 reflects the utilization of the effects of gravity on the motion. From frames 18 to 28, vertical motion of the center of the segment of concern is minimal and there is no net accelerating or decelerating force influencing this. Between frames 28 and 38 a deceleration burst occurs as the body begins to descend to the ground. This is primarily due to gravity. The forward and upward propulsion of the legs that are extending adds an acceleration component in the vertical direction which generates a force to nullify a portion of the gravitational pull acting on the body. Thus, the acceleration component a_y is virtually zero from frame 38 to around frame 42, when gravity forces again play a major

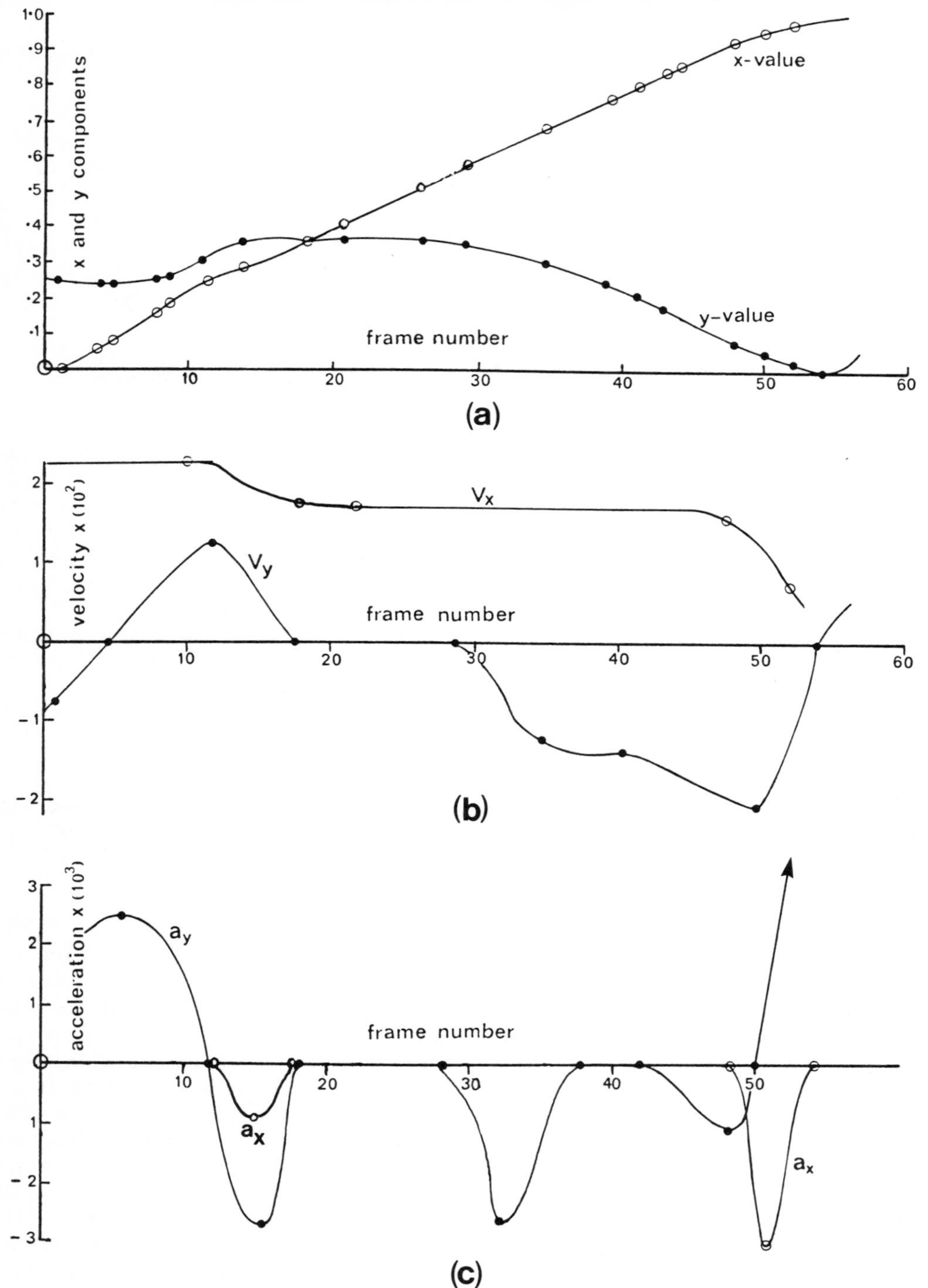

Figure 11.5. Positional, velocity, and acceleration components of "dive for height" depicted in Figure 11.4. (a) x and y positional components versus frame number; (b) x and y velocity components versus frame number, derived from (a); (c) x and y acceleration components versus frame number, and derived from (b).

role, but are countered by the reaction of the arms when the hands contact the floor. The upward thrust component and that due to rolling, lead to the substantial value of a_y beyond frame 50.

If the normalized distance scale is such that 1 unit corresponds to 4 m and the frame speed is 24 frames/s, the scales for the diagrams in Figure 11.5 must be modified. Thus, in all cases, the frame numbers must be multiplied by 0.042 (1/24) to render time in seconds. The distance scales in Figure 11.6(a) are multiplied by 4 m; the velocity scales in Figure 11.5(b) are multiplied by 4/0.042 = 95; and the acceleration scales in Figure 11.5(c) are multiplied by 2262. (Note that the peak value of the deceleration phase in Figure 11.5(c) then turns out to be 6.22 m/s^2, which is reasonable since it is less than, and close to, 9.81 m/s^2, the value for gravity alone.)

In general, attempts should be made to use equal spacings between frames of data. This will serve to facilitate subsequent data processing.

Acceleration of Body Parts

Example 1 illustrates the calculation of linear acceleration for a body component. Movement of any sort involves angular acceleration of body segments. The link segments in Figure 11.4 are seen to vary their angular dispositions in the frame sequences. The calculations of angular velocities and accelerations are performed in a fashion similar to that illustrated in Example 1 except that cognizance must be taken of the angular variations of concern.

In the analysis of gymnastic activities when the body is rotating unsupported in air or around a bar, the angles that a segment makes with one of a set of *x* and *y* axes are used and the results may be conveniently plotted on a circular graph. Figure 11.6 is made from data collected during a pike somersault performed on a trampoline. In each frame plotted, the *x* and *y* axes are drawn through the hip joint and the angles are plotted in their appropriate quadrant. Figure 11.6(a) is the graph of the trunk, while Figure 11.6(b) is that of the lower extremities.

EXAMPLE 2—Angular Velocity and Acceleration Determinations

Using the data presented in Figure 11.6(a), plot the angular velocity and angular acceleration of the trunk for a pike forward somersault if the camera is operating at 50 frames/sec.

First, a smooth graph of angular position against frame number and time (t = frame no. × 0.02 s) is plotted as in Figure 11.7. It should be recognized that the angles for frames 1 and 2 may be conveniently plotted by adding 360° to them.

As may be seen, the angular variations commence rapidly, but then slow down. It can be anticipated that there are two distinct velocity phases, an initial rapid one and then a slower one.

The angular velocity, ω, is determined by extracting pertinent slopes from the graph of angle, θ, against time and then plotting the relevant values. For convenience, the magnitude of ω, $|\omega|$, is plotted. Next, the angular acceleration, α, is derived from the ω-t curve. Its peak is sustained from frames 7 to 12. α is in fact the manifestation of a decelerating force acting on the trunk.

EXAMPLE 3—Glide Kip on Uneven Parallel Bars

Figure 11.8 depicts the angular disposition of the upper extremities in the sagittal plane of the body during the performance of a glide kip. The origin of *x* and *y* coordinates is located at the wrists and the distal end of each line represents the shoulders. The gymnast maintains elbow extension throughout the performance, one of the characteristics of skillful execution. Principal changes in angular velocity are readily seen: the movement is under way at frame F and angular velocity increases during the downward/forward movement of the body to frame I; a short period of almost no acceleration (i.e., maintaining

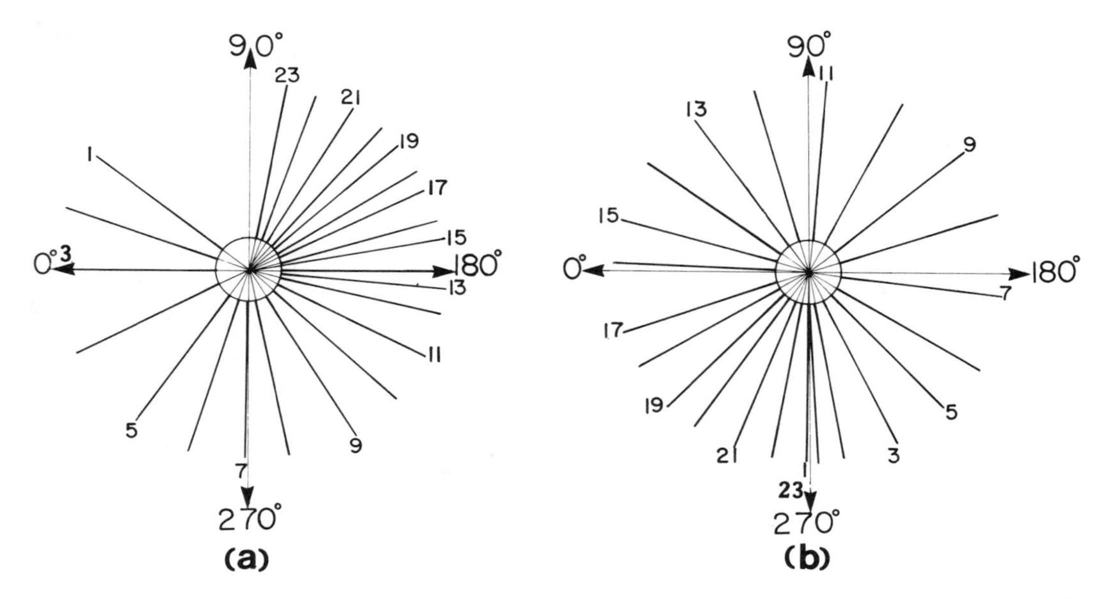

Figure 11.6. Graphs of part of a forward somersault on the trampoline. *a*, of the trunk; *b*, of the lower extremities, x and *y* axes through hip. Performer is a highly skilled university athlete. Compare the greatest angular displacement of the trunk per time, frames 1 through 5, with the greatest angular displacement of the lower extremities per unit time, frames 6 through 11. (From Reuschlein, P. L., 1962. An analysis of the speed of rotation for the forward somersault on the trampoline. Unpublished paper prepared at the University of Wisconsin.)

angular velocity) is seen in the time period between frames H and I and between frames I and J; then, the angular velocity decreases progressively until frame P, the reversal of direction of the body, when velocity is zero; near the point of reversal of direction, velocity is changing quickly and therefore acceleration is approaching its maximum for this skill; as the body now rides backward and upward toward the bar, the arms may be seen to change position rapidly at an increasing rate and are showing their greatest velocity values near the end of the movement as depicted by the changes between frames X and Y and between frames Y and Z.

The time between frames is 0.076 s: measurement of each angle of inclination of the arms provides the information necessary to plot the angular displacement of the arms against time. Such a plot reveals the characteristics of a single oscillating swing which starts at an angle of 171° at frame F, changes direction at an angle of 306° at frame P, and returns to an angle of 162° at frame Z. If the resulting plot is compared with a typical sine wave pattern, it will be seen to approximate a portion of a sine wave for a single oscillating system. The plot may have some irregular shape to it because the human body does not behave quite like a simple pendulum system. That is to say, there are forces acting within the human body in the form of muscle contractions with resulting momenta exchanges that influence the angular displacement of the arms. The first derivative of the angular displacement-time plot by comparison, closely resembles that of a cosine wave and will show the velocity changes highlighted in the above discussion. The second derivative of the displacement plot also depicts some similarity with sinusoidal variations with one burst of positive acceleration at the start of the movement for the purpose of getting under way, and another at the end of the movement, for the purpose of

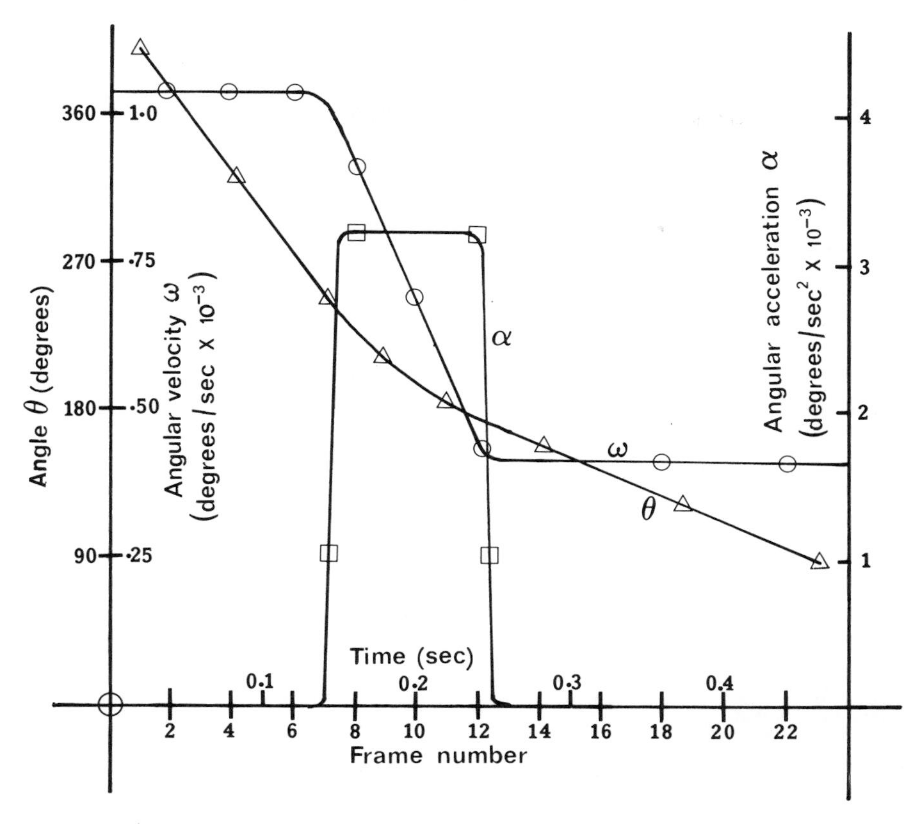

Figure 11.7. Plots of angular orientation θ, magnitude of angular velocity $|\omega|$, and angular acceleration $|\alpha|$ derived from Figure 11.6(a).

decelerating to a stop. (The interested student is invited to construct the various plots alluded to in this discussion.)

ANALYSIS OF SEGMENTAL VELOCITIES

Two examples of analyses of segmental velocities of body parts that contribute to the release velocity of a projectile are presented to familiarize the reader with the techniques involved and the problems associated with them. In any projectile skill, much of the muscular activity and resulting joint actions serve only to provide a base on which the throwing, kicking, or hitting extremity moves to project a given object; the remainder of the muscular force and resulting joint actions contribute directly to the initial velocity of the projected ball or other object. Study of films of various throwing, kicking, and striking activities have led investigators to a consensus concerning what joint actions move the kinematic chains that contribute directly to the initial velocity of the ball, javelin, etc. In the overhand throwing pattern, for example, the major joint actions involve rotation of the pelvis on one or both hips, trunk rotation around the intervertebral joints, medial rotation of the arm at the shoulder, pronation at the radio-ulnar joints, and wrist flexion. Protraction of the shoulder girdle at the sternoclavicular joint has been largely overlooked by many investigators but may play quite an important role. While the elbow does extend during an overhand throw, this action only serves to increase the radius of the arc through

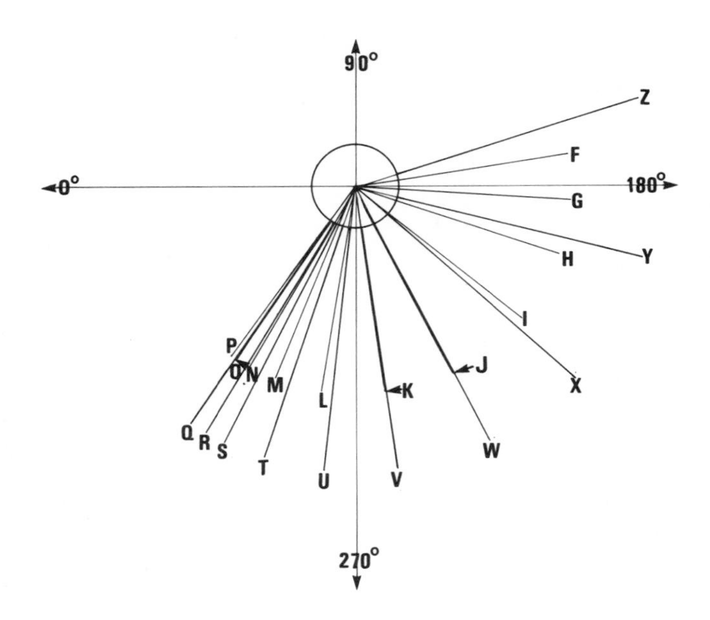

Figure 11.8. Angular displacement of the arms during a glide kip on the uneven parallel bars. x and y axes drawn through wrists. At frame P the performer is in full layout, parallel to the floor and ready to start the kip. The swing, as indicated by arms, has been decelerating from frame J and reverses as kip is performed. Note that in this case the reference position 0° is along the −x axis and clockwise movements are considered to be positive. (From Gowitzke, B. A., 1969. An analysis of levers contributing to the glide kip. Unpublished paper prepared at the University of Wisconsin.)

which hip and spinal rotation and shoulder protraction move the hand.

Investigators have long been aware of these actions, but it is only comparatively recently that a method of determining the contribution made by each of these different actions to the initial velocity of the projectile was devised first by Glassow and then Roberts (1971) in the late 1950s and early 1960s at the University of Wisconsin. Since that time more automatic and sensitive optical systems have been developed, such as the Selspot system mentioned later in this chapter. However, the principles of measurement remain the same.

TECHNIQUE

This method requires synchronized films taken from the front or back, from the side, and, whenever possible, from overhead (Fig. 11.9). (The overhead sequences increase the accuracy of all measurements of body rotation.)

Taking careful note of the sequences in Figure 11.9, one can see that the subject is wearing fins attached to the hip and shoulders. Figure 11.10 shows a subject equipped with these fins, which facilitate the measurement of trunk and pelvic rotations. The fins are made of styrofoam set in aluminum holders and the belts are elastic with Velcro fastenings. The fins are supported at the sacrum and between the shoulders. The holders and belts are designed to maintain the fin perpendicular to the sacrum and the thoracic spine. The maximal length of the fin appears as the performer poses for the scale photographs. These fins may be marked off in alternating black and white stripes or in 10-cm sections; as the body rotates, the image of the fins becomes foreshortened as they leave the picture plane.

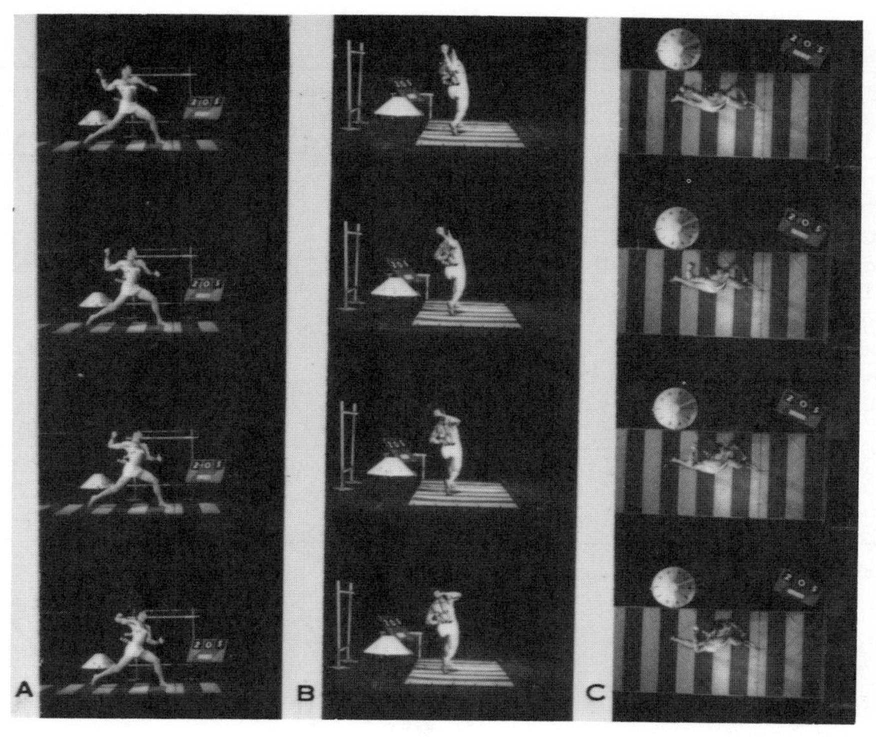

Figure 11.9. Overhand throw. Performer is photographed from (A) side; (B) rear; and (C) overhead cameras. The three films are synchronized with the aid of markings on a revolving cone. (From Blievernicht, D. L., 1967. Courtesy of the University of Wisconsin Kinesiology Laboratory and R. B. Glassow.)

The performer in Figure 11.11 also holds a measurement scale (e.g., a meter-stick) with a predetermined length—10, 20, or 30 cm—clearly marked. Similar photographs are made in each plane of view so as to calibrate the distance scales in each filmed sequence.

From Figure 11.11 it is obvious that the full length of the fin forms the hypotenuse of a right triangle while the foreshortened length forms the adjacent side. As the cosine of an angle is the ratio of the side adjacent to the angle, to the hypotenuse, it is possible to determine the angle of the trunk and pelvis from the length of their respective fins. Thus, in views such as those shown in Figure 11.10 and for the overhead camera the subject is posed in such a way that the full length of the fins is featured in the picture planes.

Figure 11.10. Scale photographs of performer wearing styrofoam fins. Side view.

Table 11.1
Overhand Throw

Segment Joint Action	Angle (a)	Angle (b)	Angular Displacement	Angular Velocity	Radians per s	PMA[a]	Linear Velocity	Contribution
				°/s		m	m/s	%
Pelvic rotation	28°	28°	0°					
Spinal rotation	17°	28°	11°	688	12	.713	8.56	26
Wrist flexion	27°(ext)	10°(fl)	37°	2313	40.4	.116	4.69	14
Shoulder (medial rotation), angle with horizontal	45°	90°	45°	2813	49.1	.177	8.69	26
Sternoclavicular	d = 5.1 cm; length clav. = 19.1 cm		16°	1000	17.6	.655	11.53	34
			Total linear velocity				33.47	
			Linear velocity calculated from Figure 11.12(d)				32.39	

[a]Projection moment arm.

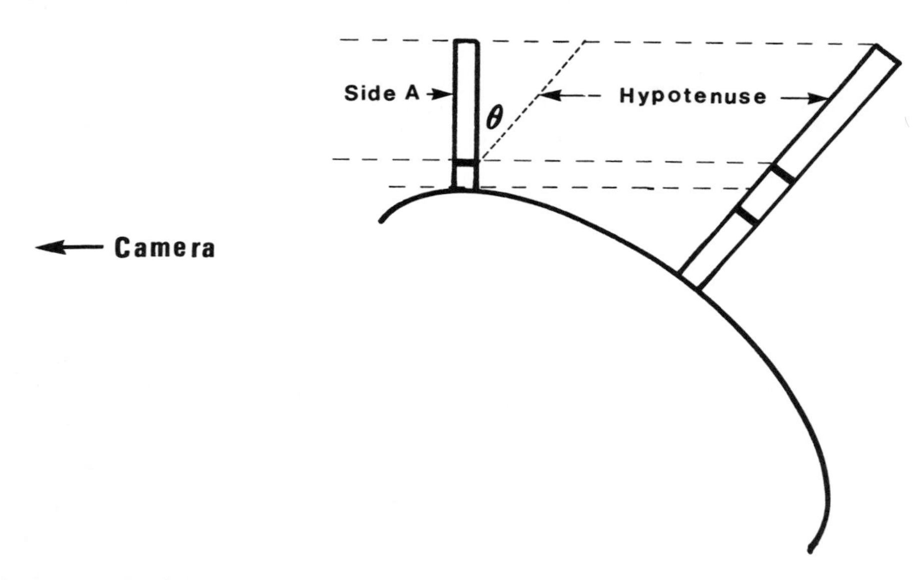

Figure 11.11. Use of fins for measuring trunk and pelvic rotation. As trunk or pelvis rotates from picture plane, fins become foreshortened. This length becomes the adjacent side of a right triangle, while actual length of fin becomes the hypotenuse of this triangle whose angle θ is the amount of rotation of the body part.

EXAMPLE 4—Analysis of Velocities of Body Segments Contributing to Overhand Throw

Tracings from several frames of film which recorded an overhand throw are shown in Figure 11.12. Note the inclusion of the scale measures. The time between frames is 0.016 sec. It is required to calculate the linear velocities resulting from all pertinent joint actions and to compare the summation of their contributions at the hand with the initial linear velocity of the ball.

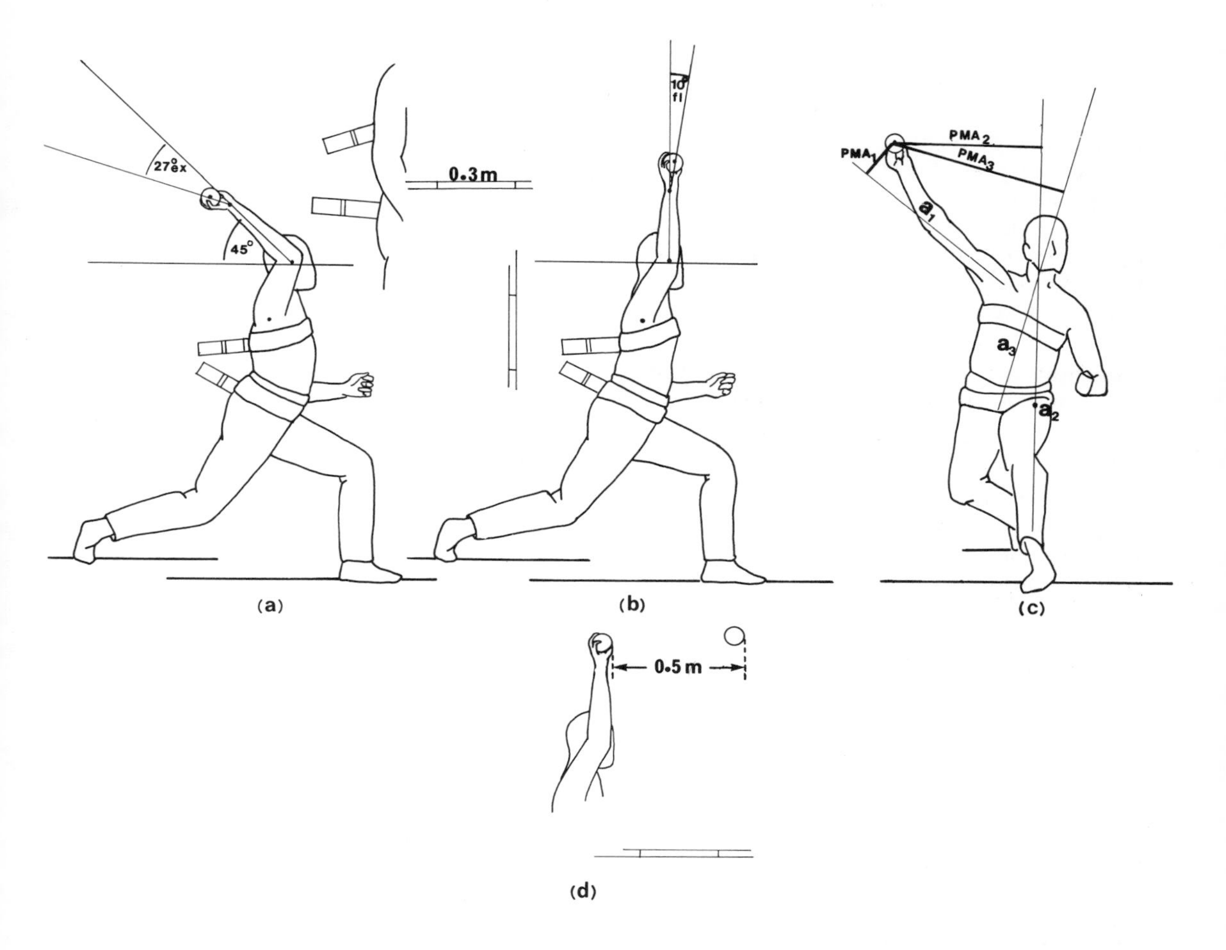

Figure 11.12. Tracings of overhand throw. (a) frame just before release, side view; (b) release frame, side view; (c) release frame, front view. Axes of rotation; a_1, at the shoulders; a_2, at the hip, a_3, of the spine. (d) distance ball has traveled from release frame to frame immediately following.

Angular Velocities of the Body Parts

At the instant of release, these velocities may be ascertained by determining their angular displacements during the last brief interval before release. The hand displacement is obtained by measuring the angle formed by the long axis of the forearm with a line drawn from the center of the ball to the wrist axis (Fig. 11.12(a) and (b); Table 11.1). The angular displacement of the forearm resulting from medial rotation of the arm at the shoulder is obtained by measuring the angle of inclination of the forearm, i.e., the angle which the forearm makes with the horizontal plane, as shown in Figure 11.12(a) and (b) (see Table 11.1).

Pelvic and Spinal Rotations

In Figure 11.12(a), the marked area of the fin between the shoulders is 21 units[a] long; in the scale, the corresponding part of the fin is 22 units, so the cosine of the angle that the trunk makes with the picture plane is 21/22 or 0.954 and the angle is

[a] On the original tracing enlargement, 1 ft measured 53/50 of 1 in, so these are the "units" referred to.

17°. Similarly, in Figure 11.12(b), that area of the fin measured 19.5 units in length so the cosine is 19.5/22 or 0.886 and the angle is 28°. The difference is 11°, which is the angular displacement of the trunk caused by spinal rotation just before the release of the ball. There is no change in the length of the pelvic fin from (a) to (b), indicating that the pelvis is held motionless just before the release.

Method of Determining the Amount of Protraction Occurring at the Sternoclavicular Joint

A tracing of Figure 11.12(a) superposed on (b) coincides with (b) to the level of the thoracic belt. (This is always the case when shoulder protraction is taking place, if there is forward motion of the entire body in relation to a fixed point during this prerelease interval, or if the entire torso shoulder protraction is not indicated.)

When the long axis of the arm in (a) is compared with that of (b), it is found that the center of the axilla moved forward 9 units (Figure 11.13). As 53 of these units equal 1 ft, there are 4.4 units to the inch. Under these circumstances the shoulder girdle has moved forward (protracted) 2 in or 5.1 cm. This subject placed his arm in the position of Figure 11.12(a), and the distance from his right sternoclavicular joint to a point in line with the center of his right axilla is measured and found to be 7.5 in or 19.1 cm. An arc with a radius of 19.1 cm is drawn and the angle subtended by a 5.1-cm chord is measured and found to be 16°.

Calculations of Linear Velocity Resulting from Joint Actions

As all angular displacements from frame to frame take place in 0.016 sec the degrees of displacement per second are easily calculated (Table 11.1). The procedure for converting degrees per second of a given joint to the linear velocity allied with the motion at an extremity, follows from Equation 3.12

Figure 11.13. Release throw. Dotted line, superimposed over prerelease throw; solid line, shows protraction of the shoulder girdle.

$$v = \omega r$$

where v is the linear velocity; ω is the angular velocity in radians/s; and r is the radius arm of concern. Thus,

$$v = \text{degrees/s} \times \frac{2\pi r}{360} = \frac{\text{degrees/s}}{57.2957} \cdot r$$

The radial length r in each case is the perpendicular distance from the center of the ball to the axis around which that particular angular displacement occurs. For the sake of convenience this distance is termed the projection moment arm or PMA. The PMA for wrist flexion is measured from Figure 11.12(b), while the other PMAs are measured from Figure 11.12(c). The PMA for the sternoclavicular joint action is parallel to that for spinal rotation and is taken as 5.8 cm shorter than that for spinal rotation (Table 11.1).

Now, all data necessary for the final calculations have been obtained from the film tracings, and only the arithmetic remains to obtain the actual linear velocity contributed by each joint action (Table 11.1). Figure 11.12(d) is a composite of Figure 11.12(b) and the frame immediately following. The only part of the post-release frame added to Figure 11.12(b) is the ball, so that the actual initial velocity may be measured. As the ball has traveled 90 units, which equals 0.518 m in the elapsed 0.016 s, this initial velocity is 32.39 m/s. This value compares extremely well with the sum of the linear velocity contributions. The last column of Table 11.1 shows the percentage of this velocity contributed by each joint action.

Comparison of these data with those from Cooper and Glassow (1976) (Table 11.2) indicate at least two weaknesses in the particular throw reflected in Table 11.1: lack of pelvic rotation and a comparatively small amount of wrist flexion. On the other hand, the Cooper and Glassow data indicate the entire absence of shoulder protraction. This absence seems rather unusual in the light of electromyographic studies which always indicate that considerable activity in the serratus anterior accompanies any forward motion of the corresponding arm.

EXAMPLE 5—Analysis of Velocities of Body Segments Contributing to Underhand Throw

Tracings from film frames of an underhand softball pitch are shown in Figure 11.14. The time between frames is 0.016 s. It is required to determine the linear velocities resulting from pertinent joint actions and to compare the summation of their contributions at the hand with the initial linear velocity imparted to the ball.

The same procedures are followed as in the overhand throw, but in this instance the characteristic shoulder action is flexion (Fig. 11.15; Table 11.3). There are, however, two problems which do not arise in the analysis of the overhand throw: first, the problem of *effective* spinal rotation when the pelvis is also in motion at the time of release; and second, that of calculating the amount of wrist extension in frame (a).

Effective Rotation of Spine and Pelvis When Pelvis is also in Motion

Whenever the pelvis is rotating on the heads of one or both femurs, the torso is naturally carried through the *same angular displacement*. Under these circumstances any one of the following three situations may be occurring.

1. The angular displacement measured at the thoracic level is **less** than that calculated for the pelvis; therefore, the spinal rotation taking place at the intervertebral joints is in a *direction opposite to* the pelvic rotation. In such an instance the amount of spinal rotation must be subtracted from the total pelvic displacement to arrive at the **effective pelvic rotation.** The spinal rotation, being in the opposite direction, makes no contribution to the speed of the ball.
2. The angular displacement measured from the thoracic fin is the same as that calculated for pelvis, so there is no rotational activity at the intervertebral joints.
3. The angular displacement resulting from spinal rotation is **greater** than that measured at the pelvis; therefore spinal rotation has been taking place and is in the same direction as that of the pelvis. In order to arrive at the actual velocity contributed by the spinal rotation, that amount of angular displacement caused by the rotating pelvis, must be *subtracted* from that measured at the thoracic level. For example, in Figure 11.14(a), the angle of the fin attached at the pelvis with the picture plane is determined to be 24° and in (b) the angle is 17°, so the difference is 7°. The thoracic fin angle in (a) is 29° and in (b) 17° so that the total displacement is 12°. However, as 7° of this displacement is due to the pelvic rotation, only 5° results from rotation at the intervertebral joints, and this amount is considered the **effective spinal rotation** (Table 11.3).

Table 11.2
Velocity Contributions[a]

	Range	Angular Velocity	Moment Arm[b]	Linear Velocity[b]
	°	°/s	m	m/s
Hip rotation	20.6	824	0.646	9.29
Spinal rotation	9.6	384	0.838	5.46
Shoulder rotation	38.5	1540	0.082	2.22
Wrist flexion	60.0	8571	0.149	22.35
Total linear velocity				39.48

[a]From Cooper, J. M., and Glassow, R. B., 1976. *Kinesiology*, second edition. St. Louis: The C. V. Mosby Company.
[b]Converted to S.I. units.

Table 11.3
Underhand Softball Pitch

Segment	Angle (a)	Angle (b)	Angular Displacement	Angular Velocity	Radians per s	PMA[a]	Linear velocity	Contribution
				°/s		m	m/s	%
Pelvic rotation	24°	17°	7°	438	7.64	0.482	3.68	16.3
Spinal rotation	29°	17°	12° 5°[b]	313	5.46	0.378	2.06	9.2
Shoulder flexion, angles of inclination	95°	85°	10°	62.5	10.91	0.762	8.31	37.0
Wrist flexion	22°(ext)	20°(flex)	42°	2625	45.81	0.125	5.73	25.5
Sternoclavicular protraction	d =2.3 cm; length clav. = 19.1 cm			438	7.64	0.354	2.70	12.0
			Total linear velocity				22.48	
			Linear velocity calculated from Fig. 11.14(b)				22.29	

[a]Projection moment arm.
[b]Effective spinal rotation: 12° − 7° = 5°.

Table 11.4
Velocity Contributions[a]

	Range	Angular Velocity	Moment Arm[b]	Linear Velocity[b]
	°	°/s	m	m/s
Hip rotation	12	400	0.442	3.08
Spinal rotation	6	200	0.488	1.70
Shoulder flexion	22	733	0.762	9.75
Wrist flexion	30	3750	0.107	6.98
Total linear velocity				21.51

[a]From Cooper, J. M., and Glassow, R. B., 1976. *Kinesiology, second edition. St. Louis: The C. V. Mosby Company.*
[b]*Converted to S.I. units.*

Determination of Amount of Wrist Extension just before Release

This problem arises because the arm is laterally rotated and the forearm supinated (Fig. 11.14(a) and (c)) so the amount of wrist extension cannot be measured from tracing (a) alone, as was possible in the overhand throw. This necessitates a second front-view tracing, that of the prerelease frame, Figure 11.14(c). Examination of this tracing shows that there is considerable radial deviation as well, and that the volar surface of the forearm is not fully in the picture plane, i.e., the forearm is facing obliquely forward and to the left of the thrower (i.e., a supinated forearm and lateral rotation at the shoulder). Consequently, if the distances from the wrist axis to the center of the ball in tracings (a) and (c) are taken as the side adjacent (frame (a)) and the hypotenuse (frame (c)), it is possible to determine (or at least make a good estimate of) the degrees of wrist extension. In (a) the distance measured 12.5 units and in (c) it measured 13.5 units, so that the cosine is 0.926 and the angle is 22° (Table 11.3).

From the data in Table 11.3 and from similarly derived data, the investigator may draw conclusions as to the efficacy of the various joint actions in contributing to the throw. Comparison of the data in Table 11.3 with those from Cooper and Glassow (1976) in Table 11.4 indicates that this throw was somewhat faster and that shoulder flexion makes the greatest contribution to the initial velocity of the ball.

Value of Determining Segmental Velocities

From Examples 4 and 5 it is evident that the analyses undertaken can pinpoint any particular weakness or strength in a given performance.

Movement analyses demonstrate that the body parts that exert the greatest moment of *resistant* force, either as a result of weight, length of moment arm, or both, are those parts that initiate the forward motion of a throw, hit, etc. On the other hand, the part or lever lightest in weight and with the shortest moment arm (least moment of resistant force) is the last segment to be moved. Thus, in a throwing or striking skill the forward rotation of the pelvis on one or both hip joints may be initiated even before the upper extremity completes the backswing phase of the throw or hit. The wrist snap and/or finger extension occur as the ball is released or hit. Ideally, all angular displacements accelerate to the desired maximal velocity at the instant of release or impact so that optimal results are achieved. When maximal velocity and distance are desired as in a throw from the outfield, a javelin or discus throw, or a golf drive, this early action of the proximal joints results in a "crack the whip" effect which, aided by the more distal joint actions, contributes significantly to the speed at which the object leaves the hand or striking implement.

This summation in optimal sequence was originally introduced by Morehouse and Cooper (1950) and is reproduced here in Figure 11.15, where each curve represents a plot of the changing linear velocity contributed by the hand(s) or foot by the joint action it represents. In theory each joint action should impart maximal linear velocity at the instant of release or impact (Fig. 11.15(a)). However, it is quite possible that only one joint action may be out of phase, e.g., the wrist snap in a hitting skill, or that several actions are not timed properly (Fig. 11.15(b) and (c)). Heretofore, there has been no means of determining just how adequate the coordination and/or timing of a skill may be, except by the end result. Now by using the technique of segmental velocities, it may be possible to determine the instantaneous linear velocity of the hand(s) or foot contributed by any joint action at any given point on the recorded data. By this means the instant of greatest velocity resulting from a given joint action may be ascertained. Comparison of this velocity with that of the release or impact may reveal the degree of coordination and timing of the performance.

This discussion assumes that all dis-

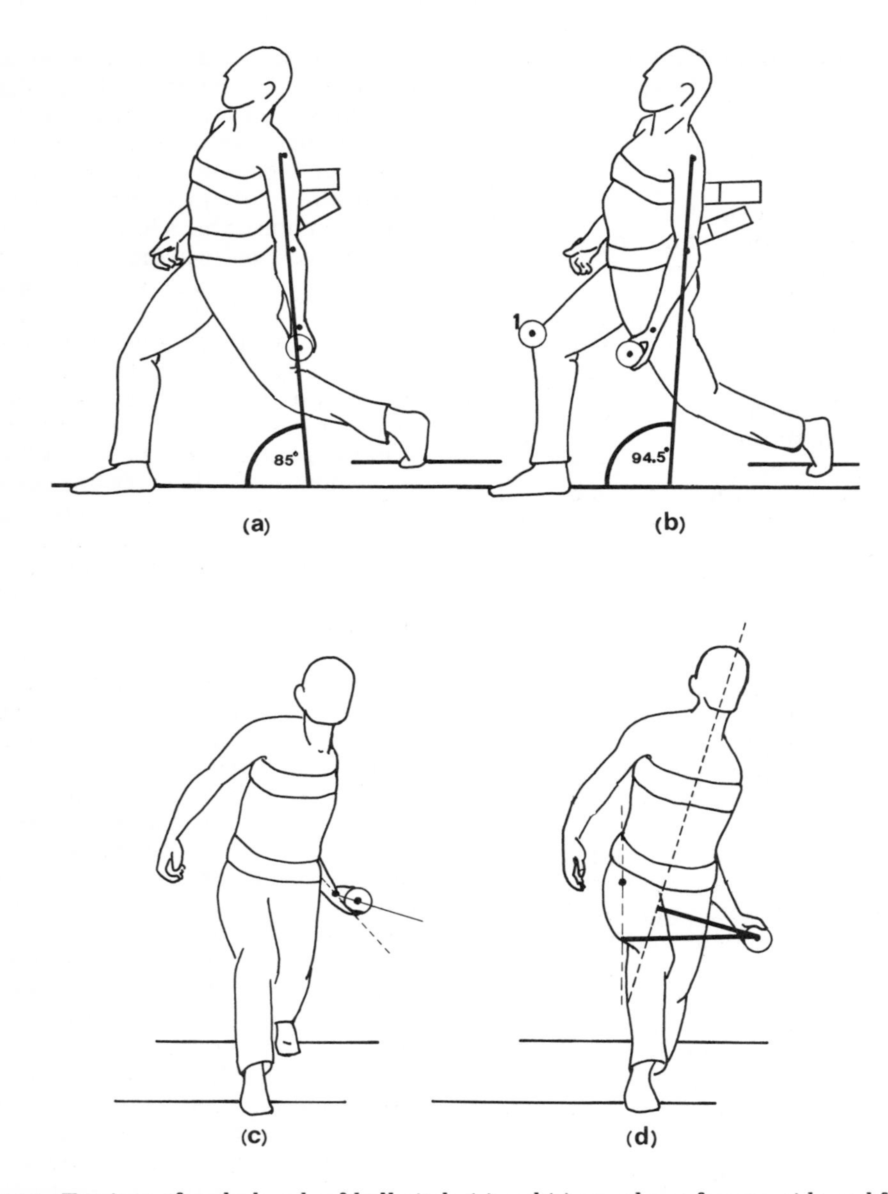

Figure 11.14. Tracings of underhand softball pitch. (a) and (c) prerelease frames, side and front; (b) and (d) release frames, side and front. 1, location of ball next frame after release.

placements, velocities, and accelerations are measured in the same plane. When multiplanar measurements are made, such as from the sagittal and frontal views of Figure 11.14, special care must be taken not to confuse and directly add velocities occurring in different planes of motion.

Much controversy surrounds this concept of summation of velocities because a direct summation, implied by Figure 11.15(a), suggests that each joint action imparts maximum velocity at the instant of release. We know, for example, that some joint actions abruptly stop (or even

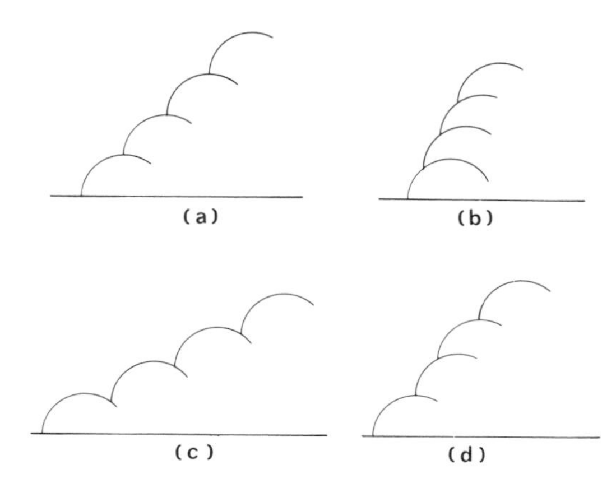

Figure 11.15. Schematic diagram of summation of segmental velocities. (a) maximal velocity of each segment at release/impact; (b) peak segment velocities not reached at release/impact; (c) peak segmental velocities passed at release/impact; (d) one segment out of phase. (After Morehouse, L. E., and Cooper, J. M., 1950. *Kinesiology*. St. Louis: C.V. Mosby.)

reverse) prior to release, thus accelerating segments distal to them commensurate with the abruptness of the stop, i.e., impulse. At best, summation of segmental velocities is a method which expresses "ballpark" figures sufficiently accurate to note some trends which may be useful in analyzing the technique of a skill.

ON INSTRUMENTATION

Technological innovation has played a vital role in promoting studies in kinesiology. In what follows, emphasis is on instrumentation mainly for studying locomotor function, but the possible ramifications for wider applications should be readily apparent to the serious student.

Capture of Body Motions

The basic ideas concerning the capture and subsequent presentation of kinesiological information are by no means a product of this decade. Our advances hinge upon technological developments and innovation that already have occurred, particularly over the last century. At the turn of the century the French physiologist Marey studied facets of human locomotion aided by his own advancements in photography. He possessed a locomotion cart used to track an ambulating subject and in which the necessary film processing could subsequently be performed. He produced diagrams reflecting the trajectories of the head, shoulder, hip, knee, and ankle in the sagittal plane.

Since that time, technological advances have allowed greater measurement precision, expanded the number of movement parameters that may be measured, and greatly siimplified the task of obtaining and processing kinematic data. Early advances involved principally photography techniques, but in the last two decades, electronics and computers have revolutionized kinematic and kinetic studies of motion.

There are several different methods of obtaining kinematic information during movement studies, with each method employing different technology. These methods may be separated into those employing imaging, goniometry, or accelerometry.

Imaging Techniques

This is the methodology used historically to study human movements. Simple two-dimensional photography falls into this category, with the developed film being analyzed to obtain the desired data. Although the types and levels of technology vary widely in imaging studies, the methodologies employed are quite similar. Whatever technology is employed, the objective is to track and capture the movement, in two- or three-dimensional space, of conveniently selected and marked body sites. Sites usually considered most convenient are the bony prominences close to the joints.

To measure motion in its entirety requires measurements in three dimensions. However, this is not necessary for studies where motion in only one plane is of interest. Three-dimensional studies of motion require at least two synchronized recording cameras, each simultaneously recording movement in a different plane of motion. Each relevant body site must be in view of at least two cameras at all times. The measurement planes may be orthogonal to each other and correspond to the plane of Figure 11.1, or they may be user-selected with movement information in any required plane obtained by using computer processing and an appropriate transformation program. Although the most general and flexible movement study system employs three-dimensional measurement, we will present only two-dimensional or single-plane methods because of ease of presentation and discussion.

Imaging systems may employ either fixed or moving cameras. In locomotion studies, the moving camera method is often employed. As an example, consider a subject walking along a walkway with a sagittal plane camera mounted on a cart moving parallel to the path of the subject and at the same speed. This arrangement allows precise measurement of the anatomical site locations for the full length of the walkway. If only a single, fixed camera is used, the same precision is limited to a very short segment of the walkway. The type of movement being studied and the precision required dictate whether the cameras should be fixed or mobile during the study. In using photographic methods, a marked background or reference grid is required so that views of the moving subject may be related to the environment in which the subject moves. When using fixed cameras, simple background marking schemes enable adequate referencing and scaling of photographed information.

The most common imaging technique for obtaining kinematic information has been cinematography. In this, 16-mm movie cameras capable of operating up to 500 frames per second are used to photograph the moving subject. The type of camera, film, and frame rate required are determined by the motion studied, whether the movements are very rapid, at a distance, etc. Body sites are indicated by visible markers, and a suitable level of illumination may have to be provided by special lighting, especially at high frame rates. Following film processing, the location of these sites in the plane of recorded motion (horizontal and vertical coordinates) must be determined for each frame. Several devices such as the Vanguard Motion Analyzer have been developed to facilitate this data extraction task. The coordinates may then be entered into a computer program and the appropriate kinematic and/or kinetic parameters calculated using suitable programs. Stick figure diagrams such as those of Figure 11.17 may also be generated. Current technology makes it possible to extract the site coordinate information from the film "on-line" to a microcomputer system using a high-resolution digitizing tablet.

Videotaping

In recent years, videotaping has become a very popular replacement for movie cameras. Current video cassette recording (VCR) equipment makes it possible to acquire, store, and display high quality images of movement, providing the better models of cameras and VCRs are employed. The high-quality VCRs have excellent stop-frame capabilities, which may be used to digitize marked body site locations. The current generation of video disk equipment is even more suitable for single-frame analysis, and the volume of data that may be stored and retrieved is unequaled.

Timing Devices for Cinematography and Videotape

One of the problems facing a user of cinematography (or videotape) who also wants to simultaneously record other movement-related data, such as EMG or force signals, is synchronizing the two data sets. Various timing devices have been designed to synchronize film views as well as providing the timing and duration of recorded human movements.

Blievernicht (1967) presented the

design of a cone-shaped rotating device for synchronizing three cameras in the three cardinal planes (Fig. 11.9). The cone, driven preferably by a synchronous motor, rotates at 60 rpm and two time scales (each having 50 equal divisions) were marked around the surface of the cone each at a different level. A number appeared at every fifth division progressing from 0 through 9. The two scales were marked 90° out of phase so that two floor-level cameras set perpendicular to each other recorded the same number on film. From overhead the cone appears flat. Two stationary pointers (one for each scale) set at right angles to each other ensure that pertinent readings are made from records derived from an overhead camera. Each scale division represents 1/50 s and therefore by interpolation, time may be estimated to 1/100 s.

Figure 11.16 shows a synchronizing device based on the same principle that also allows three-plane viewing. This device affords a 0.04-s resolution and produces sequences of electrical pulses to enable synchronization of EMG records and the like.

Automatic Imaging Systems

One of the problems associated with using cinematography or other photographic imaging methods is the considerable effort and time delay required for extracting the body site coordinate information from the films. Although this may be acceptable for kinesiology or biomechanics laboratories studying normal or athletic movements, the delay and cost in staff time may be prohibitive in a clinical laboratory where abnormal movement patterns are assessed or used in diagnosis. Considerable effort has therefore been expended in designing and constructing automatic imaging systems.

One of the earlier efforts used videotaping rather than film (Winter et al (1972b)). A major advantage of this system is the fact that the data collected are immediately available for processing. Film, on the other hand, requires special arrangements for rapid development, and errors and omissions are discovered too late for corrective action.

Body markers consisting of halves of ping-pong balls covered with reflective tape and filled with low-density polyfoam are attached at anatomical landmarks with double-sided adhesive tape. Background markers consisting of larger discs of reflective tape provide a spatial reference for the TV recording system, which utilizes 525 lines. The subject is tracked using a television camera and monitor mounted on a cart at about 3 m to the side. The lighting is adjusted to provide a very high contrast image, which enables a one-bit conversion of this data into the computer (1 for white, inside a marker, 0 for black, outside a marker.).

The TV recording is replayed on a videotape recorder and converted into digital format via a TV-computer interface. Computer programs reduce the converted data, cluster the points from each marker, and calculate the absolute coordinates of the geometric centers of the markers. At the same time corrections are made for parallax error.

This system made excellent use of the TV technology available at that time. As well, in addition to considerably speeding up the data reduction process, it also had two features essential for a clinically acceptable system: (1) minimal encumbrance to normal gait, with easy and quick preparation of subjects who walk on a regular floor surface in a normally lit environment; and (2) sufficient stride analysis to allow stride-to-stride variations to be assessed. Current VCR technology has made this approach attractive again and several centers are developing similar technology.

Storing only the coordinates of the markers extracted from a video signal, Cheng (1974) used a normal television camera as the optical sensor. In this, the x - y data are transferred to a PDP 11/10 computer under program control. It is claimed that two points on the same scanning-line can be detected only if they are sufficiently far apart (1/6 of the picture width).

Andrews and Jarrett (1976) developed a multiple camera system interfaced with a PDP-12 computer. A maximum of five horizontal coordinates may be generated on any one television line.

Bruegger and Milner (1978) reported on a system that enables virtual on-line scaled displays of tracked body motions. In this system they utilize a charge-coupled device image sensor composed of a matrix of individual light-sensing elements. The subject is instrumented at anatomical landmarks by bright incandescent lamps whose light falls on a charge-coupled device when focused by a lens system. The device is scanned electronically by an on-line computer, and the illuminated sensing elements are determined. These data are used to establish the coordinates in space of the body markers.

Although the above automatic systems are interesting from a development perspective, they have only been used in the development teams' laboratories. Consequently, their impact on motion studies in general has been small. During the last decade, however, several commercially available systems have been developed. In common with the above systems, they all require online computers for both data collection and processing. This does not add significantly to their cost since current inexpensive microcomputers have sufficient power for nearly all motion analysis data acquisition, processing, and display. An additional advantage of hav-

Figure 11.16. Synchronizing and timing device for two perpendicular film views with electromyography or other related movement recordings. (Constructed by J. Moroz, Technician, Physical Education Department, McMaster University.)

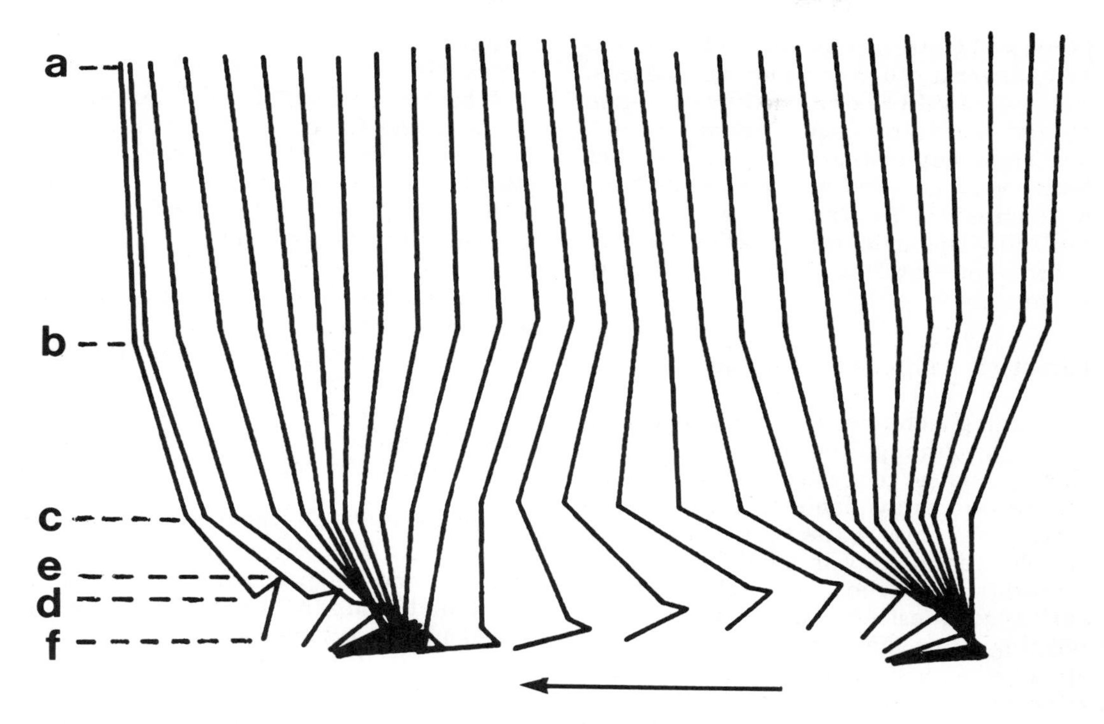

Figure 11.17. Stick figure diagram of a three-year-old normal girl walking at her natural cadence. Selspot markers have been placed at (a) shoulder, (b) greater trochanter, (c) knee, (d) malleolus, (e) heel and (f) fifth metatarsal.

ing on-line computer collection of motion data is that other analog data such as EMG and force signals may be sampled and acquired synchronously, removing the need for timing devices such as that shown in Figure 11.16. The ready integration of motion, EMG, and force data makes such systems extremely attractive for both basic, normal, and clinical movement studies.

One of the first commercial systems, the Selspot (Selcom (1976)), uses a silicon photosensor capable of transducing the coordinates (in a plane) of light-emitting diodes (LEDs) placed at anatomical landmarks. Multiple LEDs may be tracked by sequentially switching them on at very high rates. Frame rates in excess of 300 frames/sec are easily achieved for two- or three-dimensional studies. The LEDs are connected through individual cables to the switching circuitry and power supply carried by the subject. A more current, less-expensive system, the Watsmart (Northern Digital, Waterloo, Ontario, Canada) has been recently developed, based on the same principles. An early failing of these commercial systems was the lack of suitable computer programs to acquire, process, and display kinematic, EMG, or other motion analysis data. Individual laboratories were required to develop their own software. However, the current manufacturers are concentrating on the computer programming as well as the hardware in an effort to produce "turn-key" motion analysis systems. Figure 11.17 shows a stick figure representation, generated using a Selspot-based system, of a three-year-old child walking in a normally lit room. Such displays and the associated position, velocity, and acceleration data, as well as joint angle trajectories, are available to the investigator while the subject is still instrumented and in the laboratory. Several other commercial systems, such as the VICON or ExpertVision, based on different technologies, are also available. These automatic motion analysis systems, when coupled with a suitable body of computer programs, are gradually replacing cinematography as the standard equipment for the study of human movement.

Direct Measurement Techniques

Imaging technology is relatively expensive and, although the most general and flexible, may exceed the human and financial resources available to many laboratories. Far less expensive and much simpler technology is available to such investigators who are asking more restricted questions and do not require imaging data. These instruments produce analog electrical signals that are related to movement variables and can be displayed using a chart recorder or acquired directly by computers employing analog to digital converters.

Footswitches. Perhaps the simplest studies of human locomotion may be performed utilizing switches which can signal the contact of the extremities with the ground. If studies are to be undertaken in a fixed place, it is fairly easy to lay down a length of aluminum plate 1.5 mm thick, 45 cm wide and several meters long to serve as one contact of a switch, the other contacts being metallic members attached to the subject's footwear or even to the bare feet. Reasonably economic and useful footwear such as tennis or running shoes can be simply fitted with metallic contacts. Some characteristics of walking which can be delineated are the stance and swing phases when heel and toe contacts are used. Increasing the number of contacts enables the sequences of different components of footfall to be studied. Several footswitch responses may be recorded on one recording channel by placing different resistors in series with each footswitch lead, having the current in the footswitches pass through a fixed resistor across which the recorder is connected.

A battery or a suitable d.c. power supply energizes the complete circuit. Utilizing this method, Milner and Quanbury (1970) studied aspects of control in human walking. In this work, timing relationships as a function of speed and pace period are dealt with for several normal subjects. An interesting observation is that the speed of walking is determined by both frequency of stepping and pace length such that these two components are linearly connected. It seems that neuro-musculo-skeletal system defects may be characterized by deviations from this particular control law.

Brandstater et al. (1983) investigated the gait of hemiplegic subjects using only footswitches and the measurement of walking speed and found that the average walking speed and the symmetry of the single limb support durations are highly correlated with the degree of hemiplegia.

Other footswitch designs are currently being used. One type involves the fabrication of insoles containing pressure-sensitive, resistive materials under the foot areas of interest. Others employ spring-loaded mechanical switches that close when sufficient pressure has been applied. All of these switches can produce electrical signals similar to those of Fig. 11.20. Footswitches should affect the locomotion as little as possible while still being robust enough to withstand the high forces exerted on impact during gait. Currently, thin, pressure-sensitive, film resistors on a plastic backing such as those used for computer or other control keyboards offer many exciting advantages. They are very thin and may be placed easily on the bare feet without perturbing the gait. At present, no type of footswitch is recognized as being the standard, and development of better foot switches continues as better materials and technologies become available.

Measurement of Walking Speed. As stated previously, walking speed is an important descriptor of locomotion. Many parameters such as joint ranges of motion are related to the speed of walking. Therefore, it should always be measured when studying human locomotion. When imaging techniques are employed, the walking speed for each stride may be directly obtained from the velocity of the anatomical marker such as at the pelvis or the shoulder. However, the simplest method of obtaining average walking speed is by placing two adhesive strips on the walkway perpendicular to the plane of progression at a known distance apart, and using a stop watch to determine the time required by the ambulating subject to cross the two markers. An automatic version of this technique employs two infrared light beams placed a known distance apart, at approximately chest height, and an electronic elapsed time counter. The accuracy of this technique increases with increasing distance between markers or light beams. However, it must be stressed that only average walking speed may be measured this way. For most normal or stable abnormal locomotor studies, this is fortunately quite acceptable.

Electrogoniometers. If joint angles are of interest primarily, goniometers may be employed which measure the angle between any two adjacent body segments. Both two- and three-dimensional goniometers have been developed and are currently being used. Figure 11.18 shows an exoskeletal electrogoniometer (after the design of Lamoreux (1971)) fitted to a subject. These electrogoniometers may also be fitted at the ankles and hips. Electrogoniometers that measure sagittal and transverse plane rotation also have been developed and are commercially available. As well, this basic design has been modified to measure upper-limb angular motion. A commercially available system that utilizes polarized light, appropriate sensors, and special electronics also enables remote goniometric records to be obtained.

The electrogoniometer outputs are sampled at a suitable frame rate (e.g., 100 frames/s) by a digital computer which also samples footswitches used to determine the contacts of the foot (heel and toe) with the ground. After performing simple calculations on the data, displays such as that of Figure 11.20 may be generated. These instruments may also be directly connected to a chart recorder and multistride joint angle and footswitch trajectories recorded together with suitable calibration signals. Although electrogiometers are inexpensive relative to imaging techniques and may record multistride data during a walk, they possess several disadvantages. Goniometers measure relative rather than absolute angles; this may be insufficient where absolute angles are required, such as the degree of inclination of the trunk. Additionally, they may prove cumbersome to gait. especially in pediat-

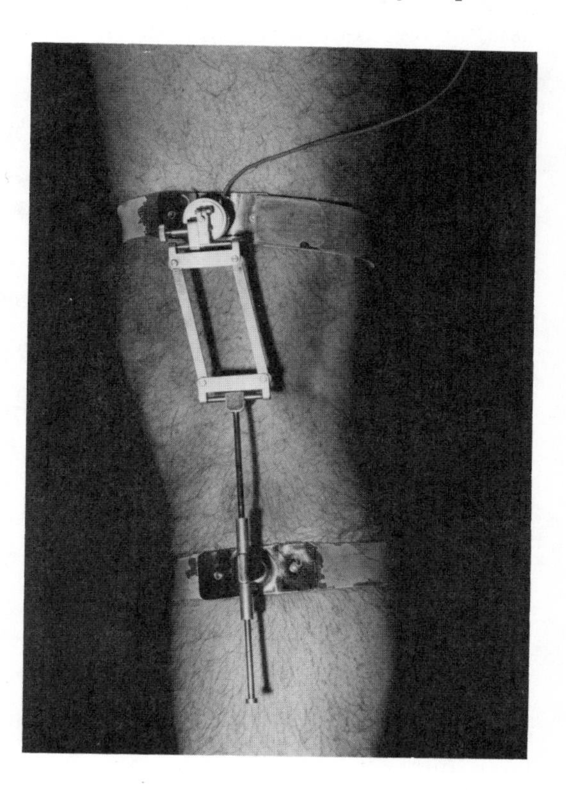

Figure 11.18. Electrogoniometer. The electrogoniometer is fitted to measure knee-joint angle variations in the sagittal plane. The mechanical linkages used ensure that only movements in the plane of concern affect the precision electrical potentiometer that is used to transduce joint angle.

ric studies. However, their low cost and ease of use probably outweigh their disadvantages in most clinical settings.

Accelerometers. Accelerometers which are compact and responsive to accelerations imposed upon them are commercially available. They may be used to detect linear and angular acceleration components. Interestingly, Newton's law dealing with force, $F = ma$, provides the basis for the construction of the accelerometers most commonly used in kinesiology. Often, a piezoelectric crystal has attached to it a weight whose inertia produces a force which acts on the crystal to generate an electrical voltage in the crystal proportional to the acceleration. (There are also piezoresistive accelerometers.) By using an electronic integrator whose input receives the output of an accelerometer, it is possible to arrive at a measure of the velocity of the accelerated object. A further integrating stage leads to the acquisition of displacement data.

Moments of Inertia of Limb Segments. Accelerometers, when used in the "quick-release" method, enable the accurate calculation of the moments of inertia of different body parts (Carlsöö (1972)). The method hinges on the equation $T = I\alpha$ where T is the total moment of force with respect to the fixed joint axis and is comprised of the moment of gravity and the moment of muscular force; I is the total moment of inertia incorporating that of the body segment and the gauge's attachment device to the body; and α is the angular acceleration of the body segment. Figure 11.19 illustrates the method for obtaining the value of I for the lower leg and foot. The cord, C, attached at the ankle connects to a force gauge, FG, so that the cord tension can be measured. The movement axis of the knee, K, is held fixed and an angular accelerometer, A, is attached to the lower leg. The subject is asked to extend the knee and thereby produce an isometric contraction of the quadriceps. A sudden release (quick release) of the cord attachment, without the subject's knowledge, results in a forward rotation of the lower leg as it extends. Taking

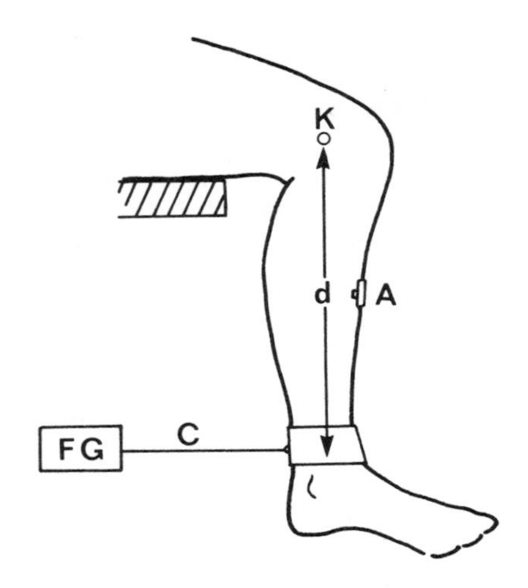

Figure 11.19. Quick-release method for determining the moment of inertia of the lower leg.

moments about the axis K leads to the required result. The moment, M produced by the musculature is $M = Fd$, where d is the perpendicular distance from K to the line of pull of the cord C. It can be confirmed electromyographically that the subject is unable to relax the muscles at the instant of release. The moment of force due to gravity is conveniently made zero by ensuring that the gravitational force acts directly under the joint axis. It therefore follows that $T = M = Fd = I\alpha$, i.e., $I = Fd/\alpha$.

This result may be corrected for the moments of inertia of the attachments, their values simply being subtracted from I.

Comparison of Methods

Although much effort and ingenuity have been expended in developing new kinematic measurement technology, little attention has been paid to determining the reliability and validity of subject measurements using this technology. When validity and reliability are considered at all, they usually concern the accuracy and precision of measurements made on test jigs, i.e, the hardware accuracy and reliability. Currently, the precision and accuracy of three-dimensional automatic systems such as Selspot and Watssmart are being addressed using very complex mathematical calibration schemes. It certainly is desirable to have as accurate and precise a measurement system as possible. However, equipment cost and utility are also very important factors and may force the sacrifice of some of the achievable accuracy and precision. As well, hardware accuracy is not directly translatable to measurement validity when such measurements are made using surface instrumentation mounted on human subjects.

Stanic et al. (1977) compared cinematography, Selspot, and electrogoniometry from the standpoint of application, data collection, and utility, but made no attempt to compare their accuracy and reliability. Winter (1979), in his book on biomechanics, also compared the current methods of obtaining motion data without determining their relative measurement accuracies. It is therefore difficult to directly compare results reported by laboratories using different motion analysis systems. De Bruin et al. (1984) determined the validity and reliability of sagittal plane knee joint angle measurements using Selspot and electrogoniometers. Measurements were made over the normal range of knee motion for ten subjects and simultaneous x-rays were taken to provide a "gold" standard of comparison. The effects on validity and reliabliity of reinstrumenting each subject were also addressed. Their results showed that, although both systems were accurate within one degree when measuring test jig angles, the electrogoniometer and Selspot systems measured, on average, only 70% and 90% of the true sagittal knee joint angle, respectively. For both systems, the average error was linearly related to the joint angle, allowing one to apply a correction factor to the measured values. These errors result from the movement of soft tissue over the skeleton as the joint angle changes. However, even following correction, errors up to ten degrees were still present for both systems. These resulted in part from misapplication of the Selspot markers over the bony landmarks and errors in calibration of the electrogoniometer. The Selspot results also apply to cinematography or other photographic imaging systems since the errors resulted from marker application to the subject and not from the measurement system itself.

The sagittal knee joint angle is one of the easiest to measure. Other body joint angles with the markers closer together may have even more measurement error. This is especially true of measurements in other planes of movement where marker-skin movement over the underlying skeleton presents an even greater problem. Therefore, a note of caution should be applied when specifying the accuracy required in the measurement system or present in measurements of body motion.

Presentation of Data

Numeric values may be derived for many gait parameters such as average walking speed, stride length, range of joint motion, and stance/swing phase ratio. These may be expressed in tabular form. However, patterns of movement or time histories are best presented graphically; to misquote the well-known proverb, "a picture is worth more than a thousand values presented in a table." This is especially true in the clinic environment where ready interpretation of data is absolutely essential. Numerous display possibilities exist including:

1. Trajectory plots–*x* versus *y* for each body marker
2. Stick figure representations such as Figure 11.17.
3. Position, velocity and acceleration for each marker versus time (*x*, *y*, or magnitude).
4. Limb and joint angles versus time or versus phase of walking such as Figure 11.20.
5. Angle-angle diagram, e.g., knee angle versus hip angle at corresponding instants of time.

When studying joint motion during

locomotion, the joint angles, along with the footswitch record, are usually displayed versus time. When a computer is being used to acquire and process the gait data, the joint angles may be displayed as a function of the gait cycle, where the gait cycle is from heel strike to heel strike as indicated by the footswitch record. As well, if multistride data are collected, the average joint angle trajectories may be calculated and displayed. Figure 11.20 shows the hip, knee, and ankle joint angles versus gait cycle for the left and right sides of a 23-year-old male walking at 4.24 km/h. The solid trajectories are the averages of five strides, while the dashed lines are the individual stride joint angles. The data were collected using electrogoniometers similar to that shown in Figure 11.18.

The footswitch records, which are the topmost traces, indicate *swing phase* when they are at baseline, *heel only* contact for the next lowest level, *forefoot* contact for the next level, and *foot flat* for the highest level. The normal sequence of events would be *heel strike* as indicated by the first vertical line crossing the trajectories, *heel only, foot flat, forefoot only, toe off* as indicated by the second line, and *swing phase*. The display shows 100% of the gait cycle starting from the first vertical line to the end of each trace, with the last 50% of the gait cycle repeated at the beginning to provide trajectory continuity at each phase of the gait cycle. The short vertical lines within each footswitch record indicate the period of single limb support. This figure shows that joint angle trajectories are very repeatable during normal locomotion with very little deviation between the results for each stride and the average trajectories. Stable pathological

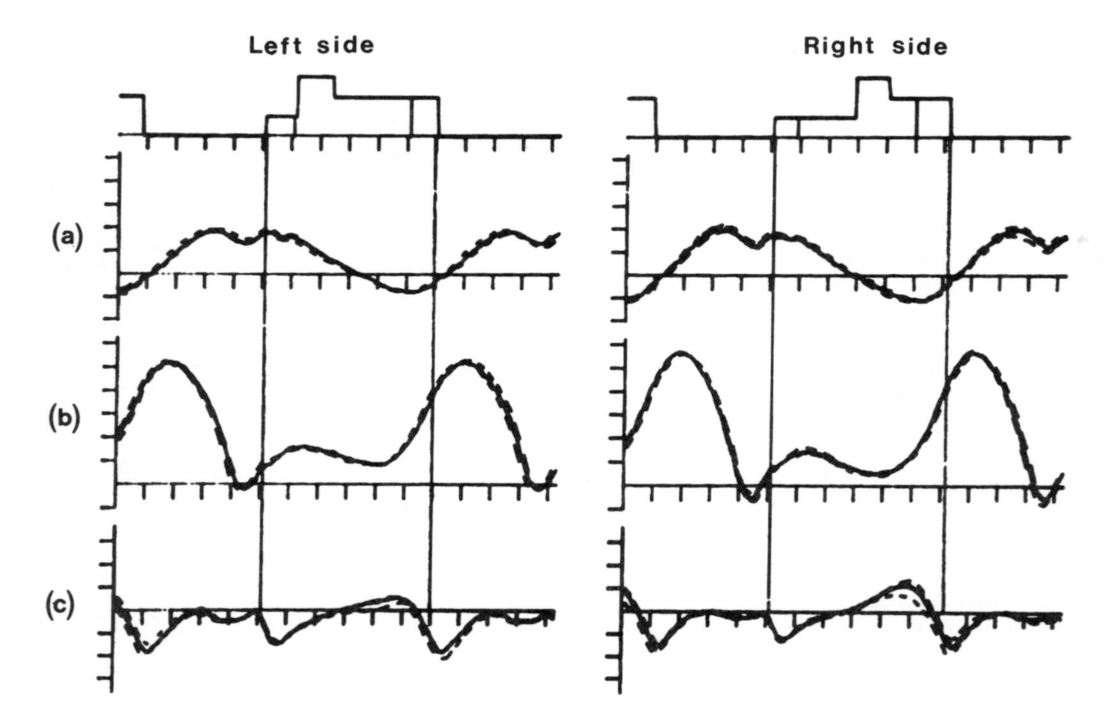

Figure 11.20. Left and right (a) hip, (b) knee, and (c) ankle joint angles versus gait cycle for a 23-year-old male subject walking at 4.24 km/hr with a stride length of 1.43 m and stride duration of 1.22 s. 150% of the gait cycle is shown, with heel strike indicated by the first vertical line. Average (solid line) and individual stride (dashed line). Horizontal hash marks are 10% of gait cycle. Vertical hash marks are 10° of joint angle.

gaits such as for stroke, cerebral palsy, and amputee patients, also demonstrate highly repeatable kinematic patterns. Angular velocities and accelerations may be obtained from the joint angle versus time data records by successive differentiation. However, differentiation usually increases signal noise and successive low-pass filtering must be used.

Rather than displaying them against the time of the gait cycle, gait variables may also be plotted against each other to amplify the degree of mutual dependence, or to produce easily recognizable patterns. Grieve (1968) suggested joint angle-angle diagrams as a useful method of displaying gait data. Figure 11.21 shows an angle-angle diagram relating the knee angle to the hip angle during each instant of the gait cycle for a normal eleven-year-old girl walking at her natural cadence. Since walking is a cyclic process, a closed loop results. The reason for the flexion-extension sequence of the knee immediately following stance initiation while the hip progressively extends is to ensure that the shock of heel strike (particularly for higher speeds and hence higher kinetic energies) upon the joints in the lower extremity is relieved by being absorbed by the musculature.

Figure 11.22 shows the hip-knee angle-angle diagrams for a nine-year-old girl with spastic diplegia. She is considered mildly affected. The angle-angle diagram is nearly normal in shape with a reduced range of knee motion and a slight knee flexion contracture, especially on the right side. It should be stressed that such diagrams should not be analyzed in detail to determine what is happening at each joint during the gait cycle. Displays such

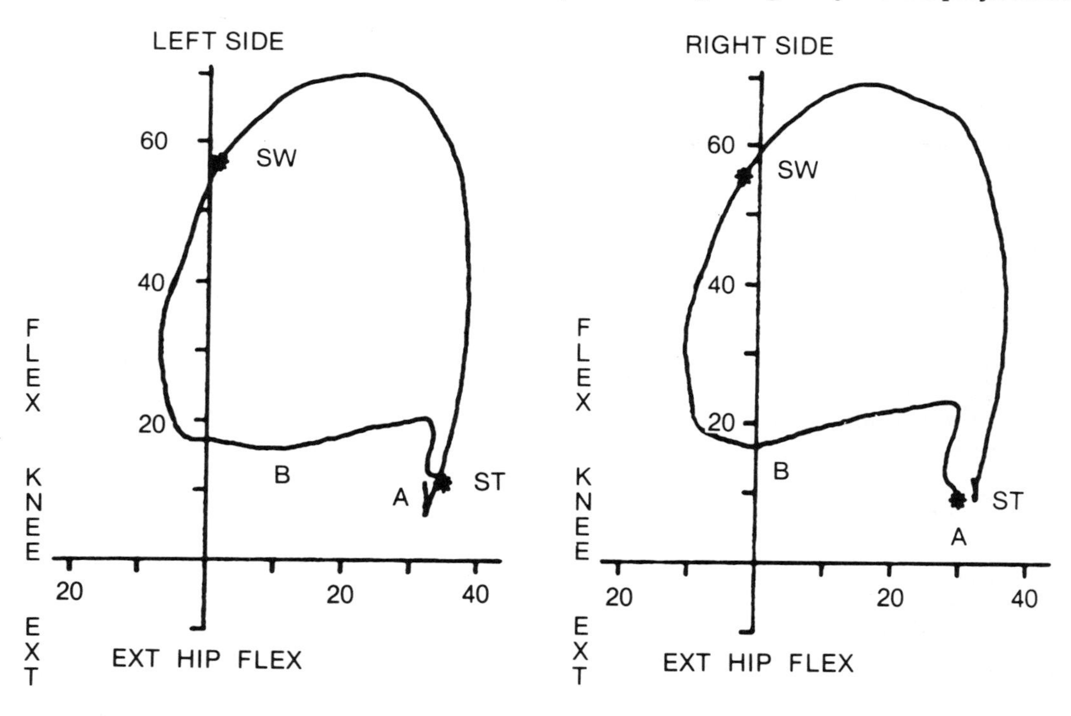

Figure 11.21. Hip-knee angle-angle diagram for a normal 11-year-old girl walking at her natural cadence. ST = initiation and SW = termination of stance phase. Axes are marked in degrees of joint angle. The origin of the axes represents neutral anatomical joint angles, i.e., 0° for hip and 180° for knee. (From deBruin, H., Russel, D.J., Latter, E., and Sadler, J.T.S., 1982. Angle-angle diagrams in monitoring and quantification of gait patterns for children with cerebral palsy. *Am. J. Phys. Med.* 61:176.)

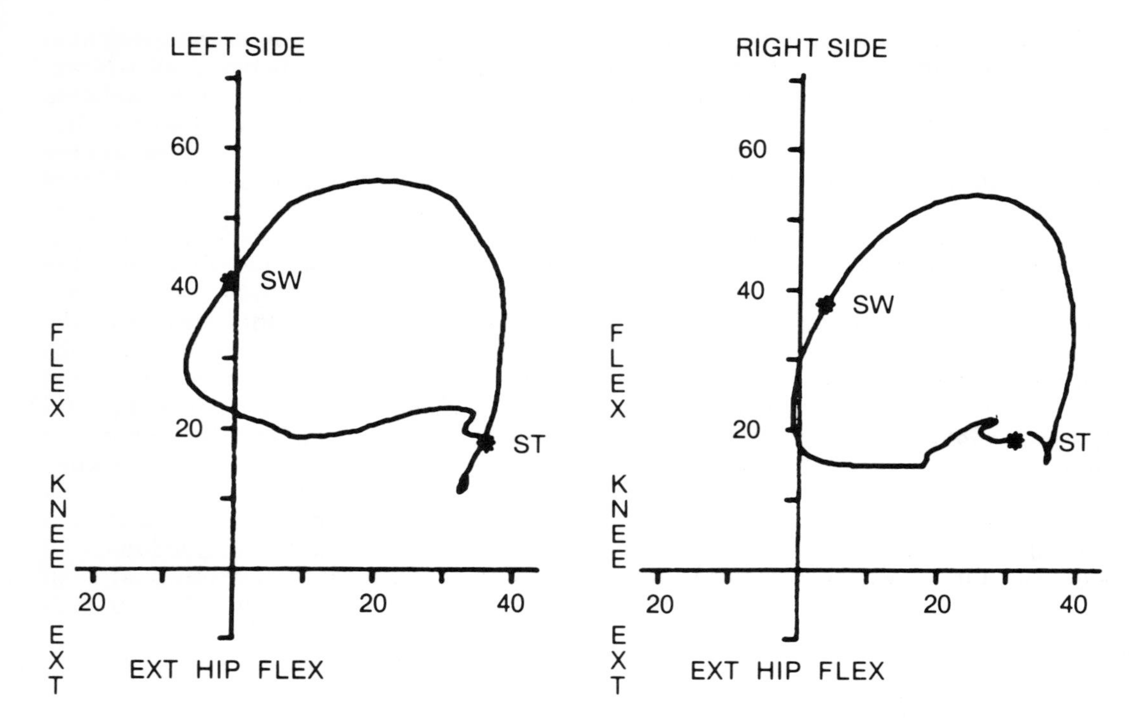

Figure 11.22. Hip-knee angle-angle diagram of a nine-year-old with spastic diplegia. ST = initiation and SW = termination of stance phase. Axes are marked in degrees of joint angle. The origin of the axes represents neutral anatomical joint angles, *i.e.*, 0° for hip and 180° for knee. (From deBruin, H., Russel, D.J., Latter, E., and Sadler, J.T.S., 1982. Angle-angle diagrams in monitoring and quantification of gait patterns for children with cerebral palsy. *Am. J. Phys. Med.* *61*:176.)

as Figure 11.20 are more appropriate for that purpose. Rather, they should be used to investigate joint synergies such as flexion-extension synergies occurring following brain injury and for easy "signature" recognition.

Ground Reaction Forces

In this chapter, we have concentrated on the acquisition and display of kinematic data relating to various human movement tasks. However, to obtain more detailed knowledge of the initiation and control of a particular movement, we require a kinetic analysis or the analysis of the internally and externally applied forces. To measure externally applied forces requires instrumenting the environment such as parallel bars, etc., with force-measuring transducers. We will limit ourselves to external forces applied to the feet during walking, running, standing, or other tasks involving foot contact with the floor or ground. These forces are called **ground reaction forces**. Knowledge of these forces, together with anatomical information and kinematic data, allows us to calculate the forces acting at the various joints of the lower extremities using classical mechanics. This type of information is obviously also of much value in the design of prosthetic and orthotic devices.

Forces applied to the ground are forces which may be depicted on a right-handed Cartesian coordinate system. Interestingly, the forces usually measured are the ground reaction forces, in other words, the equal and opposite forces of Newton's third law (Fig. 11.23). During any study of these forces, it is important to adopt either one convention or the other to designate the direction of the forces involved (Rodgers and Cavanagh, 1984:85–86).

Measurement of ground reaction forces of walking or running subjects have, in recent times, been made almost entirely with the aid of walkpaths equipped with force platforms. These platforms measure the vertical force as well as two shear forces acting along the force platform surface. These shear forces are usually resolved into anterior-posterior and medial-lateral directions (Fig. 11.23). The location of the center of pressure of the foot may be thought of as the application "point" of the resulting three-dimensional force vector (Fig. 11.24). In reality, force is distributed over a diffuse area, and the center of pressure describes the centroid of the pressure distribution at an instant in time (Rodgers and Cavanagh, 1984:86).

The foot is supported over changing amounts of foot surface area with different pressures at each part as the movement task is performed. To measure these individual pressures or vertical forces, special shoes and small pressure-measuring transducers have been designed. Although it is extremely attractive to measure the vertical forces under different areas of the foot during locomotion, with the capability of obtaining multistride results, these devices are not generally

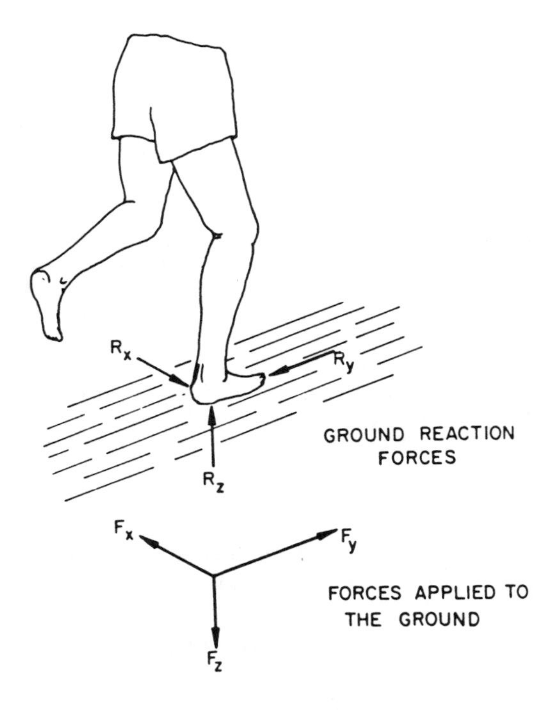

Figure 11.23. Usual conventions for the effect of ground reaction forces on the body: positive F_x, lateral (to both feet); positive F_y, backwards; positive F_z, upward. (From Rodgers, M. M., and Cavanagh, P. R., 1984. Glossary of biomechanical terms, concepts, and units. *Physical Therapy, 64:* No. 12:85.

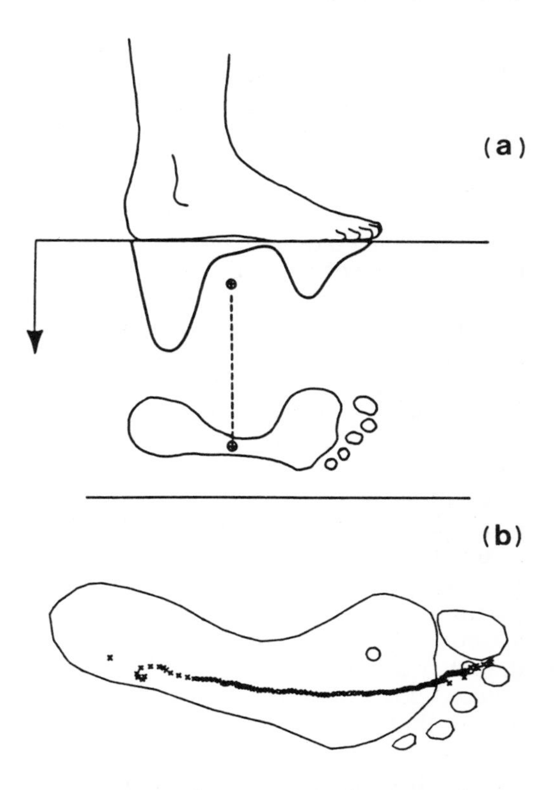

Figure 11.24. The meaning of center of pressure (C of P). (a) A theoretical distribution of pressure during standing. When pressure exists under both the heel and ball of the foot, the C of P will be in the midfoot region, which in itself is not bearing much pressure. (b) During walking, the C of P can be plotted as it moves under the foot. (From Rodgers, M. M., and Cavanagh, P. R., 1984. Glossary of Biomechanical terms, concepts, and units. *Physical Therapy, 64:* No. 12:86.

used because of their cost, reliability, and accuracy. However, this is an area of continuing development.

Many modern systems are derived from the design of Cunningham and Brown (1952). In essence, a stiff, flat metallic plate is supported at the four corners by instrumented pylons. The surface of a current modern force platform made by Advanced Medical Technology, Inc. (AMTI), measuring 46 cm by 51 cm, uses metal foil strain gauges to measure the pylon forces in three dimensions. The other most popular platform in North America, made by Kistler, uses piezoelectric transducers rather than strain gauges. These commercially available platforms, with their associated electronics and computer programs, make the measurement of ground reaction forces a standard procedure.

During the last two decades, force platforms have been used to not only study the biomechanics of normal walking and jogging, but also the effects on the gait of patients following reconstructive surgery, hip and knee joint replacements, and other orthopedic procedures. Force platform data have also been used in the analysis of mechanical energy costs and forces exerted during walking with various lower limb prostheses, orthoses, or walking aids. Currently, force platforms are also being used to investigate balance and postural control for both normals and patients with various central nervous sytem pathologies and movement disorders. For example, Riach and Hayes (1985) studied postural sway in 40 children and found that the magnitude of the sway reduces with age. The adult pattern was approached in children at approximately 7-8 years of age. Figure 11.25 shows the two-dimensional sway pattern for a typical subject. The power spectrum is also shown, which indicates the frequency components present in the sway pattern. The importance of measuring balance and postural control is becoming increasingly recognized in the assessment of geriatric subjects. The degree of balance and postural control is also a very important factor in determining the quality of gait of patients who have suffered a stroke or other central nervous system damage.

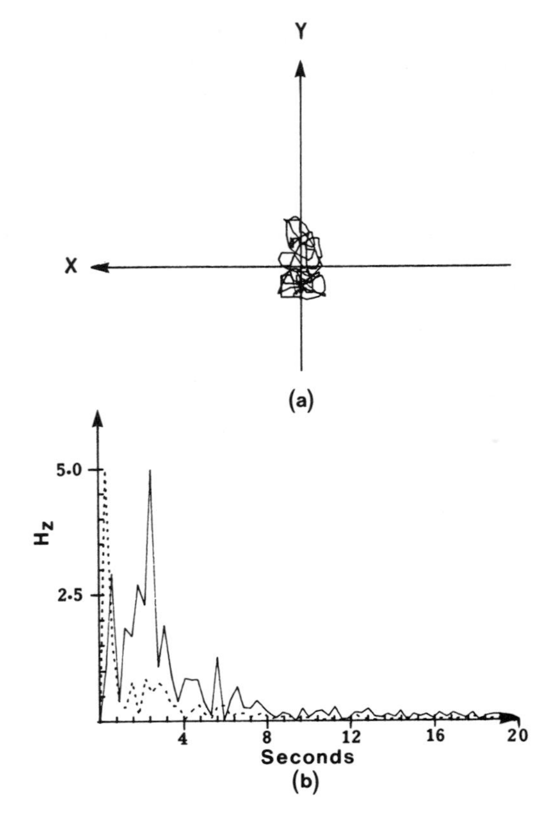

Figure 11.25. (a) postural sway in two dimensions for a child standing on a force platform; (b) frequency components of sway with solid trace for *X* direction, dashed trace for *Y* direction. (From data of Riach, C. L., and K. C. Hayes, 1985. Postural sway in young children. *Vestibular and Visual Control on Posture and Locomotor Equilibrium*, edited by Igarashi, M. and Black, F.O. Basel, Switzerland: S. Karger, pp. 232-236.)

Electromyography

Electromyography is an invaluable method for determining the patterns of activation of the muscles involved in a particular movement. The functional unit of muscle contraction is the motor unit (see Chapter 9), which is composed of a motoneuron whose cell body is located in the anterior horn of the spinal grey matter, its axon, and a number of muscle fibers innervated by that axon. Like nerve cells, muscle cells or fibers, at rest maintain electrical potential difference (voltage) across the cell membrane (see Chapter 7). When a nerve impulse traveling down a terminal branch of a motoneuron axon reaches the motor end plate, a transmitter, acetylcholine, is released from the presynaptic membrane (Fig. 7.19). This transmitter excites the postsynaptic muscle membrane and, if adequate to reach threshold, an action potential results which travels along the muscle fiber in either direction from the motor end plate to the tendons. All the muscle fibers in the motor unit are activated nearly synchronously, and the resulting summation of the individual action potentials traveling along the motor unit muscle fibers is called the motor unit action potential.

Total muscle contraction force is increased by two mechanisms: the recruitment of previously inactive motor units and increasing the discharge frequency of already active units. The myoelectric signal is the temporal and spatial integration of all motor unit action potentials detected using one or two *electrodes* from a certain volume of tissue. The myoelectric signal, when amplified and recorded, is called an **electromyogram** and the process of obtaining, processing, and analyzing electromyographic (EMG) signals is called **electromyography**.

Electrodes

For the study of human movement EMG signals may be obtained using either surface or intramuscular electrodes usually in pairs (bipolar). The amplitude and bandwidth of the EMG signal are not only determined by the electrophysiological sources and their distances from the electrodes, but also by the types and sizes of the electrodes used and their interelectrode spacing. The usable signal amplitude and bandwidth range from 10 μV to 5 mV and 10 Hz to 10 KHz respectively, depending on the electrode configuration. However, even for fine wire intramuscular electrodes, most signal energy lies below 1 KHz and the amplitudes within 2 mV peak to peak.

Surface electrodes are attached to the skin over the muscle segment under study. Surface electrodes are used to study the activity of the entire superficial muscle. The electrode spacing determines the recording or pick-up volume of tissue, with smaller spacings resulting in more selective recordings. Surface electrodes are usually of the recessed type, with electrode paste filling the cavity to provide skin contact and reduce electrode impedance. Commercial electrodes may be a disposable type such as electrocardiograph (ECG) electrodes or a reusable type with a plastic housing and double-sided adhesive tape collar. They range in diameter from 2 mm to 10 mm for the "active" part of the electrode. Silver-silver chloride (Ag-Ag Cl) electrodes with a chloride paste are invariably used because of their stability and low-noise properties. Figure 11.26 shows a human leg instrumented with pairs of 3-mm Beckman surface electrodes placed over the tibialis anterior, peroneus longus, medial gastrocnemius, and soleus muscles. The interelectrode spacing is kept to 1 cm to make the electrodes more selective.

When greater selectivity is required or deep muscles have to be studied, fine wire intramuscular electrodes are employed. These electrodes are very selective and there is no crosstalk (recording from adjacent active muscles), which may occur when surface electrodes are used over smaller superficial muscles. The most useful type are after the manner of Basmajian and Stecko (1962). They consist of two fine, highly malleable, stainless steel, Teflon-insulated wires about 25 to 50 μ in diameter passed down a hypodermic needle (No. 25 or 26). The tips are bared for 1 mm and bent back a few millimeters over the needle cannula with one bent portion 1 mm longer than the other to prevent shorting of the electrical circuit. When the needle is injected into muscle and then withdrawn, the wires remain in place because the bent-back portions serve as a hook. They can then be connected to the preamplifier using screw terminals or

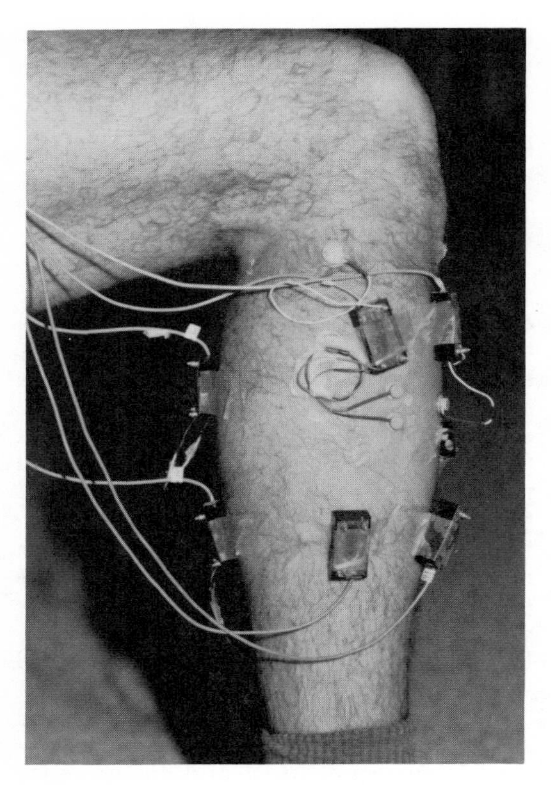

Figure 11.26. Subject instrumented with 3-mm Beckman electrodes over tibialis anterior, peroneus longus, medial gastrocnemius, and soleus muscles. Fine wire electrodes also were inserted into the tibialis anterior and peroneus muscles. Small preamplifiers can be seen attached to the skin near the electrodes to reduce noise susceptibility.

other connecting posts. Figure 11.26 shows these electrodes inserted into the tibialis anterior and peroneus longus muscles. Following use, they may be easily withdrawn. Where surface electrodes may have impedances below 1000 ohms, fine wire electrode impedances are often around 50 kilohms.

Amplification and Noise Reduction

Since the EMG signals have very small amplitudes, differential preamplifiers are used to provide signal gain and reduce or eliminate larger amplitude common mode signals such as 60 Hz power line interference, or even other biological signals such as the ECG. The amplifier input impedance must be several orders of magnitude greater than the electrode impedances to avoid signal attentuation or distortion. With modern field effect transistor (FET) input operational amplifiers, differential preamplifiers are easily constructed with sufficient signal bandwidth and gain, 90 db common mode rejection ratio and 10^{12} ohms input impedance.

Other than 60 Hz power line interference, which should have been removed by a good preamplifier, the most common sources of noise in the EMG signal are instrumentation and electrode noise, other biological signals, and motion artifact. **Motion artifact** is introduced by movements of the electrodes relative to the skin, the movement or rubbing together of electrode leads, and the flexing of connective cables during motion. Both electrode noise and motion artifact are reduced by decreasing the electrode-skin impedance. This can be accomplished for surface electrodes by swabbing the skin well with alcohol or lightly abrading it to remove the upper layers of the epidermis. The latter method is not suggested for clinical studies involving elderly patients or those with poor circulation since ulcers may develop.

Motion artifact resulting from cable flexion and other environmental noise such as magnetically coupled 60 Hz interference may be eliminated by placing the preamplifiers next to the electrodes as shown in Figure 11.24. Such amplifiers can be made very small and will offer no impediment to the movement. Finally, motion artifact, which is low-frequency artifact, may also be removed by electronically filtering out (high-pass filtering) any signals below 10 to 20 Hz. Most instrumentation noise may also be eliminated by low pass filtering the EMG signal at 1000 Hz.

Processing and Display

The amplified EMG signal, or the electrodes themselves, may be connected to the processing and recording equipment via long shielded cables. Many laboratories use telemetry to transmit the EMG and other motion-related signals to this instrumentation. Further amplification and bandpass filtering is performed by this instrumentation. The recording equipment may be a simple ink pen, multichannel chart recorder, or a higher frequency ultraviolet light (UV) recorder. The former does not have the banddwidth capability to accurately reproduce raw EMG signals, while the latter, although it has the necessary bandwidth, requires very expensive UV-sensitive paper which deteriorates with time, unless it is chemically fixed. Current modern instrumentation employs a computer to directly acquire, process, and display the EMG signals and other movement-related data. Such computers have become quite inexpensive and offer the additional advantages of more complex signal processing and display as well as signal quantification. Figures 11.3, 11.27, and 11.28 are all computer generated. Each EMG channel must be sampled at at least twice the bandwidth of the signal. It is found that low-pass filtering at 250 Hz and sampling at 500 Hz produces quite acceptable results for surface-recorded EMG signals. A bandwidth of 500 Hz and sample rate of 1000 Hz are minimally acceptable for fine wire recorded signals.

Raw EMG signals such as those shown in Figure 11.3 are quite acceptable when determining the phasic timing of activity patterns of muscles involved in a movement. Most clinical gait analyses are performed using such raw signals and visual inspection. However, the quantification of the "amount of activity" is also necessary. It is important to know not only when a muscle "turns on or off," but how much it is "on" at all times during a given contraction. Also, this allows researchers to compare results between subjects in different laboratories. The basis for most of this quantification comes from a linear detector, which is nothing more than a full-wave rectifier, which reverses the sign of all negative voltages and yields the absolute value of the raw EMG.

Figure 11.27(a) shows the raw EMG signal recorded from the tibialis anterior during normal locomotion. The full-wave rectified signal is shown in Figure 11.27(b). When the full-wave rectifier is followed by a low-pass filter, it is commonly referred to as a linear envelope detector. Such a detector may be easily constructed electronically and used to process the EMG signals prior to computer acquisition. This reduces the bandwidth of the EMG signal, depending on the cutoff frequency of the low-pass filter, and allows much reduced sampling rates of recording the processed EMG signal using a standard multichannel ink pen recorder. Nonlinear detectors such as root-mean-square detectors are used sometimes to quantify the signal.

When computers are used to perform the low-pass filtering, zero-phase lag digital filters may be employed to minimally distort the timing of the phasic pattern. Such filtering may also be performed using mid-window moving averages that calculate the average absolute EMG signal over a given time window. This time window moves continuously along the data record. It is important to note that this type of processing is often referred to incorrectly as integrated EMG. The integrated EMG, as the name implies, calculates the area under the EMG signal over a given time and gives a value of EMG in mV·s or μV·s rather than mV or μV as in the case of raw or linear envelope detected signal. Figure 11.27(c), (d), and (e) show the effect of decreasing the bandwidth of the digital low pass filter. As may be seen, linear envelope detection severely reduces the amplitude of the signal. As well, the phasic pattern is progressively smoothed with smaller details lost and, as the filter bandwidth is decreased, there is some phasic timing distortion. Also, the phasic pattern becomes more repeatable as the bandwidth of the linear envelope detector is reduced.

The use of computers augments further processing of the EMG signals and generation of appropriate displays. Such displays offer easy interpretation by the researcher or clinician so that patterns are readily detected and the most important information emphasized. Figure 11.28 shows the linear envelope-processed EMG signals (5-Hz low-pass filter) obtained from the left and right quadriceps, hamstrings, tibialis anterior, and gastrocnemius muscles of a mild left hemiplegia patient walking at 2.29 km/h. The display format is similar to that of Figure 11.22, which shows the joint angle trajectories and concurrent footswitch signals. The solid line indicates the average, processed EMG signal while the dashed lines indicate plus or minus one standard deviation from the average. These traces were obtained for eight sequential strides. As for the joint angle figure, 150% of the stride cycle is displayed, with the first vertical line indicating heel strike.

Figure 11.28 demonstrates that the bilateral patterns are nearly symmetric, hence close to normal, with essentially equal amplitudes. The EMG patterns recorded during gait are not as repeatable as the kinematic patterns, even for normal subjects. Detailed analysis of the EMG patterns should therefore be made on the average of a sufficient number of strides rather than on individual stride results. This requirement makes the use of com-

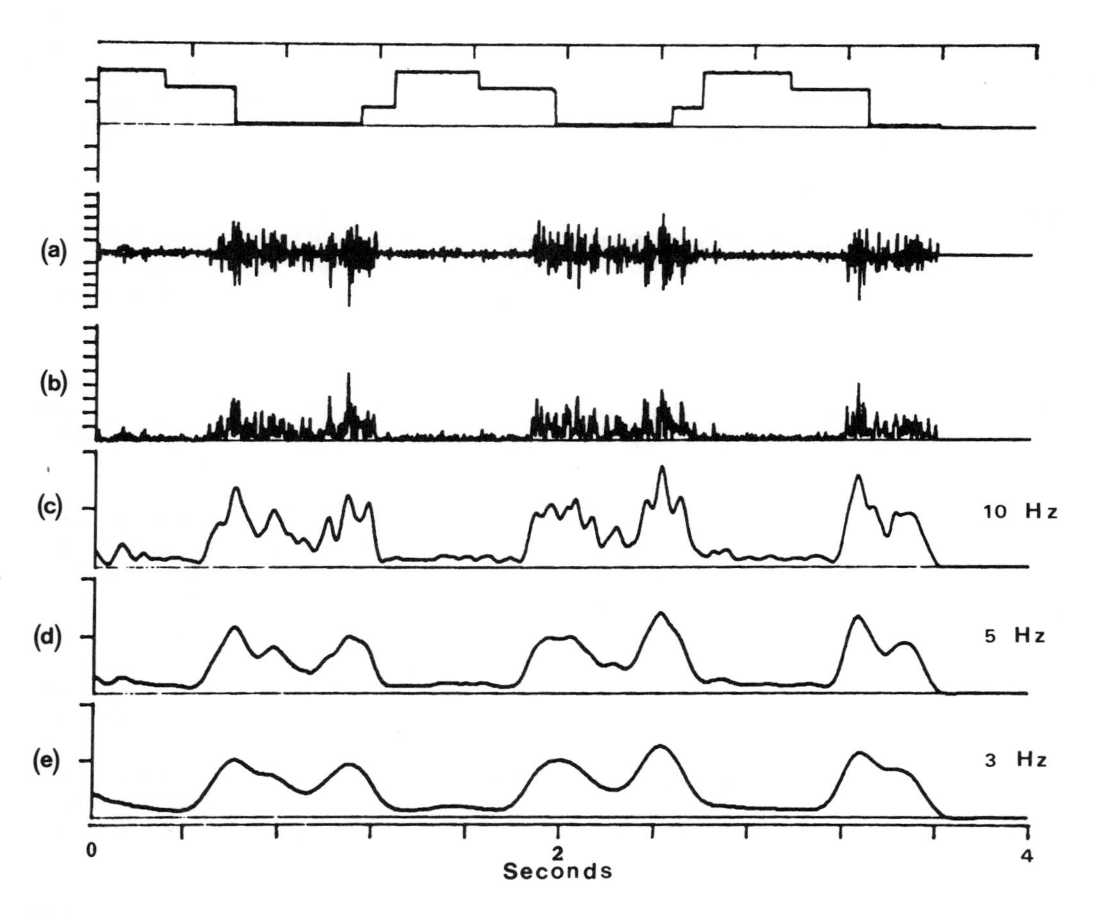

Figure 11.27. Effect of filtering the EMG signal obtained from tibialis anterior muscle of a normal subject during locomotion. The footswitch signal is at the top. (a) raw signal; (b) full wave rectified; (c), (d), and (e) linear envelope detected signal using 10 Hz, 5 Hz and 3 Hz zero-phase digital low-pass filter, respectively. Vertical axis hash marks represent 100 μV of EMG.

puters absolutely essential in analyzing muscle patterns of activity.

Once the average EMG signals are obtained, the pattern may be quantified. Computer programs can calculate the times of "turn on" and "turn off" relative to the start of heel strike for each muscle, as well as the time of peak amplitude. The peak amplitude, the area for each phasic contraction, and the averages during stance and swing also may be computed for each muscle. For many movements, it is desirable to normalize these amplitude values to those obtained for the maximum voluntary contraction (MVC) for each muscle. However, obtaining a MVC may be quite difficult, especially for brain-damaged patients. Therefore, in many clinical studies, the absolute amplitudes are reported. The pertinent question when using computers is not whether the EMG patterns may be quantified, but rather, which of the many possible parameters are most useful and sensitive.

Students interested in a more detailed presentation of electromyography are referred to Basmajian and DeLuca's *Muscles Alive* (1985) a classic that is now in its fifth edition. As well, it is appropriate to note that there are difficulties concerned with EMG terminology. The International Society of Electrophysiological Kinesiology (ISEK) published a report in August of 1980 in an attempt to alleviate some of the problems concerned with inconsistent and erroneous terms and units reported in the EMG literature. The work of this Ad Hoc Committee seems to be having an influence on the EMG research reported since that time. The brochure is entitled:

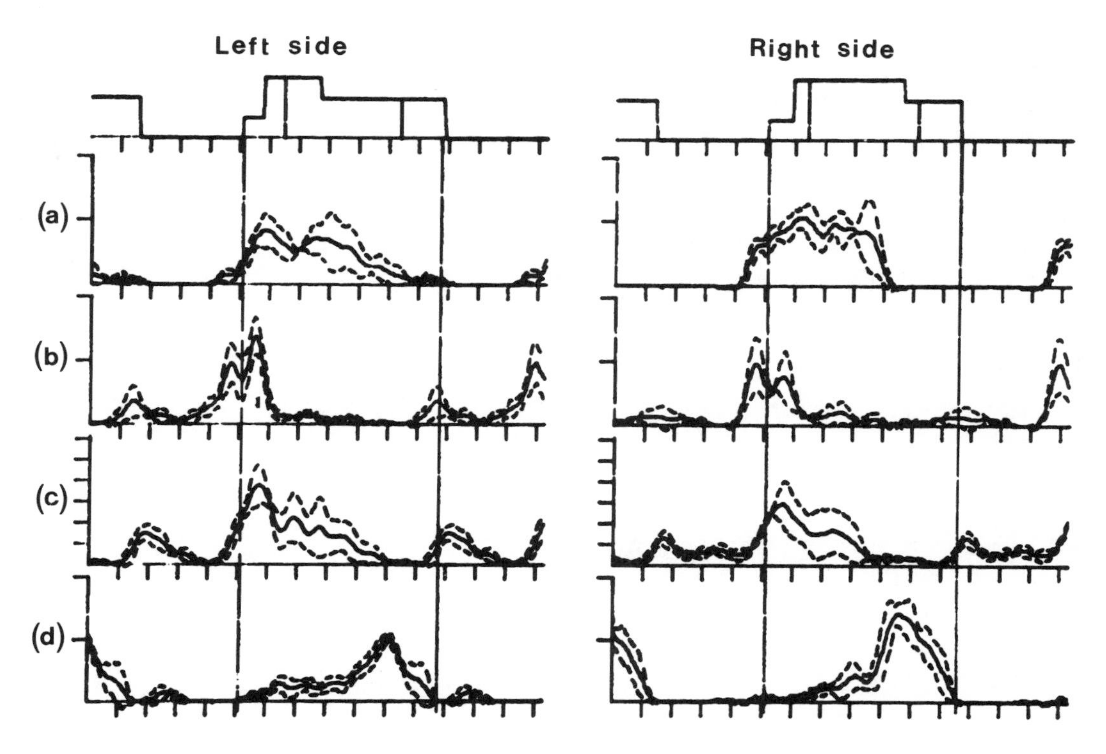

Figure 11.28. Processed EMG signals for (a) quadriceps, (b) hamstrings, (c) tibialis anterior, and (d) gastrocnemius muscles for a mild left hemiplegic subject walking at 2.29 km/hr. Footswitch signal is shown at top. Vertical axis hash marks represent 50 μV of EMG signal. Horizontal axis hash marks represent 10% of gait cycle. Solid trace is average of eight strides; dashed traces represent plus or minus one standard deviation from the average.

Units, Terms and Standards in the Reporting of EMG Research and may be obtained from the ISEK Newsletter Editor.

CONCLUSION

In this chapter, attention has been paid to methods of analysis of human movement. Pertinent approaches to the analyses rendered have been presented, and some indication has been given of data display and representation possibilities. A number of important technological aspects have been highlighted to show how studies might be facilitated, particularly by special devices and computers. No attempt has been made to be exhaustive in presenting either methodologies or technologies. The technology available for motion analysis is constantly improving and becoming more and more dependent on computer techniques.

Suggested Readings

Adal, M. N., and Barker, D., 1965. Intramuscular branching of fusimotor fibers. *J. Physiol. (London) 177:* 288.

Adrian, E. D., 1959. Sensory mechanisms, introduction. In *Handbook of Physiology,* Section 1: *Neurophysiology,* Vol. I, edited by J. Field and H. W. Magoun. Washington, D.C.: American Physiological Society.

Alexander, R. M., 1986. *Animal Mechanics.* Seattle: University of Washington Press.

American Academy of Orthopaedic Surgeons, 1965. *Joint Motion—Method of Measuring and Recording.* Edinburgh: E. and S. Livingstone, Ltd.

Andrews, B. J., and Jarrett, M. O., 1976. On-line Kinematic data acquisition. Glasgow, Scotland: Internal report, University of Strathclyde.

Appelberg, B., and Emonet-Denand, F., 1967. Motor units of the first superficial lumbrical muscle of the cat. *J. Neurophysiol. 30:* 154.

Asmussen, E., 1962. Muscular performance. In *Muscle as a Tissue,* edited by K. Rodahl and S. M. Horvath. New York: McGraw-Hill Book Company, p. 161.

Atkins, K. R., 1966. *Physics.* New York: John Wiley & Sons, Inc.

Bárány, M., 1967. ATPase activity of myosin correlated with speed of muscle shortening. In *The Contractile Process: Proceedings of a Symposium Sponsored by the New York Heart Association.* Boston: Little, Brown and Company, p. 197.

Barker, D., 1966. The motor innervation of the mammalian muscle spindle. In *Muscle Afferents and Motor Control,* edited by R. Granit. New York: John Wiley & Sons, Inc.

Barter, J. T., 1957. *Estimation of the Mass of Body Segments,* WADC Technical Report 57-260. Wright-Patterson Air Force Base, Ohio: Aero Medical Laboratory, Wright Air Development Center.

Basmajian, J. V., 1974. *Muscles Alive.* Baltimore: The Williams & Wilkins Company.

Basmajian, J. V., 1978. Personal communication.

Basmajian, J. V., and DeLuca, C., 1985. *Muscles Alive, Their Functions Revealed by Electromyography.* Baltimore: The Williams & Wilkins Company.

Basmajian, J. V., and Stecko, G. A., 1962. A new bipolar indwelling electrode for electromyography. *J. Appl. Physiol. 17:* 849.

Bearpark, R. E., 1981. Competition and the young athlete. Unpublished position paper presented to the Ontario Soccer Association.

Bell, G. H., Davidson, J. N., and Scarborough, H., 1969. *Textbook of Physiology and Biochemistry.* Baltimore: The Williams & Wilkins Company.

Bendall, J. R., 1970. *Muscles, Molecules and Movement.* London: Heinemann Education Books, Limited.

Bennett, M. V. L., 1968. Similarities between chemical and electrical mediated transmission. In *Physiological and Biochemical Aspects of Nervous Integration,* edited by F. D. Carlson. Englewood Cliffs, N.J.: Prentice-Hall, Inc.

Bessou, P., Emonet-Denand, F., and Laporte, Y., 1963. Occurrence of intrafusal muscle fibre innervation by branches of slow alpha motor fibres in the cat. *Nature 198:* 594.

Bessou, P., and LaPorte, Y., 1966. Observations on static fusimotor fibres. In *Muscle Afferents and Motor Control,* edited by R. Granit. New York: John Wiley & Sons, Inc.

Birdwhistell, R. L., 1970. Essays on body

motion communication. In *Kinesics and context,* edited by R. L. Birdswhistell. Philadelphia: University of Pennsylvania Press.
Blievernicht, D. L., 1967. Multi-dimensional timing device for cinematography. *Res. Q. 38:* 146.
Bobath, B., 1965. *Abnormal Postural Reflex Activity Caused by Brain Lesions.* London: William Heinemann Medical Books Limited.
Bodian, D., 1967. Neurons, circuits and neuroglia. In *The Neurosciences,* edited by G. C. Quarton, T. Melnechuk, and F. O. Schmidt. New York: The Rockefeller University Press.
Bowsher, D., 1967. *Introduction to the Anatomy and Physiology of the Nervous System.* Oxford: Blackwell Scientific Publishers.
Boyd, I. A., 1962. The structure and innervation of the nuclear-bag muscle fibre system and the nuclear-chain muscle fibre system in mammalian muscle spindles. *Philos. Trans. R. Soc. Lond., B 245:* 81.
Boyd, I. A., 1968. The morphology of muscle spindles and tendon organs. In *The Role of the Gamma System in Movement and Posture,* Revised edition, edited by C. A. Swinyard. New York: Association for Aid of Crippled Children, p. 2.
Boyd, I. A., and Davey, M. R., 1968. *The Composition of Peripheral Nerves,* London: E. & S. Livingstone Ltd.
Boyd, I. A., Eyzaguirre, C., Matthews, P. B. C., and Rushworth, G., 1968. *The Role of the Gamma System in Movement and Posture, Revised edition, edited by C. A. Swinyard.* New York: Association for the Aid of Crippled Children.
Boyd, I. A., Gladden, M. F., McWilliam, P. N., and Ward, J., 1975. 'Static' and 'dynamic' nuclear bag fibers in isolated cat muscle spindles. *J. Physiol. (Lond.) 250:* 11.
Brandstater, M. E., deBruin, H., Gowland, C., and Clarke, B. M., 1983. Analysis of temporal variables of hemiplegic gait. *Arch. Phys. Med. Rehabil. 65:* 583.
Broer, M. R., and Zernicke, R. F., 1979. *Efficiency of Human Movement.* Philadelphia: W.B. Saunders Company.
Brooks, V. B., 1959. Contrast and stability in the nervous system. *Trans. N.Y. Accad. Sci. Ser. II 21:* 387.
Brown, M. C., and Matthews, P. B. C., 1966. On the subdivision of the efferent fibres to muscle spindles into static and dynamic fusimotor fibres. In *Control and Innervation of Skeletal Muscle,* edited by B. L. Andrew. Dundee, Scotland: D.C. Thomson and Company, Ltd.
Bruegger, W., and Milner, M., 1978. Computer-aided tracking of body motions using a CCD-image sensor. *Med. Biol. Eng. Comput. 16:* 207.
Brunnstrom, S., 1964. Historical approach. In *Approaches to Treatment of Patients with Neuromuscular Dysfunction,* edited by S. Sattely. Dubuque, Iowa: William C. Brown Book Company.
Buchthal, F., Rosenfalck, P., and Erminio, F., 1960. Motor unit territory and fiber density in myopathies. *Neurology 10:* 398.
Buchwald, J. S., and Eldred, E., 1961. Conditioned response in the gamma efferent system. *J. Nerv. Ment. Dis. 132:* 146.
Buchwald, J. S., Standish, M., Eldred, E., and Hallas, E. S., 1964. Contribution of muscle spindle circuits to learning as suggested by training under Flaxedil. *Electroenceph. Clin. Neurophysiol. 16:* 582.
Buller, A. J., 1965. Mammalian slow and fast skeletal muscle. In *Studies in Physiology,* edited by D. R. Curtis and A. K. McIntyre. New York: Springer-Verlag Inc.
Buller, A. J., Eccles, J. C., and Eccles, R. M., 1960. Interactions between motoneurones and muscles in respect to the characteristic speed of their responses. *J. Physiol. (Lond.) 150:* 417.
Bullock, T. H., 1959. The neuron doctrine and electrophysiology. *Science 129:* 17.
Bunge, M. B., Bunge, R. P., and Peterson, E. R., 1967. The onset of synapse formation in spinal cord cultures as studied by electron microscopy. *Brain Res. 6:* 728.
Burke, R. E., 1967. Motor unit types of the cat's triceps surae muscle. *J. Physiol. (Lond.) 193:* 141.
Burke, R. E., Rudomin, P., and Kajac, F. E., 1970. Catch property in single mammalian motor units. *Science 168:* 122.
Carlsöö, S., 1972. *How Man Moves.* London: William Heinemann, Limited.
Cheng, I. S., 1974. Computer-television analysis of biped locomotion, Ph.D. Thesis. Columbus: Ohio State University.
Close, R., 1965. Effects of cross union of motor nerves to fast and slow skeletal muscles. *Nature 206:* 4986.
Close, R., 1967. Properties of motor units in fast and slow skeletal muscles of the rat. *J. Physiol. (Lond.) 193:* 45.
Close, R., and Hoh, J., 1968. The after-effects of repetitive stimulation on the isometric twitch contraction of rat fast skeletal muscle. *J. Physiol. (Lond.) 197:* 461.
Coers, C., and Woolf, A. L., 1959. *Innervation of*

Muscle. Springfield, Ill.: Charles C Thomas, Publisher.

Cooper, J. M., and Glassow, R. B., 1976. *Kinesiology*, second edition. St. Louis: The C.V. Mosby Company.

Costantin, L. L., Franzini-Armstrong, C., and Podolsky, R. J., 1965. Localization of calcium-accumulating structures in striated muscle fibers. *Science 147:* 158.

Cotton, F. S., 1932. Studies in center of gravity changes. *Aust. J. Exp. Biol. 10:* 16–34, 225–247.

Counsilman, J. E., 1968. *The Science of Swimming*. Englewood Cliffs, N.J.: Prentice-Hall, Inc.

Couteaux, R., 1960. Motor end-plate structure. In *Structure and Function of Muscle*, Vol. I, edited by G. H. Bourne. New York: Academic Press, p. 337.

Cunningham, D. M., and Brown, G. W., 1952. Two devices for measuring the forces acting on the human body during walking. *Proc. Soc. Exp. Stress Anal. 2:* 75.

Curtis, D. R., and Eccles, J. C., 1960. Synaptic action during and after repetitive stimulation. *J. Physiol. (Lond.) 150:* 374.

Damon, A., Stoudts, H. W., and McFarland, R. A., 1966. *The Human Body in Equipment Designs*. Cambridge, Mass.: Harvard University Press.

Davis, H., 1959. Excitation of auditory receptors. In *Handbook of Physiology*, Section 1: Neurophysiology, Vol. I, edited by J. Field and H. W. Magoun. Washington, D.C.: American Physiological Society.

Davis, H., 1961. Some principles of sensory receptor action. *Physiol. Rev. 41:* 391.

Dawson, P., 1935. *The Physiology of Physical Education*. Baltimore: The Williams & Wilkins Company.

DeAndrade, J. R., Grant, C., and Dixon, A. S., 1965. Joint distension and reflex inhibition in the knee. *J. Bone Joint Surg. 47A:* 313.

deBruin, H., Brandstater, M. E., Burcea, I., Plews, N., and Gowland, C., 1984. The reliability and validity of knee joint angle measurement by electrogoniometric and electrooptical systems. *Proc. of 2nd International Conference of Rehabilitation Engineering*. Ottawa, Canada, p. 501.

deBruin, H., Russel, D. J., Latter, E., and Sadler, J. T. S., 1982. Angle-angle diagrams in monitoring and quantification of gait patterns for children with cerebral palsy. *Am. J. Phys. Med. 61:* 176.

Dempster, W. T., 1955. *Space Requirements for the Seated Operator*, WADC Technical Report 55159. Wright-Patterson Air Force Base, Ohio: Wright Air Development Center.

Dempster, W. T., 1961. Free body diagrams as an approach to the mechanics of posture and motion. In *Biomechanical Studies of the Musculoskeletal System*, edited by F. G. Evans, Springfield, Ill.: Charles C. Thomas, Publisher.

Dowling, J. E., and Boycott, B. B., 1965. Neural connections of the retina. *Cold Spring Harbor Symp. Quant. Biol. 30:* 393.

Doyle, A. N., and Mayer, R. F., 1969. Studies of the motor units in the cat. *Bull. Sch. Med. Maryland 54:* 11.

Dreizen, P., and Gershman, L. C., 1970. Molecular basis of muscular contraction: Myosin. *Trans. N.Y. Acad. Sci. Ser. II 32:* 170.

Duggar, B. L., 1966. The center of gravity and moment of inertia of the human body. In *The Human Body in Equipment Design*, by A. Damon, H. W. Stoudts, and R. A. McFarland. Cambridge, Mass.: Harvard University Press.

Dyson, G. H. G., 1977. *Mechanics of Athletics*. London: Hodder and Stoughton.

Eccles, J. C., 1957. *The Physiology of Nerve Cells*. Baltimore: The Johns Hopkins Press.

Eccles, J. C., 1960. Neuron physiology, introduction. In *Handbook of Physiology*, Section 1: Neurophysiology, Vol. I, edited by J. Field and H. W. Magoun. Washington, D.C.: American Physiological Society.

Eccles, J. C., 1964. Ionic mechanism of postsynaptic inhibition. *Science 145:* 1140.

Eccles, J. C., 1965. The synapse. *Sci. Am. 212:* No. 1, 56.

Eccles, J. C., 1967. Postsynaptic inhibition in the central nervous system. In *The Neurosciences*, edited by G. C. Quarton, T. Melnechuk, and F. O. Schmidt. New York: The Rockefeller University Press.

Eccles, J. C., Magni, F., and Willis, W. D., 1962. Depolarization of central terminals of Group I afferent fibers from muscle. *J. Physiol. (Lond.) 160:* 62.

Elder, G., 1978. Personal communication.

Eldred, E., 1960. Posture and locomotion. In *Handbook of Physiology*, Section 1: Neurophysiology, Vol II, edited by J. Field and H. W. Magoun. Washington, D.C.: American Physiological Society.

Eldred, E., 1965. The dual sensory role of muscle spindles. In *The Child with Central Nervous System Deficit*. Washington, D.C., U.S. Government Printing Office.

Eldred, E., 1967a. Peripheral receptors: Their

excitation and relation to reflex patterns. *Am. J. Phys. Med. 46:* 69.

Eldred, E., 1967b. Functional implications of dynamic and static components of the spindle response to stretch. *Am. J. Phys. Med. 46:* 129.

Elftman, H., 1939. The function of muscles in locomotion. *Am. J. Physiol. 125:* 357.

Erlanger, J., and Gasser, H. S., 1937. *Electrical Signs of Nervous Activity.* Philadelphia: University of Pennsylvania Press.

Esch, D., and Lepley, M., 1974. *Evaluation of Joint Motion: Methods of Measurement and Recording.* Minneapolis: University of Minnesota Press.

Eshkol, N., and Wachman, A., 1958. *Movement Notation.* London: Weidefeld and Nicholson.

Everett, N. B., 1971. *Functional Neuroanatomy.* Philadelphia: Lea and Febiger.

Eyzaguirre, C. 1966. Properties of intrafusal muscle fibres of amphibians and mammals. In *Control and Innervation of Skeletal Muscle,* edited by B. L. Andrew. Dundee, Scotland: D. C. Thomson and Company, Ltd.

Eyzaguirre, C., 1968. Some functional characteristics of muscle spindles. In *The Role of the Gamma System in Movement and Posture,* Revised edition, edited by C. A. Swinyard. New York: Association for Aid of Crippled Children, p. 18.

Eyzaguirre, C., 1969. *Physiology of the Nervous System.* Chicago: Year Book Medical Publishers, Inc.

Fernandez-Moran, H., 1967. Membrane ultrastructure in nerve cells. In *Neurosciences,* edited by G. C. Quarton, T. Melnechuk, and F. O. Schmidt. New York: The Rockefeller University press.

Freeman, M. A. F., and Wyke, B., 1967. The innervation of the knee joint, an anatomical and histological study in the cat. *J. Anat. 101:* 505.

Frohlich, Cliff, 1980. The physics of somersaulting and twisting. *Sci. Am. 242:* No. 3, 154.

Gaddum, J., 1965. The neurological basis of learning. *Perspect. Biol. Med. 8:* 436.

Ganong, W. F., 1965. *Review of Medical Physiology.* Los Altos: Lange Medical Publications.

Gardiner, M. D., 1978. *The Principals of Exercise Therapy.* London: Bell and Hyman, Limited.

Gardner, E., 1963. *Fundamentals of Neurology.* Philadelphia: W. B. Saunders Company.

Gardner, E. B., 1965a. The neuromuscular base of human movement: Feedback mechanisms. *J. Health Phys. Educ. Rec. 36* (Oct.): 61.

Gardner, E. B., 1965b. Physiological aspects of motor learning. In Proceedings, Fall Conference of Eastern Association for Physical Education of College Women.

Gardner, E. B., 1967. The neurophysiological basis of motor learning—A review. *J. Am. Phys. Ther. Assoc. 47:* 1115.

Gardner, E. B., 1969. Proprioceptive reflexes and their participation in motor skills. *Quest XII:* 1.

Garrett, R. E., Widule, C. J., and Garrett, G. E., 1968. Computer aided analysis. *Kinesiology Rev.,* edited by A. L. O'Connell, p. 1.

Garrett, G. E., Widule, C. J., Reed, W. S., and Garrett, R. E., 1969. Human movement via computer graphics. Paper presented at the convention of the American Association for Health, Physical Education, and Recreation, Boston, 1969.

Gelfan, S., 1955. Functional activity of muscle. In *Textbook of Physiology,* Ed. 17, edited by J. F. Fulton. Philadelphia: W. B. Saunders Company, p. 123.

Gernandt, B. E., 1959. Vestibular mechanisms. In *Handbook of Physiology,* Section 1: Neurophysiology, Vol. I, edited by J. Field and H. W. Magoun. Washington, D.C.: American Physiological Society.

Gernandt, B. E., and Shimamura, M., 1961. Mechanisms of interlimb reflexes in the cat. *J. Neurophysiol 24:* 665.

Gibbs, C. L., and Ricchiuti, N. V., 1965. Activation heat in muscle: Method of determination. *Science 147:* 162.

Glassow, R. B. Personal communication, 1964.

Goldberg, J. M., and Lavine, R. A., 1968. Nervous system: Afferent mechanisms. *Ann. Rev. Physiol. 30:* 319.

Gollnick, P. D., and King, D. W., 1969. Energy release in the muscle cell. *Med. Sci. Sports 1:* 23.

Gordon, A. M., Huxley, A. F., and Julian, F. J., 1966. Variation in isometric tension with sarcomere length in vertebrate muscle fibers. *J. Physiol. (Lond.) 184:* 170.

Gowitzke, B. A., 1968. Kinesiological and neurophysiological principles applied to gymnastics. In *Kinesiology Rev.,* edited by A. L. O'Connell, p. 22.

Gowitzke, B. A., 1969. An analysis of levers contributing to the glide kip. Unpublished paper prepared at the University of Wisconsin.

Gowitzke, B. A., 1986. Muscles alive in sport. In *Biomechanics: The 1984 Olympic Scientific Congress Proceedings,* edited by M. Adrian and H. Deutsch. Eugene, Oregon: Microform Publications, p. 3.

Gowitzke, B. A., and Waddell, D. B., 1979. Technique of badminton stroke production. In *Science in Racquet Sports,* International Congress of Sports Sciences, Edmonton, Alberta. Del Mar, Calif.: Academic Publishers.

Granit, R., 1955. *Receptors and Sensory Perception.* New Haven, Conn.: Yale University Press.

Granit, R. 1975. The functional role of the muscles spindles—facts and hypotheses. *Brain 98:* 531.

Granit, R., Holmgren, B., and Merton, P. A., 1955. Two routes of excitation of muscle and their subservience to the cerebellum. *J. Physiol. (Lond.) 130:* 213.

Grant, J. C. B., and Smith, C. G., 1953. In *Morris' Human Anatomy,* edited by J. P. Schaeffer. New York: The Blakiston Company.

Gray, J. A. B., 1959. Initiation of impulses in receptors. In *Handbook of Physiology,* Section 1: Neurophysiology, Vol. I, edited by J. Field and H. W. Magoun. Washington, D.C.: American Physiological Society.

Gray, J. A. B., and Matthews, P. B. C., 1951. Response of Pacinian corpuscles in the cat's toe. *J. Physiol. (Lond.) 113:* 475.

Grieve, D. W., 1968. Gait patterns and the speed of walking. *Biomech. Eng. 3:* 119.

Grundfest, H., 1959. Synaptic and ephaptic transmission. In *Handbook of Physiology,* Section 1: Neurophysiology, Vol. I, edited by J. Field and H. W. Magoun. Washington, D.C.: American Physiological Society.

Guyton, A. C., 1972. *Structure and Function of the Nervous System.* Philadelphia: W.B. Saunders Company.

Hagbarth, K. E., 1952. Excitatory and inhibitory skin areas for flexor and extensor motoneurones. *Acta Physiol. Scand. 26:* Suppl. 94, 5.

Hagbarth, K. E., 1974. Spinal and supraspinal factors in the control of voluntary activity under normal and pathological conditions. *Scand. J. Rehab. Med. Supp. 3,* 9-18.

Hagbarth, K. E., and Eklund, G., 1966. Motor effects of vibratory stimuli in man. In *Muscle Afferents and Motor Control,* edited by R. Granit, New York: John Wiley & Sons, Inc.

Halverson, L., 1966. Development of motor patterns in young children, *Quest 6:* 44.

Hanson, J., and Lowy, J., 1960. Structure and function of the contractile apparatus in muscles of invertebrate animals. In *Structure and Function of Muscle,* Vol. I, edited by G. H. Bourne. New York: Academic Press, p. 312.

Harvey, R. J., and Matthews, P. B. C., 1961. The response of de-efferented muscle spindle endings in the cat's soleus to slow extension of the muscle. *J. Physiol. (Lond.) 157:* 370.

Hasan, Z., Enoka, R. M., and Stuart, D. G., 1985. The interface between biomechanics and neurophysiology in the study of movement: some recent approaches. In *Exercise and Sport Sciences Reviews,* edited by R. L. Terjung. New York: Macmillan Publishing Company, p. 169.

Hay, J. G., 1978. *The Biomechanics of Sports Techniques.* Englewood Cliffs, N.J.: Prentice-Hall, Inc.

Hellebrandt, F., 1956. Physiological effects of simultaneous static and dynamic exercise. *Am. J. Phys. Med. 35:* 106.

Hellebrandt, F. A., Genevieve, G., and Tepper, R. H., 1938. The relation of the center of gravity to the base of support in stance. *Am. J. Physiol. 119:* 331.

Hellebrandt, F. A., Kubin, D., Longfield, W. M., and Kelso, L. E. A., 1937. The base of support in stance. *Phys. Ther. Rev. 17:* 231.

Hellebrandt, F., Schade, M., and Carns, M., 1962. Methods of evoking TNR in normal individuals. *Am. J. Phys. Med. 41:* 90.

Hellebrandt, F., and Waterland, J., 1962. Motor patterning in stress. *Am. J. Phys. Med. 41:* 56.

Henneman, E., and Olson, C. B., 1965. Relations between structure and function in the design of skeletal muscle. *J. Neurophysiol. 28:* 581.

Henneman, E., Somjen, G., and Carpenter, D. O., 1965. Functional significance of cell size in spinal motoneurons. *J. Neurophysiol. 28:* 560.

Hicks, J. H., 1953. The mechanics of the foot. I. The joints. *J. Anat. 87:* 345.

Hill, A. V., 1965. *Trails and Trials in Physiology.* Baltimore: The Williams & Wilkins Company.

Holmqvist, B., 1961. Crossed spinal reflex actions evoked by volleys in somatic afferents. *Acta Physiol. Scand. 52:* Suppl. 1, 81.

Houk, J. C., 1979. Regulation of stiffness by skeletomotor reflexes. *Ann. Rev. Physiol. 41:* 99.

Houk, J., Henneman, E., 1967. Response of Golgi tendon organs to active contraction of the soleus muscle of the cat. *J. Neurophysiol. 30:* 466.

Houtz, S. J., and Fischer, F. J., 1959. An analysis of muscle action and joint excursion during exercise on a stationary bicycle. *J. Bone Joint Surg. 41-A:* 123.

Hoyle, G., 1970. How is muscle turned on and off? *Sci. Am. 222:* No. 4, 84.

Hubbard, J. I., and Willis, W. D., 1963. The effect of use on the transmitter release mechanism at the mammalian neuromuscular junction. In *Effect of Use and Disuse on Neuromuscular Functions,* edited by E. Gutman and P. Hnik. Amsterdam: Elsevier Publishing Company.

Hufschmidt, H. J., 1966. Demonstration of autogenic inhibition and its significance in human voluntary movement. In *Muscle Afferents and Motor Control,* edited by R. Granit. New York: John Wiley & Sons, Inc.

Hunt, C. C., 1961. On the nature of vibration receptors in the hind limb of the cat. *J. Physiol. (Lond.) 155:* 175.

Hunt, C. C., and Kuffler, S. W., 1951. Further study of efferent small-nerve fibres to mammalian muscle spindles, multiple spindle innervation and activity during contraction. *J. Physiol. (Lond.) 113:* 283.

Huxley, A. F., and Niedergerke, R., 1954. Structural changes in muscle during contraction; interference microscopy of living muscle fibers. *Nature (Lond.) 173:* 971.

Huxley, A. F., and Taylor, R. F., 1958. Local activation of striated muscle fibers. *J. Physiol. (Lond.) 144:* 426.

Huxley, H. E., 1958. The contraction of muscle. *Sci. Am. 199:* No. 5, 66.

Huxley, H. E., 1965. The mechanism of muscular contraction. *Sci. Am. 213:* No. 6, 18.

Huxley, H. E., 1969. The mechanism of muscular contraction. *Science 164:* 1356.

Huxley, H. E., and Hanson, J., 1954. Changes in cross-striations of muscle during contraction and stretch and their structural interpretation. *Nature (Lond.) 173:* 973.

Huxley, H. E., and Hanson, J., 1960. The molecular basis of contraction. In *Structure and Function of Muscle,* Vol. I, edited by G. H. Bourne. New York: Academic Press, p. 183.

Infante, A. A., Klaupiks, D., and Davies, R. E., 1964. ATP changes in muscle doing negative work. *Science 144:* 1577.

Inman, V. T., 1947. Functional aspects of the abductor muscles of the hip. *J. Bone Joint Surg. 29:* 607.

International Society of Electrophysiological Kinesiology (Ad Hoc Committee), 1980, *Units, Terms and Standards in the Reporting of EMG Research.* (Small monograph published by ISEK.)

Jackson, K. M., Joseph, J., and Wyard, S. J., 1977. Sequential muscular contraction. *J. Biomechanics. 10:* 97.

Jansen, J. K. S., 1966. On fusimotor reflex activity. In *Muscle Afferents and Motor Control,* edited by R. Granit. New York: John Wiley & Sons, Inc.

Jansen, J. K. S., and Matthews, P. B. C., 1962. The central control of the dynamic responses of muscle spindle receptors. *J. Physiol. (Lond.) 161:* 357.

Jayson, M. I. V., and Dixon, A. S., 1970. Intra-articular pressure in rheumatoid arthritis of the knee. *Ann. Rheum. Dis. 29:* 401.

Johnston, R. C., and Larson, C. B., 1969. Biomechanics of Cup Arthroplasty. *Clinical Orthopedics and Related Research, 66:* 56.

Joseph, J., 1960. *Man's Posture, Electromyographic Studies.* Springfield, Ill.: Charles C Thomas, Publisher.

Kandel, E., and Wechtel, H., 1968a. Neural aggregates in aplysia. In *Physiological and Biochemical Aspects of Nervous Integration,* edited by F. D. Carlson. Englewood Cliffs, N.J.: Prentice-Hall, Inc.

Kandel, E., and Wechtel, H., 1968b. The functional organization of neural aggregates in aplysia. In *Physiological and Biochemical Aspects of Nervous Integration,* edited by F. D. Carlson. Englewood Cliffs, N.J.: Prentice-Hall, Inc.

Karpovich, P. V., 1950. Mechanics of rising on the toes. Abstract of paper presented at the National Convention of Health, Physical Education, and Recreation, Dallas, April 1950.

Karpovich, P. V., and Manfredi, T. G., 1971. The mechanism of rising on the toes. *Res. Q. 42:* 395.

Kasvand, T., Milner, M., Quanbury, A. O., and Winter, D. A., 1976. Computers and the Kinesiology of gait. *Comput. Biol. Med. 6:* 111–120.

Katz, B., 1966. *Nerve, Muscle and Synapse.* New York: McGraw-Hill Book Company.

Keele, C. A., and Neil, E., 1965. *Samson Wright's Applied Physiology,* Edition XI. London: Oxford Press.

Kinesiology Academy, 1980. Guidelines and standards for undergraduate kinesiology, *Journal of Physical Education and Recreation 51:* No. 2, 19.

Knott, M., and Voss, D. E., 1968. *Proprioceptive Neuromuscular Facilitation.* New York: Harper and Row.

Komi, P. V., 1984. Physiological and biomechanical correlates of muscle function: effects of muscle structure and stretch-shortening cycle on force and speed. In *Exercise and Sport Sciences Reviews,* edited by R. L. Terjung. Toronto: The Collamore Press, p. 81.

Korr, I. M., Wilkinson, P. N., and Chornock. F. W., 1967. Axonal delivery of neuroplasmic components to muscle cells. *Science 155:* 342.

Lamaster, M. A., and Mortimer, E., undated. A device to measure body rotation in film analysis. Paper presented at research section of the American Association for Health, Physical Education, and Recreation.

Lamoreaux, L. W., 1971. Kinematic measurements in the study of human walking. *Bull. Prosth. Res. 10–15:* 3.

Langman, J., 1969. *Medical Embryology: Human Development, Normal and Abnormal.* Baltimore: The Williams & Wilkins Company.

Lemkuhl, D., 1966. Local factors in muscle performance. *J. Am. Phys. Ther. Assoc. 46:* 473.

Lehmkuhl, L. D., and Smith, L. K., 1983. *Brunnstrom's Clinical Kinesiology.* Philadelphia: F. A. Davis Company.

Lester, H. A., 1977. The response to acetylcholine. *Sci. Am. 236:* No. 2, 106.

LeVeau, B., 1977. *Williams and Lissner: Biomechanics of Human Motion.* Philadelphia: W.B. Saunders Company.

Lloyd, D. P. C., 1943. Neuron patterns controlling transmission of ipsilateral hind limb reflexes in the cat. *J. Neurophysiol. 6:* 293.

Lloyd, D. P. C., 1960. Spinal mechanisms involved in somatic activities. In *Handbook of Physiology,* Section 1: Neurophysiology, Vol. II, edited by J. Field and H. W. Magoun. Washington, D.C.: American Physiological Society.

Loeb, G. E., 1984. The control and responses of mammalian muscle spindles during normally executed motor tasks. In *Exercise and Sport Sciences Reviews,* edited by R. L Terjung. Toronto: The Collamore Press, p. 157.

Loofbourrow, G. N., 1960. In *Science and Medicine of Exercise and Sports,* edited by W. R. Johnson. New York: Harper and Brothers.

Lorand, L., and Molnar, J., 1962. Biochemical control of relaxation in muscle systems. In *Muscle as a Tissue,* edited by K. Rodahl and S. M. Horvath. New York: McGraw-Hill Book Company, p. 97.

Lowenstein, W. R., 1960. Biological transducers. *Sci. Am. 203:* No. 2, 98.

Luttgens, K., and Wells, K. F., 1982. *Kinesiology: Scientific Basis of Human Motion.* Philadelphia: Saunders College Publishing.

MacConaill, M. A., 1946. Studies in the mechanics of synovial joints. *Irish J. Med. Sci. 21:* 223.

MacConaill, M. A., 1950. The movements of bones and joints. *J. Bone Joint Surg. 32B:* 244.

MacConaill, M. A., 1958. Joint movement. *Chartered Physiotherapy*: 359.

MacConaill, M. A., and J. V. Basmajian, 1977. *Muscles and Movements: A Basic for Human Kinesiology.* Baltimore: The Williams & Wilkins Company.

MacDougall, J. D., Elder, G. C. B., Sale, D. G., Moroz, J. R., and Sutton, J. R., 1977a. Abstract: Skeletal muscle hypertrophy and atrophy with respect to fiber type in humans. *Can. J. Appl. Sports Sci. 2:* No. 4, 229.

MacDougall, J. D., Ward, G. R., Sale, D. G., and Sutton, J. R., 1977b. Biochemical adaptation of human skeletal muscle to heavy resistance training and immobilization. *J. Appl. Physiol. 43:* 700.

Magoun, H. W., 1958. *The Waking Brain.* Springfield, Ill.: Charles C. Thomas, Publisher.

Marchiafava, P. L., 1968. Activities of the central nervous system: Motor. *Ann. Rev. Physiol. 30:* 359.

Matthews, P. B. C., 1962. The differentiation of two types of fusimotor fibre by their effects on the dynamic response of muscle spindle primary endings. *Q. J. Exp. Physiol. 47:* 324.

Matthews, P. B. C., 1968. Central regulation of the activity of skeletal muscle. In *The Role of the Gamma System in Movement and Posture.* Revised edition, edited by C. A. Swinyard. New York: Association for Aid of Crippled Children, p. 29.

McComas, A. J., 1977. *Neuromuscular Function and Disorders.* London: Butterworths.

McLennon, H., 1963. *Synaptic Transmission.* Philadelphia: W.B. Saunders Company.

McPhedran, A. M., Wuerker, R. B., and Henneman, E., 1965. Properties of motor units in a homogeneous red muscle (soleus) of the cat. *J. Neurophysiol. 28:* 71.

Megirian, D., 1962. Bilateral facilitatory and inhibitory skin areas of the cat. *J. Neurophysiol. 25:* 127.

Melzack, R., and Wall, P. D., 1965. Pain mechanisms: A new theory. *Science 150:* 971.

Meriam, J. L., 1975. *Dynamics.* New York: John Wiley and Sons, Inc.

Miller, W. H., Ratliff, F., and Hartline, H. K., 1961. How cells receive stimuli, *Sci. Am. 205:* No. 2, 222.

Milner, M., and Quanbury, A. O., 1970. Some facets of control in human walking. *Nature 227:* 734.

Moore, J. C., 1969. *Neuroanatomy Simplified.* Dubuque, Iowa: Kendall/Hunt Publishing Company.

Morehouse, L. E., and Cooper, J. M., 1950. *Kinesiology.* St. Louis: C. V. Mosby.

Morton, D. J., 1952. *Human Locomotion and Body Form.* Baltimore: The Williams & Wilkins Company.

Mosso, A., 1884. Application de la balance a l'etude de la circulation du sang chez l'homme. *Arch. Ital. Biol.5:* 130.

Mountcastle, V. B., 1967. The problem of sensing and the neural coding of sensory events. In *The Neurosciences,* edited by G. C. Quarton, T. Melnechuk, and F. O. Schmidt. New York: The Rockefeller University Press.

Mountcastle, V. B., and Powell, R. P. S., 1959. Central nervous mechanisms subserving position sense and kinesthesis. *Johns Hopkins Hosp. Bull. 105:* 173.

Nathanson, J. A., and Greengard, P., 1977. "Second messengers" in the brain. *Sci. Am. 237:* No. 2, 108.

Noback, C. R., and Demarest, R. J., 1972. *The Nervous System: Introduction and Review.* New York: McGraw-Hill Book Company.

Nystrom, B., 1968. Effect of direct tetanization on twitch tension in developing cat leg muscles. *Acta Physiol. Scand. 74:* 319.

O'Connell, A. L., 1958. Electromyographic study of certain leg muscles during movements of the free foot and during standing. *Am. J. Phys. Med. 37:* 289.

O'Connell, A. L., 1966. Ingredients of coordinate movement. *Am. J. Phys. Med. 46:* 334.

O'Connell, A. L., 1969. *Laboratory Manual for Kinesiology.* Boston: Boston University Bookstore.

O'Connell, A. L., and Gardner, E. B., 1972. *Understanding the Scientific Bases of Human Movement,* first edition. Baltimore: The Williams & Wilkins Company.

Orlick, T. and C. Botterill, 1975. *Every Kid Can Win.* Chicago: Nelson-Hall.

Orlick, T. 1984. Evolution in children's sport. In *Sport for Children and Youths,* edited by M. R. Weiss and D. Gould. Champaign, Illinois:. Human Kinetics Publishers, Inc.: 169.

Oscarsson, O., 1966. The projection of Group I muscle afferents to the cat cerebral cortex. In *Muscle Afferents and Motor Control,* edited by R. Granit. New York: John Wiley & Sons, Inc.

Oscarsson, O., Rosen, I., and Sulg I., 1966., Organizations of neurones in the cat cerebral cortex that are influenced from Group I muscle afferents. *J. Physiol. (Lond.) 183:* 189.

Page, E. W., and Infante, J., 1973. A new instrumentation technique for gait analysis. *Digest 26th Ann. Conf. Eng. Med. Biol. 14.4:* 126.

Paillard, J., 1960. Patterning of skilled movement. In *Handbook of Physiology,* Section 1: Neurophysiology, Vol. III, edited by J. Field and H. W. Magoun. Washington, D. C.: American Physiological Society.

Palay, S. B., 1967. Principles of cellular organization in the nervous sytem. In *The Neurosciences,* edited by G. C. Quyarton, T. Melnechuk, and F. O. Schmidt, New York: The Rockefeller University Press.

Parker, D. E., 1980. The vestibular apparatus. *Sci. Am. 243:* No. 5, 120, 125.

Parsons, C., and Porter, K. R., 1966. Muscle relaxation: Evidence for and intrafibrillar restoring force in vertebrate striated muscle. *Science 153:* 3734.

Partridge, L. D., 1961. Motor control and the myotatic reflex. *Am. J. Phys. Med. 40:* 96.

Peachey, L. D., 1965. The sarcoplasmic reticulum and transverse tubules of frog's sartorius. *J. Cell. Biol. 25:* 209.

Peachey, L. D., and Porter, K. R., 1959. Intracellular impulse conduction in muscle cells. *Science 129:* 721.

Penfield, W., and Rasmussen, T., 1968. *The Cerebral Cortex of Man.* New York: The Macmillan Company.

Perry, S. V., 1960. Introduction to the contractile process in striated muscle. In *The Contractile Process: Proceedings of a Symposium Sponsored by the New York Heart Association.* Boston: Little, Brown and Company, p. 63.

Plagenhoef, S. C., 1962. An analysis of the kinematics and kinetics of selected symmetrical body actions, Doctoral dissertation. University of Michigan.

Pletta, D. H., and Frederick, D., 1964. *Engineering Mechanics, Statics and Dynamics.* New York: The Ronald Press.

Pooley, J. C., 1984. A level above competition: an inclusive model for youth sport. In *Sport for Children and Youths,* edited by M. R. Weiss and Gould, D. Champaign, Ill.: Human Kinetics Publishers, Inc.: p. 187.

Porter, K. R., and Franzini-Armstrong, C., 1965. The sarcoplasmic reticulum. *Sci. Am. 212:* No. 3, 72.

Powers, W. R., 1976. Nervous system control of muscular activity. In *Neuromuscular Mechanisms for Therapeutic and Conditioning Exer-*

cise, edited by H. G. Knuttgen. Baltimore: University Park Press.

Prives, M. G., 1960. Influence of labor and sport upon skeletal structure in man. *Anat. Rec. 136:* 261.

Ralston, H. J., 1957. Recent advances in neuromuscular physiology, *Am. J. Phys. Med. 26:* 94.

Rasch, P. J., and Burke, R. K., 1967. *Kinesiology and Applied Anatomy*. Philadelphia: Lea Febiger.

Resnick, R., and Halliday, D., 1960. *Physics—For Students of Science and Engineering*, Part I. New York: John Wiley & Sons, Inc.

Reuleaux, F., 1875. *Theoretische Kinematik: Grundig einer Theorie des Maschinenwesens*. Braunschweig: I. F. Vieweg und Sohn. (Also translated by Kennedy, A. B. W., 1876: *The Kinematic Theory of Machinery: Outline of a Theory of Machines*. London: Macmillan Publishing Company.)

Reuschlein, P. L., 1962. An analysis of the speed of rotation for the forward somersault on the trampoline. Unpublished paper prepared at the University of Wisconsin.

Riach, C. L., and Hayes, K. C., 1985. Postural sway in young children. In *Vestibular and Visual Control on Posture and Locomotor Equiilibrium*, edited by Igarashi, M. and Black, F. O., Basel, Switzerland: Karger, p. 232.

Roberts, E. M., 1971. Cinematography in biomechanical investigation. *Selected Topics on Biomechanics: Proceedings of the C.I.C. Symposium on Biomechanics*. (ed. J. M. Cooper). Chicago: Athletic Institute, 41-50.

Roberts, T. D. M., 1966. The nature of the controlled variable in the muscle servo. In *Control and Innervation of Skeletal Muscle*, edited by B. L. Andrew. Dundee, Scotland: D. C. Thomson & Company, Ltd.

Roberts, T. D. M., 1967. *The Neurophysiology of Postural Mechanisms*. New York: Plenum Press.

Rodahl, K., and Horvath, S. M., 1962. *Muscle as a Tissue*. New York: McGraw-Hill Book Company.

Rodgers, M. M. and Cavanagh, P. R., 1984. Glossary of biomechanical terms, concepts and units, *Phys. Ther. 64:* No. 12, 82.

Roebuck, J. A., Jr., 1966. Kinesiology in engineering. Paper presented at the Kinesiology Council, Convention of the American Association for Health, Physical Education, and Recreation, March 1966.

Roebuck, J. A., Jr., 1968. A system of notation for space suit mobility evaluations. *Hum. Factors 10*.

Rogers, E. M., 1960. *Physics for the Inquiring Mind*. Princeton, N.J.: Princeton University Press.

Rooney, J. R., 1969. *Biomechanics of Lameness in Horses*. Baltimore: The Williams & Wilkins Company.

Rose, J. E., and Mountcastle, V. B., 1959. Touch and kinesthesis. In *Handbook of Physiology*, Section 1, Neurophysiology, Vol. I, edited by J. Field and H. W. Magoun. Washington, D.C.: American Physiological Society.

Rushworth, G., 1968., The modification of gamma system activity in man. In *The Role of the Gamma System in Movement and Posture*. Revised edition, edited by C. A. Swinyard. New York: Association for Aid of Crippled Children, p. 48.

Sage, G. H., 1977. *Introduction to Motor Behavior: A Neuropsychological Approach*. Reading, Mass.: Addison-Wesley Publishing Company.

Schaeffer, J. P. (Editor), 1953. *Morris' Human Anatomy*. New York: The Blakiston Company.

Scott, M. G., 1963. *Analysis of Human Motion*. New York: Appleton-Century-Crofts.

Sechenov, I., 1935. *Selected Works*. Moscow: State Publishing House (1863 quote).

Seireg, A., 1969. *Mechanical Systems Analysis*. Scranton, Pa.: International Textbook Company.

Selcom, A. B., 1976. Trade Literature on the Selspot System.

Shambes, G. M., 1968. The influence of the muscle spindle on posture and movement. *J. Am. Phys. Ther. Assoc. 48:* 1094.

Sherrington, C. S., 1906. *The Integrative Action of the Nervous System*. New Haven, Conn. Yale University Press.

Simpson, J. A., 1966. Control of muscle in health and disease. In *Control and Innervation of Skeletal Muscle*, edited by B. L. Andrew. Dundee, Scotland: D. C. Thomson & Company, Ltd.

Singh, M., and Karpovich, P. V., 1966. Isotonic and isometric forces of forearm flexors and extensors. *J. Appl. Physiol. 21:* 1435.

Smidt, G. L., 1973. Biomechanical analysis of knee flexion and extension. *J. Biomechanics 6:* 679.

Smith, R. S., 1966. Properties of intrafusal muscle fibres. In *Muscle Afferents and Motor Control*, edited by R. Granit. New York: John Wiley & Sons, Inc.

Snyder, S. H., 1985. The molecular basis of

communication between cells.*Sci. Am. 253:* No. 4, 138.

Spears, B., 1966. *Fundamentals of Synchronized Swimming.* Minneapolis: Burgess Publishing Company.

Spencer, J. D., Hayes, K. C., and Alexander, I. J., 1984. Knee joint effusion and quadriceps reflex inhibition in man. *Arch. Phys. Med. Rehabil. 65:* 171.

Stanic, V., Boyd, T., Valencic, V., Kljajic, M., and Acmionic, R., 1977. Standardization of kinematic gait measurements and automatic pathological gait pattern diagnostics. *Scand. J. Rehab. Med. 9:* 95.

Stein, J. M., and Padykula, H. A., 1962. Histochemical classification of individual muscle fibers of the rat. *Am. J. Anat. 110:* 103.

Stein, R. B., 1974. Peripheral control of movement. *Physiol., Rev.* 54:215-43.

Steindler, A., 1973. *Kinesiology of the Human Body.* Springfield, Ill.: Charles C. Thomas, Publisher. Stern, J. T., 1971. Investigations concerning the theory of 'spurt' and 'shunt' muscles. *J. Biomechanics. 4:* 437.

Stern, J. T., 1971. Investigations concerning the theory of 'spurt' and 'shunt' muscles. *J. Biomechanics.* 4:437.

Stockmeyer, S. A., 1967. An interpretation of the approach of Rood to the treatment of neuromuscular disfunction. *Am. J. Phys. Med. 46:* 900 (NUSTEP Proceedings).

Stockmeyer, S. A., 1970. Personal communication.

Stratford, P., 1982. Electromyography of the quadriceps femoris muscle in subjects with normal knees and accutely effused knees. *Phys. Ther. 62:* 279.

Stratford, P., Agostino, V., Brazeau, C., and Gowitzke, B. A., 1984. Reliability of joint angle measurement: a discussion of methodology issues. *Physiotherapy Canada. 36:* No. 1, 5.

Swearingen, J. J., 1949. *Determination of the Most Comfortable Knee Angle for Pilots.* Report No. 1, Biotechnology Project No. 3-48. Oklahoma City; Civil Aeronautics Medical Research Laboratory.

Swett, J. E., and Bourassa, C. M., 1967. Short latency activity of pyramidal tract cells by Group I afferent volleys in the cat. *J. Physiol. (Lond.) 189:* 101.

Szent Gyorgyi. A., 1953. *Chemical Physiology of Body and Heart Muscle.* New York: Academic Press.

Szent, Gyorgyi. A. G., 1960. Proteins of the myofibril. In *Structure and Function of Muscle,* Vol. II, edited by G. H. Bourne. New York: Academic Press, p. 1.

Tasaki, I., 1959. Conduction of the nerve impulse. In *Handbook of Physiology,* Section 1. Neurophysiology, Vol. I, edited by J. Field and H. W. Magoun. Washington, D.C.: American Physiological Society.

Terry, R. J., and Trotter, M., 1953. In *Morris' Human Anatomy,* edited by J. P. Schaeffer. New York: The Blakiston Company.

Thompson, R. F., 1967. *Foundations of Pghysiological Psychology.* New York: Harper and Row.

Timoshenko, S., and Young, D. H., 1956. *Engineering Mechanics,* Edition 4. New York: McGraw-Hill Book Company.

Tribe, M. A., and Eraut, M. R., 1977. *Nerves and Muscle* (Basic Biology Course, Book 10). Cambridge: Cambridge University Press.

Tricker, R. A. R., and Tricker, B. J. K., 1967. *The Science of Movement.* New York: American Elsevier Publishing Company, Inc.

Truex, R. C., and Carpenter, M. B., 1969. *Human Anatomy,* sixth edition. Baltimore: The Williams & Wilkins Company.

Twitchell, T. E., 1965. Attitudinal reflexes. In *The Child with Central Nervous System Deficit.* Washington, D.C.: U.S. Government Printing Office.

van Mameren, H., and Drukker, J., 1979. Attachment and composition of skeletal muscles in relation to their function. *J. Biomechanics 12:* 859.

Waddell, D. B., 1986. Adapting sports to meet the needs of participants. In *Sport and Medicine,* edited by J. A. MacGregor and J. A. Moncur. London: E. and F. N. Spon, p. 71.

Waddell, D. B., and Gowitzke, B. A., 1977. Analysis of overhead badminton power strokes using high speed bi-plane cinematography. Presented at the First International Coaching Conference, Malmo, Sweden.

Walker, S. M., and Schrodt, G. R., 1967. Contraction of skeletal muscle. *Am. J. Phys. Med. 46:* 151.

Walls, E. W., 1960. The microanatomy of muscle. In *Structure and Function of Muscle,* Vol. I, edited by G. H. Bourne. New York: Academic Press, p. 21.

Ward, T., and Groppel, J. L., 1980. Sport implement selection; can it be based upon anthropometric indicators? In *Motor Skills: Theory into Practice,* edited by A. Rothstein, Vol. 4, No. 2, p. 103.

Waterland, J. C., and Shambes, G. M., 1970. Biplane center of gravity procedure. *Percept. Motor Skills 30:* 511.

Watkins, J., 1983. *An Introduction to Mechanics of Human Movement.* Lancaster, U.K.: MTP Press Limited.

Weiss, P., 1963, Self-renewal and proximo-distal convection in nerve fibers. In *The Effect of Use and Disuse on Neuromuscular Functions,* edited by E. Gutman and P. Hnik. Amsterdam: Elsevier Publishing Company.

Wells, J. B., 1965. Comparison of mechanical properties between slow and fast mammalian muscles. *J. Physiol. (Lond.) 178:* 252.

Wells, K. F., 1966. *Kinesiology.* Philadelphia: W.B. Saunders Company.

Wells, K. F., and Luttgens, K., 1976. *Kinesiology: Scientific Basis of Human Motion.* Philadelphia: W.B. Saunders Company.

Wenger, M. A., Jones, F. N., and Jones, M. H., 1956. *Physiological Psychology.* New York: Holt, Rinehart and Winston, Inc.

Werner, J., 1980. *Neuroscience: A Clinical Perspective.* Philadelphia: W.B. Saunders Company.

Widule, C. J., 1974. *Analysis of Human Motion.* West Lafayette, Ind.: Balt Publishers.

Williams, M., and Lissner, H. R., 1962. *Biomechanics of Human Motion.* Philadelphia: W.B. Saunders Company.

Willows, A. O. D., and Hoyle, G., 1969. Neural network triggering a fixed action pattern. *Science 166:* 1549.

Wilson, V. J., 1966a. Inhibition in the central netrvous system. *Sci. Am. 214:* No. 5, 102.

Wilson, V. J., 1966b. Regulation and function of Renshaw cell discharge. In *Muscle Afferents and Motor Control,* edited by R. Granit. New York: John Wiley & Sons, Inc.

Winter, D. A., 1979. *Biomechanics of Human Movement.* New York: John Wiley and Sons.

Winter, D. A., Greenlaw, R. K., and Hobson, D. A., 1972a. A microswitch shoe for use in locomotion studies. *J. Biomech. 5:* 553.

Winter, D. A., Hobson, D. A., and Greenlaw, R. K., 1972b. TV-computer analysis of kinematics of gait. *Comp. Biomech. Res. 5:* 498.

Wooldridge, D. E., 1963. *The Machinery of the Brain.* New York: McGraw-Hill Book Comnpany.

Wuerker, R. B., McPhedran, A. M., and Henneman. E., 1965. Properties of motor units in a heterogeneous pale muscle (m. gastrocnemius of the cat). *J. Neurophysiol. 28:* 85.

Wyke, B. D., 1967. The neurology of joints. *Annals of the Royal College of Surgeons of England 41:* 25.

Appendix 1 Solutions and Answers to Problems

CHAPTER 2

The authors' description of the movements depicted on pages 40-42 Figure 2.45 follows:

MOVEMENT DESCRIPTION

BODY PART: JOINT	**Figure 2.45(a)**	**Figure 2.45(b)**
RIGHT LOWER EXTREMITY		
I-P[a] joints	Anatomical position	Extension
M-P[a] joints	Anatomical position	Extension
Ankle	Anatomical position	Plantar flexion
Knee	Slight hyperextension	Slight flexion
Hip	Slight abduction Slight lateral rotation	Slight abduction Slight lateral rotation
LEFT LOWER EXTREMITY		
I-P joints	Flexion	Flexion
M-P joints	Flexion	Flexion
Ankle	Plantar flexion	Plantar flexion
Knee	Slight hyperextension	Slight hyperextension
Hip	Flexion Slight medial rotation	Hyperflexion
VERTEBRAL COLUMN	Lateral flexion to the left Hyperextension of lumbar spine	Slight flexion of lumbar spine
CERVICAL SPINE	Slight rotation to the left	Slight dorsiflexion
LEFT SCAPULA	Protraction Rotation upward Depression	Rotation upward

Movement Description chart *continued on next page*

Movement Description chart *continued from previous page*

LEFT UPPER EXTREMITY		
Shoulder	Flexion	Abduction
	Medial rotation	Medial rotation
Elbow	Extension	Slight hyperextension
Radioulnar	Pronation	Marked pronation
Wrist	Flexion	Flexion
M-P[a] joints	Flexion (holding onto foot)	Extension
I-P joints	Flexion (holding onto foot)	Extension

[a]*Note:* I-P = Interphalangeal, M-P = Metatarsal-phalangeal, and metacarpal-phalangeal respectively.

CHAPTER 3

Wheelchair Basketball. Treating the problem as a vector problem and using Pythagorean theorem, the velocity achieved by the ball = 5.4 m/s. The angle may be calculated as the arctan of 2 divided by 5 = 22°.

Racing Skier. Using equation 3.9 to solve all problems, the answers are as follows: #1 = 8.0 m; #2 = 21.0 m; #3 = 7.0 m.

Forehand Drive. Since all vectors are traveling in the same direction, they may be added arithmetically. #1: The velocity of the hand with respect to the shoulder = 8.0 m/s; therefore, the velocity of the hand with respect to the ground = 10.0 m/s. #2: The two radii added together and multiplied by the angular velocity make the velocity of the ball with respect to the shoulder = 15.0 m/s; therefore, the velocity of the ball with respect to the ground = 17.0 m/s.

CHAPTER 4

Note: All questions are concerned with the energy analysis involving a man sitting in a wheelchair which is on a ramp.

Question 1. Find PE at the top of the ramp.

$$PE = mgh = (100)(9.81)(1) = 981 \text{ J}$$

Question 2. Find KE at the bottom of the ramp. Note: the system is conservative during the trip down the ramp. Therefore, PE + KE = a constant for all points between A and C.

At point A: PE = 981 J and KE = 0 J; ∴ PE + KE = 981 J. At point C: PE = 0 J and PE + KE = 981 J; ∴ KE = 981 − 0 = 981 J.

Question 3. Find velocity of the man/chair combination at the bottom of the ramp.

$$KE = 1/2\, mv^2 \text{ so } v^2 = 2\,(KE)/m$$
$$\therefore v = \sqrt{2(KE)/m}$$
$$v = \sqrt{2(981)/100}$$
$$v = \sqrt{19.62}$$
$$v = 4.43 \text{ m/s}$$

Question 4. Find the linear momentum P at point C.

$$P = mv = (100)(4.43) = 443.94 \text{ kg m/s}$$

Question 5. How much work has been done upon arrival at point C?

$$\text{Work} = \text{mas}$$
$$\text{Work} = (100)(9.81)(1.0)$$
$$\text{Work} = 981 \text{ kg m}^2/\text{s}^2 = \Delta \text{ KE} = \Delta \text{ PE}$$

Question 6. How long (from A) would it take to stop the man/chair combination if a 150-N force is applied in the opposite direction to the velocity? Let $F = -150$ N applied force and Δt = the time to come to a complete stop. (N.B. Force must be negative because it is applied in the opposite direction to the velocity.) Final velocity = 0 and velocity at point C = 4.43 m/s. Thus,

$$F\Delta t = mv_2 - mv_1$$
$$\therefore \Delta t = \frac{mv_2 - mv_1}{F}$$
$$\Delta t = \frac{(100)(0) - (100)(4.43)}{-150}$$
$$\Delta t = 2.95 \text{ s}$$

CHAPTER 5

Two Gastrocnemius Heads. For part (a) working one axis at a time, the horizontal components of the resultant:

$$\vec{AD}_X = \vec{AB}_X + \vec{AC}_X$$

By designating the right horizontal as line AH:

$$\begin{aligned}\vec{AD}_X &= \vec{AB} \cos \text{HAB} + \vec{AC} \cos \text{HAC} \\ &= 1000 \cos 100° + 1000 \cos 70° \\ &= 168.37 \text{ N}\end{aligned}$$

The vertical components of the resultant:

$$\begin{aligned}\vec{AD}_Y &= \vec{AB}_Y + \vec{AC}_Y \\ &= \vec{AB} \sin \text{HAB} + \vec{AC} \sin \text{HAC} \\ &= 1000 \sin 100° + 1000 \sin 70° \\ &= 1924.50 \text{ N}\end{aligned}$$

We know from Pythagorean theorem:

$$\begin{aligned}\vec{AD} &= \sqrt{\vec{AD}_X^2 + \vec{AD}_Y^2} \\ &= \sqrt{(168.37)^2 + (1924.5)^2} \\ &= 1931.85 \text{ N}\end{aligned}$$

Finally, the angle of the resultant $\vec{AD}$:

$$\begin{aligned}\theta &= \arctan \Sigma \text{ Y}/\Sigma \text{ X} \\ &= \arctan 1924.50/168.37 \\ &= 85°\end{aligned}$$

Working question (b) in similar fashion, the magnitude of $\vec{AD}$ = 2318.40 N and its angle with respect to the horizontal is also 85°. Finally, in question (c), the magnitude of $\vec{AD}$ = 2125.67 N and its angle = 83.6°.

Problem 1. (a) By introducing the soleus into the gastrocnemius problem, another vector is introduced and is called $\vec{AE}$. The method of solution remains the same, and the horizontal components:

$$\vec{AD}_X = \vec{AB}_X + \vec{AC}_X + \vec{AE}_X$$
$$= 255.53 \text{ N}$$

The vertical components:

$$\vec{AD}_Y = \vec{AB}_Y + \vec{AC}_Y + \vec{AE}_Y$$
$$= 2920.69 \text{ N}$$

Using Pythagorean theorem and trigonometric identities, the resultant vector $\vec{AD}$ has a magnitude of 2931.85 N and a direction of 85° with respect to the horizontal.

(b) Using the same procedure for part (b), the resultant magnitude is 3125.46 N and its direction is 84.06°.

Problem 2. The problem can be solved with a simple equilibrium approach as the weight of the body being supported may be considered as the resistance and the abductors the effort in a lever system. The axis is at the head of the femur and lies in the sagittal and transverse planes. The torque of the effort is the product of the force required from the abductors and the moment arm designated d_1, while the torque of the resistance is the product of the body weight and the moment arm designated d_2.

For the normal situation in which $d_1 = 6$ cm, the force F required to balance the lever = 1400 N. In the second situation where $d_1 =$ only 3 cm, the force F required doubles to 2800 N. This is, of course, a first-class lever, and the reaction force on the head of the femur is directed upward with respect to Figure 5.26; its magnitude is 2100 N for the normal case and 3500 N for the shortened femur neck.

Problem 3. (a) This is a third-class lever because the length of the resistance moment arm (whether it be for the weight of the forearm and hand or for the weight in the hand) is greater than that of the effort moment arm. Also, the forces for effort and resistance do not work in the same direction (true of both second and third class levers).

(b) The resistance torque is the sum of the torque of the forearm and hand plus the torque of the weight in the hand.

$$T_R = (66.72)(.3759) + (18.24)(.1854)$$
$$= 28.46 \text{ Nm}$$

(c) First, the only muscle which can possibly be involved is the m. brachioradialis since both m. biceps brachii and m. brachialis are innervated by the musculocutaneous nerve.

$$\Sigma M_f = 0$$
$$\Sigma Mf_{CCW} = \Sigma Mf_{CW}$$
$$\Sigma Mf_{Effort} = \Sigma Mf_{Resistance}$$
$$EF \times EMA = \Sigma R$$

But what is length of EMA? This information can be obtained by first obtaining the angle the muscle makes with the lever, and then using that information to deduce the length of the moment arm.

$$\theta = \arctan 7.62/26.42$$
$$\theta = 16.1°$$
$$EMA = 7.62 (\cos 16.1°)$$
$$EMA = 7.32 \text{ cm}$$

Then:

$$EF = T/EMA$$
$$= 388.80 \text{ N}$$

(d) The effort force in the X direction may be designated EF_X, and in the Y direction

EF_Y; similarly, RF_X and RF_Y for resistance. Also, the sum of the reaction forces for both X and Y may be ΣJ_X and ΣJ_Y.

$$
\begin{aligned}
EF_X &= EF\ (\cos 16.1°) \\
&= 373.55\ \text{N} \\
RF_X &= 0 \\
\Sigma\ J_X &= EF_X - RF_X \\
&= 373.55\ \text{N acting to the right (see Figure 5.27)} \\
EF_Y &= EF\ (\sin 16.1°) \\
&= 107.82\ \text{N} \\
RF_Y &= 66.72 + 18.24 \\
&= 84.96\ \text{N} \\
\Sigma\ J_Y &= EF_Y - RF_Y \\
&= 22.86\ \text{N acting downward (see Figure 5.27)}
\end{aligned}
$$

Using Pythagorean theorem, the resultant joint force is 374.25 N. Using the arctan function, the angle of its application is 356.50° with respect to the right horizontal.

Problem 4. (a) The formula shows the same quantities within both brackets, and since $TPT = TQT$, the formula may be written in a more simplified form:

$$P\text{-}F \text{ compression force} = 2\ (TPT \times \cos\ (180-30)-18)\ /\ 2$$

For 30°:

$$
\begin{aligned}
P\text{-}F \text{ compression force} &= 2\ (406.74 \times \cos 66°) \\
&= 330.87\ \text{N}
\end{aligned}
$$

For 90°:

$$
\begin{aligned}
P\text{-}F \text{ compression force} &= 2\ (406.74 \times \cos 44.5) \\
&= 580.21\ \text{N}
\end{aligned}
$$

(b) Since pain usually correlates with the patella-femoral force, it would be expected that the patient would have a lesser amount of discomfort at 30°.

Problem 5. (a) First, solve for E at 30° of knee flexion.

$$
\begin{aligned}
T_E = E \times EMA &= 135\ \text{Nm} \\
E &= 135/.0488 \\
E &= 2766.39\ \text{N}
\end{aligned}
$$

Solve for R, given that $RMA = 30.48$ cm. If the system is in static equilibrium, resistance torque = effort torque = 135 Nm.

$$
\begin{aligned}
R \times RMA &= 135\ \text{Nm} \\
R &= 135/.3048 \\
R &= 442.91\ \text{N}
\end{aligned}
$$

Referring to Figure 5.28, a shear vector is added (Fig. 5.30) so that its orientation is in the negative direction.

$$
\begin{aligned}
\Sigma\ F_X &= 0 \\
\Sigma\ F_X &= \text{E} \sin 18° - \text{R} - J_X \\
J_X &= 2766.39 \times \sin 18° - 442.91 \\
J_X &= 411.95\ \text{N}
\end{aligned}
$$

Substitute J_X back into original equation and check the sign.

$$\Sigma\ F_X = \text{E} \sin 18° - \text{R} - (+411.95)$$

The 411.95 N is a negative force in the equation which means that it will have a shear force in the direction of the shear vector shown in Figure A1.1. The direction of the J_X vector indicates a force that will result in tension in the anterior stabilizers of the knee joint.

At 90°, with *RMA* of 30.48 cm, $E = 3409.09$ N, $R = 442.91$ N, $J_X = -383.41$ N. When the shear force is substituted back into the formula, it becomes a positive force and therefore a force which will result in tension in the posterior stabilizers.

At 30°, with *RMA* of 15.24 cm, $E = 2766.39$ N, $R = 885.83$ N, $J_X = -30.97$ N, so tension on posterior stabilizers.

At 90°, with *RMA* of 15.24 cm, $E = 3409.09$ N, $R = 885.83$ N, $J_X = -826.33$, so tension on posterior stabilizers.

Problem 6. The problem is solved by setting up a moment equation about the heel. First, convert therapist's mass to a force.

$$F = mg$$
$$F = 50 \text{ kg} \times 9.8 \text{ m/s}^2$$
$$F = 490 \text{ N}$$

Then, locate the distance from the heel to the center of mass.

$$\text{Level of center of mass} = 58.8\% \text{ of the body height}$$
$$= .588 \times 1.6$$
$$= 0.94 \text{ m}$$

Locate the distance from the shoulder to the heel.

$$\text{Level of glenohumeral joint} = 85.6\% \text{ of the body height}$$
$$= .856 \times 1.6$$
$$= 1.37 \text{ m}$$

Establish moment equation about the heel.

$$\Sigma M_M = 0$$
$$\text{Let } T = \text{traction force}$$
$$(T \times 1.37 \times \sin 50°) - (490 \times 0.94 \times \cos 75°) = 0$$
$$T = 113.59 \text{ N}$$

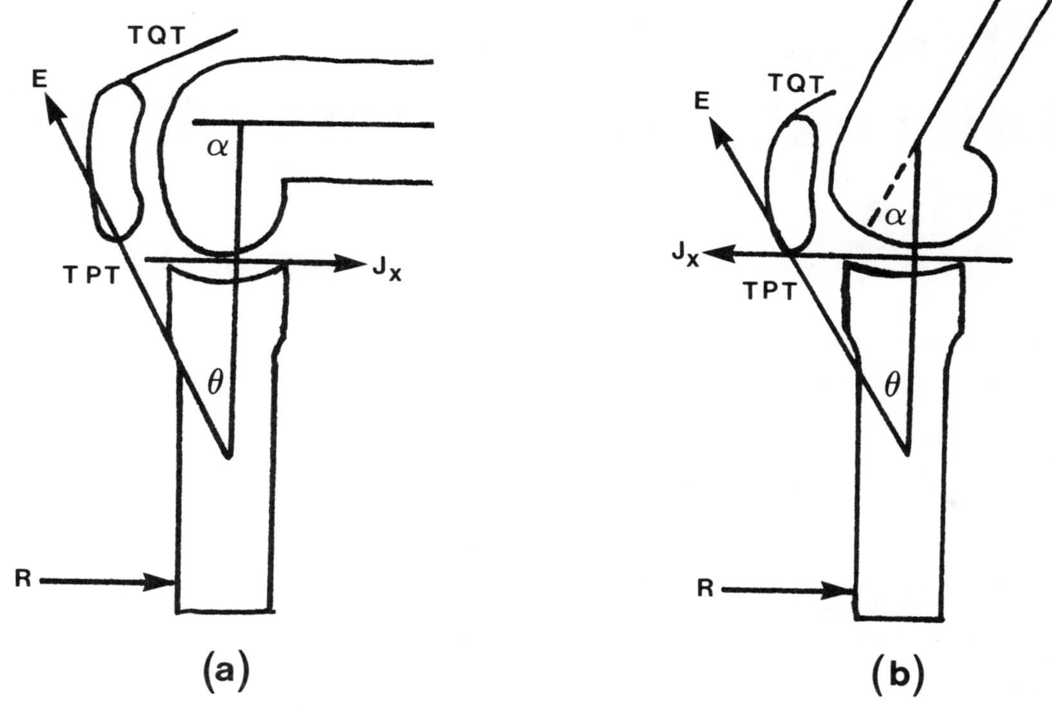

Figure A1.1. Schematic of solution for problem 5. (a) knee joint at 90° flexion; (b) knee joint at 30° flexion.

Appendix 2
International System of Units

The International System of Units or SI System (from the French Systeme Internationale D'Unites) derives from the metric system of physical units and was designated in 1960 at the 11th General Conference on weights and measures. Six basic units were adopted and defined; and other units are derived from them. They are the meter, kilogram, second, ampere, degree, kelvin and candela, for length, mass, time, electric current, temperature and luminous intensity, respectively.

Table A2.1 reflects some elemental and some derived SI units and symbols.

Table A2.2 renders a number of conversion factors relating typical United States units and those in the SI system.

Table A2.1
Some Elemental and Derived SI Units and Symbols

Quantity	SI Units: Unit	Formula	Symbol
Length	meter		m
Mass	kilogram		kg
Time	second		s
Plane angle	radian		rad
Solid angle	steradian		sr
Acceleration	meter/sec^2	m/s^2	
Area	square meter	m^2	
Density	kilogram/cubic meter	kg/m^3	
Energy	joule	N·m	J
Force	newton	kg·m/s^2	N
Frequency	hertz	s^{-1}	Hz
Kinematic viscosity	square meter/second	m^2/s	
Momentum	kilogram-meter/second	kg-m/s	
Power	watt	J/s	W
Pressure	newton/square meter	N/m^2	
Stress	newton/square meter	N/m^2	
Velocity	meter/second	m/s	
Viscosity	newton-second/square meter	N·s/m^2	
Volume	cubic meter	m^3	

Table A2.2
Conversion Factors for Typical U.S. and SI systems

inches × 25.4	=	millimeters
feet × 0.3048	=	meters
yards × 0.9144	=	meters
miles × 1.60934	=	kilometers
square inches × 6.4516	=	square centimeters
cubic inches × 16.3871	=	cubic centimeters
gallons[a] × 0.00378541	=	cubic meters
pounds (avdp) × 0.453592	=	kilograms
pounds (avdp) × 4.4498	=	newtons
horsepower × 0.7457	=	kilowatts

[a] The Imperial gallon is larger than the U.S. gallon and the relevant conversion factor is 0.004546.

Appendix 3 Trigonometric Functions

Given the right triangle ABC
Let:

angle BAC be known as θ (the Greek letter theta)
side AC be known as side A or the side adjacent to θ
side BC be known as side O or the side opposite θ
side AB be known as H or the hypotenuse of the right triangle

There are certain constant relationships for these three sides, regardless of the size of the triangle, which are dependent solely on the size of the angle θ. These relationships are expressed as:

$$\text{sine } \theta = \frac{\text{side}}{\text{opposite}} \text{ or } \sin \theta = \frac{O}{H}$$

$$\text{cosine } \theta = \frac{\text{side}}{\text{adjacent}} \text{ or } \cos \theta = \frac{A}{H}$$

$$\text{tangent } \theta = \frac{\text{side}}{\text{opposite}} \text{ or } \tan \theta = \frac{O}{A}$$

$$\text{cotangent } \theta = \frac{\text{side}}{\text{adjacent}} \text{ or } \cot \theta = \frac{A}{O}$$

Following from these relationships:

$$\sin \theta = \cos (90 - \theta)$$
$$\tan \theta = \sin \theta / \cos \theta$$

It is also possible to show that $\sin^2 \theta + \cos^2 \theta = 1$.

(Values for these functions for any angle are easily obtained using mathematical tables or hand-held calculators with these functions available.)

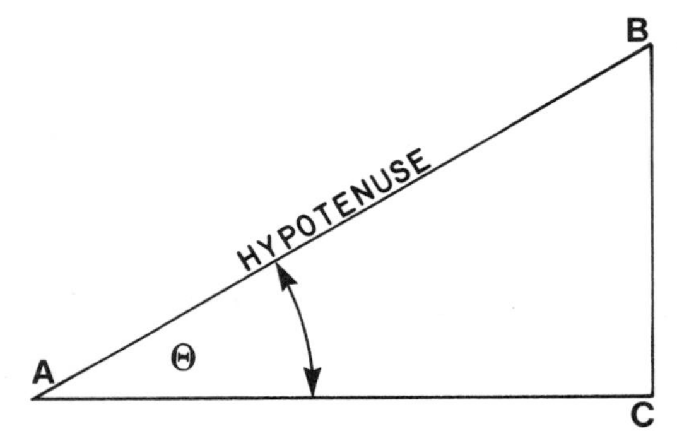

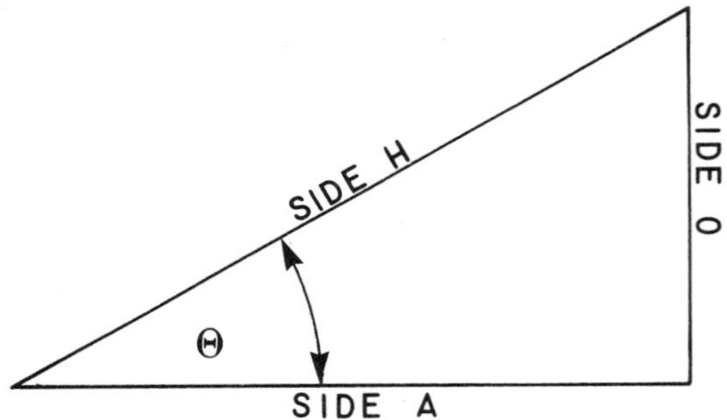

Figure A3.1. Right triangle, *ABC* θ, angle *BAC*; *AC*, the side adjacent (side *A*) to θ; *BC*, the side opposite (side *O*) to θ; the hypotenuse (side *H*) of the triangle.

Appendix 4
Body Segment Data

Table A4.1*
A. Average Segment Characteristics of Three Cadavers Dissected by Braune and Fischer (1889)

Part	Weight	Percentage of Total Body Weight	Center of Gravity Location with Respect to Joint Axes for Each Limb, Expressed as Percentage of Limb Length	
			From above	From below
	kg			
Upper arm (each)	2.127	3.3	47.0	53.0
Forearm (each)	1.335	2.1	42.1	57.9
Forearm and hand	1.872[a]	2.9	47.2[b]	52.8[b]
Hand (each)	0.533	0.85		
Upper leg (each)	6.793	10.75	43.9	56.1
Lower leg (each)	3.025	4.8	41.95	58.05
Foot (each)	1.067	1.7	43.4[c]	56.6[c]
Lower leg and foot	4.127[a]	6.5	51.9[d]	48.1[d]
Head, neck, and trunk, minus the limbs	33.990	53.0		
Head	4.440	6.95	Average locations not given. See individual cadaver data.	
Torso	29.550	46.3		
Entire body	63.85	100.0		

[a]Combined weights of several segments were slightly larger than the sum of the individual components because of loss of material during sawing.
[b]Percentage of length from cubital axis to lower edge of flexed fingers.
[c]Percentage of length from front to rear of foot.
[d]Percentage of length from knee axis to sole of foot.

Table A4.1 *continued*

B. Average Segment Characteristics of Two Cadavers Described by Braune and Fischer (1892)

Part	Weight	Percentage of Total Body Weight	Center of Gravity Location with Respect to Joint Axes for Each Limb, Expressed as Percentage of Limb Length	
			From above	From below
	kg			
Upper arm	1.51	2.92	45.9	54.1
Forearm and hand	1.30	2.54	46.0	54.0
Upper leg	5.78	11.23	43.4	56.7
Lower leg	2.32	4.53	42.4	57.6
Foot	0.95	1.88	41.7	58.3
Lower leg and foot	3.28	6.42	53.4	46.6
Head and trunk	26.12	51.31	58.1	41.9
Entire body	51.28	100.00		

*From Damon, A., Stoudts, H. W., and McFarland, R. A., 1966. *The Human Body in Equipment Design*. Cambridge, Mass.: Harvard University Press, Table 69, p. 171.

Table A4.2
Average Segment Characteristics of Eight Cadavers[a]

Part		Weight	Percentage of Total Weight
		kg	
Upper arm	R	1.614	2.77
	L	1.536	2.63
Forearm	R	0.954	1.64
	L	0.914	1.57
Hand	R	0.388	0.67
	L	0.383	0.66
Forearm and hand	R	1.343	2.30
	L	1.297	2.22
Entire upper extremity	R	2.973	5.09
	L	2.875	4.93
Upper leg	R	5.756	9.86
	L	5.812	9.95
Lower leg	R	2.741	4.69
	L	2.732	4.68
Foot	R	0.832	1.42
	L	0.872	1.49
Lower leg and foot	R	3.592	6.16
	L	3.625	6.21
Entire lower extremity	R	9.481	16.25
	L	9.408	16.13
Trunk minus limbs		33.626	57.61
Both shoulders		6.174	10.58
Head and neck		4.610	7.90
Thorax		6.763	11.58
Abdomen plus pelvis		16.395	28.09
Sum of segment weights		58.363	100.0

[a]From Dempster, W. T., 1955. *Space Requirements for the Seated Operator.* WADC Technical Report 55159. Wright-Patterson Air Force Base, Ohio: Wright Air Development Center.

Table A4.3
Location of Centers of Gravity of Body Segments[a]

Segment or Part and Reference Landmarks	Distance from Center of Gravity to Reference Dimension Stated as Percentage
Hand (position of rest), wrist axis to knuckle III	50.6% to proximal (wrist) axis
Forearm, elbow axis to wrist axis	43.0% to proximal (elbow) axis
Upper arm, glenohumeral axis to elbow axis	43.6% to proximal (glenohumeral) axis
Forearm plus hand, elbow axis to ulnar styloid	67.7% to proximal (elbow) axis
Whole upper limb, glenohumeral axis to ulnar styloid	51.2% to proximal (glenohumeral) axis
Foot, heel to toe II	A ratio of 42.9 to 57.1 along heel to toe distance establishes a point above which the center of gravity lies; the latter lies on a line between ankle axis and ball of foot
Lower leg, knee axis to ankle axis	43.3% to proximal (knee) axis
Thigh, hip axis to knee axis	43.3% to proximal (hip) axis
Leg plus foot, knee axis to medial malleolus	43.3% to proximal (knee) axis
Whole lower limb, hip axis to medial malleolus	43.4% to proximal (knee) axis
Head and trunk minus limbs, vertex to transverse line through hip axes	60.4% to vertex (top of head)
Head and neck, vertex to seventh cervical centrum	43.3% to vertex
Shoulders/thorax[b] centrum L_1 to centrum T_1	53.6% to first thoracic centrum
Abdominopelvic mass (lower trunk), centrum first lumbar to hip axes	59.9% to centrum first lumbar

[a]From Dempster, W. T., 1955. *Space Requirements for the Seated Operator.* WADC Technical Report 55159. Wright-Patterson Air Force Base, Ohio: Wright Air Development Center. See Figure A.1. Appendix A.
[b]Adapted from the above.

Table A4.4
Specific Gravity of Body Segments[a]

Body Segment		Cadaver Number								
		14815	15059	15062	15095	15097	15168	15250	15251	Mean
Trunk minus limbs		1.04	1.05	1.05	1.04	1.03	1.02	1.02	1.00	1.03
Trunk minus shoulders		1.05	1.06	1.00	1.05	1.06	1.03	1.00	1.00	1.03
Shoulders		1.00	1.04	1.02	1.05	1.03	1.04	1.05	1.05	1.04
Head and neck			1.12	1.13	1.12	1.10	1.10	1.10	1.11	1.11
Thorax			0.81	0.94	0.92	1.01	0.91	0.95	0.90	0.92
Abdominopelvic			1.00	1.00	1.00	1.02	1.03	1.03	1.00	1.01
Entire lower extremity	R	1.06	1.07	1.07	1.07	1.04	1.06	1.04	1.06	1.06
	L	1.06	1.07	1.05	1.07	1.05	1.06	1.05	1.06	1.06
Thigh	R	1.04	1.06	1.04	1.05	1.05	1.05	1.04	1.05	1.05
	L	1.04	1.06	1.03	1.05	1.04	1.05	1.05	1.05	1.05
Leg and Foot	R	1.05	1.11	1.08	1.11	1.07	1.09	1.06	1.07	1.08
	L	1.09	1.11	1.08	1.12	1.07	1.10	1.08	1.08	1.09
Leg	R	1.10	1.12	1.08	1.11	1.08	1.09	1.07	1.09	1.09
	L	1.11	1.12	1.08	1.11	1.08	1.10	1.09	1.09	1.09
Foot	R	1.02	1.17	1.08	1.17	1.12	1.11	1.01	1.07	1.09
	L	1.02	1.17	1.07	1.15	1.14	1.13	1.05	1.09	1.10
Entire upper extremity	R	1.10	1.25	1.07	1.10	1.07	1.10	1.10	1.09	1.11
	L	1.11	1.09	1.08	1.11	1.08	1.10	1.10	1.09	1.10
Arm	R	1.09	1.07	1.06	1.01	1.07	1.09	1.10	1.09	1.07
	L	1.09	1.07	1.06	1.01	1.07	1.08	1.10	1.08	1.07
Forearm and hand	R	1.11	1.18	1.10	1.01	1.12	1.13	1.11	1.13	1.11
	L	1.09	1.14	1.10	1.16	1.11	1.12	1.12	1.13	1.12
Forearm	R	1.14	1.12	1.11	1.20	1.08	1.12	1.14	1.10	1.13
	L	1.14	1.15	1.11	1.18	1.09	1.14	1.14	1.05	1.12
Hand	R	1.14	1.41	1.05	1.33	1.09	1.15	1.13	1.08	1.17
	L	1.14	1.27	1.05	1.28	1.12	1.09	1.10	1.09	1.14

[a]Calculated from mass data on cadaver parts and volumetric data derived from immersion in water. (From Dempster, W. T., 1955. *Space Requirements for the Seated Operator*. WADC Technical Report 55159. Wright-Patterson Air Force Base, Ohio: Wright Air Development Center.)

Table A4.5
Segmental Fractions of Body Weight According to Somatotype[a]

Body Type	Rotund	Muscular	Thin	Median (Air Force)
Head and neck	0.07900	0.07900	0.07900	0.07900
Shoulders-thorax	0.17103	0.18097	0.17361	0.16947
Abdomen-pelvis	0.21507	0.22753	0.21985	0.21309
Thighs	0.29776	0.27376	0.27476	0.28624
Legs	0.09832	0.09374	0.10510	0.10114
Feet	0.02288	0.02680	0.03192	0.02730
Arms	0.07132	0.07010	0.06384	0.07490
Forearms	0.03494	0.03724	0.03856	0.03642
Hands	0.00968	0.01086	0.01336	0.01244

[a]If a subject seems to match any of these types, the appropriate data may be used to calculate the location of his center of gravity by the segmental method. (Calculated and adapted from data presented by Dempster, W. T., 1955. *Space Requirements for the Seated Operator.* WADC Technical Report 55159. Wright-Patterson Air Force Base, Ohio: Wright Air Development Center.)

Table A4.6
Regression Equations for Computing the Mass (in kg) of Body Segments[a]

Body Segment	Regression Equation[b]	Standard Deviation of the Residuals
Head, neck, and trunk	$= 0.47 \times TBW + 5.4$	±2.9
Total upper extremities	$= 0.13 \times TBW + 1.4$	±1.0
Both upper arms	$= 0.08 \times TBW + 1.3$	±0.5
Forearms plus hands[c]	$= 0.06 \times TBW + 0.6$	±0.5
Both forearms[c]	$= 0.04 \times TBW + 0.2$	±0.5
Both hands	$= 0.01 \times TBW + 0.3$	±0.2
Total lower extremities	$= 0.31 \times TBW + 1.2$	±2.2
Both upper legs	$= 0.18 \times TBW + 1.5$	±1.6
Both lower legs plus feet	$= 0.13 \times TBW + 0.2$	±0.9
Both lower legs	$= 0.11 \times TBW + 0.9$	±0.7
Both feet	$= 0.02 \times TBW + 0.7$	±0.3

[a]From Barter, J. T., 1957. *Estimation of the Mass of Body Segments.* WADC Technical Report 57-260. Wright-Patterson Air Force Base, Ohio: Aero Medical Laboratory, Wright Air Development Center.
[b]TBW, total body weight.
[c]$N = 11$; all other, $N = 12$.

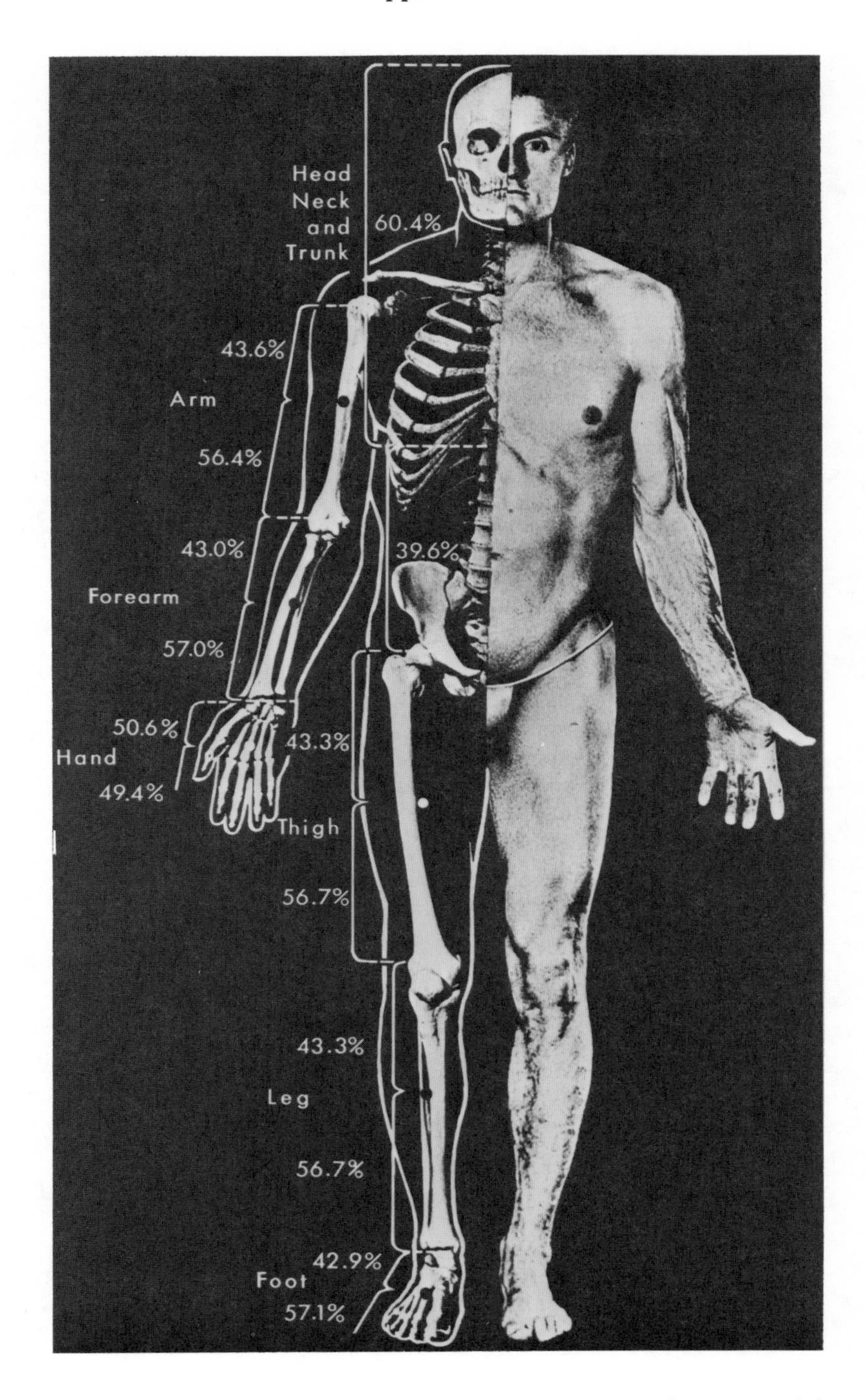

Figure A4.1. Link boundaries (at the joint centers) and percentage of distance of the centers of gravity from link boundaries. (From Williams, M. and Lissner, H. R., 1962. *Biomechanics of Human Motion*, Philadelphia: W. B. Saunders Company, Fig. A-1, p. 132.) See Table A4.3.

Index